Modern Observational Physical Oceanography

Modern Observational Physical Oceanography

Understanding the Global Ocean

Carl Wunsch

PRINCETON UNIVERSITY PRESS

PRINCETON AND OXFORD

Copyright © 2015 by Princeton University Press

Published by Princeton University Press, 41 William Street, Princeton, New Jersey 08540

In the United Kingdom: Princeton University Press, 6 Oxford Street, Woodstock, Oxfordshire OX20 1TW

press.princeton.edu

Cover art © elic/Shutterstock

All Rights Reserved

ISBN 978-0-691-15882-2

British Library Cataloging-in-Publication Data is available

This book has been composed in Minion Pro with Helvetica Neue Extended and Scala Sans (SJP)

Printed on acid-free paper. ∞

Printed in Canada

10 9 8 7 6 5 4 3 2 1

In memory of Henry M. Stommel (1920–1992)
whose spirit and ideas so dominate this subject.

Au Canard

Contents

Preface

This book is directed primarily to beginning graduate students in physical oceanography and to working scientists in allied fields seeking some understanding of what the science teaches us about the behavior of the fluid ocean. My main motivation has been to provide an introduction conveying what the observational revolution of the past thirty years has taught us—a revolution that is primarily about the ocean as a time-varying system. The field does have a number of good textbooks outlining—primarily—the essential theoretical side of the subject and having varying breadth and depth (Gill, 1982; Pedlosky, 1996; Vallis 2006; Huang, 2010; Olbers, Willebrand, and Eden, 2012). For a graduate student, or a scientist trying to understand oceanic biology or chemistry or their role in climate, however, no adequate introduction exists depicting the qualitative behavior of the fluid as now perceived—beyond the exhortation to read hundreds of sometimes difficult and contradictory papers. Continued dependence on an outmoded description of the fluid ocean has become an obstacle to progress.

The attraction of writing a book that is primarily about the theory is plain: the author can write something like: "Assume the motions have periods between a and b, and length scales between c and d. Let the dynamics be linear, hydrostatic, and have a flat bottom, etc. The governing equations are then ...." A complete, sometimes, elegant deductive product then follows. In contrast, understanding observations involves all of that theory, and a great deal more: Do the observed motions actually lie between a, b and c, d? Is the seafloor sufficiently simple that the real topography can be ignored? Are the motions demonstrably linear and hydrostatic? How much noise is there in the data (some always is)? Is it negligible or dominant? Most likely the beautiful theory explains some, but not all, of what is observed. How does one distinguish them? What about the unexplained part?

What I have tried to do here is to sketch the major elements of modern physical oceanography, with an emphasis on observations and what they seem to say. Physical oceanography is a subdiscipline of fluid dynamics, and its history shows the need for a strong and continuous coupling between theory and observation (experiment). The equations of fluid motion are so rich in possible solutions that theory without observations tends to diverge from realism; conversely observations are uninterpretable without a strong theoretical framework for their analysis. Observing the ocean is remarkably difficult: it is corrosive, opaque to electromagnetic radiation, produces biological fouling of instruments, imposes pressures over 600 atmospheres, is extremely large and time varying, and near the surface imposes rapid and strong mechanical cycling (wave forces). Existing understanding and depiction of the fluid ocean and its climate implications are perceived through the distorting lens of a very specific set of practical observations. A long list of important oceanic phenomena are known that previously existing theory could have predicted but didn't[1] and for which a theory was put in place only after the observations became available. Nonetheless, the narrative is simpler with a theoretical framework for context.

One of the more troubling developments in physical oceanography over the last thirty years has been the ever-greater separation of scientists from the data they are scrutinizing. Years ago, there was the silly mantra that "real oceanographers make their own observations," and it was used to both separate the club of seagoing scientists from their more theoretically inclined colleagues and disparage meteorologists who, as they were faced with a global system, had come to rely heavily on government-operated weather-observing systems. Of course, oceanography encountered climate and became global, and the science became much more sophisticated. The result has been that proportionally ever-fewer oceanographers still make their own observations and the subject has come to much more closely resemble the situation in meteorology. This development has two consequences: we know enormously more about the ocean than the individual scientists of thirty years ago could even dream of, and we have a new generation of scientists with little understanding of how complex observations can be and the consequent possibilities for misinterpretation.

With the rise of computers, the maturing of dynamical systems theory, and other powerful theoretical tools, it is easy to forget that physical oceanography and climate remain fundamentally observational subjects. Modelers spend much time comparing their results but sometimes lose sight of the idea that comparison to observations is the test of a model's skill—not whether it reproduces the same results as all the other models.

In recent years new elements have arisen to complicate the problem of distinguishing scientific fact from mere hypothesis or rationalization or description of untested model results. Much ocean physics, as it influences the climate system, is "slow," spanning decades and centuries. Direct observations over long time periods are extremely rare. With rising interest in climate change and the consequent growth of a tabloid science (epitomized by *Nature, Science*, their would-be competitors, and media-camp followers), much speculative storytelling, far outstripping the observational record, has come to dominate, and even corrupt, many aspects of the science.

Mathematics is the natural language of fluid dynamics, and its employment in oceanography and climate is essential. Writing a book about these subjects without mathematics is somewhat like writing one about Chinese poetry for readers who cannot read that language. Nonetheless, I have tried to keep the mathematics to a minimum, mainly describing what it says rather than deriving it. What theory is presented here is intended primarily to make the observations intelligible, to provide insight, and not to be rigorous or particularly accurate.

What is *not* known is emphasized relative to what is essentially "fact." Most textbooks tell the reader what is thought to be known and true. But the unknowns and the puzzles are much more interesting than what everyone agrees on. In any case, a serious student of ocean

---

[1]Examples where the theory was available but not applied until the right observations appeared includes the existence of the equatorial countercurrent and undercurrents; fine and microstructure in the assumed smooth temperature and salinity profiles; the intense high-latitude barotropic variability; the near-universal internal wave spectrum; the ubiquitous internal tide, etc.

physics should read this book alongside one of the more theoretical textbooks mentioned above. Whether the presentation here falls between the two stools of too little or too much theoretical framework will be judged by the reader.

Readers will also notice many omissions and nonuniformities of coverage. This subject is now so large that a comprehensive coverage by a single author may no longer be possible. Among many other interesting, important topics, only tangential attention is paid to the coastal ocean: the mixed-layer, marginal seas including the Arctic, Mediterranean, and Caribbean, the Southern Ocean, sea ice or high latitudes generally, and near-equatorial dynamics. Paleoceanography is discussed only where it provides some useful context for modern change. In the last analysis, the material reflects mainly my own interests over the past fifty plus years and thus is perhaps best regarded as a personal statement as to what has seemed most intriguing.

Acknowledgements

This book was written over a period of years at the Massachusetts Institute of Technology; Balliol College and the Department of Physics, Oxford (supported by the George Eastman Professorship); and Harvard University. Over those years, I had generous research support from the US National Aeronautics and Space Administration and the National Science Foundation. Very helpful comments and important corrections, for which I am grateful, were provided by W. Sturges, D. J. Baker, C. Garrett, E. Firing, K. Brink, and J. Pedlosky. I received many useful suggestions, not all of which I found it possible to exploit, but none were ignored. No one else bears any responsibility for remaining errors, eccentricities, and incoherence.

Carl Wunsch
Cambridge Massachusetts
March 2014

Modern Observational Physical Oceanography

CHAPTER 1

Introduction

Study of the ocean circulation is a problem in fluid dynamics. Traditionally, however, descriptions of the oceanic general circulation have begun with pictures of the large-scale temperature, salt, and oxygen and other chemical tracer properties of the deep sea. This approach rests on good historical and logical grounds: until recent times, the only properties measurable on a global basis were these scalar "tracers." Furthermore, their overall distributions have proved remarkably stable in time, and in turn that has made it possible to combine data over many decades to achieve global pictures from shipboard measurements.

In contrast, this book begins with an emphasis on the time-varying flow field as observed from a variety of modern instruments. The more traditional discussion of the time-average properties of velocity, temperature, and salinity is postponed. These latter are to be set into a context more relevant to an observer coping with a changing velocity field. Conventional pictures showing the large-scale temperature, salinity, and related distributions led to the concept of the ocean circulation as a quasi-geological phenomenon, with little or no change occurring either spatially or temporally. In the process, sometimes it was forgotten that the ocean is a fluid, and not a series of slabs sliding over one another unrelated to the equations of physics. As long as the study of the circulation was primarily of interest to academic physical oceanographers, the consequences of this distortion were of little practical consequence. Today, however, the circulation is widely regarded as an essential element in the understanding of the climate system and as a dominant factor in such politically charged phenomena as global change, sea level, and biological variations. But misconceptions concerning the very character of the circulation generate unrealistic programs for climate forecasting, observing the ocean, interpreting the record of past climate, and a host of related practical issues such as the management of fish populations.[1]

The term "oceanography" historically denoted a descriptive science, paralleling "geography"—with its heavy emphasis on terrain, crops, economic assets, regional particulars,

---

[1] A conspicuous example is the folklore postulating that the Gulf Stream can "turn off."

etc.[2] That traditional beginning is today recalled in "descriptive oceanography," to distinguish it from the wider subject employing the dynamical equations with much mathematics. Every region, depth, season, and probably year in the ocean is distinct from all others. A very large and growing literature exists depicting the elements and eccentricities of many geographical regions. Most of that subject is omitted here—rather, the focus is on those elements that can be understood in a more global context, because of their generality or exceptionality. But the reader must understand that no clear distinction exists between the regional- and global-scale descriptions, be it verbal or mathematical, and too much should not be made of the division.

Physical oceanography can no longer be encompassed in a single manageable volume, and I make no claim to being expert in more than a fraction of it. References are provided that should permit a reader interested in pursuing a subject in greater depth to do so by starting with the various papers and books cited. No serious attempt has been made to provide a historically correct attribution to the originator of an idea, and when a reference is given, unless explicitly stated otherwise no implication is intended that it refers either to the first, or even the most important, discussion. These references might be regarded as the analog of navigational beacons: they are neither the channel nor a shoal, but indicators of where those are to be found. Parts of the field are undergoing rapid development as I write, with new papers appearing weekly. Obsolescence in a book must be expected, with the navigational markers being more like bread crumbs in a world of birds and rainfall. Modern electronic search tools now permit easy access to both the earlier and later literature. Occasionally, a historical sketch is provided where it enables a better understanding of some concept.

My intention has been to make the book self-contained if not comprehensive; specific references to the fluid dynamics literature (e.g., Tritton, 1988; Kundu and Cohen, 2008) and to the more theoretical textbooks noted in the preface are provided so that the reader can locate a fuller derivation, a wider discussion, or illuminating applications. Much useful material can be found in the recent compendium of Siedler et al. (2013); like most multiauthor collections (there more than seventy), it is neither easily digested nor without internal contradictions.

By employing "boxed" discussions and appendices, I have tried to make the basic concepts, borrowed from a wide variety of subfields, at least heuristically sensible and have provided references for anyone who would like to know more. Thus sketches are provided of the singular value decomposition, the Radon transform, Bessel functions, etc. Within the text, in many cases, results are simply stated; in others, where the derivation is particularly easy or interesting or illuminating, it is at least sketched. I do not claim to have been consistent. The ocean and climate are nonlinear systems, a property one must always remember. Nonetheless, this book leans almost completely on linear mathematics on the grounds that most intuition and insight are built that way, and as has been found across the sciences, linear analyses often have skills well beyond their formal domain of validity.

Only elementary statistical methods are employed: sample means and variances, spectral estimates, etc.—just enough to get by on, given the existence of useful handbooks dealing with a variety of powerful techniques. Historically, oceanography and climate have almost never raised issues in which very fussy statistical tests were required—if apparent signals were so weak as to require powerful tests, they usually proved unimportant compared to much more

---

[2] Soviet Union scientists, in particular, made an attempt to substitute the more logical "oceanology" as the correct parallel terminology to the scientific subjects of biology, geology, etc., but the label was never accepted in the West; see Hall, 1955; Carruthers, 1955.

conspicuous, and still unexplained, signals. Many statistical methods exist for extracting weak signals from noise. In practice, however, oceanographic and climate measurements are usually subject to such basic problems as calibration drifts, sampling distribution changes, unknown external contributors, small sample size, and poorly understood statistical characteristics (e.g., they are never truly Gaussian, never truly statistically stationary) such that results dependent upon the use of elaborate methodologies should continue to be regarded as very tentative unless subjected to careful study of their sensitivity to the underlying assumptions. Common sense is useful. For example, if a process is obviously non-Gaussian, don't use ordinary statistical tests that assume it is normally distributed. The future belongs to the Bayesians, but as these methods have not yet broadly been used in the ocean literature, no explicit use of them is included.

Organization of a book such as this presents a conundrum. Discussion of instruments and measurements is almost impossible without some understanding of how the data are used—and that requires some theoretical background. But much of the theory is not very compelling without an understanding of what is measurable. Ocean variability is not interpretable without knowledge of the time-mean circulation—suitably defined—and that in turn is determined in part by the variability. A linear narrative is thus not possible—leading to a need for parallel and iterative discussions; the reader can expect to jump around among chapters. The book opens with a description of measurement methods, followed by a qualitative description of both the time mean and variability. A chapter sketching the variability theories leads to one discussing observations. Later chapters then turn attention to the more traditional ideas about the time-mean circulation.

TERMINOLOGY

Many scientists are impatient with discussions of terminology ("It's just semantics"). But precise language is an essential shorthand. Furthermore, unnecessary jargon is a serious obstacle to the exchange of ideas within the field, and much more so with neighboring disciplines. Muddled thinking is often most immediately apparent in the choice of language. Anyone who has worked in oceanography for awhile will have had the experience of reading through a paper or sitting through a talk before recognizing that one's growing bewilderment and confusion resulted from some careless or unorthodox use of the label for a concept. Examples abound. For example the term "barotropic velocity" has at least six different incompatible definitions, and a very large amount of unnecessary confusion has ensued by their sometimes unthinking, undefined use. A number of examples are discussed in Appendix C.

AN OPEN MIND

Most textbooks are directed at explaining to their readers the facts of the subject. One interpretation of the central role of science is in its overthrowing what "everyone knows." In the grand scheme of things, everyone once knew that the Sun orbited around the Earth, that the geological record was the result of the Noachian Flood, and that species were immutable. Many scientists share the human need for near-religious faith in what they "know" about the world, to the point that dogmatism becomes a major obstacle to understanding. In physical oceanography, as with all fields, many examples exist of somewhat plausible ideas being converted into a kind of faith-based science. The advice by Chamberlin (1890) to *always* maintain multiple scientific hypotheses remains most sensible.

CHAPTER 2

Observing the Ocean

Compared to the atmospheric sciences, oceanography has a special flavor because measurements are so difficult. Radio waves do not propagate through the sea at useful frequencies. Among other problems, (1) no analog exists of the capability in meteorology of measuring cloud velocities and temperature profiles from space; (2) information cannot be sent back to shore electronically in the way weather balloons transmit data to observers at the surface; (3) from a ship, ocean surface properties alone can be measured, or an instrument must be placed physically at the depth where the observation is required; and (4) if long-duration measurements are needed, the instrument must be kept in place, internally recording the changing values being sensed.[1]

The problem of observation is further complicated by the high pressures at depth: each 10 m of ocean depth increases the ambient pressure by about one atmosphere of pressure; seawater is corrosive everywhere; biological fouling is a major problem in the upper regions; and surface instruments are mechanically stressed by thousands of wave cycles over weeks and months. A somewhat brief description of many (far from all) of the major measurement technologies is given here because so much of what is believed understood of the ocean is, and has always been, filtered and distorted through the prism of the available observational tools.

Because of the present intense interest in climate trends, questions of whether the ocean has been getting warmer or fresher or moving faster or slower over the last decades and centuries, or even whether its volume is changing, are very conspicuous. Some understanding must be gained of how the measurement technology, its coverage, and its accuracy have evolved through time.[2] Deacon et al. (1971) and Peterson et al. (1996) provided useful and

[1] Instruments that autonomously periodically return to the surface for satellite communications have also recently become available. But only under special circumstances do they return to a fixed location.

[2] Any history of oceanographic instrument development would necessarily acknowledge the major role played by the US and other navies. Perceived military needs, as well as the great expense and time required to shepherd an idea from laboratory prototype to something routinely deployable by nonspecialist observers, have often meant that only a naval organization could afford it. Military needs have also commonly dictated the space and time

interesting summaries of observations dating from antiquity to the late nineteenth century. More extended accounts of measurement methods of their era can be found in von Arx 1962, Baker 1981, Heinmiller 1983, Laughton et al. 2010, Robinson 2004, and Martin 2004, the latter two for satellites. Here only a sketch is provided to give of the flavor of the subject and as background for understanding of some of the difficulties observational oceanographers have had, and continue, to face. Many instrument types and concerns are revisited later in discussions of the meaning of the observations for the physics.

Oceanographers now often download "data" from the Web, sometimes failing to recognize how remote the numbers are from the original measurements. Tabulated values commonly are derived from very complicated measurement systems and are the result of complex manipulation and processing of the observations. An outcome can be theories or models of nonexistent phenomena—or the opposite, the failure to detect significant physical processes.

2.1 SHIPS

Ships have been the major platforms for observations of the ocean since antiquity. Without them, no deep-sea measurements would have existed. Published histories of the evolution of boats and ships abound; platform evolution from human to wind to engine power is well known. What is relevant for understanding oceanographic observations are that they have been, and remain, expensive and slow. Depending upon the region, time of year, and adequacy of the vessel, they can produce extreme discomfort in scientific observers of varying intestinal fortitude. The first recognizable oceanographic expeditions are sometimes traced to Edmond Halley (1698), who measured tides and the geomagnetic field at sea (Cook, 1998), or to the *Challenger* expedition of 1872. Even at the outset, the financial requirements for ship use were so onerous that only a government agency could sponsor the work—in both these cases, it was the UK Royal Navy. For awhile in the early twentieth century, wealthy amateurs with their own seagoing yachts (the Prince of Monaco, Fridtjof Nansen, Henry Bryant Bigelow) made significant contributions to the subject, but that era came to an abrupt halt with the professionalization that occurred following World War II.[3]

In the modern world (2014), costs remain a major factor. Although inflation renders all such numbers ultimately nugatory, a modern vessel capable of crossing an ocean basin would cost about $30,000/d to operate, not including the cost of the scientists. The best modern vessels steam at about 12 knots (about 22 km/h) and thus require many days to cross an ocean basin—even without stopping to make measurements. With the ocean now known to be changing significantly day by day, and with few scientists willing to spend months and years at sea as they did hundreds of years ago, the era of the ship as the *fundamental* sampling platform in physical oceanography is over, although a need for both oceanographic ships and seagoing scientists will always exist. (The situation still remains different in biological and chemical oceanography.)

---

distributions of observations, sometimes leading to intractable problems of false apparent ocean variability from the shifting sampling distributions.

[3] Astronomy, as a science, benefits in many ways, both direct and indirect, from the activities of amateurs in producing data interpretable by the observer but often also of intense interest to the professionals. Unhappily, physical oceanography has never been able to provide many such opportunities.

2.2 NAVIGATION

Observations at sea are useful only if their location is known. The accuracy required depends directly on the type of measurement and its purpose. For Benjamin Franklin, the knowledge that water temperature within the Gulf Stream was warmer than outside it, provided a very useful insight, and large uncertainties in latitude and longitude were tolerable. For a more contemporary scientist attempting to calculate the horizontal derivatives of the density field, $\partial\rho/\partial x$, so as to determine the velocity between measurements separated by 10 km, a 1 km error in the distance can totally confound the calculation. Historically, and practically, the problem of horizontal position has been treated separately from that of determining the depth of an instrument, and thus "navigation" here refers to the problem of determining latitude and longitude at sea.

Because of its implications for commerce, the military, exploration, and science, elaborate discussions exist of the history of navigation. Sobel (1996) provided a well-known popularization of the romantic story of Harrison's marine chronometer. But oceanographers have always been intensely interested in obtaining and using the most precise available navigation systems, and a summary will have to suffice. See Bowditch, 2002, for a history and a guide to many of the available methods.

Before the Second World War, navigational techniques had hardly changed in their fundamentals since the invention of the marine chronometer. Determining position at sea depended upon the compass, the ship's log (a speed-measuring device), the determination of the positions of astronomical bodies with sextants and related instruments, elaborate numerical tables, and the very real skill of human navigators.[4] World War II brought systems related to the invention of radar, including LORAN, Decca, Omega, and the like, which permitted all-weather navigation, at least over large areas. These worked a revolution for oceanography in regions such as the North Atlantic where coverage was good. But LORAN was never global, and putatively global systems like Omega were not very accurate. Radar itself provided accurate line-of-sight navigation relative to detectable objects such as the shore, other ships, and drifting buoys.

The advent of Earth-orbiting satellites in the late 1950s eventually brought about the Transit satellite system, which was operated by the US Navy (see Bowditch, 2002) and provided accurate but highly sporadic temporal coverage, and later the revolutionary Global Positioning System (GPS), and soon perhaps, rival non-US networks. Beyond GPS, improved horizontal position accuracy is unlikely to have a major further impact on shipboard oceanography, although measurements will always exist where even greater accuracy and precision can be used. Quotidian navigational skills have eroded to the ability to push a button—one of the ways in which electronics has distanced end users from familiarity with observations.

2.3 THE PREELECTRONICS ERA

Electromagnetism as a potentially powerful observation tool was recognized very early: in 1832 Faraday made a failed attempt to use induced voltages in telegraph cables and had

---

[4]Some appreciation for the remarkable individual skills that developed can be obtained from the account of the finding of South Georgia by Frank Worsley from an open boat in the Southern Ocean—during one of the Shackleton expeditions (Huntford, 1986).

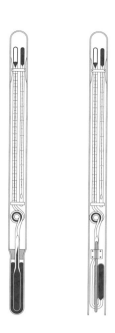

Figure 2.1: Unprotected (left) and protected (center) reversing thermometers and their auxiliaries. When inverted (top right) the mercury in the bulb is unable to drop. The auxiliaries are used to determine the corrections for the ambient temperature at which the thermometers are read on deck after reversal. (H. U. Sverdrup, M. W. Johnson, and R. H. Fleming. *The Oceans: Their Physics, Chemistry, and General Biology.* 1st ed. © 1942. Reprinted by permission of Pearson Education, Inc., Upper Saddle River, NJ.)

described a hypothetical system for measuring the flow in the English Channel (see Longuet-Higgins, 1949). But before about 1970, most measurements were made by purely mechanical means—because vacuum tube electronics and recording mechanisms available before then proved far too unreliable to work in the marine environment or to survive preparation and launch from rolling, pitching vessels with vibrating decks, high humidities, and the sometimes seasick scientists handling them. The problem of observing the ocean without recourse to electronic (or even electrical) methods led, over the years, to a collection of clever methods, some of which were widely adopted.

REVERSING THERMOMETERS AND NANSEN BOTTLES

The most impressive measurements of this era were those directed at determining the temperature and salinity at depth from ships. (Salinity, S, is the fraction of mass from salt content in seawater.) The methodologies were in place before 1880 and are worth a brief discussion because they were the backbone of physical measurements until well into the 1970s; they are a wonderful example of scientific ingenuity, and their use directly controlled the understanding and depiction of the ocean for 100 years.

Two devices are required. The first, and more remarkable one, is the *reversing thermometer*, whose major feature is a small capillary in the mercury-containing bulb that prevents the mercury from exiting when the thermometer is inverted (see Fig. 2.1). The second is the *Nansen bottle* (an evolved form of earlier designs), a hollow tube that is attached to a wire from one point above and one below the tube and then lowered into the ocean. The lower attachment point is a screw-on clamp containing an axle about which the bottle can pivot. The upper attachment point is another, releasable, clamp, which when struck by a weight (the *messenger*) sliding down the wire from above, releases the clamp so that the bottle swings upside down by the lower pivot (see Fig. 2.2). While the bottle pivots, a flange at each end emerges and seals off the water in the tube, making it available then for recovery and laboratory analysis and simultaneously releasing another messenger. When the bottle swings upside down, it

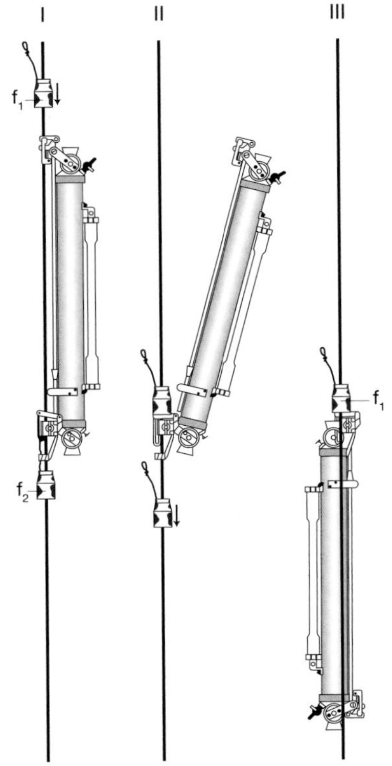

Figure 2.2: The Nansen bottle and how it pivots when the top coupling is released by a falling messenger weight (f_1) thus both closing the bottle flanges and releasing a second messenger (f_2) to trip any bottle mounted below. The three stages are shown: I, as lowered; II, in the process of inverting; III, inverted as hauled back onto the deck. (Used with permission from G. K. Dietrich, K. Kalle, W. Krauss, and G. Siedler [1980] *General Oceanography, An Introduction*, 2nd ed. English translation of the 1975 German edition. New York: Wiley.)

inverts the reversing thermometer attached to the bottle by a frame, allowing the mercury outside of the bulb to fall to what is now the bottom and reflecting the temperature at which the reversal took place. The thermometer can then be read on deck, in the inverted position. The system is completed by using two reversing thermometers—one *protected* against the ambient pressure where the reversal takes place; the other is an *unprotected thermometer* and is subject to the water pressure—which squeezes excess mercury out of the bulb, making it read artificially high. The difference between the protected and unprotected temperatures can be then converted into a pressure. From the hydrostatic relationship, which gives the relationship between water depth and pressure, the physical depth of the measurement can be calculated. A third, auxiliary, thermometer is used to correct for ambient room temperature. (Sverdrup et al., 1942, give a brief history; McConnell, 1980, describes predecessor methods.[5])

A *cast* was made by having the research vessel steam to the desired location, and a wire with a heavy weight on the end was lowered on a winch. Nansen bottles with reversing thermometers (sometimes with only a protected one) were placed on the wire at intervals from 50 to 500 m depending upon the water depth and the available resources. A messenger was attached at the base of each bottle except for the last one. The entire string of bottles (sometimes more than 30) was lowered to thousands of meters in the ocean, with the scientist allowing extra wire out for the catenary formed in a weighted cable strung out from a drifting ship. After waiting some minutes for the thermometers to equilibrate, the observer would

[5] Maximum-minimum thermometers preceded development of the reversing thermometer by some decades. They provided the major measurement method during the *Challenger* Expedition. See Roemmich et al., 2012.

launch a messenger down the cable that would trip the first bottle. When it pivoted, the first bottle would release its messenger, which would then trip the second bottle, etc., all the way to the bottom. Allowing for the finite transit time of the messengers, the whole string of now inverted, hanging bottles would be brought back onto the deck, samples drawn for salinity and other chemical analysis, and the thermometers read. The whole process could take many hours depending upon the number of bottles, the water depth, and things going wrong (many things could!). After years of evolution and experience, this system produced data of impressive accuracy, with pressures believed accurate to about 0.2 to 0.5% of the depth and temperatures, in expert hands, reproducible (precise) to about 0.002–0.01°C (Warren, 2008 discusses differences of opinion.)

Salinity measurements were reviewed by Millero et al. (2008). Measurements from the Nansen bottle samples were first usually done by chemical titrations and later by electronic measurements of water conductivity, with strenuous efforts made to retain consistency because both the technology and even the definition of salinity continues to change (Feistel et al., 2008; Millero et al., 2008). The IOC (2010) technical report describes the latest accepted definitions and procedures. An early salinity chart is shown in Fig. 2.3. Oxygen and nutrient measurements, which are ancillary to the measurement of temperature and salinity and used by physical oceanographers primarily as tracers, appeared very early (see, e.g., the plates in Wüst, 1978, one of which is shown in Fig. 2.4). Methods and their calibration and the accuracies and precisions have changed greatly over the years and continue to do so; the reader is referred to Millero's book (2006) for what is primarily the subject of the physical chemistry of seawater.

THE BATHYTHERMOGRAPH

Part of the great expense of the Nansen bottle/reversing thermometer method was the necessity of keeping a ship at a nominally fixed location for many hours or even days. Thus an urgent search existed for methods useful from a ship underway, a search that led to the invention of the *bathythermograph* (BT; see Schlee, 1978, for a description of the dispute over who invented what). A BT is a torpedo-shaped device lowered by winch from a moving ship. The torpedo is quite heavy so that it sinks despite its forward speed and cable tether, albeit the achievable depth from an oceanographic vessel never exceeded about 200 m. A bellows to measure pressure was combined with a thermocouple for temperature, which was scratched onto a smoked-glass slide as the instrument sank through the water. The instrument was then brought in by rewinding on the winch, the smoked-glass slide extracted (one problem was finding a coating that would not wash off in seawater), and the temperature as a function of depth read and recorded. These instruments appeared just before the Second World War and were intensively used by the military because the acoustic field governing submarine hunting or hiding is largely defined by the temperature distribution. The depth achievable by, for example, a destroyer moving at 30 knots is, however, much less than that from an oceanographic ship moving at 10–12 knots. Although widely employed, the data were of restricted use, owing to their relative inaccuracy, in addition to the severe depth limitation, and the absence of salinity information. BTs could also be very disagreeable to use—because catching a heavy instrument before it crashed into the vessel as it came out of the water from a rolling ship was hard on hands, fingers, bystanders, and the BT itself.[6]

[6] "A murderous instrument." H. Stommel (in Hogg and Huang, 1995).

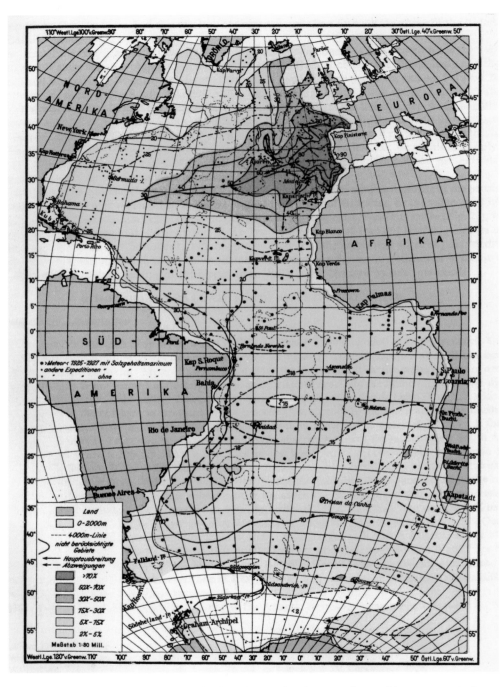

Figure 2.3: The salinity (o/oo) at mid-depth in the North Atlantic as depicted by Wüst (1978) from data obtained from the R/V *Meteor* in the 1930s. The major feature is the immense salt "tongue" emanating from the Mediterranean Sea at about 1200 m. As a local maximum it can be traced around the world in modern data. The depth of the charted values is not constant, varying from about 1000 m at the latitude of Labrador, to about 2700 m west of the Cape of Good Hope (see his Insert XII). The arrows shown on this and similar charts are Wüst's inference of what he called "spreading" of the water (*Ausbreitung* in German), a term chosen to avoid distinguishing between velocity and mixing components of fluid properties.

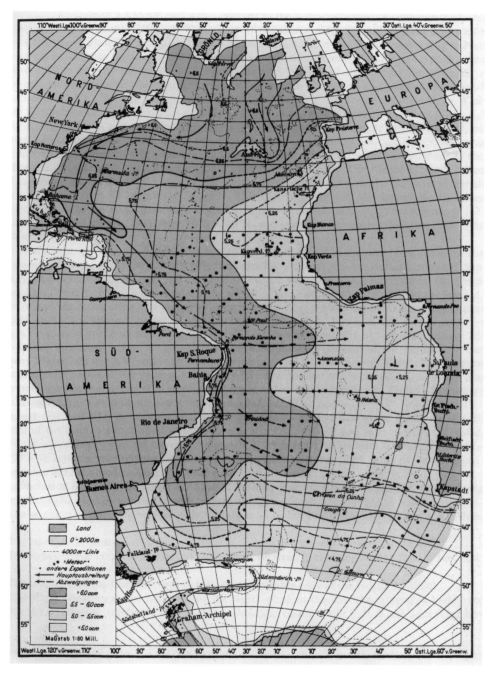

Figure 2.4: The oxygen concentration (milliliters per liter) in the deep Atlantic Ocean at the same locations as in Figure 2.3. (From Wüst, 1978.)

Figure 2.5: The von Arx current meter as used on board ship. The instrument is related to the better-known Ekman instrument, which had many variations. In this configuration, dials were read on deck with the instrument lowered to desired depths. (From W. S. von Arx [1962]. *An Introduction to Physical Oceanography*. Reading, MA: Addison-Wesley.)

CURRENT METERS

Velocity-measuring devices, current meters, that could send signals up conducting cables to the ship from which they had been lowered appeared very early in the history of the subject. Many clever designs appeared, but the most commonly used instruments were variants of propellers mounted at right angles to each other and combined with an orienting magnetic compass. Spring-wound strip-chart recorders and visual reading of dials on deck were the usual means of recording (see Fig. 2.5). For many purposes, these instruments had useful accuracies and precisions. The great difficulty was that, in contrast to temperature and salinity measurements, the velocity field appeared to vary wildly from minute to minute to hours and even days (if one could afford to keep a ship in one place that long), so the meaning of the measurements was obscure. Velocity measurements in the ocean never became routine until the invention of modern electronics. One remarkable exception was the work of Pillsbury (1891), who, using an anchored ship, made a set of current meter observations over a period of years in the Gulf Stream off Florida (described by Stommel, 1965, 8–9).

SURFACE DRIFTERS

Surface-drifting objects have been used from antiquity to determine the circulation. Bottles with messages requesting the finder to notify the sender have been deployed for hundreds of years. Bumpus (1973) summarized the configurations used at that time. Several difficulties arise in the interpretation of surface drifters, if located, including windage—the effects of the wind on objects extending above the sea surface—and slippage of the water around the drifting object. In addition, they were mostly picked up on beaches, and the time between grounding and the finding by some passerby was unknown. Thus velocities could not be computed, only nominal trajectories, given the start and end points.

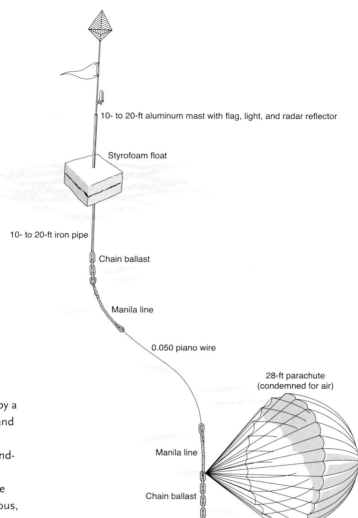

10- to 20-ft aluminum mast with flag, light, and radar reflector

Styrofoam float

10- to 20-ft iron pipe

Chain ballast

Manila line

0.050 piano wire

28-ft parachute
(condemned for air)

Manila line

Chain ballast

Figure 2.6: An older design of a surface drifter (or drogue) tracked by a ship using radar or optical means and a parachute deployed in the water. Such designs would have a high windage, and opening of the parachute requires slippage between it and the surrounding water. (From Anonymous, 1968.)

Some extension above the sea surface is desirable—so that the instrument can be seen or tracked and/or so that it can radio its position to direction finders on ships or land. Many existing designs attempt to minimize windage and slippage. After World War II, surplus parachutes were available cheaply from the military, and a number of drifters employed them in an effort to trap the moving water so that it would carry the attached instruments with it (Fig. 2.6). Later, it was recognized that without slippage, the parachute would not deploy, and so there was an inevitable systematic error. More recent designs carry GPS receivers and other tracking devices and do not use parachutes.

TIDE GAUGES

The longest time series (serial in time measurements) of any oceanographic variable come from tide gauges. These are water-level gauges, whose records are usually dominated by the

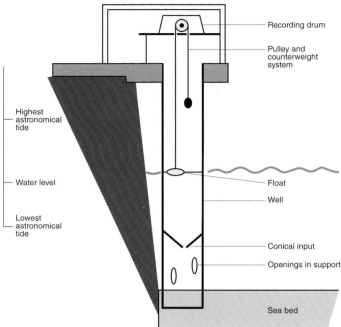

Labels in figure: Recording drum; Pulley and counterweight system; Highest astronomical tide; Water level; Lowest astronomical tide; Float; Well; Conical input; Openings in support; Sea bed

Figure 2.7: A typical Kelvin tide gauge showing the tide staff and float. (Adapted from D. Pugh [2004]. *Changing Sea Levels: Effects of Tide, Weather and Climate.* New York: Cambridge University Press. Copyright 2004 David Pugh. Reprinted with the permission of Cambridge University Press.)

tides. In the earliest form, they were graduated rulers (called *tide staffs*) strapped to the side of a pier, and the values read one or more times per day by a human observer. They gradually were replaced by floats in vertical pipes, known as *stilling wells* (see Fig. 2.7), which were open at the lower end or had holes in the side and which damped the high-frequency wave noise that otherwise made reading a ruler difficult. Eventually, spring-wound clocks were developed to control strip-chart recorders, in what are known as Kelvin tide gauges. This system freed the observer from having to visit several times per day, although observers were still required to clean the system and to calibrate it—generally by using a nearby tide staff. Records from such devices extend in a number of cases for more than 100 years, although the vicissitudes of harbor construction, wars, storms, etc., led to data gaps and serious calibration troubles. Note that the *nilometer* (see Fig. 2.8 and http://www.waterhistory.org), of ancient origin, was essentially a tide staff, although it did not provide an oceanographic record except insofar that it represented an element in the freshwater input boundary condition.

COMPUTERS

The reversing thermometer and Nansen bottle were the technological backbone of physical oceanography from its beginnings to about 1970. In the following decades, it was the transistor and its descendants as low-power integrated circuits that completely remade the observational methods and greatly enhanced understanding. Digital computers, built with integrated circuits, have become so pervasive both in permitting the handling of vast volumes of data and in the production of numerical model representations of the ocean, that it is easy to forget how much of recent progress is primarily due to the availability of the underlying, out-of-sight technology. Before the era of the transistor, and well into the period of digital computers based upon vacuum tubes, oceanographers had relied, along with the wider sci-

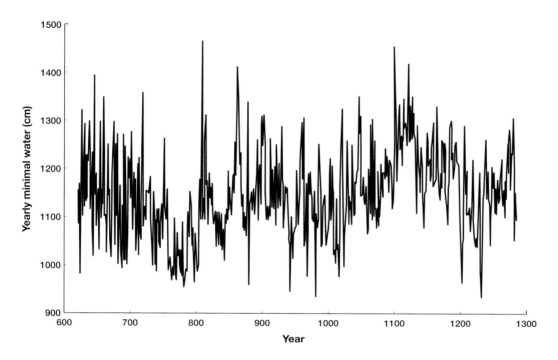

Figure 2.8: Part of the nilometer record of low water each year in the Nile at Cairo. It is probably a unique time series of such length and one derived from a staff. Years denote those of the common era (CE). The apparent long-term fluctuations are typical of climate records generally. Construction of the Aswan Dam in the twentieth century ended the utility of the nilometer—an example of the problems encountered in obtaining long records and their intrinsic noisiness. (From http://www.waterhistory.org/histories/cairo.)

entific community, primarily on analog computing techniques. Ishiguro (1972) provided a summary, and F. Ursell (in Laughton et al., 2010) described the heroic efforts then required to obtain the Fourier transform of a record, which today can be done in a fraction of a second with one line of computer code.

METEOROLOGICAL VARIABLES

The ocean is set moving and maintained in motion by astronomical bodies (the tides) and the transfer of momentum, enthalpy (heat energy), and moisture between the ocean and the atmosphere. Contributions exist (discussed in Chap. 3) from geothermal heating and mass flux through the seafloor and from freshwater injection from the land and ice fields. The ocean and its circulation cannot be understood without an adequate description of the meteorological fields above the sea. Apart from observations at a handful of islands, the only possibilities for determining the open-ocean fields were from ships. These measurements began in the earliest seafaring days with purely visual, but highly expert, estimates of wind force and direction and rainfall, as recorded in ships' logs by watch officers. The Beaufort scale of wind speed comes from this era, and because sailing ship crews were seriously interested in the

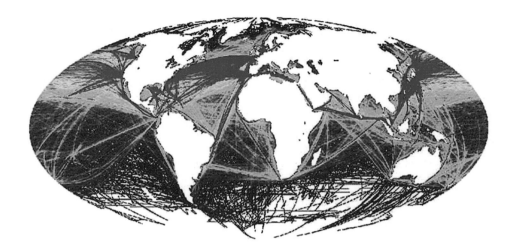

Figure 2.9: Shipping lanes of the world. These have changed considerably over the years with the opening of the Panama and Suez Canals and the shift from wind power to steam. Colors represent a qualitative measure of density from blue (low) to red (high). Inhomogeneity of coverage is a typical problem of ocean observations because of the very great spatial variations in ocean physics and variability. (From Halpern et al., 2008, online material.)

wind, the standard of reporting was very high.[7] Cardone et al. (1990) discussed the changing technologies. See the paper by Kent at al. (1993) for a more general discussion of shipboard observing system accuracies, with the important caveat that they relied heavily on models to provide a standard. Conversion from wind velocity to the force acting on the ocean (the stress) is an important problem that remains incompletely solved (Risien and Chelton, 2008).

A serious issue remains today: most ships stick to major shipping lanes (Fig. 2.9), and thus reports are confined to narrow strips and are not representative of the world ocean. Rainfall is notoriously sporadic, and the representativeness of shipboard rain observations is difficult to evaluate. Other quantities, such as relative humidity (from wet- and dry-bulb temperatures), were much less frequently reported, as were such quantities as cloud cover (which untrained observers have difficulty estimating). Calculation from available shipboard measurements of important quantities such as evaporation rates are, to this day, the subject of research. More generally, ships sensibly avoid regions of bad weather, and thus, as expected, high-latitude winter-time observations are scarce, as are reports from hurricanes, and the data sets are biassed toward fair weather.

2.4 THE ELECTRONICS ERA

The invention of the low-power integrated circuit revolutionized oceanography as it did almost all of science. In addition to giving rise to the now-ubiquitous computer, about 1970 it became possible to make sophisticated instruments that could run unattended for months and

---

[7] The scale began as a description of wind force as it influenced the sails of a frigate and was later converted to anemometer equivalents.

years and could reliably record the resulting time series for eventual recovery. Measurements became ever cheaper. The process has culminated thus far in satellite measurements from space, free-drifting floats, self-propelled "gliders," and remote acoustic technologies. Change is ongoing.

Because so much of the history of modern physical oceanography is bound up with the new technologies, the writers in the volume edited by Jochum and Murtugudde (2006) devote much space to it. Many interesting instruments exist, but only a fraction can be described here, enough to give the flavor of some of the more common technologies. The early days of the electronics era are described in more detail by Baker (1981) and the essay by J. Crease in Laughton et al. 2010 is a good description of the transitional era.

MOORED CURRENT METERS AND PROFILERS

Current meters, as placed in mooring lines as in Fig. 2.10, are widely regarded as a mature technology—they have been in use for decades, and a great many oceanographic inferences have been based upon these records. Nonetheless, despite the best efforts of scientists and engineers, problems remain. Figure 2.11 shows scatter diagrams of speed and direction from three different types of current meters as well as a comparison between two of the same type placed close together. The magnitudes of these differences are surprising, and subtle data analysis must be done with an understanding of the conditions under which records might be considered reliable. Certain issues appear repeatedly in the use of moored current meters:

1. In low-current regimes, mechanical rotors can stall for extended periods. Acoustic instruments may not encounter sufficient backscattering particulates in the water. These problems appear as zero-velocity intervals and bias the results.

2. Instruments are most commonly mounted on moorings that do not penetrate the surface (subsurface moorings; see Fig. 2.10). In high currents, such moorings can lean over by hundreds of meters or more (depending upon how far from the seafloor the mooring extends), meaning that the measurement is not being made at the nominal instrument depth, a particular problem during the most energetic flows. More generally, unless extremely expensive and uncommon rigid mounting structures are used, all instruments move to a greater or lesser degree during the measurements, which cannot be regarded as strictly obtained at a fixed position. Hogg (1991) discussed approximate correction methods. The extent of this problem has to be analyzed on a case-by-case basis.

3. Years ago it was found that surface waves and other high-frequency movement of the surface floats on surface-buoyed moorings tended to "pump" the speed-measuring device on certain current meter designs even at great depth (Gould and Sambuco, 1975), leading to spuriously high kinetic energies. The problem has been generally dealt with by designing new types of current meters less subject to the problem (the *vector-measuring current meter*, VMCM), using subsurface moorings, and/or using mooring types capable of decoupling the near-surface instruments from the rest of the mooring (slack elements near the surface). Figures such as 2.11 show, however, that the consistency-calibration problem is not fully solved, with disconcertingly large differences between speeds and directions determined by different current meter types even on the same subsurface mooring (see Hogg and Frye, 2007). Hendry and Hartling (1979) documented a quite unexpected measurement error.[8] These records

---

[8] A magnetostrictive pressure-induced effect biasing the compass directions.

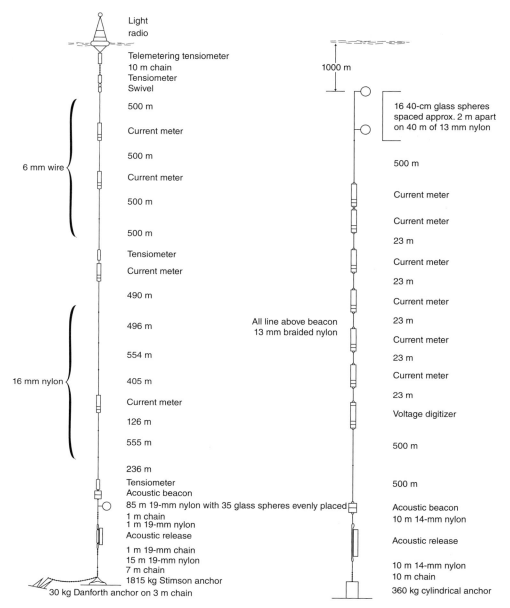

Light
radio

Telemetering tensiometer
10 m chain
Tensiometer
Swivel

500 m

Current meter

500 m

Current meter

500 m

500 m

Tensiometer
Current meter

490 m

496 m

554 m

405 m

Current meter

126 m

555 m

236 m
Tensiometer
Acoustic beacon
85 m 19-mm nylon with 35 glass spheres evenly placed
1 m chain
1 m 19-mm nylon
Acoustic release

1 m 19-mm chain
15 m 19-mm nylon
7 m chain
1815 kg Stimson anchor
30 kg Danforth anchor on 3 m chain

6 mm wire

16 mm nylon

1000 m

16 40-cm glass spheres
spaced approx. 2 m apart
on 40 m of 13 mm nylon

500 m

Current meter

Current meter

23 m

Current meter

23 m

Current meter

23 m

Current meter

23 m

Current meter

23 m

Voltage digitizer

500 m

500 m

Acoustic beacon
10 m 14-mm nylon

Acoustic release

10 m 14-mm nylon
10 m chain
360 kg cylindrical anchor

All line above beacon
13 mm braided nylon

Figure 2.10: A typical surface mooring (left) and one with no surface expression (right). Subsurface moorings became much more common after it was realized that wave action on the surface float of taut moorings coupled into the subsurface current meters giving spuriously high speeds (Gould and Sambuco, 1975). Subsurface moorings became possible only with the invention of the *acoustic release*—a device capable of releasing the anchor from the mooring cable, upon an acoustic command from a surface ship, when the upward tension on the instrument is typically very strong (thousands of Newtons). Battery lifetime in releases is one of the limiting factors in mooring duration. Special buoyancy elements capable of surviving at depths of thousands of meters had also to be developed—typically heavy glass spheres. Instruments in the mooring line include those for measuring the ocean (current meters, temperature sensors, and the like) and those included for engineering purposes (tensiometers, accelerometers). (From Fofonoff and Webster, 1971.)

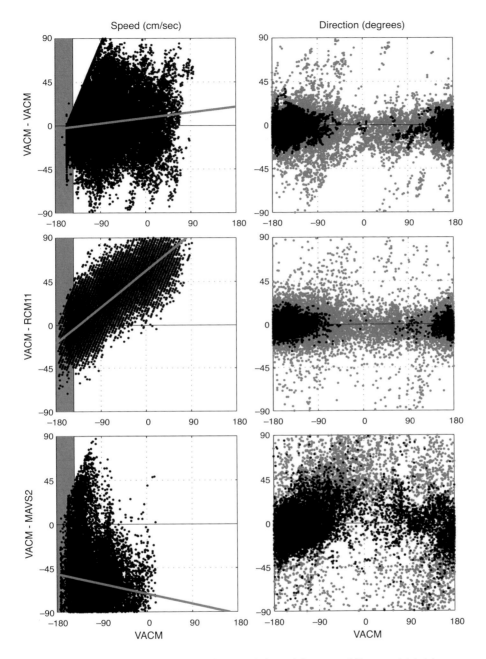

Figure 2.11: Scatter diagrams for speed difference (left) and direction difference (right) between two vector-averaging current meters (VACMs; upper panels) on the same mooring, between one of the VACMs and a rotor-and-vane instrument (RCM11; middle panels), and between a VACM and an acoustic instrument (MAVS2; bottom panels). Instruments were separated by about 10 m in the vertical. Darker points in the direction-difference plots indicate when the speed is greater than 5 cm/s while the lighter ones are for speeds less than this value. Directions for speed differences with magnitude greater than 4 cm/s are not displayed. (From Hogg and Frye, 2007, which should be consulted for much more detail.)

now reside in easily accessible databases but need to be used with an understanding of their limitations.

Devices that can measure velocity, temperature, oxygen, and other properties while moving vertically up and down a moored line have also been developed. Some of them (e.g., see Krishfield et al., 2008) have been specially designed to operate under sea ice.

DRIFTERS

Apart from the time-of-arrival issue, the main concerns with surface-drifting objects were the contributions from wind effects, the degree of slippage relative to the water, and the actual trajectories and their timing. Even today, wind effects remain a concern, particularly in high-wind regimes, but the timing issue has been largely resolved by tracking instruments from satellites; all of the intermediate positions and times are available, in addition to those at the beginning and end (see Niiler et al., 1995). Many of these surface drifters today carry temperature, salinity, and atmospheric-pressure sensors.

FLOATS

By the early 1960s, the technologies of batteries, clocks, transducers, and pressure cases had advanced to the point that devices could be built that would float at pre-determined pressure depths and emit acoustic pulses at known times, fulfilling the vision of Swallow (1955) and Stommel (1955).[9] Scientists on a ship at a known geographical position could lower a listening device (a *hydrophone*) and by determining the delay between known time of emission and time of arrival, find the distance to the float. By then steaming to a new position, and assuming that the float moved little in the interim, they could determine a second or third distance and then use simple circular navigation (Rossby and Webb, 1970) to produce a float position.

These early *Swallow floats* have evolved greatly. Among other changes, escape from the need to keep a ship at sea was engineered by the use of moored hydrophones to record times of arrival, with the navigation being done back in the laboratory after the mooring was recovered. Eventually, instead of sound sources being on the floats and the hydrophones moored, their positions were interchanged, so that a comparatively small number of moored, expensive acoustic sources could be heard by a large number of passively listening floats (albeit the internally recording floats had to be recovered or had to eventually surface and relay the data by satellite). At low enough frequencies (requiring comparatively large acoustic sources), the presence of the acoustic waveguide could be exploited (see Fig. 2.12), leading to far greater acoustic ranges. Such instruments with sound sources are generally known as SOFAR floats,[10] and when the sound source is moored, are known as RAFOS floats (reversed spelling of SOFAR; see Rossby et al., 1986). One further evolved version—one not employing acoustics but with which only the time and position of launch and return to the surface are obtained, and temperature and salinity profiles are found as the float sinks to depth and returns to the surface—are called (Profiling) Autonomous Lagrangian Circulation Explorer, or (P)ALACE floats, and are discussed further below with the Argo program.

---

[9]By convention, a *drifter* is a freely moving surface-confined instrument, and a *float* is one that moves in the ocean at depth. "Drifters float, and floats sink" (Anonymous).

[10]The waveguide is known as the *SOFAR channel* (SOund Fixing And Ranging), as an analog of RADAR).

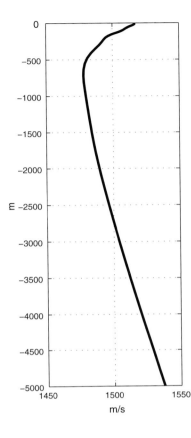

Figure 2.12: The sound speed near 140°W, 34°N in the eastern subtropical Pacific Ocean. The profile, determined from temperature and salinity via an empirical formula, is characteristic of much of the global ocean with a distinct minimum in the vicinity of 1000 m. The great importance of such profiles is that the minimum gives rise to a *waveguide*, the SOFAR channel, for very efficient long-range sound propagation. This effect is exploited by SOFAR and RAFOS floats, as well as in acoustic tomography. (See Medwin and Blue, 2005 for general discussion of ocean acoustics.)

Concerns about the degree to which floats at fixed pressures do follow water parcels—that is, whether they provide Lagrangian measurements—have led to versions that dynamically adjust to maintain, for example, fixed temperatures or densities (see Rossby et al., 1986). Others (e.g., Voorhis, 1970) have been fitted with vanes so as to rotate as fluid moves vertically past the float. Analysis of even truly Lagrangian measurements is inherently more difficult than for Eulerian ones at fixed points, as the former necessarily mix space and time variations, whereas the latter (almost) cleanly separate them. The papers by Davis (1991), D'Asaro (2003), and Gould (2005) are good starting points.

HYDROGRAPHIC AND VELOCITY PROFILING DEVICES

Hydrography in physical oceanography is the measurement and interpretation of temperature and salinity measurements.[11] Electronic instruments capable of continuous profiling of temperature and salinity as functions of measured pressure started to become available in the early 1960s (see Baker, 1981). These were first labelled STD (salinity-temperature-depth) devices, and later, CTD (conductivity-temperature-depth) instruments. A salinity-measuring device might be thought better and more sophisticated than one that only measures conductivity, but the calculation of salinity from conductivity is sufficiently complicated that it quickly

---

[11] In navies, "hydrography" historically has meant instead the determination and charting of coastlines and water depths.

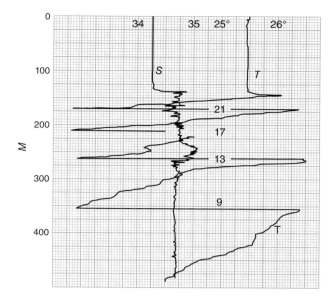

Figure 2.13: The small-scale structure (now known as *finestructure*) that first became apparent only when continuous profiling instruments were developed. Yet smaller scales, visible with later instruments, are now known as *microstructure*. Numerals 21, 17, etc., indicate resetting of the temperature scale to keep it on the strip-chart recorder. Temperature and salinity scales are indicated at the top in this photograph of the original analog recording and are subject to a number of later corrections. (Used with permission from H. Stommel and K.N. Federov [1967]. "Small scale structure in temperature and salinity near Timor and Mindinao." *Tellus* 19: 306–325. Tellus/Co-Action Publishing.)

became clear that "offline" computer power could do a much better job than could be done within the instrument itself. The early data were widely regarded with skepticism, because they showed a multiplicity of small scale structures (Fig. 2.13) that were at odds with the smooth profiles that had always been drawn through the vertically sparse data obtained from reversing thermometers. Doubts also arose because empirical estimates of turbulent mixing (eddy) coefficients were so large that the observed small-scale structures would not have been able to survive very long after the unknown events generating them had occurred. Skepticism gradually eroded as it was shown that the structures were reproducible from different instruments, persisted in many cases over large horizontal distances and times, and as the theories of what came to be called *finestructure* and *microstructure* emerged to rationalize the observations. The former are structures roughly between one and tens of meters, and the latter anything below about 1 m.

This technology eventually gave rise to a whole family of profiling instruments, not just for temperature and salinity, but also for quantities such as oxygen concentration, and ultimately the velocity field (see Fig. 2.14). Vertical resolution is now on the centimeter scale (Osborn and Cox, 1972). Many free-falling profiling devices have been developed, including those based upon electromagnetic field variations (Sanford et al., 1978). Dependence on ships or moorings means that measurements remain comparatively rare.

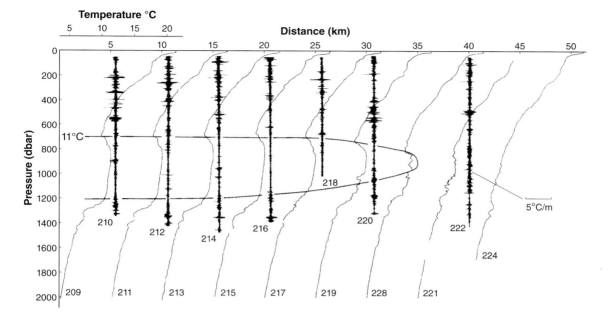

Figure 2.14: The velocity microstructure intensity, measured in terms of the oceanic dissipation, ε, overlain on a series of simultaneous CTD temperature profiles. numbered 209, 210, etc. Largest values of $\varepsilon > 10^9$ W/kg. (ε is defined in Eq. 3.33). Characteristic "bursts" are seen on scales down to the molecular. (From Oakey, 1988.)

PALACE/ARGO FLOATS

Beginning about 2002, the technology of both floats and profiling instruments had improved to the point that the worldwide oceanographic community began deploying instruments on free-drifting floats. In a program now called Argo (Roemmich et al., 2009), approximately 3000 profiling (PALACE) floats are in the water at any one time. They are launched from a ship and typically descend to 1000 or 2000 m, where they drift for, commonly, 10 d. After that time, they ascend to the sea surface, measuring temperature and salinity as they go. At the sea surface, they remain long enough to upload their measurements and a GPS-determined position to a communication satellite. They then descend to their resident or parking depth and begin another cycle. Figure 2.15 shows the positions of deployed floats that successfully reported in March of 2011.

Discussions are underway of extending the float capability to the full water depth, and its technical feasibility is not in doubt: cost is the major factor. A serious energy penalty exists as well as a requirement for much stronger pressure housings and more demanding calibration. This program is an important step toward freeing oceanographers from the straitjacket of having to use ships for in situ measurements. In many ways, the Argo floats are a revolutionary equivalent of the weather-balloon capability long available to meteorologists. Note, however, that oceanographic structures are laterally much smaller than atmospheric weather systems, and the Argo network remains thin compared to much oceanic structure, even if the existing population of about 3000 can be maintained. As with radiosondes, reliable worldwide calibration is an ongoing problem.

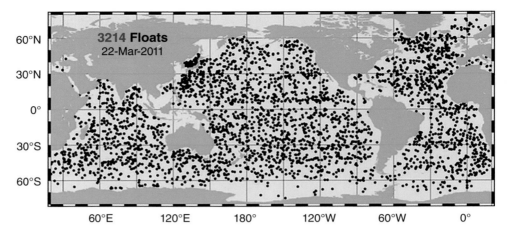

Figure 2.15: The surface positions of the Argo floats reporting temperature and salinity profiles in March 2011. The coverage is extraordinary compared to historical ship-based measurements. A major concern remains with the absence of observations below 1000–2000 m and the tendency of floats to cluster in certain regions and to avoid others. Care must be taken both with calibration and the interpretation of the measurements in the presence of the strong eddy field. (From the Argo project website.)

ACOUSTICS

Acoustic techniques are attractive because sound can travel long distances within the fluid ocean—in contrast to the behavior of electromagnetic signals. Generally speaking, the lower the frequency, the greater the range. But in contrast to the electromagnetic situation, penetrating acoustic frequencies and wavelengths are useful in practice. Apart from the acoustic depth sounder, which has a very long history, the most widespread use has been in navigation, as in the float methods already described.

A general methodology used over distances ranging from centimeters to hundreds of meters is based upon acoustic Doppler (AD) techniques. The fundamental measurement is the frequency shift observed from a transmitted sound pulse that is back-reflected from particulate matter in the water column. The time delay of the back-scattered signal gives the range to the point of measurement, and the velocity of the particulate matter is inferred from the degree of Doppler shifting. Acoustic Doppler techniques have been implemented on current meters (see Hogg and Frye, 2007, for references), in profiling instruments mounted on moorings (e.g., Doherty et al., 1999), in devices mounted on underway ships and capable of providing measurements to a few hundred meters, and on lowered instruments (LADCP). Underway measurements need to be carefully corrected for ship motion (see Joyce, 1982; Flagg et al., 2006) and all applications require an adequate supply of suspended particulates.

Acoustic tomographic methods are an analog of medical tomography and involve the inference of temperature and velocity structures from the behavior of the travel times of acoustic rays or normal modes through volumes of ocean. As such, they represent a form of remote sensing in the same way that a radiologist makes inferences about the interior of a patient from the behavior of X-rays or ultrasound. Munk et al. (1995) described these methods in detail. Their more widespread use awaits the engineering development of cheaper acoustic sources having adequate bandwidth and power and of better exploitation of the natural background sources of earthquakes, animals, ice breakup, and ships.

Another similarly under exploited and promising acoustic method is that based upon techniques arising from seismic surveying of the seafloor. High-frequency acoustic signals emitted by a transiting ship that also tows arrays of hydrophones are used for oil exploration and more basic science. Traditionally, marine seismologists suppressed measurements of reflection returns from within the water column arising from density contrasts, but it is known that the entire water column can be imaged that way (Holbrook, 2003). Now needed is the design of acoustic sources and towed-array configurations best suited to the water-column problem—and those would not necessarily coincide with the ones used for subseafloor oil exploration.

THE EXPENDABLE BATHYTHERMOGRAPH

Beginning in the 1970s the very much depth-limited and cumbersome BT was replaced by an expendable version (XBT). Because of the instrument's use in antisubmarine warfare, methods were developed for wrapping extremely thin, fragile, coated, conducting wire that would unspool as the device dropped, essentially in free fall, from a moving ship. A measured temperature signal was sent up the wire to an on-deck recorder—initially analog, today digital. By calibrating the fall rate, the elapsed time could be converted into a measurement depth. These devices could operate as deep as 800 m, and soon many thousands of profiles were being obtained every year. No special onboard-winches are required, untrained observers can use them, even automatic launchers and recorders have been employed, and the measuring ships can travel at full speed through the water.

XBT measurements came to dominate measurements of upper-ocean temperatures (salinity-measuring versions were also developed but were never much used). With the end of the Cold War, the number of such measurements began to decline rapidly, and the upper ocean is, in some regions, less well sampled today than it was decades ago (see, e.g., Fig. 2 of Abraham et al., 2013). Furthermore, antisubmarine warfare primarily involves very large upper-ocean thermal anomalies (via the speed of sound). High-accuracy calibration was thus never required nor done. As interest in climate change has arisen, the data have been used for determining oceanic changes at precisions and accuracies far beyond the calibration standards. The result has been a seriously confused science, and careful examination has shown large, systematic, and space- and time-varying bias errors in the data (e.g., Gouretski and Koltermann, 2007; Reverdin et al., 2009, DiNezio and Goni, 2011, Abraham et al., 2013). This story is not untypical of the history of oceanography, with events such as the Cold War having a major influence on what was measured and where, then followed by the application of the data to problems outside of their original intended or calibrated use.

BOTTOM PRESSURE GAUGES

Surface barometers provide great insight into the behavior of the atmosphere, even in the absence of any measurements at height. The equivalent measurement at the bottom of the ocean might have been thought to be equally illuminating. But the oceanic measurement is far more difficult: Two of the major problems are that instead of dealing with one atmosphere of pressure, values can exceed 400 atmospheres, and that the instrumental altitude (the water depth) is not known even close to the required accuracies (1 cm vertical uncertainty corresponds to an important oceanic signal). Nonetheless, efforts were devoted to deploying instruments that could measure bottom-pressure *fluctuations* at the centimeter or smaller level, even though the absolute pressure was of no use. Vibrating string instruments were first used and later replaced by more stable vibrating crystals (e.g., Irish and Snodgrass, 1972; Wearn and Larson, 1982). Both of these sensors were very sensitive to temperature as well as to pressure, and so an accurate thermometer was also required. These instruments are still in use and have proven valuable for the study of tides and other high-frequency phenomena, but their creep under the high pressures and subsidence into the seafloor have generally precluded the study of low-frequency signals (see Filloux, 1971). The rise of satellite altimetry (taken up below) has made the sea-surface height measurement a better analog of meteorological barometric data. Very recently, satellites that can measure time-dependent gravity have begun to permit the inference from space measurements of changing ocean-bottom pressures.

ELECTROMAGNETIC METHODS

As already noted, the idea of using voltages induced in electrical cables by the moving overlying conducting fluid dates back to Faraday. Luther et al. (1991) described a successful, regional example using purpose-laid instruments. The most important application of this method thus far has been the measurements of the transport of the Gulf Stream in the region where it is called the Florida Current, which lies between Florida and the offshore Bahama Islands (discussed below). Some efforts have been made to use abandoned telephone cables on ocean-basin-wide scales, but the data have proven almost uninterpretable.

METEOROLOGICAL MEASUREMENTS

Sophisticated meteorological observing systems are today carried by some ships and report vector wind, relative humidity, barometric pressure, etc., automatically (see Kent et al., 1993). More common, however, have been devices read on the bridge by the officer on watch and hand recorded in the deck log. Many issues arise: improperly mounted anemometers can be shielded behind ship structures from some wind directions; busy mates can mistakenly add, rather than subtract, the ship's motion to the apparent vector wind; east longitudes can be entered into logbooks as west, etc. All lead to problematic reports. As was true in the preelectronics era, determining the physics of the relationship between the windfield measured at some distance above the sea surface, conventionally 10 m, and the actual stress experienced by the underlying ocean remains a difficult and very active area of research.

In modern weather forecasting, global data sets of all types (from weather balloons, surface observers, satellites, moorings) are combined with global atmospheric models to form an estimate of the three-dimensional atmospheric state, nominally every 6 h. These estimates are then used to make forecasts out to several days. At the intervals when the data are introduced into the model, the result is known as an *analysis*, in which observational data is combined in an approximate least-squares fit to the model.

The long time series available from the atmospheric analyses are commonly used to seek fluctuations and trends. But atmospheric model computer codes changed greatly over the decades, and concern arose that false trends might appear both because of the constantly shifting model and the methods used to combine models with the data sets. This consideration led to the construction of what meteorologists call *reanalyses* in which for intervals typically of 50 to 100 years, a model and the methodology are held fixed (see Kalnay, 2003), although the observing system changes radically. The accuracy and physical consistency of these estimates is a major concern and is discussed in Appendix B.

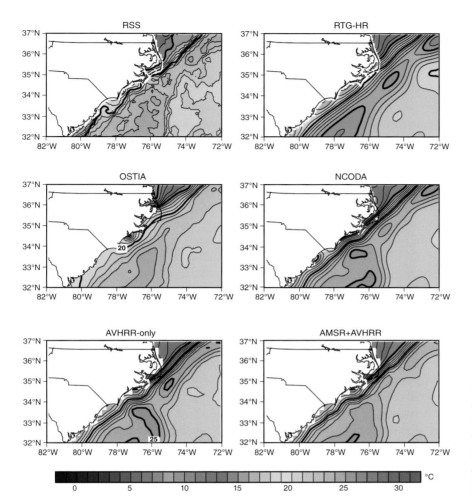

Figure 2.16: The variations among different analysis products of SST near the South Carolina coast on 1 January 2007. See the reference for the meaning of the acronyms. (From Reynolds and Chelton, 2010.)

2.5 THE RISE OF SATELLITES

Earth-orbiting satellites opened the possibility to obtain truly global measurements of the ocean for the first time. Satellites do not, however, evade the problem that radio and light waves at useful frequencies cannot penetrate the interior beyond a few meters. Consequently, with the exception of the gravity missions, the only satellite data of use are those from oceanic surface fields. In the years following the launch of Sputnik in 1957, satellite measurements of passive infrared and microwave emissions from the sea surface were used to determine surface temperatures and, secondarily, ocean color (of central importance in biological processes). Because sea-surface temperatures (SST) are so indirectly connected to the water column below, they have tended to remain primarily of interest to meteorologists, who employ them as lower boundary conditions on the overlying atmosphere. They have, however, had great qualitative importance in convincing oceanographers of the complexity of surface flow and thermal patterns. Reynolds and Chelton (2010) discuss the ambiguities that underlie SST (see Fig. 2.16). In the summertime, particularly, it may be a "skin-temperature," changing substantially within millimeters of the sea surface. (See fig. 1 of Donlon et al., 2002.)

Several books exist summarizing methods of satellite observations of the ocean (e.g., Stewart, 1985, Baker, 1990, Robinson, 2004, Martin, 2004) and thus spaceborne data are given short shrift here, with only a brief summary of altimetric, wind, and gravity measurements.

ALTIMETRY

Satellite measurements had only a peripheral influence on oceanography until the rise of accurate altimetry at the end of 1992. The principle of this measurement is simple.[12] First, define the shape of the sea surface if the ocean were brought to rest with a constant density: this surface coincides with a gravitational equipotential called the *geoid*, whose height is $N_g(\lambda, \phi)$, where ϕ is latitude, and λ, longitude, ignoring slow time-dependent terms and tidal disturbances to it. Flying a radar on a spacecraft over the ocean determines the distance, $r_s(\lambda, \phi, t)$, of the spacecraft from the sea surface via the round-trip travel time from the spacecraft to the surface. From knowledge of the distance, R, from the center of the earth to the spacecraft, the height of sea surface relative to the center is, $S_a(\lambda, \phi, t) = R(\lambda, \phi, t) - r_s(\lambda, \phi, t)$. Then

$$\eta(\lambda, \phi, t) = S_a(\lambda, \phi, t) - N_g(\lambda, \phi) \tag{2.1}$$

would be the height of the sea surface relative to the geoid—the dynamical variable of interest in determining the fluid motion. Making the generally excellent assumption of hydrostatic balance, $\rho g \eta$ would be the pressure field at the sea surface—evaluated at the geoid. While η is a sea-surface property, the pressure field determined from it is dynamically coupled to the entire water column—unlike other surface fields. This relationship is discussed in more detail later in the context of the oceanic circulation.

Without here going into the details of altimetry (see Stewart, 1985; Fu and Cazenave, 2001; Robinson, 2004; Martin, 2004), suffice it to say that its use becomes a major challenge when it is recognized that for measurements on the basin and larger scales, S_a or η must be known with an accuracy approaching 1 cm from a spacecraft orbiting at over 1000 km above the surface of the Earth—an accuracy of one part in 10^8. Several major corrections must be made to the observations before they are quantitatively useful for describing oceanic variability (see Fig. 2.17). To a good first approximation, the ground track under the satellite (Fig. 2.18) is governed by the Earth spinning underneath the orbital plane, which slowly precesses. The consequence is a complicated space-time sampling of the sea surface, whose properties differ with the particular spacecraft inclination and repeat period. (See Fig. 3.31 for the result after 10 d for one type of mission.) To obtain the accuracies required to produce a scientifically useful measurement, a considerable list of primary corrections must be made. An incomplete list includes

1. Orbiting spacecraft see a space-time structure dominated by the deviation of the Earth from a sphere. As listed in Appendix D, the polar radius differs from the mean equatorial one by about 21 km and, as measured along the orbit, is by far the dominant component; it is subtracted. Variations in N_g exceed 200 m overall (see Fig. 3.2), and errors in its estimates translate into errors in η, and vice versa.

2. Open-ocean tides have amplitudes exceeding 1 m. The altimeter itself was the first true, nearly global tide gauge, and subject to understanding of the "alias" periods discussed in

---

[12] Wunsch and Stammer, 1998.

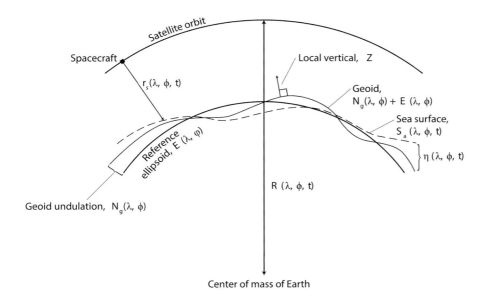

Figure 2.17: A definition sketch of altimetric measurement geometry. Height, *r*, above the surface is about 1300 km. The dynamically important variable is the lateral slope of η, the difference between the sea-surface height S_a and the the geoid height $N_g + E$. Sometimes the presence of the reference ellipsoid, *E*, is suppressed in writing; see also Chelton et al. (2001). (From Stammer and Wunsch, 1998.)

Appendix A, open-ocean tides are now known to an accuracy of about 1 cm. *Shallow water tides can be far larger than in the open ocean and remain less well-determined, and altimetric data must be used cautiously in regions shallower than 500–1000 m (see Vignudelli et al., 2010).*

3. Atmospheric water vapor content varies and changes the speed of light through the atmosphere. Water vapor content must be measured and can introduce further errors.

4. The height and orientation of the spacecraft are determined by a combination of measurement (GPS navigation devices on the spacecraft, laser tracking from the ground, and special systems such as DORIS[13]) and the relativistic equations of orbital motion. Residual orbital errors map into a complicated space-time varying error in the calculated height of the sea surface.

5. Ionospheric variations change the speed of light. This problem is dealt with in some altimetric systems by direct measurement—employing two distinct radar frequencies in the altimeter.

6. A strong, but not universal, tendency exists for the sea surface to respond nearly statically to atmospheric pressure changes, with an approximately 1 cm/millibar (1 hPa) change in the atmospheric load, and this response is usually called the *inverted barometer effect*. The correction is imperfect because the ocean can respond dynamically at periods longer than a

---

[13] Doppler Orbitography and Radio-positioning Integrated by Satellite.

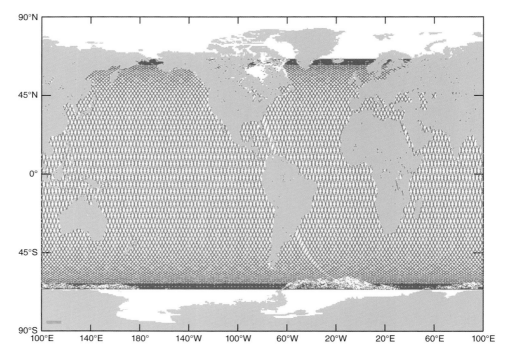

Figure 2.18: The ground track built up over 10 d by the TOPEX-POSEIDON and Jason altimetric spacecraft. On any given day, measurements are widely separated spatially as the Earth rotates under the slowly precessing plane of the satellite's orbit. Note that no coverage from these mission types exists at the highest latitudes. (From the website of the European Space Agency, ESA.)

few days, will not respond at all at periods shorter than a few hours, and does not respond to atmospheric pressure changes that are uniform over the ocean. "Uniform over the ocean" may, in practice, refer to large basins rather than to the truly global, as basins cannot instantaneously exchange mass. Fig. 3.36 shows the estimated 6-year time-mean atmospheric pressure over the oceans with values of tens of millibars or tens of centimeters of sea-surface height equivalent. Shorter-period temporal variations are much larger.

7. Sea-surface scattering phenomena affect the radar pulse in several ways, leading to the electromagnetic bias correction. One effect is the tendency for large waves to have steeper crests than troughs, and the latter return a stronger reflected radar signal. The raw pulse thus sees "too deep" relative to the local mean sea surface. Complicated electromagnetic effects also occur upon reflection from a moving conducting fluid. These effects are generally dealt with by analyzing the detailed radar pulse return, using aircraft-based calibration tests, and developing algorithms for defining the mean surface.

8. General aliasing of high-frequency motions. Undersampled signals have an apparent frequency content sometimes radically different from the correct one. The central TOPEX/POSEIDON[14] and follow-on Jason mission measurements have had a repeat period of 10 d,

---

[14] TOPEX is an acronym of Ocean TOPography EXperiment. POSEIDON is a dual acronym created by Michel Lefebvre: Premier Observatoire Spatial Étude Intensive Dynamique Ocean et Nivosphere (sic), or Positioning Ocean

thus producing a *Nyquist frequency* (the highest visible frequency; see Appendix A) of 1 cycle/20 d. Motions with time scales shorter than 20 d necessarily appear at longer periods, and the conspicuous semidiurnal and diurnal tide signals in the altimeters at periods from about 60 d to 6 months and longer are the clearest evidence of this effect (see Tab. 7.1). Space-time aliasing, in which both incorrect spatial and temporal structures appear, has been considered, mainly in the abstract.[15] The effect has not been adequately quantified because the spatial aliases are a direct function both of the poorly known space-time structure at high frequencies and of the details of how the motions are spatially filtered in the data reduction systems.

9. At accuracy levels approaching 1 cm, a host of smaller contributions can be important for some problems. These range from the effects of the Chandler wobble of the Earth's rotation axis (Ch. 6) to residual uncorrected tides, seasonal biases caused by the changing incidence of sea ice at high latitudes, the assumption that the center of mass of the Earth remains fixed, and concerns that the GPS system has internal drifts. Some of these are particularly significant in attempts to determine global mean sea level change.

SCATTEROMETRY

The second kind of satellite data with a major impact on the determination of the ocean circulation have been from scatterometers (see Stewart, 1985; Risien and Chelton, 2008), which provide a nearly direct measurement of the major atmospheric forcing of the ocean—the wind stress acting at the sea surface—freeing oceanographers for the first time from the sporadically distributed and often poorly calibrated shipboard anemometer records (see Fig. 2.19). The basis of the measurement is a determination by radar backscatter of the amplitude and orientation of the small-scale ripples—which theory and observation show are the scale at which momentum is transferred from the atmosphere to the ocean. These measurements, like altimetry, are neither simple nor perfect. Spatial sampling follows the complex surface pattern implied by any orbiting spacecraft, so that several days are required to build up a global picture. Meteorological analyses and reanalyses now employ these data in concert with other atmospheric data.

Conversion of the radar backscatter signal into the force acting on the sea surface is a complex, and not wholly solved, problem because it involves the physics of the ocean surface boundary layer, the most complicated part of the entire system, as well as the scattering behavior of an electromagnetic wave from a moving, conducting surface. Measurement nominallly detects the stress exerted on the sea surface; but for practical reasons, calibration is in terms of a conversion into wind velocity, with a subsequent (usually) conversion back into stress. Risien and Chelton (2008) described some of the methods used.

GRAVITY-MEASURING SATELLITES

The direct inference of changes in ocean bottom pressure from satellite data became a reality with the Gravity Recovery and Climate Experiment (GRACE) mission, which began in 2002 (usable data came later). Two heavy masses are orbited, one chasing the other around the Earth, at as low an altitude as is practical (Tapley et al., 2004). With lasers or radars, precise

---

Solid Earth Ice Dynamics Orbiting Navigator.
    [15] Wunsch, 1989; Greenslade et al., 1997; Tai, 2006.

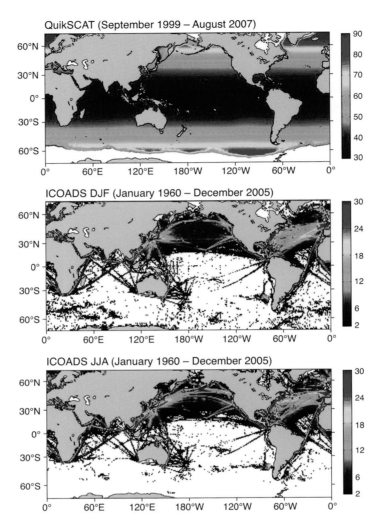

Figure 2.19: In the top panel, values are the average number of measurements from the satellite QuikSCAT in each $1/4° \times 1/4°$ grid cell/month over the 8 years 1999–2007. In contrast, the middle panel shows the number of ship observations per $1° \times 1°$ cell in December, January, February (DJF) in the 46-year period 1960–2005. The bottom panel is the same available ship observations except during the months of June, July, August (JJA). The near absence of southern hemisphere data is striking as is the bias toward ship tracks in the northern hemisphere. No data exist from ice-covered regions. (After Risien and Chelton, 2008.)

measurements can be made of the relative velocity of the two masses—which will fluctuate as they encounter disturbances in the gravity field at the orbital altitude.[16]

---

[16] Another gravity-measuring satellite with a different design and directed primarily toward the time-mean geoid rather than the time dependence is called GOCE (Gravity Field and Steady-State Ocean Circulation Explorer); see Knudsen et al., 2011.

The gravitational field at any satellite position $\mathbf{r} = (r, \lambda, \phi)$, where r is the radius, ϕ the latitude,[17] and λ the longitude, is calculated from gradients of the Newtonian potential,

$$\varphi_G(r, \lambda, \phi, t) = G \iiint \frac{\rho(r', \phi', \lambda', t)}{\mathbf{r} - \mathbf{r}'} dV, \tag{2.2}$$

where $\mathbf{r}'$ is the position vector of any mass contributing to the field at $\mathbf{r}$. The integrals are taken over the entire Earth volume, V, including the solid, fluid, and gaseous parts. G is the gravitational constant. Time changes in the gradients of φ_G are ascribed to time changes in ρ. Because the ocean bottom pressure, $p_b(\lambda, \phi, t)$, is the integral (through the hydrostatic relation) of all the mass above any point λ, ϕ, changes in bottom pressure will influence φ_G, and hence the relative velocities of the two spacecraft, and one can "invert" the resulting estimates of the gradient φ_G for the density changes, $\Delta\rho$. In conventional undergraduate physics problems, φ_G is computed from a known ρ—a "forward" problem. Determining ρ is thus an *inverse problem*, because the properties of φ_G are measured and the goal is to determine the ρ that produced the measurements (see Appendix B). Because the gravity field at any given location is an integral over all of the disturbing masses, globally, the result will not be a unique solution. In the process of solving the problem, strong assumptions are sometimes made about the local versus global character of $\Delta\rho$.

If the ocean responds only as an *inverted barometer*, compensating any increase of decrease in atmospheric pressure by moving the sea surface downward or upward, no net mass disturbing signal exists, because the ocean and atmosphere contributions cancel. But to the degree that the ocean fails to respond, atmospheric loads will produce measurable bottom-load (pressure) signals. Local heating and cooling do not change the bottom pressure, but wind-induced movements in the water-column height, as well as the addition and removal of freshwater, will produce seafloor signals. The largest nontidal signals seen in GRACE data are the annual changes in ground water stored on land, and these fluctuations must be corrected to extract the comparatively small oceanographic ones.

ELEPHANT SEALS AND OTHER ANIMALS

One of the more interesting developments of the past few years has been the ability to put sensors and navigation equipment (GPS) on diving animals, as in the elephant seals of Figures 2.20 and 2.21 (see Charrassin et al., 2008; Roquet et al., 2013).[18] Over 300,000 profiles from these animals already (2014) exist, including numerous ones from under Antarctic sea ice where previously little information has been obtainable. Where this kind of intriguing capability of ocean observations will ultimately lead is obscure at the present time. Animals are naturally sensitive to water temperatures, salinities, oxygen content, and pressures, and some have highly sophisticated navigation and information-transmission capabilities. Perhaps advances in biological engineering will produce animals bred with purely biological sensors and navigation and communication devices to produce data accessible to scientists. At the moment that hope is science fiction.

---

[17] Colatitude is measured from the north pole and is 90° minus the latitude.

[18] The developments occurred because of concerns that acoustic tomography measurements might unduly disturb marine mammals. While those signal strengths showed no measurable biological effects, the instrumented-animal data proved useful both for understanding mammal behavior and for providing environmental information.

Figure 2.20: A southern elephant seal on the beach at South Georgia with head-mounted instruments including temperature, salinity, position, and depth devices, plus a transmitter for relaying profile information to a satellite communication system. (Credit: Michael Fedak/SEaOS.)

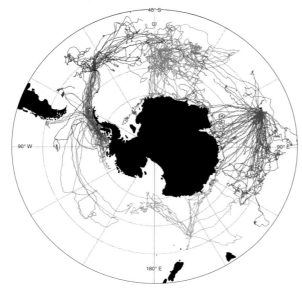

Figure 2.21: As of 2010, the paths taken by diving elephant seals, along which vertical profiles have been obtained. Nesting islands are color coded. Many more data have been obtained in the intervening time. (From F. Roquet, pers. comm., 2010.)

2.6 INTERMEDIATE- AND LONG-DURATION MEASUREMENTS

Achieving long-duration measurements remains a central, very important goal for oceanographers, particularly when it comes to understanding climate change. It is thus useful to examine the few existing extended records to understand some of the problems of their use. Because the near-global coverage by high-accuracy altimetric satellites began in 1992 and from profiling floats about 2002, the records of most interest are, at the time of writing (circa 2014) anything longer than about two decades. A few problems are generic to all such measurements and two warrant special mention: (1) calibration and (2) space-time sampling.

Many measurements begin as casual ones for purely local practical reasons, and only if sustained for some time do they excite interest as perhaps showing long-term (climate or otherwise) shifts. But many aspects change, including the technology, locations, the measurement time of day, and harbor construction, which can take place nearby. Often these changes are unrecorded, and calibration is thought to be unimportant, because the desired signals are so strong. Years later, scientists looking for slow fluctuations and weak trends may be frustrated by these problems.

Space-time sampling is a major concern both in trend determination and when data are used to form large-scale averages, perhaps even global ones. Data confined to shipping lanes or to the northern hemisphere or to summer seasons alone will likely, if averaged, be unrepresentative of the correct large-scale space-time mean. Aliasing errors, where space and time sampling is too infrequent to properly render frequency and wavenumber structures, are ubiquitous and rarely negligible.

TIDE GAUGES

Although tide gauges have undergone major changes in technology along with everything else, the advent of electronics had a less dramatic impact on them than it did on most other measurement methods. Data are still commonly acquired by installations attached to piers and other structures, and so the instruments can be tended nearly continuously by observers, who can also calibrate the observations several times per day. Older measurements, obtained by reading tide-staff levels, were generally carefully made and calibrated relative to landmarks, but major issues of temporal aliasing exist in the preinstrumental records. Sea level records extend back to the seventeenth century (in the Netherlands), and many records from the nineteenth century based upon variants of the Kelvin tide gauge were obtained and provided time-continuous data. Because global absolute levels were impossible to determine, local absolute gauge heights were rarely known before the GPS system became available.

The very oldest Dutch records are not useful for scientific purposes because they were made for practical purposes in regions of intense human modification of the shoreline. The record from Brest, France, dates from the early eighteenth century (Wöppelmann et al., 2008; see Fig. 2.22). Calibration hundreds of years later of such records is difficult. Beginning about 1900, the number of long records becomes, relatively speaking, numerous, although most of them were obtained in major northern hemisphere ports lying on the continental margins (Douglas et al., 2001; Church et al., 2010; see Fig. 12.15).

CABLED OBSERVATORIES

Prior to the invention of low-power recording devices and developments in high-energy density batteries, the idea of running cables into the deep sea was very attractive. Cable-laying ships were in wide use by telegraph and telephone companies, and instruments mounted in the deep ocean on cables connected to land would simultaneously avoid several serious problems: adequate power and data storage were available, as was very accurate time-keeping, and observers could see the data without any delay.

The most determined effort to construct an oceanic "observatory" was probably that of Stommel (1954), who concluded that Bermuda would be an ideal site: it was within easy flying or voyaging time from US East Coast ports; it appeared to be in a representative part of the subtropical oceans; the island slopes are steep, so that very long cable runs to reach deep water would not be necessary; and the local Bermuda Biological Station could provide logistical support.

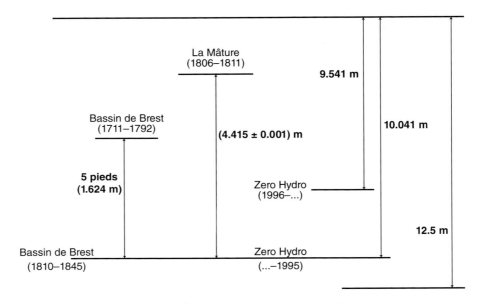

Figure 2.22: The changing *datum* (reference zero) at Brest, France, which has to be reconstructed, imperfectly, through time. The change in positions and in measurement units from prerevolutionary feet (*pieds*) and inches is also indicated. (From Wöppelmann et al., 2006.)

Stommel's observatory was abandoned after about two years and although another short-lived cable was installed later,[19] no extended record exists, except for an offshore hydrographic station time series that is ongoing. Because the United States and Canada have recently set out to establish an expensive global infrastructure of cabled observatories, it is worth briefly reviewing the reasons why Stommel and his successors did not pursue the idea.

The Stommel observatory consisted of a temperature-measuring cable (Fig. 2.23), plus a Nansen-bottle reversing-thermometer hydrographic station further offshore to be measured at monthly intervals, a series of surface drifters tracked by radar from the island, and the tide gauge at the Bermuda Biological Station, which had an independent existence.

Aside from a great number of practical difficulties (e.g., leaky cables and bad weather precluding measurement at the hydrographic station), much of the data proved uninterpretable.[20] With the benefit of sixty years of hindsight, it is evident that apart from some utility in observing internal waves,[21] the low frequencies were dominated by the eddy field, which was then unrecognized. Data obtained at Bermuda are typical only of Bermuda and cannot be readily extrapolated to anywhere else, even in the North Atlantic. The hydrographic station sampling was later increased to (nominally) every two weeks to reduce what was an obvious aliasing of high frequencies. Ship-based hydrographic casts, subject as they are to various funding and logistic vicissitudes over the years, are the one element of Stommel's vision that survives. Ultimately it must be replaced by automatic methods not dependent upon having human observers physically present.

---

[19] Wunsch and Dahlen, 1970.

[20] Stommel's comment about the drifters was "false mean flows from … rectification. They repose in published obscurity" (H. Stommel, in Hogg and Huang, 1995, vol. 1).

[21] Haurwitz et al., 1959, reinterpreted by Wunsch, 1972.

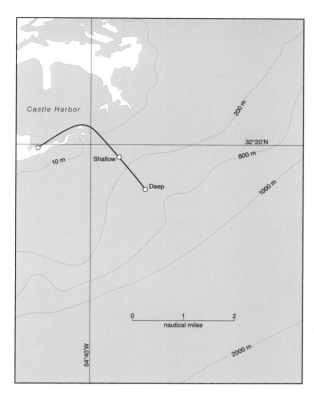

Figure 2.23: The position of the bottom-laid cable deployed by H. Stommel in the early 1950s on the southeast Bermuda slope. Positions of the two thermistors are indicated. (From Haurwitz et al., 1959.)

The great expense of cables and their maintenance and the limited number of locations where the length of the required cable is manageable means that such measurements are unlikely ever to be common. Restricted numbers of locations present a serious obstacle today for scientists attempting to describe a global fluid, which as will be seen, contains many distinct geographical regimes.

TELEGRAPH- AND TELEPHONE-CABLE DATA

A number of attempts were made to use abandoned telegraph and telephone cables for ocean transport fluctuations, but observations were rarely interpretable or sustainable. A uniquely fruitful application of this technology comes from the Florida Straits, where the Gulf Stream near Miami is trapped between the coast of Florida and the offshore Bahama Bank. Although it had been found that current meanders were generating misinterpreted low-frequency signals, a major calibration campaign (Baringer and Larsen, 2001; Meinen et al., 2010), intermittently repeated, has permitted inference of the volume transport of the Gulf Stream in this area. In Figure 2.24 note the gaps and the large, and probably random, variations. Apart from a small annual cycle, the record appears dominated by stochastic noise and with no apparent trend.

OCEAN WEATHER SHIPS

During and after World War II, ships were placed on aviation routes both to assist downed aviators and to provide weather information along the route. After some time, oceanographic measurements were begun on many of them, and sometimes they made observations en route

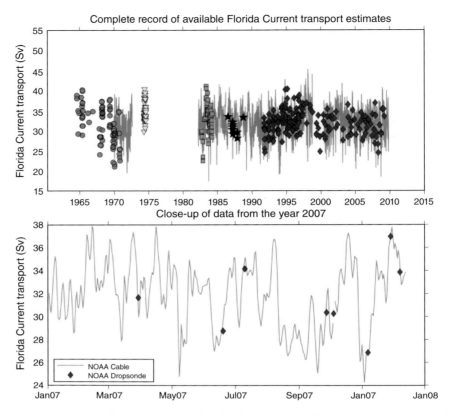

Figure 2.24: (Upper panel) An estimate of the Florida Current (Gulf Stream) transport, extending the record back into the 1970s. No qualitative change is apparent. (Lower panel) One year of transport values depicted on an expanded scale. Note the considerable range and the probability of strong aliasing if appropriate averages are not obtained. Diamonds indicate direct measurements for calibration. See the reference for an explanation of the symbols denoting calibration measurements. NOAA is the National Oceanic and Atmospheric Administration. A dropsonde is a transport measuring device. (From Meinen et al., 2010.)

to and from their station. With the advent of jet aircraft and satellite weather observations, the rationale for maintaining expensive ships in place was no longer compelling, and so these observations, despite their demonstrated utility, were terminated. Freeland (2006) discussed Station P (Papa; Fig. 2.25a). The longitude-time sampling (of the upper ocean) along its transit line is shown in Figure 2.25b, demonstrating the increased sampling frequency with time that was imposed as the temporal variability came to be appreciated. The last weather ship, Station M (Mike), was removed in 2010[22] and their data are now of significance only for the past.

OTHER LONG-DURATION RECORDS

Few other extended oceanographic records exist. Great efforts have gone into compiling ship reports of sea-surface temperatures, some of which originate in the seventeenth century. These

[22]Schiermeier, 2010.

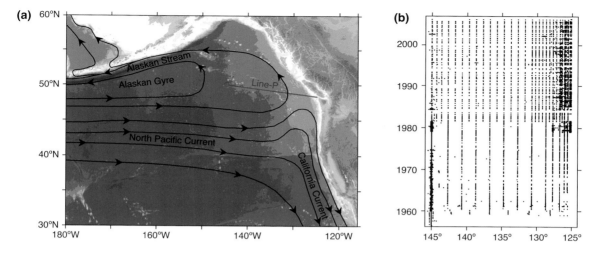

Figure 2.25: (a) The line covered by weather-ship Station P on the way to and from its nominal position. Black lines are a schematic of the surface flow. (From Freeland, 2006.) (b) Longitude-time coverage from Ocean Station P. Notice how the density of measurement increased with time as the true variability came to be recognized. (After Freeland, 2006.)

data have led to lengthy discussions of how to correct for changing technologies and, to the extent they are successfully applied, to atlases of ocean sea-surface temperature (see Folland and Parker, 1995; Kennedy et al., 2011). Some of the corrections are not very convincing, and acceptance of the results sometimes seems naive, or the "triumph of hope over experience."

PANULIRUS STATION—BERMUDA

The monthly hydrographic station off the south shore of the island was included in Stommel's Bermuda Observatory, and was named the "Panulirus Station" after the original Bermuda Biological Station vessel. After monthly sampling was understood to be inadequate, the rate was nominally increased to semimonthly, albeit the series was broken by bad weather and, for much longer periods, by lost equipment and funding. The record continues today (e.g., Joyce and Robbins, 1996; Phillips and Joyce, 2007). Some aliasing must still exist even at the higher rate, although it does not appear to have been quantified.

THE CLIMATOLOGICAL ATLASES

Systematic hydrographic measurements (temperature, salinity, depth) have been made since the middle 1800s. Data, particularly from the last fifty plus years, have been combined several times into global climatological atlases of temperature and salinity, the best known of which have been that of Levitus et al. (1998) and its evolved successors, the *World Ocean Atlas* (Locarnini et al., 2006), that of Gouretski and Koltermann (2004), and R. Curry's *Hydrobase* (available via the Woods Hole Oceanographic website). Some others are confined to specific regions (Arctic Seas, Southern Ocean, North Atlantic Ocean).

The most important issue with these atlases concerns the data distribution through time (see Fig. 2.26). Hydrographic observations are extremely nonuniformly distributed

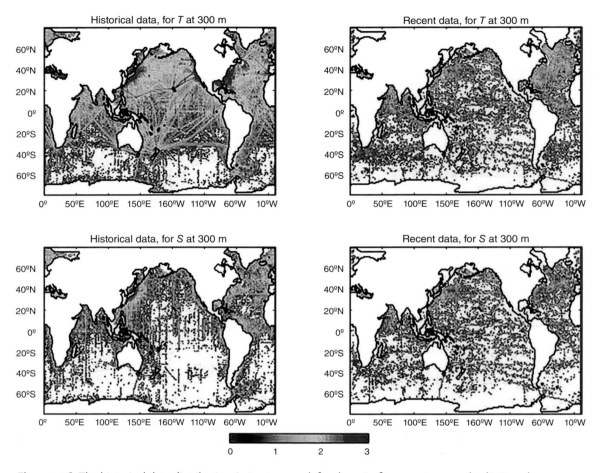

Figure 2.26: The historical data distribution (prior to 2001, left column) of temperatures and salinities above 300 m as available in the early 2000s where more than three observations exist in a $1° \times 1°$ box. The $\log_{10}$ of the number in the archives is shown. The right column shows the equivalent coverage in the interval since about 1990. The mean ocean depth is about 3800 m, and the data density falls off sharply with depth. (From Forget and Wunsch, 2007.)

throughout the ocean both spatially and temporally over the decades, and a clear bias is present, particularly at high latitudes, toward measurements obtained primarily in local summertime. (High-latitude winter observations from ships are understandably very scarce.) The southern hemisphere is grossly undersampled. Almost as important and troubling are the issues of calibration, as technologies and the care with which scientists and technicians made the measurements changed.

Figure 2.27 shows the parts of the world ocean in 5° squares where (white regions) Worthington (1981) concluded that as of 1978, at least one(!) useful hydrographic measurement extended to the abyss. Black areas had *no* useful measurement, even to mid-water depths. In the plot, the Southern Ocean seems to have remarkably good coverage, but much of the data comes from a single ship, the R/V *Eltanin* surveys, over a very limited span of years (Gordon and Molinelli, 1982). Interpreting such sampling as producing accurate time averages is a major leap of faith.

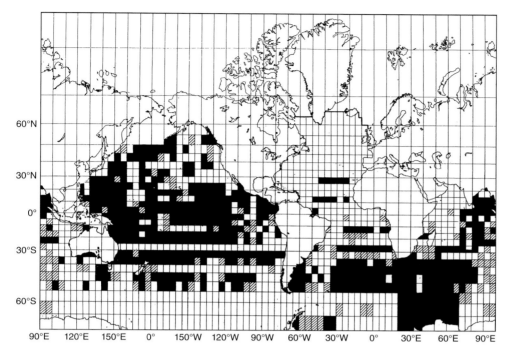

Figure 2.27: Historical temperatures and salinity measurement distribution to 1978. White blocks are 5° squares with at least *one* top-to-bottom hydrographic station regarded by Worthington (1981) as of high enough quality to be useful. Cross-hatched areas had at least one station to an intermediate depth. Black areas had no station regarded by him as adequate. That a single station could be regarded as conceivably adequate sampling reflects the era in which the ocean was widely regarded as essentially static and unchanging. A different rendering of the sampling issue can be seen in Figure 2.26.

Worthington (1981) had stringent rules for quality control of the data; in particular, he rejected most salinity measurements not made by a small club of institutions and individuals whom he regarded as qualified to make the observations. (See the discussion by Warren, 2008.) Later authors relaxed, sometimes greatly, the criteria for acceptance of data, introducing the possibility of major systematic errors. In the more than three decades since Worthington's compilation, many more high-quality measurements have been obtained, and a similar map constructed today would be much more nearly uniformly "white." But in determining low-frequency ocean variability, such as ocean heat-content changes, the paucity of older high-quality global coverage is the major limitation.

The climatological estimates and atlases raise a number of specific concerns. Typically, they provide a gridded value at every one of the three-dimensional grid points in their domain, whether or not the data distribution is adequate to form a useful average. Although often some form of reliability value is provided, those are often incomplete reflections of the data. For example, small sample numbers may well produce a small variance; if that variance is used as a measure of uncertainty, it is a misleading indicator.

A very demanding use to which the atlases have been put has been trend determination in quantities such as oceanic heat or freshwater content. Data sets are divided into time intervals

(5 years, 25 years, etc.), and averages are calculated over those intervals and then examined for changes, or a straight-line or higher-order polynomial fit continuously. Here all of the issues of data distribution and technology change come to the fore: (1) Some decades have many fewer data than others, and averages taken over intervals having very different sample numbers will *necessarily* produce apparent time differences. (2) Technology changes can readily introduce apparent trends from subtle or unexpected causes. Abraham et al. (2013) discussed the general problem of systematic errors in XBT data. Gouretski and Jancke (2001) noted a large number of Nansen casts where the recorded bottle depth corresponded to that nominally intended. Their inference is that either no protected/unprotected pairs of reversing thermometers were available to determine the actual depth, or that the scientists simply failed to calculate the actual depth. Large numbers of such mislabelled depths can produce artificial temperature and salinity trends. (3) Unless both the signal and the noise structure are spatially uniform, even if the numbers of sample values were constant in time, changes in their spatial distribution necessarily lead to apparent changes whose magnitude must be estimated to determine the significance of the result.

Water samples stored for long periods before the salinity measurement is made can undergo significant evaporation unless special precautions are taken. Comparing salinities determined in shore laboratories months after they were acquired with recent measurements made immediately on board ship after the data were collected can produce spurious apparent freshening.

SATELLITE MEASUREMENTS OF RADIATION

The oldest records are infrared measurements of the sea-surface—interpreted as sea surface temperature (SST)—but they were troubled by cloud cover, aerosols, and unstable, not-well calibrated, sensors. Microwave-band measurements of SST of useful accuracy were gradually obtained and, unlike infrared data, are largely impervious to clouds. Satellite records of sea-surface temperature, color, and ice cover now extend over several decades. Perpetual concerns are calibration, changed sampling rates, spatial coverage, and of course, funding. The textbooks on ocean-observing satellites all devote much space to the measurement of SST.

FLOATS

Owens et al. (1988) reported on one float trajectory extending over 9 years (Fig. 2.28). Gaps in acoustic coverage led to breaks in the measured trajectory. Generally speaking, float records of longer than a year or two are rare. The Argo system is beginning to produce multiyear data sets, but because the floats regularly cycle between the surface and a kilometer or more depth, the computed trajectories are very noisy. Argo data are primarily used for the temperature and salinity profiles obtained as the floats move between the abyss and the surface, rather than providing traditional float trajectory information.

TRACERS

Passive tracers (those not affecting the density of seawater) have been introduced into the ocean both deliberately and inadvertently. The nuclear weapons testing that took place in the atmosphere beginning in 1945 reached a crescendo in the early 1960s and then diminished abruptly with the signing of the Partial Test Ban Treaty. Those tests introduced a variety of

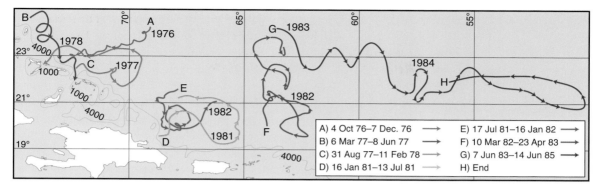

Figure 2.28: The trajectory of a SOFAR float tracked over nine years. Gaps result from lost acoustic coverage. (From Owens et al., 1988.)

products into the atmosphere, some quite nasty, but which did not exist naturally in significant amounts in the ocean. The radioactive isotope of hydrogen, tritium, ^{3}H, has a half-life of about 12.5 years and was tracked as it entered into the ocean over several decades until it had decayed so far as to become undetectable (Broecker et al., 1986); see Fig. 2.29. The pathways traced out by its initially increasing concentration do give at least a qualitative impression of oceanic water trajectories (Fig. 2.30). Other tracers arising from the bomb tests include carbon-14 (^{14}C, radiocarbon) and the stable helium-3 (^{3}He) decay product of tritium, which is often used to calculate tracer "ages," a subject requiring special treatment (see p. 342). Other artificially generated substances, such as the chlorofluorocarbons (Freons), can also be used. In recent years, artificial tracers such as SF_6 have been deliberately introduced in small areas to study small-scale ocean stirring and mixing (Ledwell et al., 2000; "stirring" and "mixing" are technically different processes and will be taken up later.)

2.7 EXPERIMENTS AND EXPEDITIONS

Deployment strategies of instruments that were used over the years are an important element in the evolving understanding of the ocean, and to a great extent its history becomes that of the entire subject.[23] An oversimplified summary would be that two parallel efforts have always characterized the field: (1) regional exploration using ships, and much later, the deployment of arrays of moored instruments and floats; and (2) basin and global reconnaissances and then, in the very recent past, attempts at sustained global coverage. Somewhat typical of the first kind of efforts were the northern North Atlantic description by Helland-Hansen and Nansen (1920) in the early days, and the Mid-Ocean Dynamics Experiment (MODE

---

[23] Almost completely lost in most modern accounts of oceanography are the excitement and romance of work at sea. Early oceanography was part of the heroic age of exploration that began with Christopher Columbus. The connection began to fade only with the arrival of travel by jet aircraft, the appearance of satellite data, and the Internet. Deacon, 1971, is a standard account of the history to 1900. Physical oceanography since about 1960 has been covered by Jochum and Murtugudde (2006), and numerous interesting individual reminiscences and oral histories exist.

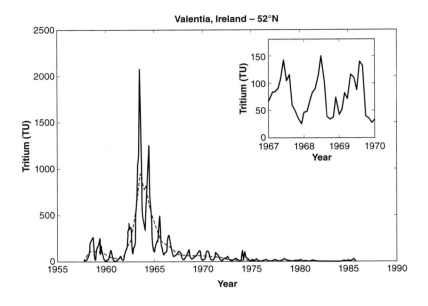

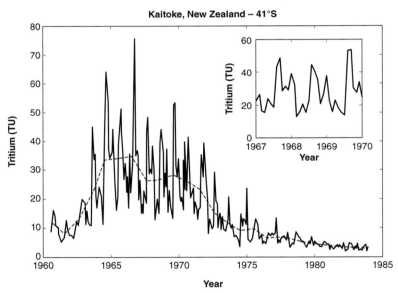

Figure 2.29: The tritium (^{3}H) content in tritium units (TU) in rain from the northern (top panel) and southern hemisphere (bottom panel). Insets show the years 1967–1970 in more detail. Notice the much wider scale in the northern hemisphere, which saw the dominant rainout of this hydrogen isotope. Almost all of the tritium came from atmospheric nuclear weapons tests and reached a clear maximum in the northern hemisphere in the early 1960s. With a 12.5-year half-life, ^{3}H does not occur naturally in the ocean, and because it becomes part of the water molecule, it is a passive tracer of water movement. 1 TU is the number of tritium atoms in a population of 10^{18} hydrogen atoms. (From Doney et al., 1992.)

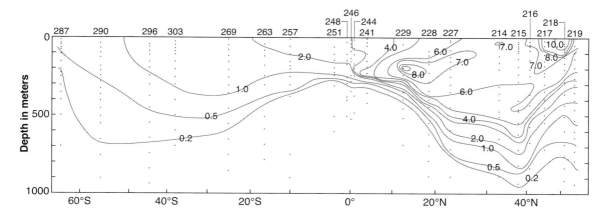

Figure 2.30: A section showing the tritium concentration (TU) in the North Pacific: from the Aleutians to 69°S along a meridian varying between 170°E and 165°W in about 1973. Station numbers are shown at the top. The most striking feature is the near-surface confinement (a contrast with some other parts of the ocean) and the strong increase toward the north. (From Fine et al., 1981.)

Group, 1978) in latter times. An entire library can be filled with the still-growing literature on the oceanography of special sub-regions of the oceans—all of which differ. High-latitude oceanography is a current focus.

Conventionally, modern oceanography is dated from the British *Challenger* Expedition, which circumnavigated the world over several years in the 1870s with a degree of sampling apparently not ever repeated by a single ship. In the 1920s, the German research vessel *Meteor* conducted a hydrographic survey of the Atlantic Ocean (see Wüst and Defant, 1993). During the International Geophysical Year (IGY), the United Kingdom and the United States resurveyed the North Atlantic (Fuglister, 1960); the International Indian Ocean Expedition was mounted to obtain long hydrographic lines there (Wyrtki et al., 1971); and the R/V *Eltanin* covered the Southern Ocean (Gordon and Molinelli, 1982).

The availability in the 1970s of self-contained in situ instruments led to diminishing interest in large or regional-scale hydrographic surveys in favor of a series of localized "experiments" directed primarily at physical processes.[24] A very incomplete list of the latter and the associated programs would include the coupled ocean-atmosphere ENSO phenomenon (Tropical Ocean–Global Atmosphere, TOGA), coastal upwelling (Coastal Upwelling Ecosystems Analysis–CUEA), and geostrophic eddies (Mid-Ocean Dynamics Experiment, MODE). These later programs often combined many of the new current meters, drifters, floats, profilers, etc., with classical, but localized, hydrographic surveys.

Toward the end of the twentieth century, the growth of interest in climate (by definition a global problem), the increasingly robust and long-lived electronic devices, and the availablility of satellite-borne instruments anticipated to produce useful measurements of sea-surface height, winds, time-dependent gravity, and sea-surface temperature led to the first global

---

[24]Oceanographers almost never carry out experiments in the classical sense of measuring a controlled system. The term "experiment" has come to mean any set of field measurements, sometimes with the implication of testing a hypothesis.

oceanographic program, the World Ocean Circulation Experiment (WOCE).[25] That program encompassed traditional shipboard hydrography, but on a global scale, combined with many of the array of available electronic devices and satellite altimeters and other spaceborne measuring devices. Many of the results from WOCE are used throughout this book. Out of WOCE and successor programs have come the instruments that continue to provide near-global coverage, including the satellites, the Argo floats, and, as always, shipboard hydrographic surveys and their ancillary measurements.

In the abstract, an important oceanographic problem concerns devising the best *future* mix of instruments and their deployment strategy for providing the most useful information. Ocean instruments are expensive, and their use confronts difficult logistical challenges, thus justifying an investment in understanding the scientific benefits and costs of different instrument mixes and deployment strategies. Elements of this optimization problem have been discussed in the literature, but the subject is not pursued here, in part because of its complexity on several levels. Fundamentally, without agreement on what the specific goals and purposes are, optimization of a field deployment is either impossible or not meaningful. Instrument types are not fungible—moored array groups skilled in that work will not decide to focus instead on using floats; not flying an ocean-observing satellite does not lead to more ships. Deciding that modelling skill can substitute for direct measurements can lead to intense scientific controversy. We leave the subject here.

2.8   THE FRONTIER: DURATION

As spatial coverage of the ocean has come to be at least intermittently nearly complete, the frontier of observational difficulties has shifted to the problem of *duration*: the lack of long records is one of the most fundamental problems in physical oceanography and its related climate and biological sciences. Once a time series has been established, many things militate against its sustenance: (1) Absent an unexpected event, the cost of extending a 50-year record for another 10 years often appears much less compelling than measuring some other variable with a new instrument for 3 years. (2) Technologies inevitably shift, and a time series measured with an "old-fashioned" method runs the risk of appearing archaic and so is not pursued, even if the technology remains useful. A decision to replace a still-useful older technology (e.g., reversing thermometers on Nansen bottles were replaced by CTDs) requires considerable knowledge and insight, an interest in the often difficult intercalibration problems and in the trade-offs of ease of use and cost against possibly greater precision and accuracy. (3) An old scientific rule of thumb is that a record is worth reanalyzing when it doubles in length. Thus a scientist with a 30-year data record has little personal incentive to fight for funding and calibration to extend the record by another 10 years, important as that might be to the wider community. (4) Measurements become seemingly routine and are turned over to technicians to maintain—people who will not themselves use the data and who have no stake in nor understanding of the necessity of tight quality control; deterioration is almost inevitable. Keeling (1998) provided an interesting perspective on the vicissitudes of maintaining long-duration measurements. He was measuring atmospheric CO_2 concentrations, but the problems are generic.

---

[25] An informal account of its history is in Wunsch, 2006b, and Siedler et al. (2013) attempted a science summary.

2.9 OTHER GENERIC OBSERVATION ISSUES

MODELS

Occasionally some scientists, ones typically devoted to making observations, will proclaim that they "never use models"—only "data." But even the simplest measurements involve a model of some sort and hence a series of often complex assumptions. A commonplace example is the mercury thermometer, which requires conversion of a properly calibrated length into a temperature value. XBT data interpretation is directly dependent upon a model of the fall rate.

Ocean models are often conceptual. Consider a sequence of temperatures measured at 300 m depth at a temporal separation of 1 h. Interpretation of the changes recorded would be radically different for a scientist who believed she was studying internal gravity waves; one who thought she was seeing changes in mixed-layer depth; or one studying moving frontal structures. The "model" is not necessarily a numerical one—it is nevertheless a complex and sophisticated world view and is a presence in any usable measurement.

SAMPLING

The concept of aliasing, described more completely in Appendix A, has already been invoked several times and is a constant concern. For example, space agencies like to fly satellites that maintain Sun synchrony—returning to each point on the Earth's surface at a fixed time of day. In those orbits they are able to remain always illuminated by the Sun and avoid the use of expensive and heavy battery packs, but daily samples can incur large errors (e.g., Gentemann et al., 2003).

ERROR ESTIMATES

No perfect measurements exist and error estimation is often the critical step in an analysis. Determining errors can be extremely difficult and is an effort that calls attention to the weaknesses of a result, sometimes even demonstrating that a hard-won result is not useful. Consequently, some authors just ignore the entire subject.

Often the errors are the essence of the science: For example, the first person to calculate the transport of the Gulf Stream near Florida could have usefully inferred that it was "about 28 Sv,"[26] which tells a reader that it is unlikely to be 3 or 300. But a second investigator who reports that it is about 33 Sv is hiding essential information: if a realistic error estimate produced 33 ± 5 Sv, a reasonable inference is that the first reported measurement was indistinguishable from the new one. If, however, the new report was 33 ± 1, then a third investigator might wonder if the discrepancy between first and second measurements was a real one, or simply that the first measurement had a large, but unreported error bar. If the third measurement is 26 ± 2 Sv, questions arise about differing methodologies, seasonal cycles, sampling errors, etc. (The ± notation usually denotes what is called the standard error derived from the standard deviation of an ordinary average. For Gaussian processes it corresponds to roughly a 68% confidence interval—see Appendix A. More cautious authors will use error bars of two standard deviations, which is roughly a 95% confidence interval, and will check the

---

[26] As Appendix D notes, 1 Sverdrup, abbreviated as Sv, is defined as a volume transport of 10^6 m^3/s and for seawater is approximately a mass transport of 10^9 kg/s, with about a 3% error. It is not a Système Internationale (SI) unit.

near-Gaussian assumption.) See Lanzante, 2005, for a discussion of the problem of deciding on (in)consistency of estimates.

In measurements generally, a critical distinction lies between the ideas of accuracy and precision. The former refers to absolute standards, the latter to the ability to detect changes, be they spatial or temporal. Thus a reversing thermometer might have an absolute calibration accuracy of ±0.01°C, but a trained observer might (and sometimes did) claim to be able to read it repeatedly to within ±0.003°C. If true, *changes* considerably smaller than the accuracy can be detected. Improved accuracy in an instrument is not always useful: if a thermometer could be calibrated to ±0.0001°C, reports of measurements to that accuracy would not be of much interest if repeating the measurement 5 min later or at 1 km separation would always show changes much greater than the instrumental accuracy. Accuracy and precision must be understood and used in the context of the magnitudes and character of the noise, whether instrumental or natural. (The terminology is itself sometimes not precise, as *changes* are sometimes quoted as having an accuracy determined by the precision.)

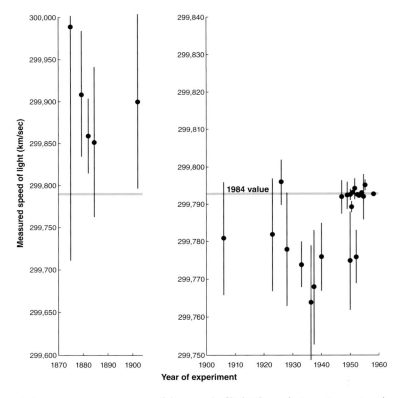

Figure 2.31: Laboratory measurements of the speed of light through time. Assuming the most recent values are correct (?), it is notable that numerous very careful measurements under controlled laboratory conditions are not consistent within one standard error with that value. (Under conventional assumptions, about 35% of the values should lie outside one standard deviation.) Did the speed of light change through time? Or were there systematic errors unknown to the experimentalists? Why do the values cluster over decades? Are much less well-controlled oceanographic measurements immune to such difficulties? (From Henrion and Fischoff, 1986.)

Much more difficult to cope with are the inevitable *systematic*, or *bias*, errors, which can arise from a very large number of causes. Their existence means that no matter how many independent measurements are made, their average will differ from the true value. Figure 2.31 is a well-known example (Henrion and Fischhoff, 1986; see also Morgan et al., 1990) that shows the measurement of the speed of light through time and represents the generic difficulties. If the most recent value of the speed of light shown is taken to be correct, then far too many of the earlier measurements are inconsistent with their estimated value at one standard error. The clustering of high and low values is striking. A conundrum must be faced: cosmological theories exist requiring that the speed of light be time variables, and evidence for such changes would be very exciting. But most physicists would likely instead conclude that unknown and uncorrected systematic errors were present in the apparatus, despite the best efforts of specialized scientists working under extremely carefully controlled conditions who were aware, consciously or otherwise, of values being obtained in other laboratories. Uncontrolled oceanographic measurement problems are far more difficult to parse than this one.

Systematic errors can have many causes: physics omitted from a model of the data; failure to recognize instrument drift; sampling distributions (e.g., ocean data obtained only in the summer time); temporal or spatial aliasing; unconscious knowledge of the "right" answer, etc. Kunze et al. (2006) described an example where ocean mixing rates were overestimated by a factor of 10 because of a failure to understand instrument noise characteristics. Sturges (1974) described a historically important conflict between geodesists and physical oceanographers, one pointed out by R. B. Montgomery: the seemingly very crude oceanographic measurements of sea-surface slope along meridional coast lines conflicted in sign with the supposedly far more accurate geodetic measurements on land. The explanation lay with identification of a systematic error in land geodetic surveying, one that arose from atmospheric light-refraction effects incurred because theodolite users try to work with the Sun at their backs. Eternal vigilance and an open mind are required.

A NOTE ON NUMERICAL VALUES

Oceanographic science is largely about numbers, and many are quoted in this book. Without claiming complete consistency, I try to employ numbers in the scientific convention that provides some information about their accuracy. Thus a statement that a transport is about 170 Sv, without an error bar, should be interpreted as implying an uncertainty of *about* 10 Sv, sometimes written as ±5. A statement that the transport is 173.4 Sv implies an accuracy of about 0.5 Sv. A recent paper quoted a calculated energy transfer rate of 0.204 exajoules (EJ), implying a spurious accuracy of ±0.005 EJ. The value should have been written "as about 0.2 EJ." Psychologists know about the "principal of conservation of digits," the unwillingness to surrender any calculated numbers, no matter how irrelevant they might be.

An Interlude

Understanding the nature of the fluid ocean is nearly impossible without the use of basic data analysis and representation tools. A subset of these tools would include probabilistic and statistical concepts, their combination with Fourier analysis, time-domain representations, the ideas of minimum variance estimation, "objective mapping," and numerous other pieces of machinery. Terms like "spectrum," "spectral estimate," "narrow-band process," "white noise," "confidence interval," "Q," etc., are essential to making sense of what is observed and understood. Because of the central importance of these ideas to everything that follows, Appendix A has been provided to at least outline some of the most central tools. Readers who are unfamiliar with them are urged to read that appendix now, before proceeding, and to refer to it as needed.

A comparatively strong emphasis is placed on Fourier and related methods because nature often divides disparate physics by time scale (more so than by space scale), and the ability to interpret frequency domain analyses is essential.

CHAPTER 3

What Does the Ocean Look Like?

This chapter is a bit dry (no pun intended), as its purpose is to provide some basic geographical and descriptive information required to make sense of the fluid motions of the ocean. The first part is largely descriptive in the conventional, pictorial sense. The second part will represent a discontinuity for many readers: it sets out the mathematical tools—equations and concepts—in the language necessary for quantitative descriptions and goes beyond the mere display of pictures or of wiggly lines.

3.1 SOME PHYSICAL DESCRIPTIONS

Where possible, the emphasis here is on global characteristics. Every location in the ocean is different at least in detail from any other, and the temporal variability renders the oceanic flow unique at any instant.[1] Describing global ocean characteristics is not easy, because it involves measurements in physically diverse regions through varying seasons, decadal shifts in properties, changing instrument technologies, and high-frequency and high wavenumber structures whose significance is not always obvious.

Because of the near-total reliance on ships, most oceanography until relatively recently was necessarily regional in nature, and attempts at global syntheses of any type remained rare.[2] Much of the description has been of the supposed time-average ocean, with the implicit assumption that *large-scale fuzzy snapshot = time average*, an inference only now breaking down. The somewhat vague term "general circulation" of the oceans is commonly used to refer to the putative time average (or the variability about that average). No clear definition of the spatial scale to which it applies is available, but it would commonly apply to long-lived features changing horizontally over distances on the order of 100 km and larger.

---

[1]"You can't step twice into the same river" (Heraclitus) can also be said of the ocean. Much of the recent oceanographic literature might be satirized as having titles such as "El Niño signals in the spring-time temperatures at mid-depths in the Gulf of Codfish from 1980–1995." Such results may be important locally but do not obviously lend themselves to global generalizations.

[2]The work of J. L. Reid of the Scripps Institution of Oceanography over many decades is noteworthy in this regard. Older textbooks such as those by Sverdrup et al. (1942) and Dietrich et al. (1980) described global water masses and a crude notion of the flow field. Talley et al. (2011) provide a recent update.

An impressive array of very diverse oceanic and meteorological phenomena are touched on in this chapter and those that follow. Going beyond descriptions and discussions of the very specific elements and synthesizing such disparate elements into an understanding of the ocean circulation is an overarching goal of physical oceanography. Achieving that goal requires combining what is known about all of these things with the best understanding of the governing dynamical and thermodynamical (and sometimes chemical or biological) laws.

3.1.1 Gravity and the Shape of the Earth

The Earth is nearly a spheroid (an ellipsoid having two equal equatorial radii), but one whose mean equatorial radius, $a_e = 6378$ km, is about 21 km greater than the polar one, $a_p = 6357$ km (Lambeck, 1980; see Fig. 3.1). This equatorial bulge is a consequence of Earth rotation history, and its stability and size carry important information about the properties of the planetary interior. *Geodesy* is the science of determining the shape of the Earth on all scales. In combination with the geophysical explanation for the observed structure, it is an interesting and sophisticated subject with many ramifications (Munk and Macdonald, 1960; Lambeck, 1980; Hofmann-Wellenhof and Moritz, 2006), including a surprising number that involve physical oceanography and climate.

Dynamical consequences in the fluid ocean of the deviation of the Earth from a perfect sphere have not been observable. Kinematic effects do occur in the study of long-distance acoustic or surface-swell propagation—that is, along ray trajectories where precise paths can be determined.[3] Satellite altimeter measurements are also dominated by the polar flattening.

a = 2, c = 1

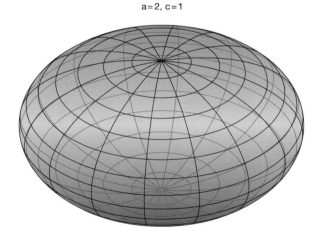

Figure 3.1: The Earth is a flattened spheroid with a polar radius about 21 km shorter than the equatorial one. A spheroid is an ellipsoid in which the two equatorial radii are equal, and it is a good approximation for the Earth. The surface is denoted $E(\lambda, \phi)$. (From Wolfram Alpha.)

For oceanographic purposes, the equipotential of special interest is called the *geoid* and might be defined as the one to which the ocean would conform if it were brought to rest and its density made uniform at some definition of its mean value. That value of the equipotential is then continued over land.[4] Let the average density of the ocean be $\bar{\rho}$, and let $r(\lambda, \phi)$ be any radius vector from the center of the Earth as a function of latitude ϕ and longitude λ.

---

[3] Munk et al., 1995, p. 336.

[4] Historically, the geoid was supposed to be the equipotential corresponding to global mean sea level. As will be seen, the value of the global average sea level is not well-determined, is partially ambiguous, and is apparently changing with time. The nonlinearity of the equation-of-state of seawater means that the density averaging must be very carefully defined. Pragmatically, however, spatially uniform errors in N_g have no dynamical effect on the

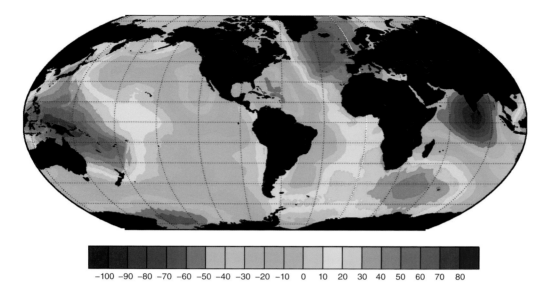

−100 −90 −80 −70 −60 −50 −40 −30 −20 −10 0 10 20 30 40 50 60 70 80

Figure 3.2: The estimated elevation, N_g, of the geoid (equipotential) relative to the reference spheroid E. Contours are in meters. $S_{tot} = E + N_g$ would be the surface topography of an ocean brought to rest with a uniform density. The particular estimate shown is derived from the GRACE satellite mission (see Reigber et al., 2005). Subsequent observations will improve this estimate, but they are unlikely to be visibly apparent on the global scale.

Then an ocean at rest with this uniform density would have a surface shape relative to the center of the Earth of

$$r(\lambda, \phi) = S_{tot}(\lambda, \phi), \tag{3.1}$$

where S_{tot} is called the *geoid undulation*, or *elevation* (the geoid itself—in energy units—is the value Φ_g, which is taken to be constant, of the gravitational equipotential on S_{tot}). If the solid earth were a perfect spheroid with uniform density ρ_e, then $S_{tot}(\lambda, \phi) = S_{tot}(\phi)$ would also be a spheroid. Define $E(\lambda, \phi)$ as the best-fitting reference ellipsoid, and the *reference geoid height* is

$$r(\lambda, \phi) = E(\lambda, \phi) + N_g(\lambda, \phi). \tag{3.2}$$

so that the shorthand "geoid-height" refers now to the anomaly $N_g(\lambda, \phi)$. In the physical oceanographic context, the somewhat improper term "geoid" alone usually refers to the potential height anomaly, not to the gravitational potential value.

Earth's surface features and interior are not, however, of uniform density; bulges and depressions exist—corresponding to mountain ranges, ocean-floor features, and density anomalies deep within the Earth. The geoid elevation is known to deviate from the reference ellipsoid elevation by ±100 m, as depicted in Figure 3.2, the most notable features being, for example, the Sri Lanka low and the high over the northern North Atlantic Ocean.

---

ocean circulation estimates because only the lateral gradients appear in the equations of motion. Consistency of definition does arise when combining data from different satellites or instrument types into a global estimate. In times of major sea level change, the entire concept becomes amorphous.

One recent estimate of N_g is the EGM2008 (Earth Gravity Model 2008; see various websites), which is said to be complete to spherical harmonic order $n = 2160$ (roughly wavelengths of $40000/2160 \approx 19$ km), although it has a complex error as a function of location and wavelength. This estimate has already have been superseded by the use of later measurements, but they differ only in short-wavelength details not detectable by eye in displays such as that in Figure 3.2.

3.1.2 Topography

The area of the oceans is about 3.6×10^{14} m^3, and the volume is about 1.4×10^{18} m^3. Areas shallower than about 200 m occupy approximately 7% of the total, while their volume contribution is negligible (Menard and Smith, 1966). Figures 3.3 and 3.4 depict the topography as estimated from modern acoustic sounding and satellite data. Extended regions of comparatively flat topography exist, separated by areas of intense three-dimensional structures, including the mid-ocean ridge system. The ridge system has a two-dimensional large-scale structure but is made up of many thousands of three-dimensional seamounts and other structures, perhaps most clearly depicted in the semiqualitative Heezen and Tharp painting of the seafloor best seen online.

Figure D.1 indicates the oceanic areas within fixed-depth intervals. Most oceanic depths lie between 3 and 5.5 km, with the most common values at just over 5 km. The mean oceanic depth, including the continental shelves, is about 3800 m, with a maximum (in trenches) of about 7000 m. With an Earth radius of about 6700 km, the mean thickness of the ocean is about 0.1% of the total. If the Earth is depicted as a near sphere, the ocean is an extremely thin skin at the surface, with major consequences for its flow dynamics.

To generalize about oceanic topography is not simple: it is spatially very inhomogeneous, having complex anisotropies on all scales. Much oceanic fluid-dynamics theory exists for motions over flat topography, but these regions occupy only a fraction of the oceanic area. A fringe of shallow water (the continental shelves) rings the deep ocean, with water depths near 100 m but with greatly varying widths (see Fig. 3.5 and compare the region of the Argentine shelf with that off California), different depths, and a gradually changing and steepening gradient as one progresses from deep into shallow water.

Considerable uncertainty remains about important details of the seafloor on small scales. Figures 3.3 and 3.4 are based in part upon satellite altimetric measurements of the sea surface (Sandwell and Smith, 2001; Becker et al., 2009). Topographic features disturb the local geoid, and measurements by altimeters of the sea-surface shape can be exploited to infer the underlying topography. Inferences of topography from the geoid cannot, however, produce a unique solution, and significant remaining errors exist in the charts displayed. Sometimes only statistical descriptions of seafloor topography are needed, and estimates exist (Goff and Jordan, 1989), but the regimes are very diverse.

Topographic features generate both kinematic and dynamic effects on water movement. The kinematic effects are most obvious: comparatively narrow passages between deep ocean basins permit the exchange of water between basins. Over time, the existence of such exchanges can greatly modify the temperature, salinity, and chemical properties of the ocean. Consequently, the omission of some small feature in an ocean model, or the inadvertent but erroneous inclusion of a nonexistent exchange path, can destroy the water temperature and salinity properties in an otherwise perfect model.

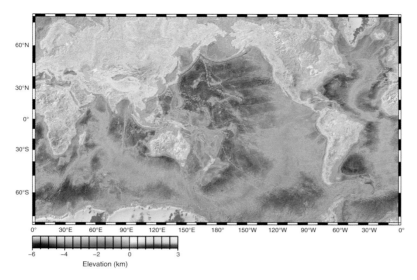

Figure 3.3: Ocean and land topography. The mid-ocean ridge system and the so-called abyssal plains are most prominent, but changes occur on all scales with a great variety of amplitudes. Various websites provide the underlying data and its extensions into shallow water and the Arctic Sea. (From Smith and Sandwell, 1997.)

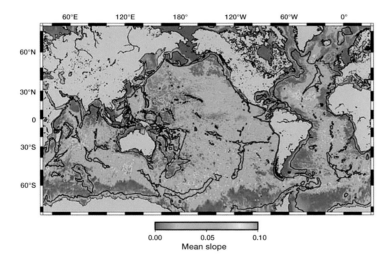

Figure 3.4: Slopes of bottom topography as estimated by Wessel et al. (2010) using both ship sounding data and satellite altimetry. Values of the slope are very important in ocean dynamics, and can differ greatly from, for example, abyssal plains to continental margins and ridges. The spatial resolution of this chart is not very clear but is probably best viewed as having been smoothed over about 10 km. Many low-elevation seamounts are believed missing because satellite altimeters do not detect deep-ocean features of elevation less than about 1000 m and ship tracks remain sparse. The 3000 m contour is shown as the black line.

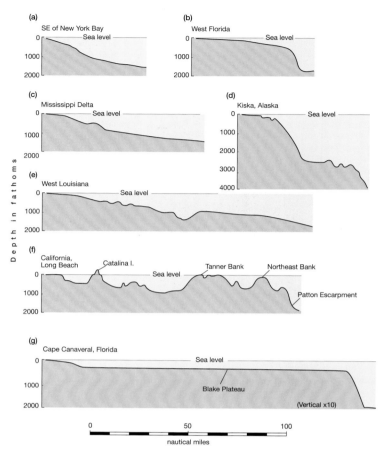

Figure 3.5: Various continental boundary configurations. Note the very large vertical exaggeration. The shapes of the continental margins have a profound influence on the fluid motions that can take place there and commonly control the exchange of properties between the coastal regions and the open sea. 1 fathom ≈ 1.8, 1 nautical mile ≈ 1.85 km. (After Shepard, 1973.)

3.1.3 Water Movement

Suppose a current meter capable of measuring the two horizontal components of velocity and the local water temperature as a function of time is placed for 800 h at an unremarkable part of the North Atlantic Ocean at a depth of 600 m (see Fig. 3.6; the water depth is about 4600 m). What do the data show? Figures 3.7–3.11 display the answer in a variety of forms. Figure 3.7 shows the velocity field sampled at hourly intervals. A striking, energetic motion on time scales of a few hours appears in both the temperature and velocity fields, but they show little visual resemblance. If the records are averaged in time over a day, much of this high-frequency movement disappears. Figure 3.8 shows the flow in the form of a *stick plot* in which the daily average vectors point in the geographical direction $[u(t), v(t)]$. Displacements calculated from daily averages of $u(t)$ and $v(t)$ over the entire record are shown in Figure 3.9. This representation corresponds to a hypothetical particle having the velocity measured at the

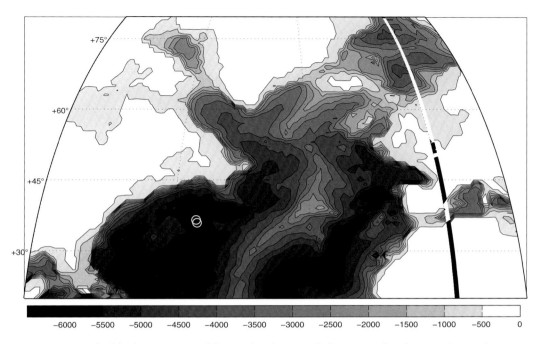

Figure 3.6: Smoothed bathymetry (m) of the North Atlantic with the two circles showing the nearby positions of the current meter records displayed in Figures 3.8 to 3.11. Note the loss in the chart of the small-scale features, particularly on the mid-ocean ridge.

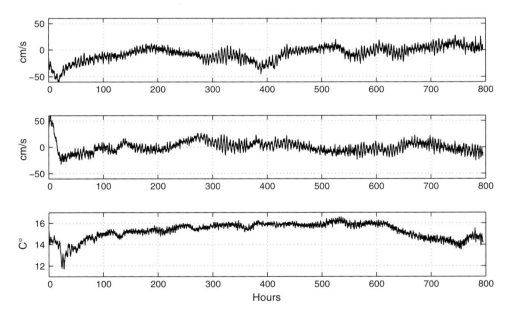

Figure 3.7: Two components of velocity (cm/s) and the temperature from 15 min interval samples from an instrument at 600 m near 36°N, 55°W as in Figure 3.6. The measurements are dominated by the variability with no significant nonzero time average.

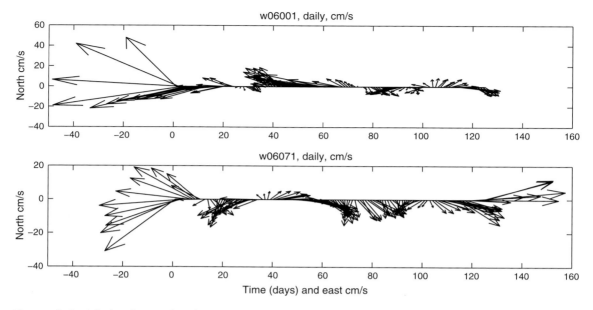

Figure 3.8: A stick plot showing hourly average vectors from the current meter record in Figure 3.7 and another at the same depth from a mooring about 68 km away. Notice the change in scales.

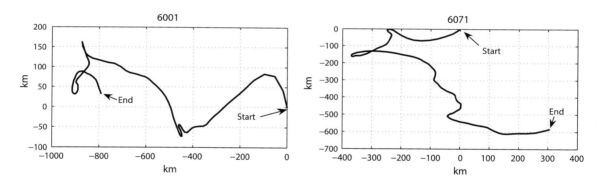

Figure 3.9: The accumulating sum of the x,y displacements of a fictitious particle residing at the position of the current from the daily averages of the same records as in Figure 3.8. Note the different scales and the radically different net displacements from moorings separated by only about 1° of distance at the locations in Figure 3.6.

position of the current meter—it is hypothetical because any real particle will have moved away under the measured velocity. The same figure shows a corresponding plot from a second mooring located about 68 km away: the strikingly different result from there illustrates the great complexity of velocities in the ocean—in contrast to the very large scale structures in temperature and salinity. Had the technology available to oceanographers in the mid-nineteenth century permitted observations like these, the theoretical structure of the subject likely would have evolved very differently.

Some insight into the complex structure of the velocity records can be obtained by breaking the time variability down into frequency bands with the spectral methods from

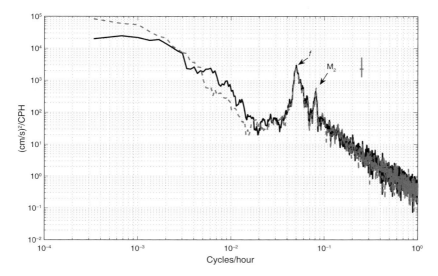

Figure 3.10: The power density spectral estimates from the 15 min samples of the current meter record shown in Figure 3.7. The solid curve is for the east component (u), and the dashed one is for the north (v). The spectra are essentially identical for all frequencies greater than the Coriolis parameter f but diverge at lower frequencies. f and the evident M_2 tidal constituent are marked on the velocity spectral plots. Temperature has a somewhat similar spectrum (not shown) but without the very conspicuous inertial wave frequency. The vertical bar is an approximate 95% confidence limit for the locally changing continuum part of the spectrum (it is not appropriate for frequency regions of sharp peaks).

Appendix A. Figure 3.10 displays the power density spectral estimate for the two components of velocity measured at 15 min intervals. Here a conspicuous energy maximum at a period near 20 h is the *inertial (wave) peak*, whose radian frequency is very close to $2\Omega \sin \phi \doteq f$ where ϕ is the latitude and Ω is the radian rotation frequency of the Earth. Figure 3.11 shows the two components $v(t)$ versus $u(t)$ where the contributions are restricted to the band of periods lying between about 25 and 18 h, a band including the inertial peak. (The appearance of a spectral peak does *not* imply periodic motions.) The motions occurring when $2\pi s > f$ are called *internal waves* or *internal gravity waves*.

Frequencies of $2\pi s < f$ are an amalgam of what are sometimes called mesoscale or geostrophic eddies and more general motions at low frequencies. No low-frequency cut-off is known—except that the temporal variations in the ocean probably extend out to the age of its creation several billion years ago.

Figure 3.12 displays power densities for a mooring near the Mid-Atlantic Ridge at 27°N at three different depths. Depending upon the frequency and the instrument depth, different approximate power laws appear, superimposed upon which are various maxima (peaks) of varying statistical significance, shape, and physics.

A somewhat different exploration of the flow field in the ocean is obtained from the neutrally buoyant floats, whose movement produces results such as those seen in Figures 3.13 and 3.14. The major characteristic of the somewhat tangled trajectories represents an important qualitative statement about the nature of oceanic flows and their interpretation.

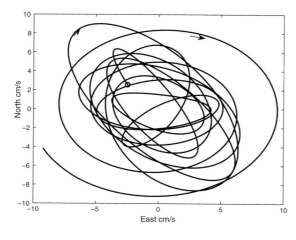

Figure 3.11: A *hodograph* plot (*v* versus *u*) of the velocity components in the record of Figure 3.7 from the frequency components lying in the range of periods between 25 and 18 h. This band encompasses the inertial frequency with a period near 20 h. The velocities are dominantly clockwise as the simplest theory of inertial oscillations (see Ch. 4) would require. Motions are *not* periodic but instead are a narrow-band random process.

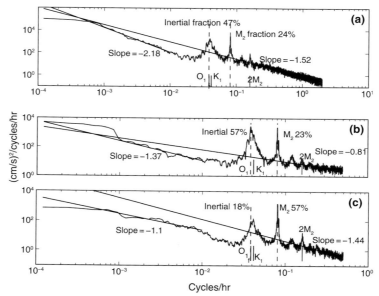

Figure 3.12: Kinetic-energy spectral estimates for instruments on a mooring over the Mid-Atlantic Ridge near 27°N (see Fu et al., 1982, for the original data). (a) Instrument at 128 m. (b) Instrument at 1500 m, and (c) 3900 m (near bottom). The inertial, principal lunar semidiurnal M_2 and diurnal O_1, K_1 tidal peaks are marked, along with their percentage of kinetic energy lying between f and the highest frequency estimate. Least-squares power-law fits for periods between 10 and 2 hours and for periods lying between 100 and 1000 hours are shown. The approximate percentage of energy of the internal wave band lying in the inertial peak and the M_2 peak are noted. Percentages were computed from the Parseval relationship. In most records, the peak centered near f is broader and higher than the one appearing at the M_2 frequency. When f is close to the diurnal frequency, it is also close to half the frequency of M_2, and some interesting resonant triad interaction physics can apply. Some spectra show the first overtone, $2M_2$, of the semidiurnal tide. The geostrophic eddy band is greatly reduced in energy near the bottom, as is the inertial band, presumably because of the proximity of steep topography. Note the different axis scales. (From Ferrari and Wunsch, 2009.)

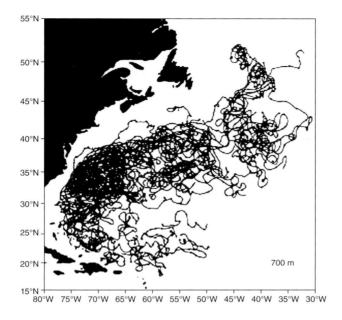

Figure 3.13: Float trajectories at a nominal 700 m pressure level in the western North Atlantic every 30 d. These are sometimes known as *spaghetti diagrams*. The floats were launched at a variety of positions in the western North Atlantic and tracked for varying time durations. (From Owens, 1991.)

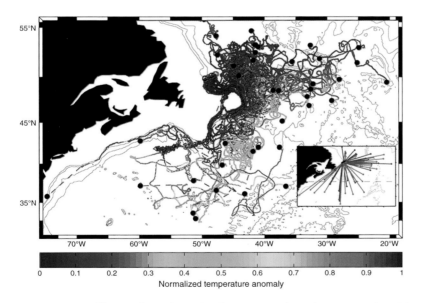

Figure 3.14: Movement of floats released into the deep western boundary current (DWBC; see Fig. 11.9) at nominal depths of 700 and 1500 m near 50°N. The results are notable both because the floats failed to move into the subtropical gyre in the DWBC and for the complexity of the individual pathways. Colors indicate the temperature *change* along the float trajectories. (From Bower et al., 2009.)

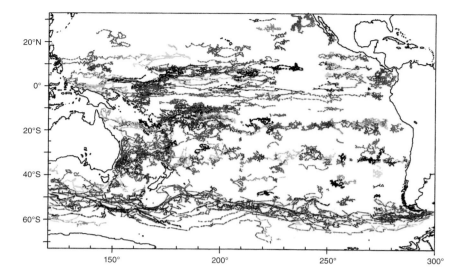

Figure 3.15: Depiction of a large-scale flow in the ocean from floats at depth. Each float has a different color. In this region, a largely zonal velocity emerges, on top of which is seen smaller-scale tangles. The result is at odds with theories predicting a largely meridional interior flow. The relationship between motions measured at fixed points by instruments such as current meters and those undergone by instruments nominally moving with the fluid itself is at the heart both of the differing Eulerian and Lagrangian descriptions and of the nature of oceanic turbulence. (From Davis, 1998, which used ALACE floats; cf. Fig. 2.28.)

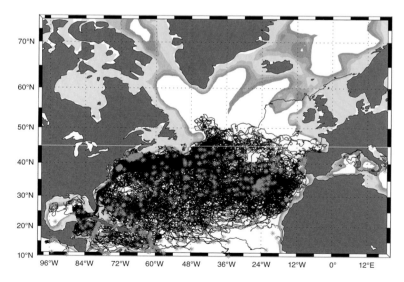

Figure 3.16: Surface-drifter trajectories in the subtropical gyre. Notice how few instruments moved into the poleward regions: all but one remained in what is called the subtropical gyre. Deployment took place at many different locations within the gyre, and tracking was conducted over varying time durations. (From Brambilla and Talley, 2006.)

Such pictures are often called "spaghetti diagrams"—a good descriptor of the motions they depict. On a larger scale, visual patterns do emerge: for example Figure 3.15 shows the strong zonality of the Pacific Ocean flow at depth.

Drifters at the sea surface are simpler and cheaper instruments compared to neutrally buoyant floats, and so many more such records exist. Figure 3.16 shows a series of trajectories of a relatively large number of drifters in the North Atlantic. The tale of their movement is complex, and the existence of that complexity, and not its details, is an important part of descriptive oceanography.

Figure 3.17 displays an estimate of the ratio of the surface kinetic energy of the time variability to that of the time mean as obtained from satellite altimetry.[5] The latter is over-estimated because the averaging time used was short (7 years) and the low-frequency variability contaminates the longer-term mean. Variability kinetic energy in the ocean is one to three orders of magnitude greater than that of the record-length time mean. These flow properties are visible in the current-meter, drifter, and float data. How the intense small-scale variability and the large-scale patterns manage to coexist is a central question.

Even the concept of a time-average velocity, independent of an arbitrarily chosen averaging time, becomes obscure if the ocean is constantly changing on all time scales out to its entire lifetime. A few places do display what appears to be a robust time average *over the record length* relative to the variability. One of those rare records is shown in Figure 2.24. That it is hardly typical can be seen by comparing it to Figure 3.18, in which prolonged direction reversals occur in the supposedly permanent North Atlantic deep western boundary current (DWBC).

That the oceanic flow is dominated by large-scale, slowly changing flows was a convenient fiction for many years. For anyone interested in studying oceanic properties, be they physical, chemical, or biological, the most important qualitative message is that *the ocean is in constant, rapidly changing motion, everywhere.*

3.2 GROSS THERMAL AND SALINITY PROPERTIES

By about 1870, ocean scientists had largely solved the problem of observing—by purely mechanical means—the three-dimensional oceanic temperatures and salinities to a useful accuracy and precision, and they began what was a century of exploration to map these oceanic fields. The major discovery was that, when depicted from coarse spatially sampled data, a large-scale nearly permanent temperature and salinity field is always present (see Figs. 3.19–3.24). This discovery dominated all aspects of oceanography, theoretical and observational, until comparatively recently and is still a central feature of the oceanographic literature. Some notion of the structure of that quasi-permanent background is necessary for interpreting the variability.

---

[5]The calculation uses the spatial slope of the altimetric measurement of height, η, as computed from Equation (9.29).

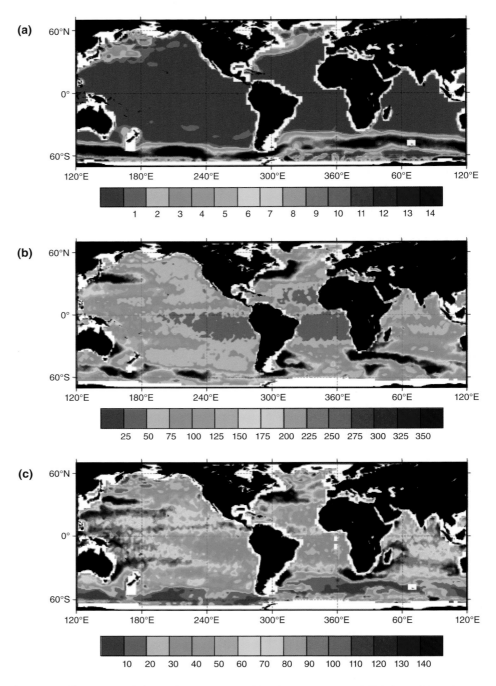

Figure 3.17: The top panel shows the 7-year mean kinetic energy times sin² ϕ from altimetric satellite measurements, the middle panel from the temporal variability about that mean, and the bottom panel is their ratio. The geostrophic relationship, Equation (3.20) was used. Although because of the short record the time mean kinetic energy is likely overestimated, it is a small fraction of the total kinetic energy almost everywhere except in the Southern Ocean, where the ratio is close to one. (From Wunsch, 2002.)

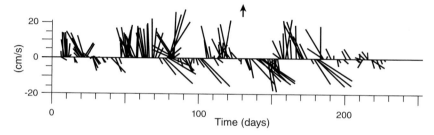

Figure 3.18: A stick plot showing the magnitude and direction of daily averages in a current-meter record from an instrument set into the North Atlantic deep western boundary current (DWBC) near Cape Hatteras. Time axis is in days. *NB: values plotted as positive flows are directed to the southwest, the direction of the permanent DWBC.* The periods of prolonged reversal show how apparent mean flows, even in major current systems, have values dependent upon the time and duration of measurement. (From Pickart and Watts, 2000.)

HYDROGRAPHIC STATIONS

In the days, which extended well into the 1970s, when Nansen bottles and reversing thermometers were used to sample the ocean, the resulting sparse vertical samples were connected by simple smooth curves. When the continuously profiling STD and CTD instruments became available and produced results like the one shown in Figure 2.13, considerable consternation resulted, with some researchers inferring the presence of instrumental defects. Today, the small-scale features known as finestructure and microstructure are taken for granted and their study has become a subject in its own right.

Individual profiles such as that shown in Figure 3.19, are summations not only of the large-scale thermohaline structure, but also of all other physics present, including internal waves, mesoscale eddies, near-surface fronts, internal tides, etc., all of which will be taken up later. In a few special regions, steplike features (Fig. 3.20) appear in both temperature and salinity and persist for very long times. Their existence is believed to be a consequence of a *double-diffusive instability*, which arises from the radically different values of the molecular diffusion of heat and salt—a rare instance in which molecular processes have a visible macroscopic consequence (see Schmitt et al., 2005).

SECTIONS

For most of the history of the science, the large-scale circulation of the oceans was inferred from *hydrographic sections*, or just *sections*—the two-dimensional profiles along ship tracks of temperature and salinity obtained from Nansen-bottle–reversing-thermometer measurements and then from CTDs. Figures 3.21–3.24 are representative of two modern versions, one pair being primarily zonal across the North Atlantic Ocean, the other meridionally down the central Pacific Ocean. Many hundreds of such sections have been published in papers, special atlases (both printed and now electronic) and textbooks. Only within the last twenty five years have measurements extending to the seafloor became the norm rather than special cases.

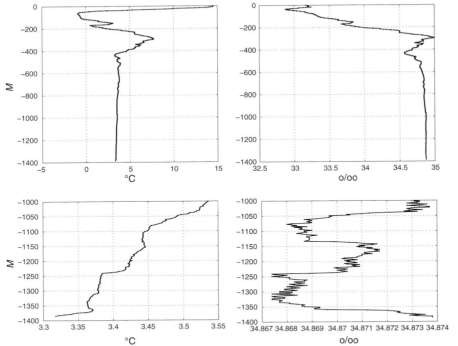

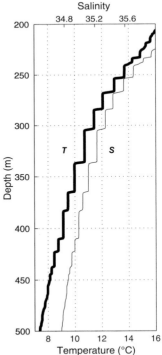

Figure 3.19: A CTD (conductivity-temperature-depth) profile from the North Atlantic Ocean extending to about 1400 m. The position was at 43°N, 51°W in July 1997. The two left panels are temperature, with the lower 400 m expanded, and the two right panels display the corresponding salinities as computed from conductivity and temperature. The finestructure, particularly in salinity, is marked, and there may well be remnants of instrumental problems. Profiles such as this are in distinct contrast to those sampled with water bottles and with smooth curves hand-drawn between sparse points.

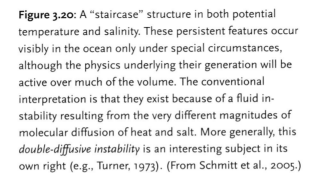

Figure 3.20: A "staircase" structure in both potential temperature and salinity. These persistent features occur visibly in the ocean only under special circumstances, although the physics underlying their generation will be active over much of the volume. The conventional interpretation is that they exist because of a fluid in-stability resulting from the very different magnitudes of molecular diffusion of heat and salt. More generally, this *double-diffusive instability* is an interesting subject in its own right (e.g., Turner, 1973). (From Schmitt et al., 2005.)

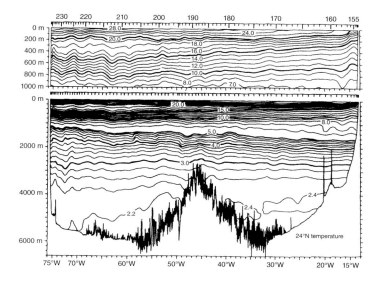

Figure 3.21: Potential temperature as a function of depth (vertical axis) and longitude (horizontal axis) across the Atlantic at about 25°N. The Gulf Stream does not appear, being off to the west beyond the land barrier of the Bahama Bank. Note the gradual shoaling of the upper-ocean isotherms to the east and the comparatively flat interior structures. The *thermocline* is the depth range of maximum rate of change. (From Roemmich and Wunsch, 1985.)

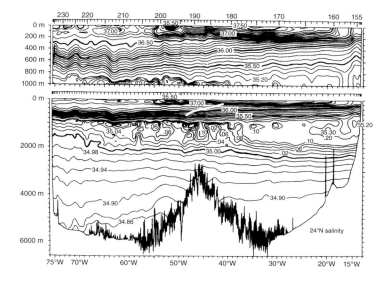

Figure 3.22: A salinity section (in o/oo) corresponding to Figure 3.21. The intrusion of high-salinity water coming from the Mediterranean Sea is particularly pronounced. Figure 3.21 shows that the density of the high-salinity water is compensated for by the presence of higher temperatures relative to the neighboring fluid.

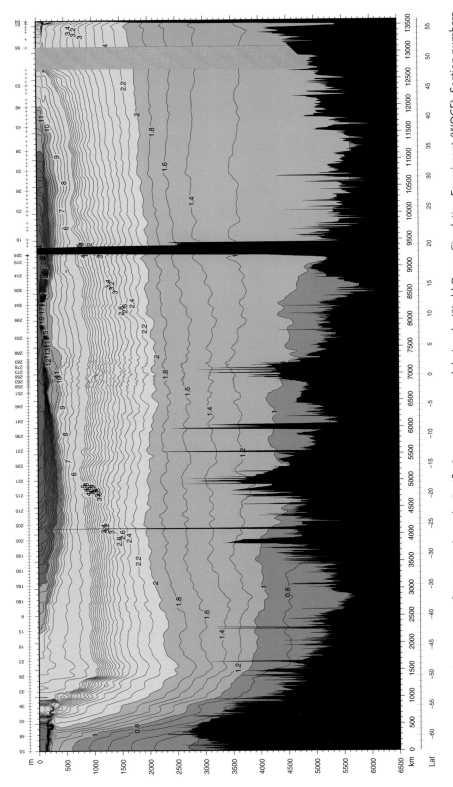

Figure 3.23: A potential-temperature °C section down the Pacific Ocean as measured during the World Ocean Circulation Experiment (WOCE). Section numbers and positions are at the top and distances in kilometers are at the bottom. Notice the very cold water near bottom on the southern end, but rising toward the surface as land is approached and where ice cover precludes shipborne measurements. Contour interval varies with depth. (From Talley, 2007.)

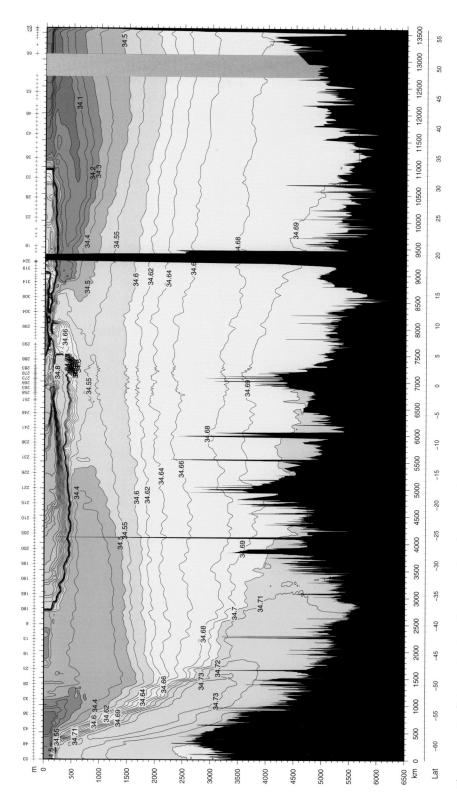

Figure 3.24: A salinity section (nondimensional on the practical scale) down the Pacific Ocean as measured during the World Ocean Circulation Experiment (WOCE). The very cold water near bottom on the southern end in Figure 3.23 is also saltier than its surroundings.

A number of extended reviews of these sections and how they have been interpreted and used are available (Warren, 1981; Reid, 1981; Talley et al., 2011). The large-scale contoured features were recognized from the earliest days as being stable and reproducible from measurements separated by years and decades. Most conspicuous in the temperature sections is the rapid decrease in values at several hundred meters in a region called the *thermocline* and with corresponding changes in salinity called the *halocline*. Explaining the existence, depth, and rate of change of these properties has long been a central goal of general circulation theory. More generally, the weaker gradients at depth and the sometimes complex structures near-surface above the thermocline represent important elements of oceanic fluid dynamics. Much of general circulation descriptive oceanography concerns the origin of the visually striking features, for example, of the relatively cold and saline water visible at the southern end of the section in Figures 3.23 and 3.24 or the very different structure near the surface at low latitudes in the same section. The meaning of the much-smaller-scale features was historically far less clear, and they were regarded as a noise to be ignored.

The second major use of the sections was in the computation of density, $\rho(x, y, z, t)$, from the observed temperature and salinity through the equation-of-state (EOS). From the equations of motion described later, the spatial derivatives of ρ can be used to make inferences about the velocities perpendicular to the section in the *dynamic method*. But because velocity depends most directly not upon ρ, but upon its spatial derivatives, the generally neglected small-scale features in the sections usually dominate the flow field, sometimes radically modifying the understanding of the physics. That the temperatures and salinities at fixed locations are strongly time variable (for example, Figs. A.9 and A.10) also influences the section interpretation.

WATER MASSES

Long ago, it was noticed from the hydrographic section data that temperatures and salinities in the ocean tend to lie within very narrow ranges. Figures 3.25 and 3.26 display histograms of the temperature and salinity of the water volumes. Regions of the ocean where temperature, T, and salinity, S, change little but occupy large volumes or have extreme values are known as *water masses* and have been given geographically evocative names such as North Atlantic Deep Water, Pacific Intermediate Water, Antarctic Bottom Water, etc. (Worthington, 1981); the list is long.

The largest volume of water in the ocean occupies a range in T-S space between about 34 and 35 in salinity and 0 and 3°C in temperature. The mean temperature is about 3.5°C and the mean salinity is about 34.8, values that need explanation.[6] Other scalars such as nitrate or oxygen concentrations are also used to "tag" water masses, but these are not discussed here.

DEGREES OF STRATIFICATION

A useful measure of the vertical stratification is the *buoyancy frequency*, N, defined as

$$N(\lambda, \phi, z)^2 = -\frac{g}{\rho(\lambda, \phi, z)}\left[\frac{\partial\rho(\lambda, \phi, z)}{\partial z} - \frac{\partial\rho(\lambda, \phi, z)}{\partial z}\bigg|_{adiabatic}\right] \quad (3.3)$$

$$= -\frac{g}{\rho(\lambda, \phi, z)}\frac{\partial\rho(\lambda, \phi, z)}{\partial z} - \frac{g(\lambda, \phi, z)^2}{c_s(\lambda, \phi, z)^2},$$

---

[6] See Lund et al., 2011.

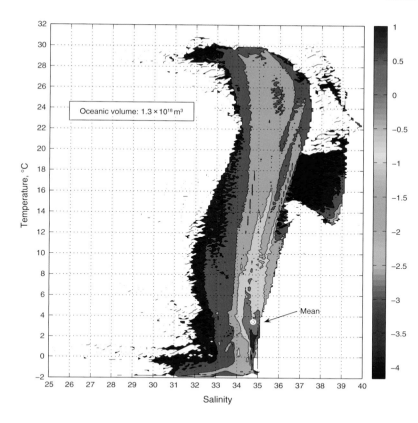

Figure 3.25: Logarithm (base 10) of the relative volumes occupied by potential temperature and salinity in intervals of 0.1°C in temperature and 0.1 in salinity. Most of the water in the ocean is cold, at 2–3°C around 34.7 in salinity; The white circle marks the mean of about 3.46°C and 34.78 in salinity. Values are taken from a state estimate, as described in Appendix B.

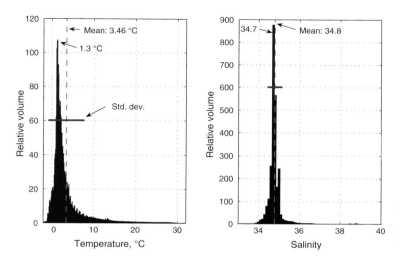

Figure 3.26: One-dimensional histograms of temperature and salinity volumes showing the narrow distribution about the mode. The standard deviation of volume-weighted temperature is about 4.3°C, and for salinity about 0.37.

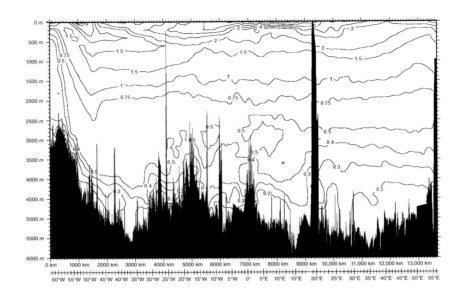

Figure 3.27: The calculated buoyancy frequency $N(\phi, \lambda, z)$ in cycles/hour in a section down the middle of the Pacific Ocean. Much of the smaller scale structure is ephemeral. (From Munk et al., 1995.)

where here $\rho\,(\lambda, \phi, z)$ is an estimate of the time-average local density. Lateral derivatives of ρ are assumed to be negligible compared to the vertical one. $N\,(\lambda, \phi, z)$ is sometimes known as the *Brunt-Väisälä* frequency, but the more physically informative name is preferable. The quantity $(g/c_s)^2$ where c_s is the speed of sound, arises from the adiabatic density gradient—and the term subtracts out the density change owing to sea water compression under the high static pressure at depths. In seawater, $c_s \approx 1500$ m/s.[7] Depending upon the accuracy required, g/ρ can, for convenience, be replaced by $g/\bar{\rho}$, with $\bar{\rho}$ constant.

Should $N^2 < 0$, the density increases upward, and the water column would be statically unstable—a rare occurrence. For a stably stratified ocean ($N^2 > 0$), a fluid parcel hypothetically displaced vertically will tend to oscillate about its initial position with the period $2\pi/N$ (demonstrated on a laboratory scale in Larsen, 1969). Figure 3.27 displays a typical section. Below the very surface layers, the total vertical variation is by a factor of 5 to 10. Some properties (e.g., wave amplitudes and vertical displacements) tend to depend upon the square root of $N(z)$—see the discussion of the WKBJ approximation in Chapter 4—and thus are less sensitive to this structure.

Figure 3.28 shows the global spatial changes at three different depths—a reflection of the changing basic stratification with position. Because ρ is a complicated function of temperature,

[7] Munk et al., 1995; their Appendix B is an acoustic atlas of the ocean.

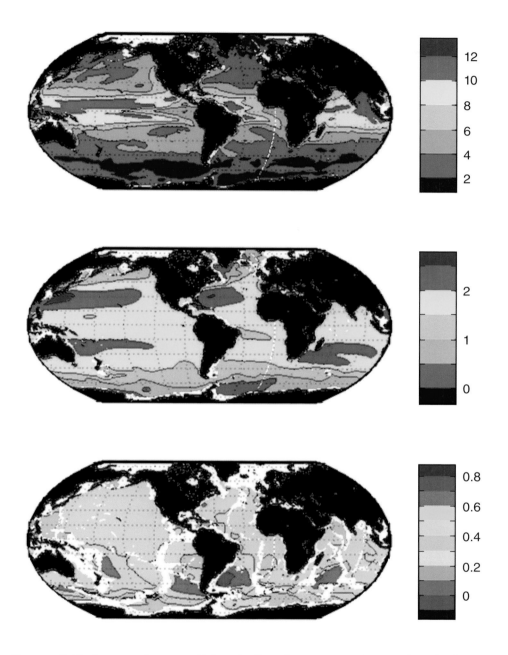

Figure 3.28: The buoyancy frequency, N, in cycles/hour from a state estimate at three depths (top to bottom): 117.5 m, 847.5 m, 3450.5 m. Considerable lateral heterogeneity is apparent, with the high-latitude reduced values particularly being conspicuous, along with the near-surface equatorial maxima. Note the high-southern-latitude structures in the abyss—features closely tied to topographic changes. (From Wunsch, 2013.)

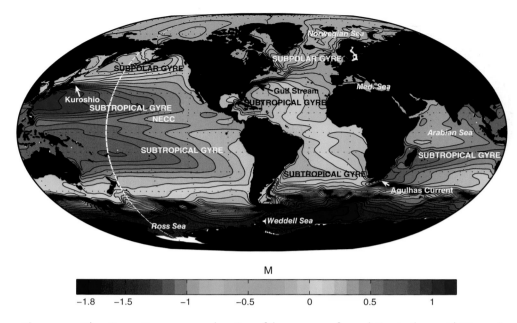

M

-1.8   -1.5   -1   -0.5   0   0.5   1

Figure 3.29: The 16-year time-average elevation of the ocean surface relative to the geoid (Fig. 3.2) in meters from a global state estimate. A few of the named features of the ocean circulation are indicated. Physics demands that the part of the surface flow in geostrophic balance should follow these contours (Eq. 7.13). The non-geostrophic component, however, does *not* conform to that restriction and can exceed the geostrophic component in magnitude. Geostrophic components dominate below about 200 m and will only slowly vary with depth. A useful comparison is with Figure 3.30, which shows an estimate of the integrated top-to-bottom flow, with the difference being an indication of how much the flow does vary with depth.

salinity, and pressure, precise values in weakly stratified water can depend upon exactly which the equation-of-state is being used with observed values and how its derivatives are taken.[8] The presence in observations of temporal disturbances in the stratification leads to the use of various spatial and temporal averages, each having their own problems.

Decades were required to conduct a sparse temperature-salinity survey of the global ocean. Fortunately, the large-scale temperature and salinity structure of the ocean is stable, so measurements from time spans of decades could be patched together and contoured. Renderings of the global-scale fields now exist, including for example, those of Reid (1981), Forget (2010), and several climatological atlases. The website at http://www.ewoce.org/ (in 2013) provides an interactive atlas based upon most of the recent data of this type.

---

[8] See Millard et al., 1990, and King et al., 2012, for a discussion of the formally rigorous calculation of N^2. Natural variability and instrumental noise levels considerably complicate the discussion.

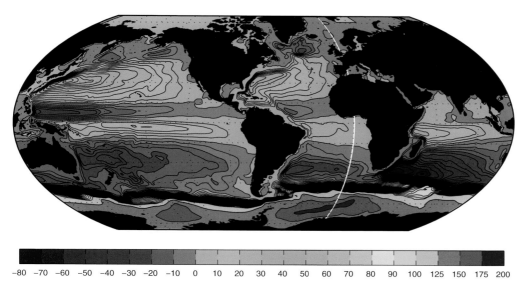

-80 -70 -60 -50 -40 -30 -20 -10 0 10 20 30 40 50 60 70 80 90 100 125 150 175 200

Figure 3.30: The top-to-bottom transport stream function (Eq. 4.60) from a 16-year time average. Units are 10^6 m³/s (Sv). Most of the features of Figure 3.29 are visible but are quantitatively different owing to the considerable flow contributions arising from the deep ocean.

Climatology estimates are presented on uniform grids but differ in how they do space-time weighting, their degrees of quality control, and how they extrapolate to sometimes large unmeasured regions in both space and time. Space and/or time averages of temperature done at fixed depths, z, can be qualitatively different from those calculated on a space- and time-varying density surface, $\rho(x, y, z, t)$. Ambiguities of interpretation exist, fundamentally, because space-time averaging of the governing equations produces complex kinematic and dynamic nonlinear terms. The resulting differences between the various estimated averages becomes a major problem for studies of the ocean on decadal and longer periods.

3.2.1 A Global View of the Surface Ocean

Figure 3.29 is a time-average estimate of the surface elevation of the ocean relative to the geoid synthesized from a combination of many different types of ocean observations. The range of elevation in the 16-year time average is close to 3 m. Explanation is needed to rationalize the magnitudes and the distances over which the elevation changes. Physics demands that the flow field at the sea surface closely follows the contours of elevation (except very close to the equator) and that the speed of movement is inversely proportional to the separation of the contours (Eq. 3.20 below); thus the figure provides some feeling for the major oceanic flow structures at and near the surface. Although the surface flow is similar to that found at depth, abyssal flows do differ from those higher in the water column. For comparison, Figure 3.30 displays the vertically integrated flow (the transport stream function) corresponding to the same state as in Figure 3.29.

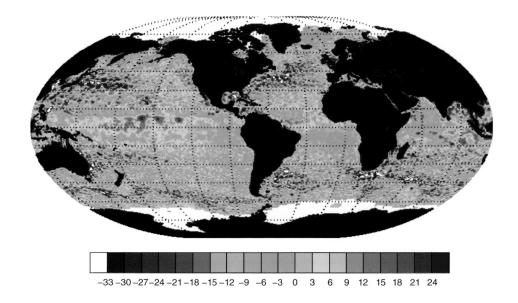

Figure 3.31: A slightly blurry "snapshot" of the sea-surface elevation anomaly in centimeters from 10 d of the combined TOPEX/POSEIDON and ERS-1 altimeter missions. Because of the mapping process, significant spatial detail below about 300 km has been suppressed. The background time average differs slightly from that in Figure 3.29 because the latter uses both more data than just the altimetry and a numerical model of the ocean. The range of disturbances here is about 1.5 m. It occurs over distances of a few hundred kilometers and, for balanced flows, implies geostrophic velocities greatly exceeding those computed from the time average. The temporal variance of η is shown in Figure 3.32. (From the AVISO (Archiving, Validation and Interpretation of Satellite Oceanographic Data) center product; see Le Traon et al., 1998.)

In contrast, Figure 3.31 shows the disturbance (anomaly) in the time-mean elevation as estimated from satellite altimeter data over a single 10 d period. The spatial scales of the anomaly are evidently far smaller than in the time-average and reach ±1 m over those scales (the figure has a saturated color bar). Because the speed is inversely proportional to the contour separation, the velocities and kinetic energies in the anomaly greatly exceed those of the time mean. The strong spatial inhomogeneity of these features, whose RMS is shown in Figure 3.32, is apparent.

3.2.2 The Atmospheric Forcing Structure

The ocean and atmosphere are in direct contact at the sea surface, and they exchange momentum, angular momentum, energy, enthalpy, freshwater, and gases such as carbon dioxide and oxygen.

Much of the structure of the atmosphere is determined by the sea-surface temperature patterns, and the large overall reservoirs of freshwater and carbon that occupy the ocean. In turn, many properties of the ocean are consequences of the atmospheric state. The system is a coupled one, so ultimately, separate discussions of the atmospheric and ocean circulations are not possible.

(a)

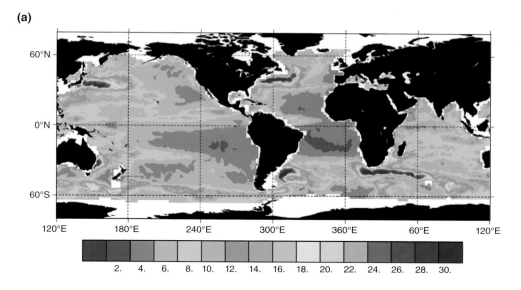

(b)

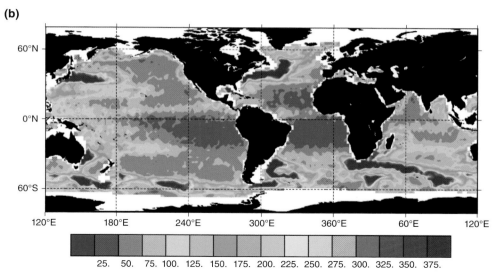

Figure 3.32: (a) An estimate of the RMS surface elevation, η, in centimeters as seen in four years of altimetric data. (b) An estimate of the kinetic energy per unit mass, cm^2/s^2, in the ocean, derived from the altimetric variability and the geostrophic relationship in the equations in (3.20). The results have been multiplied by $\sin^2\phi$, where ϕ is the latitude, so as to suppress the equatorial singularity implied by geostrophic balance. Compare to Figure 3.17b from 7 years of data. Much longer records are now available, but the results differ only in details. Conversion from η to velocity via $\nabla\eta$ is dependent upon the spatial scale used to define the gradient, and the kinetic energy is somewhat *underestimated* here. (From Wunsch and Stammer, 1998.)

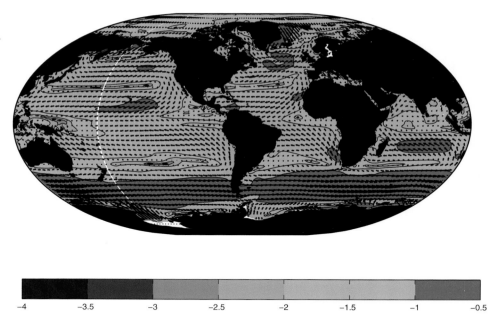

Figure 3.33: The estimated 18-year average *stress* exerted on the ocean by the winds as adjusted from a meteorological "reanalysis" by the European Centre for Medium Range Weather Forecasts (ECMWF) to be consistent with oceanic properties (reanalyses are discussed in Appendix B). Plotted every 4° of latitude and longitude. Colors (and vector lengths) give the base-10 logarithm of the strength in N/m^2. The clearest features are the very strong westerlies in the Southern Ocean and high-latitude North Atlantic. Isolated bands of weak stress are most conspicuous at mid- and low latitudes. The relationship between the wind and the stress exerted by it is a nonlinear transformation discussed later, but in general, stresses are large where winds are large. The large-scale features of dominantly zonal westerly and easterly winds are apparent. In some places (the monsoonal areas), strong annual and semiannual variations from the time mean occur, and the daily synoptic wind systems dominate the patterns away from the low-latitude trade-wind areas.

In practice, however, the complexities of the ocean and atmosphere are so great that simplifications are required to understand the physics. One major simplification is to regard the atmospheric state as known, and ask how the ocean responds to it. Such an approach is the traditional one, and it has great power. But then no answer is provided to the question of why the atmospheric state is what it is, and whether a different ocean response would produce a very different atmosphere? Answering questions like this is a working definition of climate studies: the understanding of the fully coupled system. Here, the atmospheric state will be treated as nominally known.

WINDS

Figure 3.33 depicts the time-average vector wind *stress* (the force per unit area) over the ocean from 18 years of atmospheric estimates. These values are derived from a combination of meteorological observations, a weather-forecast numerical model, and adjustments to the meteorology required to produce a realistic ocean circulation.

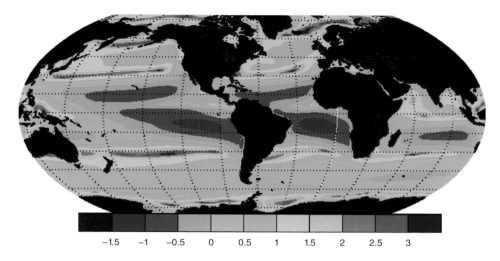

−1.5   −1   −0.5   0   0.5   1   1.5   2   2.5   3

Figure 3.34: The logarithm to the base 10 of the ratio of the kinetic energy of the variability in the wind field to that of the time mean from the reanalysis known as JRA-25 over the years 1992–2009. Estimates are from six-hourly values. Over much of the world ocean, the RMS time-dependent wind components dominate the mean winds. (After Onogi et al., 2007.)

Figure 3.34 displays the log of the ratio of the kinetic energy of the time-variable wind to that of the mean wind from 1992 to 2009 taken from the Japanese Meteorological Agency reanalysis (discussed in Appendix B). Generally speaking, the wind-field kinetic energy is dominated by the synoptic-scale time variable components at latitudes above about 25° latitude. As early mariners knew, tropical-region tradewind systems are comparatively stable, whereas high-latitude wind fields are dominated by their variability. Strong Southern Ocean winds are also relatively stable but still show very large temporal variances. Only an animation of the wind field can convey how violently changeable it is, but the reader is reminded that mid- and high-latitude synoptic wind systems shift direction and magnitude on a 3–5 d time scale. (Atmospheric estimates are far from perfect, having known major biases and other errors. See Josey et al., 2002, and Appendix B.)

Shown in Figure 3.35 is a quantitative rendering of the time scales of the wind components from the second moments of their frequency spectra.[9] Dominant scales shift from on the order of 10+ d in the trade-wind regions to 3 d in the high-latitude, synoptically dominated regions, a result that is consistent with everyday experience. Wavenumber spectra of the wind fields are not particularly well known (but see Gage and Nastrom, 1985, and Fig. 12.5).

ATMOSPHERIC PRESSURE LOADING

Variations in atmospheric pressure generate a measurable, but much weaker, response than does the wind field. Figure 3.36 displays the time-average atmospheric pressure from a meteorological analysis and its root-mean-square (RMS) temporal variability. The very strong values of the variability at high latitudes are quite marked.

---

[9] The second moment, $\sum_k s_k^2 \tilde{\Phi}(s_k) / \sum_k \tilde{\Phi}(s_k)$, is closely related to the rate of threshold crossings and extreme events. See Vanmarcke, 1983.

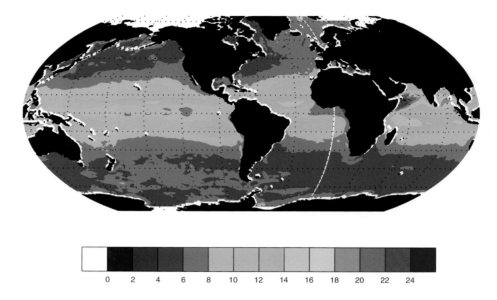

Figure 3.35: Approximate dominant time scales present in the wind forcing of the ocean (in days) derived from the normalized second-moment statistics of the wind stress spectrum. Note that the time scales are shorter than or comparable to the apparent instability growth times of oceanic motions discussed in Figure 11.6 (color bars are reversed). The dominant feature here is the massive shift toward shorter time scales at high latitudes compared to the tropics—the reverse of much oceanic behavior. Forcing was the adjusted stress from an ocean state estimate.

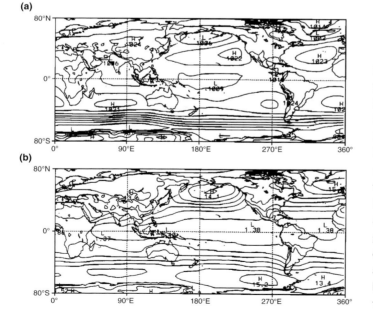

Figure 3.36: (a) The time average (6 years) of surface atmospheric pressure (millibars, mb) as estimated from the European Centre for Medium Range Forecasts (ECMWF) analysis. (b) The RMS variability (mb) about the time mean. 1 mb = 1 hPa is nearly equal to the static weight of 1 cm of seawater. Long-term shifts in these patterns have a measurable effect on regional (but not global) sea level, and the hour-to-hour changes are an important contributor to altimetric measurements of sea-surface height, η. (From Wunsch and Stammer, 1997.)

FRESHWATER

An estimate of the net annual over-ocean precipitation is depicted in Figure 3.37. The accuracy of such estimates is poorly known. To a large degree, the oceanic response depends upon the net difference of evaporation and precipitation. These involve very different physical processes and are not expected to covary in space and time in any simple way. Béranger et al. (2006; see Fig. 3.38) integrated estimates of precipitation rates around the Earth and showed very large discrepancies among them—a lower bound on the actual uncertainty. Because the atmosphere stores so little water, evaporation and precipitation must, when averaged over periods of about a year, be very nearly in balance, thus justifying the study on multiyear time scales, of their simple average difference.[10] Water storage on land does, however, vary measurably, although the magnitudes are small relative to integrated precipitation.

Direct measurements of evaporation from space have not been possible. Figure 3.39 shows an estimate, from a combination of indirect data and a meteorological model, of the *net* evaporation minus precipitation in the annual average. In some regions, the errors are believed to approach 100%.

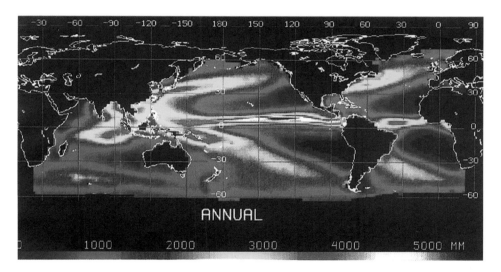

Figure 3.37: An estimate of the global over-ocean annual mean precipitation from the microwave sounding unit (MSU) on various satellites. The intense bands associated with the low-latitude tropics and the midlatitude regions of atmospheric cyclogenesis are apparent. Compare this structure to the estimated long-term evaporation minus precipitation in Figure 3.39, whose structure is markedly different. Figure 3.38 indicates major discrepancies between the MSU measurements and other estimates of precipitation. (From Spencer, 1993.)

---

[10] See Minster et al., 1999.

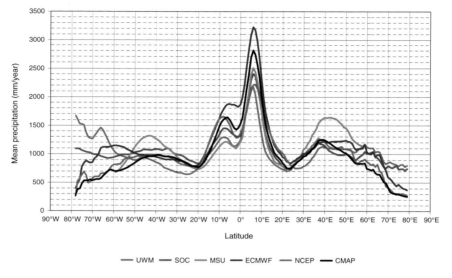

Figure 3.38: Estimates of precipitation climatologies displaying the disconcertingly large discrepancies between various estimates of mean precipitation when integrated zonally around the world. Differences are in the range of meters/year. Whether some of these estimates are qualitatively most reliable remains unresolved. 1 mm/y = 3.2×10^{-11} m/s. 1000 mm/y, if uniform over the globe, including land, is about 16 Sv. UWM: University of Wisconsin, Milwaukee; SOC: Southampton Oceanography Centre; MSU: Microwave Sounding Unit; ECMWF: European Centre for Medium Range Forecasts; NCEP: National Center for Environmental Prediction; CMAP: Climate Center Merged Analysis of Precipitation. See the reference for the corresponding spatial maps. (From Béranger et al., 2006.)

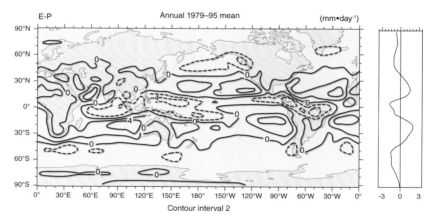

Figure 3.39: (Left panel) The net estimated multiyear time average evaporation minus precipitation (E-P) derived from both observations and a model component in a reanalysis. Zonal means are shown in the right panel. The accuracy of estimates like this one is probably low. (Right panel) The temporal standard deviation and its zonal integral. The difference of the pattern of precipitation and of E-P is an indication of the evaporation spatial structure. Compare with Figure 3.38. (From Trenberth and Guillemot, 1998.)

Table 3.1 and Figure 3.40 attempt a gross global summary of the rates at which water is added to and removed from the oceans. The largest forcings by far are the mean precipitation and evaporation. Other terms, including land-water runoff and melting land glaciers, are very much smaller.

HEAT TRANSFER

Enthalpy transfer between the atmosphere and the ocean is of importance both because of the ability of the ocean to store and transport large amounts of heat, almost all of which is then largely retransferred back to the atmosphere, and because it represents a buoyancy forcing term (along with freshwater) that is intricately involved in interpreting and controlling oceanic motions and their properties. Over the years, many methods have been used to estimate time-average and time-variable air-sea heat transfers involving sensible, latent, and direct radiative components (see for example, Kraus and Businger, 1994). Figure 3.41 displays an estimate of the time-average heat transfer as computed by Stammer et al. (2004) from atmospheric data but adjusted to be consistent with oceanic observations. Crudely, a strong latitudinal dependence exists with net heating in the tropics and net cooling at high latitudes, most conspicuously in the northern hemisphere. Major warming regions exist in the Southern Ocean. Global integrals of the net heating must nearly vanish, because estimates of global warming are less than 1 W/m^2—a very small residual of the much larger competing negative and positive values obvious in the figure.

3.2.3 The Surface Layer

The surface and near-surface ocean are complicated places. Making measurements there is particularly difficult because the sea surface is a moving boundary, one capable of exerting enormous, repetitive, mechanical stresses on any instrument placed at or near it (thousands of cycles per month). The region is one of intense biofouling, with some places being subject to changing ice cover and iceberg movement. Surface waves, which dominate the kinetic energy at the sea surface, span scales from millimeters to kilometers, with even the smallest-scale ripples being important to the transmission of stress from the wind into the sea. On a global basis, surface waves have almost two orders of magnitude more kinetic energy than do the motions of most direct involvement in the ocean circulation.[11] Even if they were only a nuisance, a noise process, they would present a serious impediment to measuring and understanding the much-weaker flows. But surface waves are capable of interacting both with themselves and with larger-scale flows to produce nontrivial effects on the general circulation (e.g., McWilliams and Restrepo, 1999). Surface gravity waves have been studied by humans for thousands of years and modern physics analysis began with Isaac Newton. Despite their familiarity and this long history, understanding remains incomplete.[12]

[11] Ferrari and Wunsch, 2009.
[12] See Craik, 1985; Jones and Toba, 2001; Janssen, 2004; Sullivan and McWilliams, 2010.

Table 3.1: Order-of-magnitude numbers for freshwater exchanges with the ocean. Various quoted units have been converted to Sverdrups (Sv) in the third column. Reliability of these numbers varies greatly (see Bahr et al., 2009, and Schanze et al., 2010). In global volume flux terms, all of these elements are dwarfed by the Ekman pumping of 100+ Sv, which also carries a salinity anomaly signature, ΔS, of near-surface waters.

Input Source	Rate (Various Units)	Sverdrups $(10^6 \text{m}^3/\text{s})$	Reference
1 mm/y global sea level water equivalent		0.01	
Global mean over-ocean precip.		12 ± 6	CMAPP website (NOAA); Xi and Arkin (1997)
Global mean land runoff	$37{,}000 \text{ km}^3/\text{y}$	1.2	Dai et al. (2009), Greenland/Ant. excluded
Groundwater discharge (percolation)	$2.2\text{–}2.4 \times 10^{12} \text{ m}^3/\text{y}$	0.07	Zekster et al. (2007); Moore (2010)
Global mean evaporation		-13	To balance precip. and runoff
Greenland climatological runoff	$100\text{–}200 \text{ km}^3/\text{y}$	$(3-6) \times 10^{-3}$	Box et al. (2004)
Antarctica climatological runoff	170 mm/y	0.07	Bromwich et al. (2004); Jacobs et al. (1992)
Net ice mass loss: Greenland	$137\text{–}286 \text{ Gt/y}$	$(4-9) \times 10^{-3}$	Velicogna (2009), Shephard et al. (2012)
Net ice mass loss: Antarctica	$69 \pm 18 \text{ Gt/y}$	$(3.3 \pm 0.6) \times 10^{-3}$	King et al. (2012)

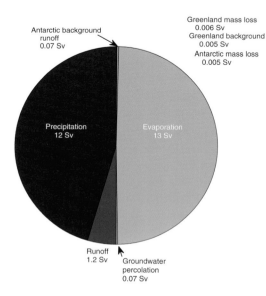

Figure 3.40: Nominal values of the freshwater injection into the ocean (same as Table 3.1) in Sverdrups. Many of these values are very uncertain. The system is dominated by the over-ocean precipitation and evaporation and probably by their temporal variations. A few values, too small to be shown in the figure are listed in the upper right.

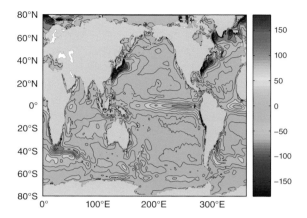

Figure 3.41: Estimated transfers of heat (W/m²) as a multiyear mean between the ocean and atmosphere. Blue areas depict regions of heat loss from the ocean, and yellow-red areas are those of ocean warming. Many different versions of such pictures exist—differing both qualitatively and quantitatively—but have similar gross patterns indicating severe heat loss over the relatively warm western boundary currents and net heating at lower latitudes. Variations about the mean can be very large, exceeding 1000 W/m² at in the western boundary current regions in winter. Note the large regions of warming in the Southern Ocean. The long-term global average is a small residual of large positive and negative numbers and is expected to be on the order of 1 W/m² or less—the global warming signature—although larger interannual and longer variations are also expected. (After Stammer et al., 2004.)

The ocean surface layer is subject to forcing by the atmosphere via winds, atmospheric pressure loads, evaporation, and precipitation. These processes also transfer gases (oxygen, carbon, etc.) between the ocean and atmosphere that are important in biology and climate. In high winds, such as those experienced in hurricanes, the air-sea interface itself can vanish, becoming a continuum of air and water.

Some of the intricacies of the other motions seen within the near-surface layer were discussed by Kraus and Businger (1994), Thorpe (2005) and others. One of the key concepts of the steady state is that of the Ekman layer, whose properties are best discussed later as part of the time-average flow.

3.2.4 Abyssal Boundary Layers

Compared to the sea surface, the seafloor is a simple boundary, because it does not usually move (and tsunami generation by earthquakes is not being discussed here). Significant fluxes through the seafloor are confined to a weak (compared to solar forcing), large-scale thermal flux and a highly localized injection of extremely hot, salty water (see Kadko, 1995). Over flat topography, a mixed-layer commonly exists near the bottom, one partially analogous to that seen near the sea surface (Fig. 3.42). The physics of the rigid horizontal boundary layer is more like the forced flow over a flat plate (albeit a rotating one), which is connected with classical turbulence theory (e.g., Munk and Wimbush, 1970; Monin and Yaglom, 1975; Armi and Millard, 1976). Interesting complexities arise when planar surfaces slope vertically through the stable stratification and become even more intricate when topographic structures

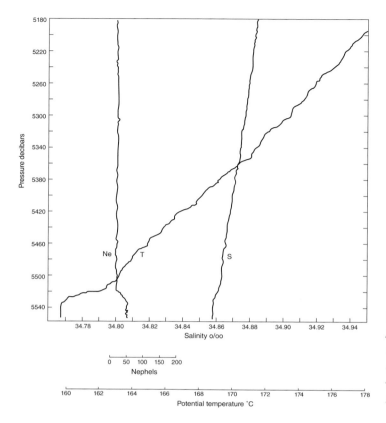

Figure 3.42: Profiles over the Hatteras Abyssal Plain of temperature, salinity, and nepheloid (suspended-particulate) concentration. The mixed layer close to the bottom is apparent. (From Armi and D'Asaro, 1980.)

are imposed upon that slope. These regions have strongly time-dependent flows, making extraction of time-mean fields extremely challenging. The ocean is heated from below by a geothermal heat flux with a global average value of about 0.1 W/m^2 with some extreme local values (Fig. 3.43). This forcing acts to drive the abyssal ocean in the sense of classical Rayleigh-Bénard convection of a fluid heated from below (Chandrasekhar, 1968). Because the oceanic heat capacity is so large however, direct conversion into kinetic energy is comparatively weak, although not everywhere negligible. The mean of 0.1 W/m^2 is not small compared to the estimated residuals of atmospheric heating of the ocean in the greenhouse warming effect of a few tenths of watts per meter-squared.

3.3 EQUATIONS OF MOTION

This part of the chapter has an entirely different flavor from what has gone before. Putting a discussion of the equations of motion into a descriptive chapter might seem perverse, but the real descriptive language is that of fluid dynamics, and at a minimum, a knowledge of the vocabulary is essential. The reader is assumed to have at least a basic understanding of fluid dynamics or else access to a good textbook on the subject. Here, the equations of motion are made postulates, as is the utility of numerous approximations to those equations.[13]

[13] Numerous textbooks on fluid dynamics exist that align with a diversity of reader tastes and levels. The texts by Kundu and Cohen (2008) and Tritton (1988) include a number of geophysical fluid examples. All of the more

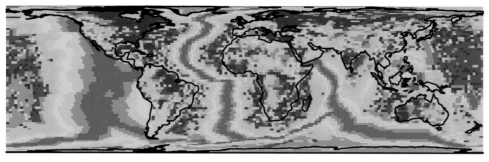

Final estimate of heat flow (mW/m²) (area-weighted median)

■ 4-50	■ 55-58	■ 61-63	■ 70-80	■ 99-129
■ 51-54	■ 59-60	■ 64-69	■ 81-98	■ 130-919

Figure 3.43: An estimate of the geothermal heating through the seafloor. A useful error estimate has also been provided; compare Pollack et al., 1993, and Emile-Geay and Madec, 2009. Oceanic averages are about 0.1 W/m², which can be compared to the magnitudes in Figure 11.18. Because the heating takes place at the bottom of the ocean rather than at the top, it cannot be assumed that the oceanic response to the abyssal forcing is negligible. (From Davies, 2013.)

Fluid dynamics is *always* the story of useful approximations—no one ever solves the full, rigorous, equations with their boundary and initial conditions. Knowing when and how to use which approximations is the challenge: A mathematical description of the flow that permitted the Wright Brothers' airplane to fly could be used to diagram the forces acting on a Boeing 747. The model might be used to determine the result if the latter were caused to stall. Few of us, however, would be likely to accept the resulting scenario as an adequate basis for buying a passenger ticket. Approximations are essential, but like many powerful tools, they can be dangerous to the user. Fortunately for aviation, far more accurate and complete analyses have been developed—and comprehensively tested—in the intervening years.

3.3.1 The Sphere

The Earth deviates measurably from a sphere, but experience shows that representing it in spherical coordinates is adequate for describing observed fluid motions. The most convenient origin is the North Pole, with the angular distance being measured southward from there by the *colatitude*, $\theta = \pi/2 - \phi$, where ϕ is the latitude (see Figure 3.44). Consequently, the meridional velocity is positive in the direction of increasing θ and is thus positive *southward*. The zonal velocity, if positive in the direction of increasing longitude, λ, renders the coordinate system left-handed.

When working in local Cartesian coordinates, however, it is near-universal practice to set the meridional velocity as positive to the north and to label it as v, with u denoting the positive velocity in the zonal direction—making it a right-handed system. Thus on the sphere, many authors use latitude, instead of colatitude, and with u, v oriented as in the Cartesian system (but not always).

theoretical textbooks listed in the Preface describe the basic hydrodynamic concepts.

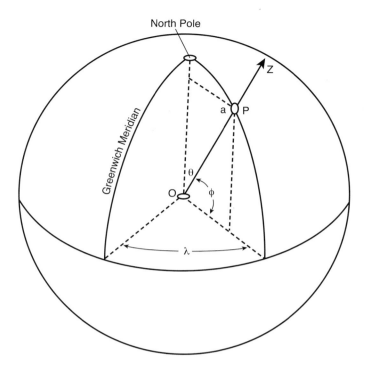

Figure 3.44: The definition of the spherical coordinate system used in the thin-shell limit with $r = a + z \approx a$. ϕ is the latitude, θ is the colatitude, and λ is the longitude measured eastward.

Here, because of the common reliance on the Cartesian approximation, latitude ϕ is used, and then a useful set of equations is as follows (compare Huang, 2010, p. 64):

$$\rho\left(\frac{du}{dt} - \frac{uv\tan\phi}{a} + \frac{uw}{a} - 2\Omega\sin\phi v\right) = -\frac{1}{a\cos\phi}\frac{\partial p}{\partial\lambda} + v\rho\nabla^2 u \tag{3.4a}$$

$$\rho\left(\frac{dv}{dt} + \frac{u^2\tan\phi}{a} + \frac{vw}{a} + 2\Omega\sin\phi u\right) = -\frac{1}{a}\frac{\partial p}{\partial\phi} + v\rho\nabla^2 v \tag{3.4b}$$

$$\rho\left(\frac{dw}{dt} - \frac{u^2 + v^2}{a}\right) = -\frac{\partial p}{\partial z} - g\rho + v\rho\nabla^2 w \tag{3.4c}$$

$$\frac{1}{a\cos\phi}\left(\frac{\partial u}{\partial\lambda} + \frac{\partial(v\cos\phi)}{\partial\phi}\right) + \frac{\partial w}{\partial z} = 0 \tag{3.4d}$$

$$\frac{\partial\rho}{\partial t} + \frac{1}{a\cos\phi}\left(\frac{(\partial\rho u)}{\partial\lambda} + \frac{\partial(\rho v\cos\phi)}{\partial\phi}\right) + \frac{\partial(\rho w)}{\partial z} = \rho\kappa_\rho\nabla^2\rho \tag{3.4e}$$

where,

$$\frac{d}{dt} \equiv \frac{\partial}{\partial t} + \frac{u}{a\cos\phi}\frac{\partial}{\partial\lambda} + \frac{v}{a}\frac{\partial}{\partial\phi} + w\frac{\partial}{\partial z}, \tag{3.5}$$

$$\nabla^2 = \frac{1}{a^2\cos\phi}\frac{\partial}{\partial\phi}\left(\cos\phi\frac{\partial}{\partial\phi}\right) + \frac{1}{a^2\cos^2\phi}\frac{\partial^2}{\partial\lambda^2} + \frac{\partial^2}{\partial z^2}.$$

The longitude is λ, the radius is a, and the radial position is $r = a + z$ (see Fig. 3.44), u, v, w are the velocities in the increasing λ, ϕ, r directions, respectively, and p is the pressure. ν, κ_ρ are the *molecular* coefficients of kinematic *viscosity* and density *diffusivity* respectively, which multiply the viscous and diffusive terms. In addition to the use of spherical coordinates, Equations (3.4a)–(3.4e) contain other approximations. Among them, the radius vector has been written as $r = a + z$, where a is a mean radius of the sea surface, and z is a very small deviation; z is neglected with respect to a, except where a derivative in r has to be taken. In effect, the *thin shell approximation* is being used; it should be noted that the mean depth of the ocean, $\bar{h}$, is very small relative to a, $(\bar{h}/a \approx 7 \times 10^{-4})$.

DENSITY

Because ρ is a nonlinear function of T, S, it is common to write separate equations for each of the form in Equation (3.4e), with individual diffusivities, κ_T, κ_S respectively. Because $\kappa_T \approx 1.4 \times 10^{-7}$ m^2/s, $\kappa_S \approx 1.5 \times 10^{-9}$ m^2/s, differing by two orders of magnitude, treating the diffusion of a single density variable can be inadequate in some problems (e.g., as used in explaining Fig. 3.20). ρ is then a quantity calculated diagnostically from T, S.

An equation-of-state relating the density to T, S, and pressure, p, is needed and is written generically as

$$\rho = \rho\,(T, S, p)\,. \tag{3.6}$$

No complete theory for the dependencies of ρ exists—it is constructed from laboratory measurements, resulting in a complex empirical expression, but one that has been improved over many decades, a subject reviewed by Millero et al. (2008) and the IOC (2010).

Because adiabatic compression of water increases its temperature along with its density, in the deeper parts of the ocean a distinction must be made between the in situ temperature and the *potential temperature*, which is reduced relative to the in situ value by the amount of compressional heating. Various formulas are given alternatively with in situ or local and potential temperatures, so the definitions must be checked. Corresponding to the potential temperature is the calculated *potential density*, the value being referred to a specific pressure value, often the sea surface $(p = 0)$.

Oceanographic analyses employ various versions of the density field. Unless otherwise stated, the densities here are in situ ones, being higher than their potential density counterparts. For notational purposes, $\rho\,(\lambda, \phi, z, t)$ is the instantaneous in situ density at all points and all times. $\bar{\rho}$ is used as a volumetric- and time-mean density field, with the averaging being either regional or global depending on the context. $\rho_0\,(\lambda, \phi, z)$ is used to denote the density within the water column in an ocean ostensibly at rest. Absent an argument t, density is treated as time independent. Again, a profile can represent the ocean anywhere from a single point to a global average and is almost always understood to represent a large scale. $\rho'\,(\lambda, \phi, z, t) = \rho\,(\lambda, \phi, z, t) - \rho_0\,(\lambda, \phi, z)$ is the anomaly of density relative to the rest state.

Because the mean density of seawater differs little from a nominal value near 1030 kg/m^3, the specific choice of ρ—in whatever form, instantaneous or time- or space-averaged—is often not very important in observations. The exception occurs when the density is multiplied by gravity, g. If $\Delta\rho$ represents the spatial and temporal anomalies about any reference value, the *buoyancy forces*, $-g\Delta\rho$, are not small compared to the other terms retained in the equations of motion. Discussions of the effects of compressibility over large volumes and large times also demand very careful definitions of ρ.[14]

Because the numerical value of ρ varies only slightly throughout the ocean, it is an oceanographic convenience, if somewhat sloppy, to write interchangeably about volume fluxes ($v \times$ area) and mass fluxes ($\rho\, v \times$ area), the values differing by 10^3 (with about a 3% error). Thus the oceanographers' volume-flux unit of 10^6 m^3/s (1 Sv) is used and is for many purposes interchangeable with units of 10^9 kg/s, a mass flux rate, although if 3% errors are not tolerable, the interchange is not acceptable.

For some dynamical problems a rough linearization is useful:

$$\rho = \bar{\rho}\left(1 - \alpha_T T + \beta_S S\right), \tag{3.7}$$

where $\bar{\rho} \approx 1038$ kg/m^3; $\alpha_T = 5 - 30 \times 10^{-5}/°$C, and $\beta_S \approx 0.8 \times 10^{-4}$ (Thorpe, 2005, p. 375), and for which the ranges and approximation signs are necessary because the linearization depends upon the particular reference density $\bar{\rho}$. Pressure effects are being absorbed into the constants α_T, β_S. α_T is the *coefficient of thermal expansion*. The widely used name for β_S—the *coefficient of haline contraction*—should be suppressed, because salt is never added to or removed from to the ocean, except on geological time scales.[15] Here it is called the *haline density coefficient*.

Equations (3.4a–e) also employ what is usually called the *traditional approximation*, in which only the component of the Earth's spin-axis vector perpendicular to the local sea surface appears. Considerable theoretical justification exists for adopting it, and it has proven highly useful. Like all approximations, it does sometimes fail (e.g., van Haren, 2008; Stewart and Dellar, 2011). The two major physical regions where it is particularly suspect are the equator, where the vertical component of the rotation vector vanishes, and in places and depths where $N < f$, so that stratification effects are minimal. For present purposes, it is just one more approximation to be re-evaluated should contrary evidence appear. Miles (1974) is a useful starting point for a discussion of its range of validity.

In many oceanographic and related problems, motions are sufficiently weak to permit the neglect of product terms in the equations, such as $(v/a)\,\partial u/\partial\phi$ in the momentum equations. This assumption leads to linearization of the equations. Often linearization is carried out for disturbances in an ocean at rest, in which a basic stratification exists that is treated as a function of z (the radius) alone, ignoring any lateral variations. Let $\rho_0(z)$ be the local density profile of a resting ocean. Then a steady hydrostatic pressure, p_0, would exist:

$$0 = -\frac{\partial p_0(z)}{\partial z} - g\rho_0(z), \tag{3.8}$$

[14] Nonlinearities in the equation-of-state can produce some unexpected and puzzling effects. For example, consider two equal water volumes with different temperatures and salinities—(T_1, S_1), (T_2, S_2)—but having the same density. If they are mixed, the new volume, with nominal temperature and salinity $[(T_1 + T_2)/2, (S_1 + S_2)/2]$ may be *denser* than either of the parents, as a result of a process called *cabelling*. Gille (2004), and Schanze and Schmitt (2013) discussed such effects and, specifically, how they may influence sea level change. A specialized literature exists on the interesting and important question of whether surfaces exist within the ocean, that are related to but different from those of constant in situ density and for which lateral movement of a water parcel occurs without any work being done with or against local gravity. Such surfaces are usually called *neutral surfaces* (see, e.g., Klocker et al., 2009; IOC, 2010).

[15] Determining whether adding salt to an already saline fluid causes the volume to expand or contract is a problem in physical chemistry, one depending upon the starting salinity and temperature distributions.

which can be integrated to

$$p_0(z) = g \int_z^0 \rho_0(z)\, dz.$$

(3.9)

In the rest state, the sea surface is assumed to be $z = 0$, that is, also a spherical shape. g is being treated as a constant. Latitude variation of g dominates because it is usually combined with the rotational centrifugal force of the Earth and absorbed into the reference equipotential, S_{tot}, whose nonsphericity continues to be ignored.

Lateral variations of ρ are in practice important, but because $\rho_0(\lambda, \phi, z)$ has a relatively large vertical derivative, a useful linearization of Equations (3.4a–e) is,

$$\bar{\rho}\left(\frac{\partial u}{\partial t} - 2\Omega \sin \phi v\right) = -\frac{1}{a \cos \phi} \frac{\partial p'}{\partial \lambda} + \nu \bar{\rho} \nabla^2 u$$

(3.10a)

$$\bar{\rho}\left(\frac{\partial v}{\partial t} + 2\Omega \sin \phi u\right) = -\frac{1}{a} \frac{\partial p'}{\partial \phi} + \nu \bar{\rho} \nabla^2 v$$

(3.10b)

$$\bar{\rho}\frac{\partial w}{\partial t} = -\frac{\partial p'}{\partial z} - g\rho' + \nu \bar{\rho} \nabla^2 w$$

(3.10c)

$$\frac{1}{a \cos \phi}\left(\frac{\partial u}{\partial \lambda} + \frac{\partial(v \cos \phi)}{\partial \phi}\right) + \frac{\partial w}{\partial z} = 0$$

(3.10d)

$$\frac{\partial \rho'}{\partial t} + w\frac{\partial \rho_0}{\partial z} = \kappa_\rho \nabla^2 \rho'$$

(3.10e)

Now $\rho = \rho_0(\lambda, \phi, z) + \rho'(\lambda, \phi, z, t)$; $p = p_0(\lambda, \phi, z) + p'(\lambda, \phi, z, t)$. $\bar{\rho}$ is a constant space-time average of the in situ density and is used to multiply the acceleration terms in the three momentum equations. Note the use of $g\rho'$ in Equation (3.10c). (In the older centimeter-gram-second, CGS units, system, $\rho \approx 1$, and it conveniently disappeared from many of the equations—at the cost of occasional dimensional paradoxes if not properly restored.)

The use of $\bar{\rho}$ when multiplying terms such as $\partial u/\partial t$ neglects the small inertial mass differences between light and dense fluid, and hence their slightly different responses to a given force. The neglect is part of the *Boussinesq approximation*. ρ' appears in the momentum equations only where multiplied by gravity, g. Several related approximations are labelled as "Boussinesq": for example, when the fluid is treated as though it is incompressible; hence there can be no sound waves present—their speed formally going to infinity. A truly incompressible fluid is physically impossible (Bohren and Albrecht, 1998), and issues arise in defining the energetics of fluids under the Boussinesq approximation (Young, 2010). Nonetheless, the approximation is useful and often essential.

For phenomena in which the density variation is of no consequence (ρ constant), this linearized, thin-shell system simplifies further to

$$\frac{\partial u}{\partial t} + 2\Omega v \sin \phi = -\frac{1}{a \cos \phi \bar{\rho}} \frac{\partial p'}{\partial \lambda} + \nu \nabla^2 u$$

(3.11a)

$$\frac{\partial v}{\partial t} - 2\Omega u \sin \phi = -\frac{1}{a\bar{\rho}} \frac{\partial p'}{\partial \phi} + \nu \nabla^2 u$$

(3.11b)

$$0 = -\frac{\partial p'}{\bar{\rho} \partial z} - g$$

(3.11c)

$$\frac{1}{a \cos \phi}\left(\frac{\partial u}{\partial \lambda} + \frac{\partial(v \cos \phi)}{\partial \phi}\right) + \frac{\partial w}{\partial z} = 0$$

(3.11d)

The hydrostatic equation can be integrated to give

$$\frac{p'(\lambda, \phi, z)}{\bar{\rho}} = g(\eta(\lambda, \phi) - z), \tag{3.12}$$

relating the pressure field to the now nonzero surface elevation, η, and the depth. If the continuity equation is integrated from the bottom, $z = -h$, to the surface, and $w(z = 0) = \partial\eta/\partial t$ is invoked

$$\frac{\partial\eta}{\partial t} + \frac{h}{a\cos\phi}\left(\frac{\partial u}{\partial\phi} + \frac{\partial(v\cos\phi)}{\partial\lambda}\right) = 0. \tag{3.13}$$

Eliminating p' in Equations (3.11a) and (3.11b) in favor of η produces a system in u, v, η.

These equations, omitting the viscous terms, were formulated and used by Laplace at a very early date (1774). Terms proportional to Ω are named for Coriolis, who worked long after Laplace and who also had antecedents (see Stigler's Law in Appendix C). Although sometimes labelled as "fictitious" forces, they are real to an observer on a rotating Earth. Great effort has gone into analyzing the time-dependent solutions to these equations (see especially Lamb, 1932, Chap. 6; Longuet-Higgins, 1964, 1965a). The presence of the geometrical terms $\sin\phi$, $\cos\phi$ makes their solutions awkward, and insight is obtained by also analyzing them in a local Cartesian approximation.

In practical terms, the direct effects of the molecular coefficients of viscosity and diffusion in the ocean commonly do prove negligible. An example of a direct effect is the penetration depth of wind forcing, for which theory provides the characteristic depth $\sqrt{\nu/2\Omega} = \sqrt{8.9 \times 10^{-3}}$ m ≈ 10 cm. Observations, however, show that a value closer to 100 m is more correct. To cope with this discrepancy, the molecular value, ν, is commonly replaced with a much larger *eddy-viscosity*, A_V, such that the *Ekman depth*, $\sqrt{A_V/(2\Omega)} \approx 100$ m.

Justification for this substitution is based upon a form of *Reynolds analogy*. As written, the governing equations are postulated to describe only components of the flow corresponding to some "large" spatial and/or temporal scale. Superimposed upon them are otherwise undescribed turbulent *eddies*, whose effects on larger scales is assumed to be analogous to the action of molecular processes, except with much greater numerical values. Very roughly speaking, eddies might be regarded as being on the order of 100 km in horizontal radius, changing over about 1000 m in the vertical, being essentially stochastic in nature, and acting on larger-scale flows that vary over horizontal distances on the order of 1000+ km. The subscript V is used because, as a consequence of the extreme thinness of the ocean basins and the presence of strong vertical stratification, processes acting in the horizontal in the ocean are commonly very different from ones in the vertical. Thus the operator

$$\nu\nabla^2(\cdot) \to A_H\nabla_H^2(\cdot) + A_V\frac{\partial^2(\cdot)}{\partial z^2}, \tag{3.14}$$

is used, where ∇_H is the two-dimensional horizontal gradient. Other possibilities include permitting the zonal and meridional values of A_H to differ (not common) and imposing spatial variations,

$$\nu\nabla^2(\cdot) \to \nabla_H(A_H\nabla_H(\cdot)) + \frac{\partial}{\partial z}A_V\frac{\partial(\cdot)}{\partial z}. \tag{3.15}$$

The operator $\kappa_T \nabla^2$ is dealt with similarly, $\kappa \rightarrow \kappa_V, \kappa_H$. Eddy diffusivities of salt and temperature would be identical if the turbulence assumptions are correct. Determining what all of these coefficients might be and what they mean is a central part of modern oceanography; their artificial nature must be remembered. In the most general case, which is not often discussed, the turbulent-viscous-dissipation parameters can be represented by 3×3 nondiagonal tensors. Modelling studies show that the basic Reynolds analog terms are inadequate when integrated over long times, and the search for more accurate constructs continues. The most recent progress toward realistically including the effects of turbulent mixing on the larger scale has involved direct replacement of the time averages of the nonlinear advective terms such as $v/a\partial u/\partial \phi$ and other similar expressions with new forms intended to more explicitly account for the eddy influence, while still retaining some form of Equation (3.15) (see p. 309). Little or no justification exists from observations for the underlying assumptions of disjoint space or time scales nor that stirring and mixing are functions only of the local gradients.

In contrast, however, to the invocation of common turbulent mixing of heat and salt, the thermohaline "staircase" in Figure 3.20 is most easily interpreted as a direct demonstration of the presence of the different molecular coefficients. Turner (1973) discussed the very interesting general fluid property of *double-diffusion* where numerous, macroscopic effects of differing molecular diffusion rates are manifest.[16] In a steady-state system, energy and related structures in all their forms must be dissipated at the same rate at which they are introduced. In a fluid, all dissipation ultimately takes place at the molecular scale—so that momentum and kinetic energy are removed by the workings of molecular viscosity and diffusion on a millimeter to centimeter scale. The ultimate products are Joule heating of the fluid and changes in the internal energy from redistribution of salt and other dissolved products, through terms dependent upon $\kappa_{T,S} \nabla^2 (T, S)$. If a layer of saline fluid is placed at the bottom of a bucket of water of depth, $h \approx 0.1$ m, the time it takes to homogenize the salinity will be a multiple of the *e*-folding time scale of order $h^2/\kappa_S \approx 0.1^2/10^{-9}$ s, or several months. Molecular diffusivity acts only on the very small surface area of the interface between fresh and salt water. Stirring the fluid with a stick takes the saline water and moves it into long thin streaks throughout the bucket, much expanding the surface area on which the molecular diffusivity can work.

Oceanographers thus distinguish between "stirring" and "mixing": stirring is the process of generating physical structures of immense surface area so that the molecular values can operate effectively to remove structures (mix) at the required rate. An analogous phenomenon operates on flow structures via viscous dissipation. The reader is warned however, that "mixing," particularly in a modelling context, usually refers to "stirring," with true mixing being implicit only.

Despite the simplifications, including the spherical thin-shell, Boussinesq, and traditional approximations, these equations are still never employed in their full form in this book: they remain far too complicated. Much effort has been directed over the years at justifying these, and further, simplifications, so that they are no more complicated than necessary for insight into the particular situation at hand. Careful derivations and discussions in detail can be found in, for example, the books by Gill (1982), Pedlosky (1987), and Müller (2006). Numerical models use these and fuller forms of the equations, but their rendering into discrete representations involves a series of additional approximations.

---

[16] Momentum diffusion (viscosity) can be one of the properties; see Baker, 1971.

EULERIAN VERSUS LAGRANGIAN MOTIONS

The equations written here are in the *Eulerian* form—which refers to velocities at fixed points in space. Often, however, the velocities and trajectories following particular fluid parcels are of the most interest. Thus a small volume of fluid, dyed red or carrying large amounts of oxygen or carbon, moves through the fluid from one position, **r** at time *t*, to other positions, **r′** at times *t′*. The velocities at fixed points refer to different fluid parcels that do not remain there—they move on to other positions. A formulation exists (called *Lagrangian*) in which the equations are written following parcels labelled by their starting positions.

Lagrangian fluid equations are rarely used, usually being much more difficult to solve than are the Eulerian form.[17] But because the movement of fluid particles, and the properties they carry, is so important, a number of methods have been developed to calculate approximately the Lagrangian motions from the Eulerian solutions, and some of these will be discussed later. In strongly diffusive flows, the identity of fluid parcels can be lost.

3.3.2 Vorticity

Any reader of the fluid dynamics literature will recognize the constant invocation of *vorticity* and *vorticity dynamics*. As fluid dynamics textbooks show, the vorticity of a fluid is closely related to the rotational motion of the fluid particles. Consider the *circulation, C,* around a small closed loop,

$$C = \oint \mathbf{v} \cdot \hat{\mathbf{t}} dl \tag{3.16}$$

where $\mathbf{v} = (u, v, w)$ and $\hat{\mathbf{t}}$ is a unit vector tangent to the loop. Stokes's theorem of multivariable calculus shows that

$$C = \iint_A (\nabla \times \mathbf{v}) \cdot \hat{\mathbf{n}} dA = \iint_A \boldsymbol{\zeta} \cdot \hat{\mathbf{n}} dA, \tag{3.17a}$$

$$\boldsymbol{\zeta} = \nabla \times \mathbf{v}, \tag{3.17b}$$

where $\boldsymbol{\zeta}$ is the *vorticity*, and $\hat{\mathbf{n}}$ is the unit normal vector to the differential area dA. The curl, which produces the vorticity, can be calculated in any suitable coordinate system. Then in Cartesian form

$$\boldsymbol{\zeta} = \hat{\boldsymbol{\imath}}\left(\frac{\partial w}{\partial y} - \frac{\partial v}{\partial z}\right) + \hat{\boldsymbol{\jmath}}\left(\frac{\partial u}{\partial z} - \frac{\partial w}{\partial x}\right) + \hat{\mathbf{k}}\left(\frac{\partial v}{\partial x} - \frac{\partial u}{\partial y}\right), \tag{3.18}$$

$$= \hat{\boldsymbol{\imath}}\zeta_x + \hat{\boldsymbol{\jmath}}\zeta_y + \hat{\mathbf{k}}\zeta_z,$$

with the vectors in the x, y, z directions. If the circulation integral is in the x-y horizontal plane, C reduces to

$$C = \iint_A \left(\frac{\partial v}{\partial x} - \frac{\partial u}{\partial y}\right) dxdy = \iint_A \zeta_z dxdy,$$

where ζ_z is the vertical component of $\boldsymbol{\zeta}$.

Because it is directly connected to the circulation, vorticity is intimately related to the *local* angular momentum in the fluid (it is twice the fluid-parcel angular momentum), although

---

[17] In part because the mass conservation equation is highly nonlinear and particle trajectories are intrinsically complex phenomena even in flows with comparatively simple Eulerian descriptions. Bennett (2006) provides an extended treatment of Lagrangian fluid mechanics.

because fluid parcels deform, it is not identically the angular momentum, and care is required to interpret it.[18] Vorticity emerges as a center of attention for a number of reasons. Cross-differentiating Equations (3.10a) and (3.10b) produces an equation in ζ_z in which the pressure does not appear, often rendering the equations much more tractable. Pressure alone among the acting forces is primarily established nonlocally, making its determination dependent upon the entire fluid domain. Because it enters the momentum-balance equations only through the gradient, taking the curl of the equations removes it altogether. The thin-shell character of the ocean and the dominance of rotation and/or stratification in various settings render negligible many of the terms in Equation (3.18) and its more general equivalents.

Various conservation rules can be obtained for the vorticity depending upon assumptions about forcing, friction, boundaries, etc. The most fundamental of these results is the Kelvin circulation theorem (see any fluid dynamics text): that in a nondissipative fluid, circulation is a conservative quantity. In a rotating system like the ocean, the spinning Earth carries a major angular momentum component, which appears in the equations as a *planetary vorticity* and which can be exchanged with the ocean. Because of the thin-shell geometry, the *effective* rotation (the vertical component) and hence the planetary vorticity is a function of latitude. The possibility of exchange leads to the concept of *potential vorticity*—which represents a quantity able to be converted into various elements of the flow, much as potential energy can be converted into other forms, such as kinetic. In a rotating homogeneous fluid, the potential vorticity is

$$PV = \frac{\zeta(\lambda, \phi, t) + 2\Omega \sin \phi}{h(\lambda, \phi)}. \tag{3.19}$$

Absent forces and dissipation in movement, PV is conserved by parcels exchanging values of ζ, the relative vorticity, with $f = 2\Omega \sin \phi$, the planetary vorticity, and by changing the fluid depth, or thickness, h as they move.

When stratification and spatially varying large-scale background flows are included, the conservation laws can become very complex and the connection to the ordinary vorticity (Eq. 3.17a) may not be obvious. Nonetheless, in various forms these rules become the backbone of much of the theory of the oceanic general circulation.

3.3.3 How Big Are Terms?

The equations of fluid motion encompass such a vast array of phenomena both in general, and in the ocean in particular, that some systematic way must be found to decide which of the numerous terms in the full equations are likely to be of importance in any particular situation. To that end, methods based upon nondimensionalizing the problem parameters have developed. In a *scale analysis*, these methods typically involve assuming, for example, that lateral variations occur primarily over distances, L, and that velocities have magnitudes, U, etc. Doing this appropriately permits, for example, inferences that for ocean-surface waves having periods of 1 s, Earth rotation is likely to be unimportant, or that surface tension will likely be negligible on the spatial scales of an ocean basin. Textbooks discuss these methods in detail.

---

[18] Proper definition of the rotational movement plus deformation of a fluid parcel involves analyzing the strain in a contiuum. Tritton (1988, p. 81) has provided a clear discussion. Quantities such as shears, $\partial u/\partial y$, etc., are fundamentally rotational in nature.

Nondimensionalization and subsequent scale analysis are very powerful guides—but they almost never lead to absolute inferences. Thus the general circulation can be understood in large part by correctly assuming that its lateral spatial variations occur over a distance of the order $L = 5000$ km. But the important boundary current widths are less than $L \approx 100$ km. In such problems, strong observational and theoretical clues emerge—leading to the necessity of extending the original assumptions. Intuition, experience, and mathematical abilities all come into play.

By the appropriate choice of magnitudes of changes in time and space, of velocity, of density, etc., it proves possible to rewrite the fluid equations so that the velocities, lengths, etc., appearing are dimensionless. This ability proves a great convenience. Carrying out that recipe leads to the formation of nondimensional parameters, that is, ratios, that are useful in characterizing oceanic fluid flows. Given the numerous regimes of oceanic flow, many such parameters exist. The best-known nondimensional parameter, one occurring almost everywhere in fluid dynamics, is the *Reynolds number*, Re $= UL/\nu$. It is a measure of the importance of flow strength to viscosity and is a major determinant of whether a flow is turbulent or laminar. If the characteristic velocity is 1 cm/s, the length scale 1000 km, and the molecular viscosity ν, 10^{-6} m^2/s, then Re $= 10^{10}$. Sometimes such numbers are calculated using instead the eddy coefficients introduced above, thus greatly reducing the size of the Reynolds number. The legitimacy of that practice remains subject to debate. A relative of Re is the *Péclet number*, $Pe = UL/\kappa$ (where κ is a diffusion coefficient), which is measure of the importance of advection to diffusion in, for example, thermal fields.

In physical oceanography, additional important nondimensional parameters are the *Rossby number*, $Ro = U/fL$, and the *Ekman number* $Ek = \nu/fh^2$, where h is the water depth. Over much of the ocean, for fluid scales larger than a few kilometers, $Ro \ll 1$, implying a strong role for Earth rotation. It was already noticed that use of the molecular value of ν implies a much-too-small penetration depth of the theoretical wind-driven flow, leading to invocation of the eddy viscosity to increase it. Nonrotating oceans have infinite Rossby and Ekman numbers and consequently often radically different physical states. Nondimensional parameters also provide a convenient shorthand for understanding the relevance of a laboratory or numerical experiment for oceanographic application. If the laboratory Reynolds number is 1000 and the Rossby number infinite (a nonrotating system) compared to the oceanic values estimated above, strong notice is being given that there *could* be a fundamental difference in the applicable physics.

The *Prandtl number*, Pr $= \nu/\kappa$, is a measure of viscous versus diffusive effects in a flow. If the molecular values are used, Pr $= 9.3$ (appropriate for temperature). Absent conflicting information, turbulent values are often interpreted as implying Pr $= 1$, temperature or salt and momentum being mixed identically. Note that $Pe = \text{Re}\,\text{Pr}$.

In some cases, the *Rayleigh number*, $Ra = g\Delta\rho L^3/\nu\kappa$ is useful and can be extremely large in the open ocean when molecular values are used.[19] Many other such parameters appear throughout the literature. Their definitions are not necessarily unique, depending upon which fields a scientist wishes to impose externally (perhaps a temperature gradient instead of a velocity scale—the latter being implied in some cases by the former—or a horizontal rather than a vertical distance).

---

[19] As large as 10^{32} (Hazewinkel et al., 2012) if evaluated with molecular coefficients.

In most theory-based textbooks, equations and the graphs of their solutions are usually placed in nondimensional form—leading to a very convenient greater universality. Here, in part because of the wide availability of those results and the wish to convey dimensional magnitudes, many of the results are shown with specific dimensional values—with a consequent loss of efficiency and generality. In a later chapter, an analogous discussion leads to the construction of various *dimensional* time scales relevant to the ocean.

3.3.4 Geostrophy

Away from boundaries, at latitudes higher than a few degrees, and typically over distances exceeding roughly 10–100 km, a scale analysis of the governing equations shows a clear dominance of the pressure and Coriolis terms. This *geostrophic balance*, along with hydrostatic balance, is the dominant feature of ocean circulation physics, both in observations and in theory. In the spherical coordinate system, the momentum equations reduce to

$$-\left(2\Omega\sin\phi\right)v = -\frac{1}{a\cos\phi\,\bar{\rho}}\frac{\partial p}{\partial\lambda} \tag{3.20a}$$

$$\left(2\Omega\sin\phi\right)u = -\frac{1}{a\,\bar{\rho}}\frac{\partial p}{\partial\phi} \tag{3.20b}$$

$$0 = -\frac{\partial p}{\partial z} - g\rho \tag{3.20c}$$

$$\frac{1}{a\cos\phi}\left(\frac{\partial u}{\partial\lambda} + \frac{\partial(v\cos\phi)}{\partial\phi}\right) + \frac{\partial w}{\partial z} = 0. \tag{3.20d}$$

The operative word is "balance," in that these equations describe a nominal steady state in which no dissipation exists and hence there are no sources of energy or momentum and no time evolution occurs. In an unstratified ocean, Equation (3.20c) becomes

$$0 = -\frac{\partial p}{\partial z} - g\bar{\rho} \tag{3.21}$$

In spherical coordinates, the horizontal gradient of p is

$$\nabla_h p = \left[\frac{1}{a\cos\phi}\frac{\partial p}{\partial\lambda}, \frac{1}{a}\frac{\partial p}{\partial\phi}\right] \tag{3.22}$$

and thus,

$$[u, v]\cdot\nabla_h p = 0 \tag{3.23}$$

the property repeatedly invoked in this book that geostrophic flows, to a very good approximation, follow the contours of p. Equations (3.20a–d) show too, the property that flows are strongest where pressure contours are closest together, and that for any fixed pressure difference in the fluid, the rate of flow is greater at lower latitudes. p serves as an accurate *stream function* for the horizontal velocity, in the sense that if

$$u = \frac{-1}{2\Omega a\sin\phi}\frac{\partial p}{\partial\phi}, \quad v = \frac{1}{2\Omega a\sin\phi\cos\phi}\frac{\partial p}{\partial\lambda}$$

then $w \sim 0$, (cf. Eq. 3.23). Regional low and high pressures correspond to flows circulating around them as *cyclonic* and *anticyclonic* systems, respectively, which are familiar from atmospheric weather systems. The geostrophic system fails near the equator.

When employed for observations over periods of time beyond a few days, small deviations from these *balance equations*, most conspicuously in the appearance of temporal changes, are seen. Determining which of the numerous small extra terms should be kept to understand time evolution, viscous-like boundary conditions at walls, nonlinearities, etc., has been carefully analyzed. The most common system in use, not necessarily employing dissipational terms, is known as *quasi-geostrophy*, and for justification of its use the reader is referred to one of the textbooks. Other systems (semi-geostrophy, surface quasi-geostrophy) do exist and may be more appropriate depending on the circumstances.

THE TAYLOR-PROUDMAN APPROXIMATION

A useful qualitative inference from the constant density system, Equations (3.20a–d) derive from noticing that the only z-dependence in the system occurs in the hydrostatic relationship and in the continuity equation. The former is integrated once to $p = -gz\bar{\rho} + C(\lambda, \phi)$, where C is the integration constant. When substituted into the two momentum equations, u, v are seen to be independent of z. Then continuity produces

$$w = -z\left(\frac{\partial u}{a \cos \phi \partial \lambda} + \frac{\partial(v \cos \phi)}{a \cos \phi \partial \phi}\right) + D(\lambda, \phi) \tag{3.24}$$

and D is another integration constant. w is thus at most a linear function of z. If the flow is taking place between two rigid plates at $z = -h$ and $z = 0$, $w(0) = w(-h) = 0$, and the only possibility is that $w = 0$ everywhere in the water column.

In a rotating, hydrostatic system the homogeneous fluid is thus inhibited from generating vertical velocities. This result is due to J. Proudman. In a celebrated experiment, G. I. Taylor showed that when a uniform horizontal flow in a rotating fluid with a rigid lid encounters a small topographic bump, the flow would indeed go *around* and not over the bump, leaving a nearly stagnant column of fluid above the bump; the column is commonly called a *Taylor column*. The general result has become known as the Taylor-Proudman theorem, but is better called a "limit." Even with a free surface, the limit can remain an accurate one. In a stratified fluid, the effects of the Taylor-Proudman constraint are often still visible over abyssal topography, albeit diminishing with vertical distance above the bump (forming a Taylor-Proudman *cone*, instead of a cylinder), as can be shown from Equations (3.20a–d).[20]

3.3.5 Boundary Conditions

The equations used in the description of the ocean involve three components of velocity, $\mathbf{v} = [u, v, w]$: pressure, p; (potential) temperature, T; and salinity, S. Sometimes, temperature and salinity are lumped together as buoyancy, $b = -g\rho$, meaning the density, or its anomaly relative to some average, multiplied by gravity. T, S, ρ are subject to boundary conditions at the edges of the fluid, including the sides, bottom, and free surface.

At solid boundaries, such as the seafloor and continental margins, the normal component of flow must vanish, a requirement that can be written $\hat{\mathbf{n}} \cdot \mathbf{v} = 0$, where $\hat{\mathbf{n}}$ is a unit vector

---

[20] Legend tells of a fluid dynamics instructor who, wishing to demonstrate the Taylor column effect to a class, placed a small turtle at the bottom of a rapidly rotating fluid. Unfortunately for the demonstration, the turtle sensibly declined to move at all (relative to the rotating platform). Whether the object moves, or a large-scale flow is imposed on a fixed object, is immaterial to the result.

normal to the boundary. At a solid wall in a laboratory-scale fluid with molecular viscosity, the required boundary condition is *no-slip* meaning that the component of velocity is tangent to the wall, $\hat{\mathbf{t}} \cdot \mathbf{v} = 0$. $\hat{\mathbf{t}}$ is a unit vector perpendicular to $\hat{\mathbf{n}}$ and thus parallel to the wall. The no-slip condition is *not* a fundamental one but is instead empirical and one that took fluid dynamicists decades of theory and experiment to agree was correct.[21]

When eddy viscosities are used, it is not obvious that the no-slip condition remains the most appropriate one (and because of the large Reynolds number requirement, laboratory modeling of oceanographically relevant turbulent flows is very difficult). This issue arises most commonly in the use of numerical models (Adcroft and Marshall, 1998). In such models, the situation can become even murkier when higher-order terms such as $B\nabla^4 u$, are introduced, with the intention of representing the dissipation in a more spatial-scale selective way. B is a coefficient of unknown value and structure.[22] Because these rules are often quite ad hoc—that is, not systematically derived from the equations themselves—and raise the order of the partial differential equations, appropriate boundary conditions involving the derivatives of the flow field tangent to the boundary must be postulated. Their consequent detailed effects remain obscure.

The most common boundary condition for temperature and salinity (or buoyancy) at a solid boundary is that of no flux into or out of the boundary, which is treated as a perfect insulator for temperature and as impervious to salt penetration. Because the normal velocity vanishes there, the requirement becomes, for the remaining diffusive component, $\hat{\mathbf{n}} \cdot \nabla T = 0$, or equivalently for S. In the real ocean, complexities such as the flux of heat or salt or even momentum through the sloping seafloor can, under some circumstances, become significant.

The sea surface is a considerably more complicated place to impose boundary conditions, because its location is a function of the flow itself. This complexity leads to a number of approximations. Most common is the assumption that while the elevation, η, of the surface is changing and is to be determined, the boundary conditions on the flow can be imposed with only slight error at the undisturbed position, where $z = \eta = 0$—which is a linearization.

Except in surf zones and regions of extreme tides, the sea surface does not move much compared to the mean oceanic depths or to vertical motions within the water column, and sometimes it can be treated as a *rigid lid* with little effect on the dynamics. On the other hand, its movement is essential for the existence of some phenomena (e.g., surface gravity waves, the tidal rise and fall) and for the observations of some phenomena from space (internal waves and internal tides and baroclinic eddies).

Two boundary conditions apply at a moving free surface in the absence of external forces. One is the so-called kinematic condition: that the vertical velocity, w, should equal, in the linear limit, the rate of change of the surface elevation, η,

$$w\,(z = \eta \approx 0) = \frac{\partial \eta}{\partial t}. \tag{3.25}$$

The second is a condition preventing a pressure jump across the sea surface (which would lead to an infinite acceleration). Treating a constant atmospheric pressure as $p_a = 0$ (a space-time constant has no dynamical effect on the ocean), the requirement is

$$p\,(z = \eta) = 0. \tag{3.26}$$

---

[21] The subject is still not entirely closed; see Lauga et al., 2007.
[22] See Griffies, 2004.

This relationship is usually linearized about $z = 0$, where the oceanic pressure has a resting-state hydrostatic part, p_0, and a disturbance part, p', related to the water velocity. The pressure change owing to the rest value and induced by a vertical movement, η, is, to lowest order, $\eta \partial p_0 / \partial z$. Thus,

$$p'(z = 0) + \eta \frac{dp_0(z = 0)}{dz} \approx 0, \tag{3.27}$$

the *dynamic boundary condition*. A space- or time-varying atmospheric load introduces a nonzero term on the right-hand side of Equation (3.27) and is what generates the inverted barometer effect.

Other forces at the moving sea surface have to be accounted for. Consider the components of the wind-exerted force per unit area on the sea surface and denoted by $\tau = (\tau^x, \tau^y)$. *Assume further that they are in turn independent of η (including waves) or of the induced flow field;* then the appropriate boundary condition is

$$\left[\rho A_V \left(\frac{\partial u}{\partial z}, \frac{\partial v}{\partial z} \right) \right]_{z=\eta} \simeq \left[\rho A_V \left(\frac{\partial u}{\partial z}, \frac{\partial v}{\partial z} \right) \right]_{z=0} = [\tau^x, \tau^y] \tag{3.28}$$

the *stress condition*. Equation (3.28) results from very strong assumptions, but it is a useful simplification of an extremely complicated place.

Temperature (heat) transfer at the sea surface is an involved phenomenon, because several mechanisms give rise to temperature changes in fluids: (1) *Sensible heat* conduction occurs when a warm atmosphere heats or cools the underlying ocean by a diffusive process sometimes written as $K_v \partial T / \partial z|_{z=0} = q$, where q is a forcing rate. (2) Evaporation releases *latent heat*, cooling the sea surface. (3) Precipitation at an arbitrary temperature can heat or cool it, depending upon the temperature difference between the rain- or snowfall and the ambient water temperature. (4) Solar radiation, both direct at short wavelengths and reradiated at longer wavelengths, can directly heat and cool the upper ocean to varying depths—depending upon its color and opacity.

Almost no salt is exchanged between the ocean and atmosphere, so that an appropriate boundary condition might be

$$K_v \frac{\partial S}{\partial z} \bigg|_z \simeq 0.$$

But ocean salinity (salt concentration) does change—because the ocean and atmosphere exchange freshwater—not salt. Thus the correct surface boundary condition at the sea surface applies instead to the freshwater component of the circulation—a property that does not explicitly appear in the equations written above. It is a mass- or volume-flux condition that might be written as $\rho w(z = \eta) = -F(\lambda, \phi, t)$, where $F(\lambda, \phi, t) = E(\lambda, \phi, t) - P(\lambda, \phi, t)$, the net evaporation minus precipitation volume or mass-transfer rates. Most common, however, is the pretense, particularly in numerical models, that salinity *is* transferred through the atmosphere. A surface concentration boundary condition, $S(\lambda, \phi, z = \eta, t) = S_0(\lambda, \phi, t)$, or a salt-*flux* boundary condition, $K_v (\partial S/\partial z)|_z = q_S$, is used and usually linearized about $z = 0$. F is set to zero. (See Huang, 2010, for the important consequences of this substitution.)

Consistent with the assertion that almost everything in fluid dynamics is an approximation and that none of those approximations is universally appropriate, the conventional boundary conditions used in oceanography do fail under various circumstances. Sidewalls are never vertical and are commonly broken up by hills and valleys and incised with canyons and other features. Other complications at boundaries occur, including injection of freshwater (river runoff) and at depth (percolation from the continents) and numerous other effects.

3.3.6 Cartesian Approximation

The governing equations (3.4a–3.4e) are complicated for a number of very different reasons, ranging from the presence of nonlinear terms to the need for spherical coordinates. For some types of oceanic motions, a local Cartesian approximation can be satisfactory and justifiable, and in many cases, even where it fails quantitatively, it can still capture the fundamental physics of motions on a sphere. In that limit, a set of equations are written that will often be invoked under the assumption that justification has been, or could be, provided in the form of an error estimate. Define x, y, z as locally eastward, locally northward, and in the vertical. z is positive upward from the sea surface. The corresponding velocities are (u, v, w). Then Equations (3.4a–3.4e) become in this limit

$$\bar{\rho}\left(\frac{du}{dt} - fv\right) = -\frac{\partial p}{\partial x} + \bar{\rho}A_H\left(\frac{\partial^2 u}{\partial x^2} + \frac{\partial^2 u}{\partial y^2}\right) + \bar{\rho}A_V\frac{\partial^2 u}{\partial z^2} \tag{3.29a}$$

$$\bar{\rho}\left(\frac{dv}{dt} + fu\right) = -\frac{\partial p}{\partial y} + \bar{\rho}A_H\left(\frac{\partial^2 v}{\partial x^2} + \frac{\partial^2 v}{\partial y^2}\right) + \bar{\rho}A_V\frac{\partial^2 v}{\partial z^2} \tag{3.29b}$$

$$\bar{\rho}\frac{dw}{dt} = -\frac{\partial p}{\partial z} - g\left(\rho_0\left(z\right) + \rho'\right) + \bar{\rho}A_H\left(\frac{\partial^2 w}{\partial x^2} + \frac{\partial^2 w}{\partial y^2}\right) + \bar{\rho}A_V\frac{\partial^2 w}{\partial z^2} \tag{3.29c}$$

$$\frac{\partial u}{\partial x} + \frac{\partial v}{\partial y} + \frac{\partial w}{\partial z} = 0 \tag{3.29d}$$

$$\frac{d\rho}{dt} = K_H\left(\frac{\partial^2 \rho}{\partial x^2} + \frac{\partial^2 \rho}{\partial y^2}\right) + K_V\frac{\partial^2 \rho}{\partial z^2} \tag{3.29e}$$

$$\frac{d}{dt} = \frac{\partial}{\partial t} + u\frac{d}{dx} + v\frac{d}{dy} + w\frac{d}{dz}$$

and is used below. The Boussinesq approximation has been invoked, and the various eddy coefficients have been written in the most basic form as constants. Here $\rho_0\left(z\right)$, producing the rest hydrostatic pressure, has been retained along with the full p.

3.3.7 The β-Plane

Originating with Rossby (1939), β-plane approximations are special cases of the Cartesian approximation in which the Coriolis parameter $f = 2\Omega \sin \phi$ is approximated by a linear rule, $f = f_0 + \beta y$. f_0 is the value at the locally defined origin of the meridional coordinate y. In the most common use, f is treated as a constant, except where differentiation occurs (e.g., in eliminating the pressure forces by cross-differentiation in the horizontal momentum equations).

Linearizing the set Equations (3.29a–e), omitting the dissipative terms, and subtracting the rest pressure,

$$\bar{\rho}\left(\frac{\partial u}{\partial t} - fv\right) = -\frac{\partial p'}{\partial x} \tag{3.30a}$$

$$\bar{\rho}\left(\frac{\partial v}{\partial t} + fu\right) = -\frac{\partial p'}{\partial y} \tag{3.30b}$$

$$\bar{\rho}\frac{\partial w}{\partial t} = -\frac{\partial p'}{\partial z} - g\rho' \tag{3.30c}$$

$$\frac{\partial u}{\partial x} + \frac{\partial v}{\partial y} + \frac{\partial w}{\partial z} = 0 \tag{3.30d}$$

$$\frac{\partial \rho'}{\partial t} + w\frac{\partial \rho_0}{\partial z} = 0 \tag{3.30e}$$

In the last equation, the nonlinear term $w\partial\rho/\partial z$ survives because of the strength of the resting ocean stratification, $\rho_0(z)$. This equation is usefully rewritten using the buoyancy frequency

$$\frac{\partial \rho'}{\partial t} - w\frac{\bar{\rho}_0 N^2(z)}{g} = 0. \tag{3.31}$$

The great advantage of the β-plane approximation is that it captures what proves to be the most fundamental physical characteristic of large-scale fluid motion on a sphere—the influence of the meridional dependence of f—while eliminating the mathematical complexities of spherical coordinates. Justification for the approximation is given in the various textbooks (and see Longuet-Higgins, 1965a). It has sometimes been used outside of its limits of quantitative accuracy; nonetheless, the physical insights derived from it have been profound. Beta-plane solutions more generally are discussed by Lindzen (1967), and special forms are used near the poles.

3.3.8 Conservation Laws

The identification of hypothetically conserved quantities can be extremely useful, both in characterizing a system and for testing physical understanding. "Hypothetically" is invoked because achieving closed systems either analytically or numerically can prove extremely difficult. Thus while all physical systems must conserve mass, the ocean is partially open, exchanging mass not only with the atmosphere, but also with water stored on land and as sea ice and through exchanges into and from the seafloor. The Boussinesq-approximated equations conserve volume, rather than mass, and while the difference is commonly of no concern, problems can arise, particularly when calculating changes in global mean sea level.

Some of the most important of the many possible conservation laws in a fluid are those related to the conservation of fluid circulation, or vorticity, ζ, written in Cartesian coordinates in Equation (3.18). In oceanic flows dominated by Earth rotation, an exchange of vorticity takes place between the fluid and the rotating Earth, leading to some particularly useful constraints. Perhaps the simplest example of such a rule is derived from Equations (3.30a–e), the Cartesian approximation in an ocean of constant density on a β-plane. Differentiating

Equation (3.30a) by $\partial/\partial y$ and Equation (3.30b) by $\partial/\partial x$ and then subtracting produces

$$\frac{\partial}{\partial t}\left(\frac{\partial u}{\partial y} - \frac{\partial v}{\partial x}\right) - f\left(\frac{\partial v}{\partial y} + \frac{\partial u}{\partial x}\right) + \frac{\partial f}{\partial y}v = 0$$

or

$$\frac{\partial \zeta_z}{\partial t} + f\frac{\partial w}{\partial z} + \beta v = 0, \tag{3.32}$$

pressure having disappeared from the equations. Equation (3.32) is a special case of the conservation of potential vorticity in the flow. If the bottom is flat, h is taken to be constant, the flow steady, and the surface rigid, then $w = 0$ everywhere, and $\beta v = 0$, meaning that fluid cannot change latitude—because its potential vorticity would have to increase or decrease, absent any vorticity exchange with the planetary component (f) or a depth change permitting it.

That a geostrophically balanced flow does not "want" to change latitude can be seen in another way. In such a flow, fluid does not cross the lines of constant pressure (Eq. 3.23). Should the pressure surfaces have a north-south trend to them, the total amount of fluid moving between those surfaces would have to increase or decrease, depending upon whether f were decreasing or increasing, respectively. Thus either a fluid source or sink (a forcing term) must be present or the geostrophic constraint must somehow be broken. Given that large-scale north-south flows are known to be present in the ocean circulation, one or both of these mechanisms must operate.

3.3.9 Instability

Much fluid dynamics is devoted to the study and understanding of fluid instabilities. Oceanic stability theory is a focus of works such as Pedlosky's (1987). Drazin and Reid (2004) provided an account of the wider fluid dynamical aspects of the theory. Often, fluid flows, if they can be established physically, become unstable, breaking up into smaller-scale motions and perhaps ultimately appearing completely different from the original state. The consequences of this possibility are many:

1. Some possible flow fields are never observed in nature. For example, a mathematically possible solution for a steady wind field blowing over a resting ocean never occurs if the wind velocity exceeds more than an extremely small value: the slightest disturbance in the water leads to a growing motion that quickly converts the sea surface into a field of capillary and gravity waves and other motions. The path of the open-ocean Gulf Stream is forever changing, meandering, and pinching off eddies. It is unstable and mathematical: steady-state Gulf Stream solutions are never observed. Even if one were somehow established, disturbances would immediately appear. Atmospheric weather is omnipresent and is most commonly described as the result of an unstable large-scale atmospheric flow. A pot of water on a stove heated from below becomes unstable when the temperature at the bottom diminishes sufficiently rapidly upward, so that the resulting buoyancy force overcomes the damping effects of friction and diffusion. Figure 3.20 is interpreted as the product of a double-diffusive instability in a fluid stratified in both temperature and salt.

2. Instability processes can be an extremely efficient way to transfer oceanic energies from one spatial and temporal scale to another, completely different, one. Thus the large-scale

gyre circulations of the ocean, which vary over distances on the order of 5000 km and are quasi-permanent, produce an intense eddy field having spatial scales on the order of 100 km varying over weeks to months (Fig. 3.31). Quasi-steady global-scale atmospheric flows (40,000 km scale) break up into weather systems having spatial scales of thousands of kilometers and time scales of days.

The statement that instabilities generate spatial scales smaller than that of the initial flow has important exceptions, and oceanic situations exist in which small scales apparently manage to interact to produce *larger* ones. That "small"-scale weather systems can interact to produce planetary-scale atmospheric motions was one of the important post–World War II meteorological discoveries,[23] and analogous effects are known to exist in the ocean.

FINAL DISSIPATION

Existing estimates of the energy required to sustain the ocean circulation are of the order of 2 Terrawatts (TW = 10^{12} W, ~ 0.01 W/m^2 equivalent), with a gross uncertainty of about 50% (see Section 11.1.1).[24] A full understanding of the physics of the ocean circulation implies that propagation and transformation of this injected energy, and the mechanisms and regions of dissipation, would be quantitatively known. At the present time, understanding the energy transformations following injection into the circulation is the subject of much current work (e.g., Winters et al., 1995; Tailleux, 2013; Roquet, 2013).

Discussion of net dissipation requires some care, because it takes place, ultimately, at the molecular level, a space scale that is not observable except in sporadic measurements. On the molecular scale associated with high wavenumbers, the dissipation rate of kinetic energy is a shear variance written in a mathematical short-hand as

$$\varepsilon = \frac{\rho v}{2} \sum_{i=1}^{3} \sum_{j=1}^{3} \left(\frac{\partial u_i}{\partial x_j} + \frac{\partial u_j}{\partial x_i} \right)^2 \tag{3.33}$$

with v, the molecular viscosity. The u_i are the three components of velocity, (u, v, w), and x_j are, sequentially, (x, y, z). $\partial u_i / \partial x_j$ are the *rates of strain*. Analogously, dissipation of the temperature variance in this range is measured by

$$\chi_T = 2\kappa_T \left(|\nabla T|^2 \right), \tag{3.34}$$

where the gradient is taken in three dimensions. If the turbulence is really isotropic at these molecular scales, Equations (3.33) and (3.34) simplify. Analogous expressions are used for salinity, $\kappa_T \to \kappa_S$ etc. (See Thorpe, 2005, chap. 1 and 6.)

3.4 MODELS

Analytical and numerical models of the ocean serve several different functions. On the one hand, they are the means by which the fundamental physical elements have been isolated

---

[23] V. P. Starr, who was instrumental in describing and understanding this phenomenon, called it "negative viscosity," implying the appearance of negative values of A_H, A_V, in a terminology and approach that has fallen out of favor.

[24] The total energy entering the ocean in all forms is more than an order of magnitude larger including, especially, the part generating surface waves.

and understood. The system is simplified so as to understand specific mechanisms without necessarily implying realism. An outstanding example was Stommel's (1948) model of the Gulf Stream—which was based upon the β-plane with steady, linear dynamics, in a flat-bottom, rectangular ocean having vertical sidewalls, no stratification, a steady wind system, and idealized dissipation only at the bottom. That model, despite its gross simplifications, is a cornerstone of understanding of the dominating role in the ocean circulation of the meridional gradient of the Coriolis parameter.

At the opposite extreme is the use of numerical models intended to be as realistic as possible. Here the goal is a synthesis of *all* physical processes, so as to represent the real world, often for purposes of prediction. Under the right circumstances such models are extremely useful, for example, in weather forecasting, in the design of aircraft, or in the operation of a nuclear power plant. In these arenas, short-term tests can be made of the model compared to the observed outcome, be it a comparison with the actual weather a few days into the future, and/or to real tests of aircraft or power plant response to disturbances. In contrast, major difficulties are encountered with intrinsically untestable models. That is the situation in dealing with ocean responses over decades and longer where the observations are fragmentary and noisy at best or nonexistent at worst. The temptation to conflate a model calculation with the real world has rarely been resisted, and the results now haunt climate studies.

Users of models that are intended to be fully realistic face many difficulties. Such a model would be as complex to understand as the real system but have several additional problems: (1) The equations in the model will never be a fully accurate representation of the ones that nature solves. No model has perfect resolution, meaning that some processes are *always* omitted—an obstacle that nature does not face. The user must determine whether the omission of those processes is important. Even were it possible to perfectly numerically represent the assumed equations, errors always exist in computer codes (e.g., Basili et al., 1992). No evidence exists that nature ever solves an incorrect version of the Navier-Stokes or thermodynamic equations, but models surely do. (2) Solutions of perfect models would differ from the real world owing to a host of other error sources, including those in the data used to initialize the model and those used in any subsystem to provide boundary conditions. Thus even a complete "Earth system model" incorporating all elements of the climate system must still be forced by a time-varying solar insolation field, which inevitably will be partially uncertain.

The use of large-scale models is a formidable problem of numerical engineering. The problem is one of "engineering" in the sense that the underlying physics are reasonably well understood (the fluid and thermodynamic equations), and the modeller must render the system useful in practice. The analog would be a bridge builder, fully aware of the equations governing stress and strain and creep, who has to convert that knowledge into a functioning structure while taking into account the properties of real materials, costs, local logistics, careless or venal contractors, traffic loads, and unexpected geological structures. Ocean and climate modellers face a host of analogous practical difficulties ranging from the choice of numerical scheme and trade-offs in accuracy and precision against numerical efficiency, to issues arising from specific machine architectures (vector or parallel), input/output limitations, computation versus storage, adequacy of tests, coding errors, etc. Solving these problems is difficult and requires great ingenuity and skill. Models are very useful, but like all engineered structures, they have various modes of failure—some quite surprising and sometimes even catastrophic.

Much of the recent oceanographic literature is devoted not to the ocean per se, but to issues related to model representations of it (see Griffies, 2004; Losch et al., 2004). Important and interesting as those issues are, they are largely omitted here except to the extent that they directly influence the interpretation of observations.

MODEL ERRORS

Models, like observations, involve two distinct types of errors: random and systematic. As with data, random errors are much more easily understood and dealt with. Suppose a model is driven by a meteorological field, perhaps a wind-stress component, τ^x, known to contain errors, $\varepsilon_\tau(\mathbf{r}, t)$, of zero space-time mean $\langle \varepsilon_\tau(\mathbf{r}, t) \rangle = 0$. No spurious time or space-average offset from the true value exists, and it might well be argued, for example, that errors in a model-calculated velocity component, ε_u, also have zero space-time means, that is, $\langle \varepsilon_u \rangle = 0$. This inference is correct for a linear model; to the extent that a model is nonlinear the inference can fail dramatically.

But perhaps more insidious are the effects of what is called a *random walk*, something that exists even in fully linear systems. The result $\langle \varepsilon_u \rangle = 0$ is a theoretical one that can be obtained as a sample average only with a record of infinite duration. The ocean—and other elements of the system such as land ice—has an extended memory, reaching to thousands of years in some cases (e.g., Fig. 11.15). Hasselmann's (1976) stochastic model shows what can happen. Consider a one-dimensional slab "ocean" forced by an "atmosphere" producing a zero-mean, completely white-noise thermal flux, $q_H(n\Delta t)$, $\langle q_H \rangle = 0$. The ocean is a passive reservoir able to store heat such that the temperature anomaly, T, at time t is given by

$$\rho c_p \frac{dT}{dt} = q_H(t),$$

where c_p is the heat capacity, and rendered into discrete form at intervals Δt, as,

$$T(n\Delta t) = T((n-1)\Delta t) + q_H(n\Delta t), \tag{3.35}$$

all physical constants having been absorbed into q_H. Figure 3.45 shows the behavior of this random walk: although $\langle T \rangle = 0$, the system spends most of its time far away from its average.[25] If, slightly more realistically, this ocean has a finite-duration memory, Equation (3.35) can be written

$$T(n\Delta t) = \alpha T((n-1)\Delta t) + q_H(n\Delta t), \quad 0 < \alpha < 1, \tag{3.36}$$

(this equation is an example of an AR(1), discussed in Appendix A). Depending upon the magnitude of α, or its equivalent in more complex systems of this type, model averages taken over very long time intervals can be expected to, indeed *must*, have elements whose apparent mean values are far from the theoretical value of zero. These features also exist in two and three-space dimensions.

Systematic errors in models are similar to those in data and can arise from manifold causes. Many numerical schemes can be written as Taylor series expansions of the corresponding differential forms so that typically the accuracy of the numerically computed value is some power of the step-size, h_ε. Thus a fourth-order Runge-Kutta scheme (Press et al., 2007) incurs an error of order h_ε^6. Depending upon the particular numerical scheme, the error term can be random in sign, in which case it will produce a random walk, or more insidiously, it might be of one sign. If its sign is fixed and the model takes $N = 10^7$ time steps, the error is amplified 10^7 times. In a random walk, the error growth is slower, often of order $\sqrt{N\Delta t}$, but depends upon the physical dimensions and physics.

[25] If q_H is restricted to being randomly ± 1, one has the enlightening coin-flip Game of Peter and Paul (Feller, 1957). This extremely simple process is well worth exploration on a small computer: it has a "red" spectrum, is fractal in nature, and has a sample average whose uncertainty grows with the record length.

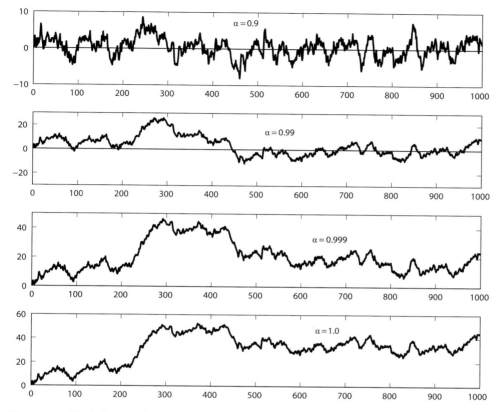

Figure 3.45: The behavior of a time series generated as $T(n) = \alpha T(n-1) + q_H(n)$, for four different values of α as shown, with $\Delta t = 1$. The white noise process, $q_H(n)$, is the same in all four cases. As α approaches 1 from below, the system memory grows, and the tendency for the time series to remain far from its true mean value of zero also grows. Note the change in scales. These simple variations of a random-walk process are readily generalized to higher dimensions and far more complicated rules. The limit, $\alpha = 1$, is no longer stationary, having a variance that grows indefinitely with duration—becoming a conventional random walk with an infinite memory.

Commonly, arbitrary error growth of either type is asserted not to affect models because "feedbacks" preclude it. A model ocean also would not be permitted to boil or freeze solid (the modeler would adjust something if it did). But regional systematic errors can grow large when a model is integrated for long periods of time. A model whose North Atlantic meridional heat transport was too low by 5%, or about 8×10^{13} W would after 10 years have misplaced about 2.5×10^{22} J of heat energy and may have erroneously melted or formed 8×10^{16} kg $\approx 8 \times 10^{13}$ m^3 of sea ice, with manifold implications for the salt distribution, albedo, convection, etc. These numbers should not be taken overly seriously, as many processes affect the formation of sea ice, including the atmospheric state, but they do show the magnitude of the difficulties.

Numerical models of the ocean always involve some typical minimum spatial resolving power. Some coarse-resolution ocean and climate models employ grids on the order of $3°$ separation in latitude and longitude and as a few as one layer, or level, in the vertical. Sometimes models are constructed using summations of functions like sines and cosines

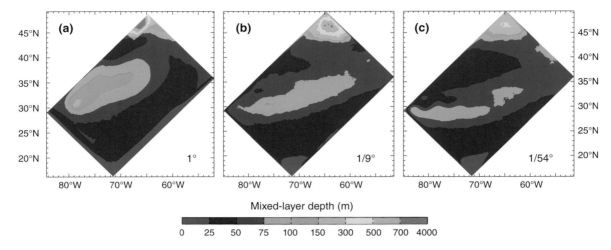

Figure 3.46: Surface mixed-layer depths computed from a regional model with three different lateral resolutions. The gross changes dependent upon the resolution, and the way in which the influence of the unresolved scales is accounted for, shows the difficulty of understanding model errors of this type. Horizontal resolutions are indicated in each panel. (From Lévy et al., 2010.)

or *finite elements* that act as a *basis* for representing the solutions. An equivalent highest wavenumber will then exist. For any smallest characteristic distance, Δr, in any direction, a fundamental question always exists of the importance of motions and structures (perhaps in temperature or bottom topography) not being represented.

To give some impression of the issues, consider Figure 3.46 where an oceanic mixed-layer depth was computed in a regional model for three different spatial resolutions, which had a disconcerting dependence on resolution. The central problem facing ocean modelers is the difficulty in representing on larger scales the undoubted influence of those smaller ones that are not resolved. This issue leads into the very large field of model *parameterization*.

At the present time, parameterization efforts are directed primarily at representing the actions of geostrophic eddies and, increasingly, of internal waves (all taken up later). Comparatively little attention has been paid to the problem of parameterizing the transport-dominating boundary currents, whose structure also remains unresolved. In some situations, the possibility of parameterizations with useful accuracy remains unclear, and it is likely that only advanced computer power will ultimately produce results sufficiently accurate for particular purposes. Parameterizations that produce realistic results at one model resolution may be quite inaccurate at another.

CHAPTER 4

Linear Wave Dynamics

4.1 BACKGROUND

An understanding of the various time-varying elements of oceanic flows requires some background in fluid physics. As with most problems, linear theories are the starting point and provide the most fundamental insights. Sometimes a linear theory does capture the predominant elements, or second best, important nonlinear elements are found to be usefully represented as perturbations around a base state. Linearization sometimes does simply fail.

The elementary theory of wave motion in a rotating, stratified fluid is sketched with the equations that appeared in Chapter 3. Although ocean-bottom topography is the source of many of the most interesting elements of the circulation, the motions in oceans with flat bottoms, both stratified and unstratified, will be explored here first.

4.2 SURFACE GRAVITY WAVES

THE BASIC CONSTRUCT

This discussion begins with a subject that is *not* a focus of this book—ordinary surface gravity waves. The rationale is that these waves are probably the most familiar of all, are the source of common intuition about wavelike behavior, and are a useful comparison for some of the unexpected and unfamiliar behavior of other types of oceanic wave motions.

In the wider wave context, however, surface gravity waves are unusual: they are *edge* or *evanescent* waves, because they are trapped at the sea surface and are thus a boundary phenomenon. They also are usually analyzed, not in terms of a wave equation, but as the solutions of a Laplace equation. (For a much fuller account, see Phillips, 1977, or Kundu and Cohen, 2008, p. 219, whom I loosely follow).

Consider a nonrotating, homogeneous fluid of constant depth, h. In the linear equation limit, in Equations (3.30a–e) set $f = \rho' = 0$, and the simplified equations become

$$\bar{\rho}\frac{\partial u}{\partial t} = -\frac{\partial p}{\partial x}, \tag{4.1a}$$

$$\bar{\rho}\frac{\partial v}{\partial t} = -\frac{\partial p}{\partial y}, \tag{4.1b}$$

$$\bar{\rho}\frac{\partial w}{\partial t} = -\frac{\partial p}{\partial z} - g\bar{\rho}, \tag{4.1c}$$

$$\frac{\partial u}{\partial x} + \frac{\partial v}{\partial y} + \frac{\partial w}{\partial z} = 0, \tag{4.1d}$$

p being the full pressure. If the flow has no vorticity (no curl), it can be written as the gradient of a potential function $(u, v, w) = \nabla\varphi$.[1] Viscous and density effects are thereby excluded. Equation (4.1d) becomes

$$\nabla^2\varphi = 0,$$

the Laplace equation with solutions

$$\varphi(x, y, z, t) = Ae^{ikx+ily-imz} + Be^{ikx+ily+imz}, \quad m = i\sqrt{k^2 + l^2} \tag{4.2}$$

(The omitted factor $\exp(-i\sigma t)$ is understood; in what follows it will appear and disappear without notice. Similarly, real and imaginary parts of complex exponentials will be taken as convenient.) A and B are constant. At a rigid bottom boundary, at $z = -h$, $w(-h) = \partial\varphi(z = -h)/\partial z = 0$, and

$$\varphi(x, y, z, t) = A\left(e^{ikx+ily+\sqrt{k^2+l^2}z} + e^{-2\sqrt{k^2+l^2}}e^{ikx+ily-\sqrt{k^2+l^2}z}\right). \tag{4.3}$$

Components of $\nabla\varphi$ are substituted for the velocity in all other equations, and setting $p = p_0 + p'$, where $p_0 = g\bar{\rho}z$ (the rest pressure), $p' = \bar{\rho}\partial\varphi/\partial t$. The two linearized conditions (the kinematic one and the dynamic boundary condition that surface-pressure perturbations vanish) must be satisfied. The first (Eq. 3.25) is

$$w = \frac{\partial\eta}{\partial t} = -i\sigma\eta = \frac{\partial\varphi(z = 0)}{\partial z}. \tag{4.4}$$

The second, already shown in Equation (3.27), is usefully derived here directly from the equations of motion. Substituting the gradient of φ for the velocities in Equations (4.1a–d), they combine to produce

$$\rho\frac{\partial\varphi}{\partial t} + p + \rho gz = 0. \tag{4.5}$$

Evaluating Equation (4.5) at $z = \eta$, using Equation (4.4) with vanishing p,

$$p(z = \eta) = 0 \implies \frac{\partial\varphi(z = 0)}{\partial t} + g\eta = 0, \tag{4.6}$$

where φ is evaluated at $z = 0$ instead of $z = \eta$.

Substituting for φ, p', η in Equation (4.6),

$$\sigma^2 = g\sqrt{k^2 + l^2}\tanh\left(\sqrt{k^2 + l^2}h\right),$$

---

[1] Not to be confused with ϕ, the latitude.

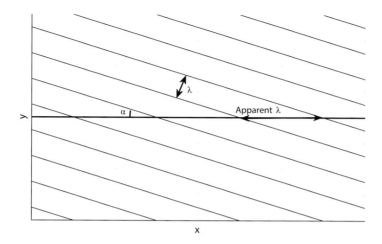

Figure 4.1: Phase propagation in a wave traveling perpendicular to the thin lines. A plane wave measured along a zonal axis will have an apparent wavelength exceeding the true one, depending upon the orientation of the wave.

the *dispersion relation*, as a result of the boundary conditions being satisfied, and thus,

$$\varphi\left(x, y, z, t\right) = A \frac{\cosh \sqrt{k^2 + l^2}\left(z + h\right)}{\sinh\left(\sqrt{k^2 + l^2}h\right)} e^{ikx + ily - i\sigma t}. \tag{4.7}$$

With the convention $\sigma \geq 0$ and the wavenumbers having either sign, in the limit as $h \to \infty$,

$$\sigma^2 = g\left(k^2 + l^2\right)^{1/2}$$

$$\varphi\left(x, y, z, t\right) = A \exp\left(ikx + ily - i\sigma t\right) \exp\left(\sqrt{k^2 + l^2}z\right), \tag{4.8}$$

decaying exponentially downward from the surface. Lines of constant horizontal phase are

$$kx + ly - \sigma t = \text{constant} \tag{4.9}$$

and move in the vector direction, (k, l), with a phase velocity

$$\mathbf{v}_p = \left(\frac{\sigma k}{k^2 + l^2}, \frac{\sigma l}{k^2 + l^2}\right), \tag{4.10}$$

whose magnitude is $\sigma/\sqrt{k^2 + l^2}$ and whose wavelength is $\lambda = 2\pi/\sqrt{k^2 + l^2}$.

In an observational context, the distinction between the true phase velocity and the apparent one is important. Consider an observer having an array of point-measuring instruments along a line as shown in Figure 4.1 in the presence of a wave obliquely incident onto the array as shown. Along the array, the *apparent* distance between wave crests or troughs is

$$\lambda_a = \frac{2\pi}{\sqrt{k^2 + l^2} \sin \alpha}, \tag{4.11}$$

where $\alpha = \tan^{-1}(l/k)$ is the angle of incidence onto a zonal line. If the difference in arrival times of successive crests is measured at any position along the lines, the apparent phase velocity is $v_{xa} = v_x/\sin\alpha$, becoming infinite as the wave approaches broadside arrival. Measurements made along one direction are only readily interpreted from knowledge of the underlying physics.

In this system, no distinction exists between x and y, and aligning the x-axis with the direction of propagation makes $l = 0$, simplifying the algebra. The phase velocity is then

$$v_{px} = \sigma/k = \sqrt{(g/k)\tanh(kh)}, \quad v_{py} = 0, \tag{4.12}$$

and as $h \to \infty$, $v_{px} \to \sqrt{g/k}$. Because of the real exponential in z in Equation (4.8), no vertical propagation occurs. The phase velocity depends upon the wavenumber (or frequency), and so different waves will move at different speeds and thus are said to be *dispersive*. A *nondispersive* system thus is one in which σ/k is a constant.

If two waves, φ_1, φ_2 of amplitude A of neighboring frequencies $\sigma, \sigma + \Delta\sigma$ are present, they will have neighboring wavenumbers $k, k + \Delta k$, and their combined potential (suppressing the z-dependence) is,

$$\varphi = A\exp(ikx - i\sigma t) + A\exp(i(k + \Delta k)x - i(\sigma + \Delta\sigma)t). \tag{4.13}$$

An interference pattern is produced as they go in and out of phase (Fig. 4.2). If Δk and $\Delta\sigma$ are very small, the *envelope* of the interference pattern moves at the rate

$$v_{gx} = \frac{\partial\sigma}{\partial k} = \frac{1}{2\sigma}\left(g\tanh(kh) + gkh\,\mathrm{sech}^2(kh)\right), \quad v_{gy} = \frac{\partial\sigma}{\partial l} = 0, \tag{4.14}$$

known as the *group velocity*.[2] The group velocity is here directed along the phase velocity, but has a different numerical value, with the phase lines appearing to move *through* the *envelope* (or, *packets*). In a dispersive wave field, the packets will form and disappear under propagation while the different frequencies and wavelengths interfere constructively and destructively as they go. Equation (4.14) simplifies usefully in the limits $kh \to 0$, $kh \to \infty$. The group velocity is commonly also the *signal velocity*—which determines the shortest time required for an observer to be able to infer that a disturbance elsewhere has occurred.

THE LONG-WAVELENGTH LIMIT

For long wavelengths, as $kh \to 0$ (the shallow-water limit), the vertical acceleration, $\partial w/\partial t = \partial/\partial t\,(\partial\varphi/\partial z)$, becomes negligible in the vertical momentum equation, and the various dispersion and energy curves simplify. A useful equation in η can be derived by integrating what is now the hydrostatic-pressure equation from the seafloor to the surface, so that $p = g\eta - gz$. Equations (3.30a–e) become

$$\frac{\partial u}{\partial t} = -g\frac{\partial\eta}{\partial x} \tag{4.15a}$$

$$\frac{\partial v}{\partial t} = -g\frac{\partial\eta}{\partial y} \tag{4.15b}$$

$$\frac{\partial\eta}{\partial t} + \frac{\partial(hu)}{\partial x} + \frac{\partial(hv)}{\partial y} = 0 \tag{4.15c}$$

---

[2] See, for example, Stoker, 1992, p. 51, or Lighthill 1978, p. 239, the latter for a physical discussion.

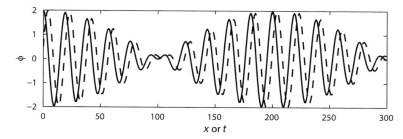

Figure 4.2: The constructive and destructive interference generated by superimposing two waves having slightly different frequencies and wavenumbers (both differing by 10%): here both are moving toward the right. The envelope of the interference pattern moves with the group velocity, which in the limit as Δk, $\Delta\sigma$ vanish is $\partial\sigma/\partial k$. Energy moves with the group velocity (Eq. 4.14). The dotted line shows the position of the disturbance after a delay of two time units.

In this limit, u, v are independent of depth, and w, which has disappeared from the equations, is a linear function of depth increasing from zero at the bottom to $\partial\eta/\partial t$ at $z = 0$. That the waves are a surface-bound phenomenon is no longer obvious.

The equation in η is

$$\frac{\partial}{\partial x}\left(h\frac{\partial\eta}{\partial x}\right) + \frac{\partial}{\partial y}\left(h\frac{\partial\eta}{\partial y}\right) - \frac{1}{g}\frac{\partial^2\eta}{\partial t^2} = 0, \tag{4.16}$$

or for a flat-bottom, periodic wave,

$$\frac{\partial^2\eta}{\partial x^2} + \frac{\partial^2\eta}{\partial y^2} + \frac{\sigma^2}{gh}\eta = 0, \tag{4.17}$$

a classical *Helmholtz Equation*.

PARTICLE VELOCITIES

Assuming A is real and taking the imaginary part of φ in Equation (4.7), then with $l = 0$, the particle velocities at a point x, z are

$$u = Ak\frac{\cosh k\,(z + h)}{\sinh\,(kh)}\cos\,(kx - \sigma t) \tag{4.18}$$

$$w = Ak\frac{\sinh k\,(z + h)}{\sinh\,(kh)}\sin\,(kx - \sigma t) \tag{4.19}$$

and define a set of *streamlines*,

$$\psi = -A\frac{\sinh\,(k\,(z + h))}{\sinh\,(kh)}\cos\,(kx - \sigma t), \tag{4.20}$$

such that $u = \partial\psi/\partial z$ and $w = -\partial\psi/\partial x$ and where the flow is parallel to the lines of constant ψ (in contrast to the representation in $\nabla\varphi$, where the flow is perpendicular to surfaces of constant φ). The surface elevation is

$$\frac{\partial\eta}{\partial t} = w\,(0), \quad \eta = \frac{Ak}{\sigma}\cos\,(kx - \sigma t). \tag{4.21}$$

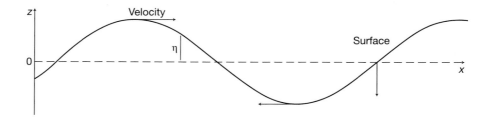

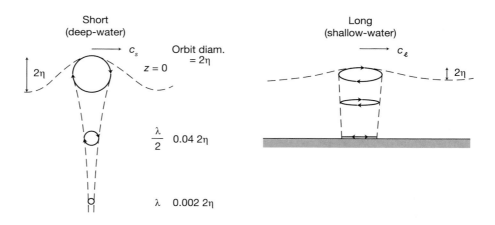

Figure 4.3: (a) The approximate velocity, pressure and (b, c) particle velocity distributions in water waves. In (b), the circles are in practice distorted by the exponential decay factor $\exp(kz)$ and thus give rise to the Stokes drift. C_s and C_l are the short- and long-wave speeds. Orbit diameters are shown at depths of 1/2 and 1 wavelength. (After Pond and Pickard, 1983.)

Equation (4.20) simplifies as $kh \to 0$ to

$$\psi = -A\frac{k(z+h)}{kh}\cos(kx - \sigma t),\tag{4.22}$$

showing as expected that u is constant with depth and w is linear.

In the general case, from Equations (4.18), (4.57), or (4.20) the flow field, and, thus, the particle motions, are ellipses with axes decaying with depth. With integration in time, the particle trajectories are obtained and it is seen that their motion is *prograde*, meaning that at the top of the ellipse the particle is moving in the direction of the phase velocity.

EULERIAN AND LAGRANGIAN MOTIONS AND STOKES DRIFT

In the two versions of fluid mechanics, Lagrangian and Eulerian representations coincide for infinitesimal motions, but for finite amplitude calculations, they differ. Consider that in the solutions Equations (4.18) and (4.19), the particle velocities decay with depth, and the trajectories are ellipses with the top of the ellipse being slightly larger than the bottom (Fig. 4.3). The forward motion of a particle at the top of the ellipse is slightly faster and extends slightly farther forward than does the backward, return motion. Thus the fluid parcel does not return to its starting position, producing a nonzero time-average Lagrangian velocity and displacement. Because the effect is *systematic* over every wave cycle, a fluid particle will, in addition to its to-and-fro motion, slowly drift in the direction of the wave, having larger drift displacements near the surface than at depth. (Beach swimmers outside the surf zone commonly notice that particulate matter in the water column tends to drift toward the shore.)

Calculation of the Lagrangian velocity can be carried out by invoking perturbation theory (here following Phillips, 1977). Let a fluid parcel be at position $\mathbf{a} = (x, z)$ at $t = 0$. Then at any future time its position is

$$\mathbf{x} = \mathbf{a} + \int_0^t \mathbf{u}_L(\mathbf{a}, t')\, dt',$$

where $\mathbf{u}_L = d\mathbf{x}/dt$ is the velocity of the particle that was *originally* at $\mathbf{a}$ and is here called the Lagrangian velocity. For small velocities and times (one wave period),

$$\mathbf{u}_L(\mathbf{a}, t) = \mathbf{u}\left[\mathbf{a} + \int_0^t \mathbf{u}_L(\mathbf{a}, t')\, dt', t\right] \approx \mathbf{u}(\mathbf{a}, t) + \left[\int_0^t \mathbf{u}(\mathbf{a}, t')\, dt'\right]\cdot\nabla_a \mathbf{u}(\mathbf{a}, t) + \dots, \quad (4.23)$$

$\mathbf{u}(\mathbf{a}, t)$ is the ordinary Eulerian velocity at the fixed point $\mathbf{a}$. On the right-hand side of Equation (4.23) are the first two terms in a Taylor series expansion about $\mathbf{a}$ of that velocity, with ∇_a being the gradient in the position $\mathbf{a}$. Longuet-Higgins (1969) usefully wrote

$$\mathbf{u}_L = \mathbf{u}_E + \mathbf{u}_S, \quad (4.24)$$

where $\mathbf{u}_E$ is the Eulerian velocity in the linear approximation and $\mathbf{u}_s$ is the *Stokes velocity*.

With a bracket denoting an average over a wave period,

$$\langle \mathbf{u}_L \rangle = \langle \mathbf{u}_E \rangle + \langle \mathbf{u}_S \rangle. \quad (4.25)$$

Using the full nonlinear equations for deep-water gravity waves, it is possible to prove that $\langle \mathbf{u}_E \rangle = 0$ exactly, below the wave troughs (Stoker, 1992), and hence the mean particle velocity is (Phillips, 1977, his Eq. 3.3.5)

$$\langle u_L \rangle = \langle u_S \rangle = \frac{\sigma k A^2 \cosh 2k(z + h)}{2 \sinh^2 kh}, \quad \langle w_l \rangle = \langle w_S \rangle = 0, \quad (4.26)$$

with amplitude decaying with depth. (The apparent infinity in $\langle u_L \rangle$ as $kh \to 0$ results from using the wave elevation amplitude, A, which in the same limit would necessarily become infinitesimal so as to prevent an infinite u.)

The Stokes drift arises because the inhomogeneity of the wave field produces small systematic changes over a wave period as a fluid parcel moves away slightly from the position of the calculated Eulerian velocity. Should the fluid particles carry properties, such as salinity, or oxygen from one region to another, they can make a significant contribution to the redistribution of oceanic properties generally. Any kind of wave motion whose magnitude is inhomogeneous in space will produce finite particle velocities, even if the mean Eulerian flow is zero. A somewhat analogous phenomenon, not discussed here, is that of *streaming*, which arises from viscous effects near boundaries.[3]

High-frequency motions, for example, water waves with periods on the order of 1 s, are thus capable of generating rectified, that is, time-mean, flows. Earth rotation dominates

[3] For example, Phillips, 1977; Craik, 1985.

oceanic time-average flows, and the consequences of the resulting Coriolis forces on Stokes drifts, should they persist for sufficiently long times, must be taken into account. At the moment, simply notice that the meaning of "time-average" is not absolute and here it is best interpreted as implying an averaging duration that is long compared to the wave periods but much shorter than $2\pi/f$.

RADIATION STRESSES

More generally, in the Eulerian framework, time-average, nonlinear terms such as $\langle u^2 \rangle$, $\langle uv \rangle$, $\langle uw \rangle$, and their spatial derivatives do not vanish. If these terms in a wave train have spatial dependencies (e.g., in a shallow-water wave over variable bottom topography), interesting and important secondary motions can be set up. In the wave context, such terms are commonly known as *radiation stresses* (Longuet-Higgins and Stewart, 1964; Phillips, 1977). A conspicuous example of their presence is the *set* effect, the spatial structure in time-average sea level that occurs where long gravity waves are breaking on a coast. Analogous effects are known for other wave types and would occur in any region where a wave field becomes inhomogeneous, with the outcome depending upon the rates and mechanisms of dissipation, including breaking.

TRANSIENT MOTIONS

Most wave properties are well understood in these linear cases from the analysis of periodic solutions. Transient, nonperiodic solutions are of very great practical importance (e.g., tsunamis) and can be represented mathematically from integrals over all frequencies or wavenumbers. For surface gravity waves in infinitely deep water, the propagation away from an initial disturbance (e.g., a stone tossed into a pond) is commonly known as the Euler-Cauchy problem and has been worked out in detail for some special cases by Lamb (1932), Stoker (1992, p. 156), and Leblond and Mysak (1978). The algebra can become quite intricate. A transient motion will leave a fluid particle displaced from its initial position because, even in the absence of a Stokes drift, the orbits are not then closed. A series of transients, for example, from sequentially dropping a series of stones into a pond, will gradually displace fluid particles from the places they originally occupied.

ENERGETICS

Consider the energy in these waves. The simplest case to analyze is the hydrostatic shallow-water system, Equations (4.15a–c) with constant ρ. Divide through by ρ, so that all results are per unit mass, multiply Equation (4.15a) by u and (4.15b) by v, and add

$$\frac{\partial\left(\left(u^2+v^2\right)/2\right)}{\partial t} = -ug\frac{\partial\eta}{\partial x} - vg\frac{\partial\eta}{\partial y} \tag{4.27a}$$

$$= -\frac{\partial\left(ug\eta\right)}{\partial x} - \frac{\partial\left(vg\eta\right)}{\partial y} + g\eta\left(\frac{\partial u}{\partial x} + \frac{\partial v}{\partial y}\right) \tag{4.27b}$$

$$= -\nabla\cdot\left(g\eta\mathbf{u}\right) - \frac{1}{2h}\frac{\partial\eta^2}{\partial t}\mathbf{u} = \left(u,v\right), \tag{4.27c}$$

using Equation (4.15c).

When the last expression is rewritten as

$$\frac{\partial \left(u^2 + v^2 + g\eta^2/h \right)/2}{\partial t} + \nabla \cdot (g\eta\mathbf{u}) = \mathbf{0}, \tag{4.28}$$

$g\eta^2/2h$ is identified as the potential energy, PE, and $\left(u^2 + v^2 \right)/2$ is the kinetic energy, KE. The term $\mathbf{e} = p\mathbf{u} = g\eta\mathbf{u}$ is the rate of pressure work, and its *divergence* generates changes in the total energy, $E = KE + PE$. Absent horizontal boundary conditions, $\mathbf{e}$ is ambiguous up to the addition of an arbitrary, divergence-free vector, $\nabla \times \mathbf{b}$, meaning that only spatial differences of $\mathbf{e}$ are physically significant. Usually interest lies with the average, $\langle \cdot \rangle$, of these quantities over a wave period.

Substitution of any of the hydrostatic wave solutions into $g \langle \eta\mathbf{u} \rangle$ shows that it can be written

$$\langle \mathbf{e} \rangle = \langle E \rangle \, \mathbf{v}_g, \tag{4.29}$$

a very general result that energy in linear waves moves with the group velocity. It is also true that

$$\langle \mathbf{e} \rangle = \langle p\mathbf{u} \rangle ,$$

and the two expressions are identical. But for more general waves, $\langle E \rangle \mathbf{v}_g \neq \langle p\mathbf{u} \rangle$, not necessarily even having the same direction; they can differ by any divergenceless vector.

If a train of surface gravity waves whose group velocity is directed toward a beach where at least partial absorption takes place through dissipative processes, then some of the energy contained in the waves is being deposited there. That the group velocity has a more central dynamical role than does the phase velocity can be understood by recognizing that phases, particularly in the presence of more than one wave, are kinematic interference patterns—which are lines of constructive and destructive summation and do not necessarily have any particular physical significance. In the presence of a time-mean shear, however, the phase velocity can become a physically important parameter (e.g., Bühler, 2009). External disturbances such as wind fields are commonly prescribed in terms of their phase velocities.

For travelling waves, potential energy reaches an instantaneous maximum at a point when $|\eta|$ is maximum, that is, at the crests and troughs, and it vanishes at points of zero elevation. Conversely, kinetic energy is minimum at the crests and troughs and is maximum at the elevation zeros. Kinetic energy is constantly being exchanged with potential energy as a function of time and position. With a bit more algebra, all of these results can be made applicable to the more general nonhydrostatic case (see Kundu and Cohen, 2008.)

Mean potential energy was defined here from the vertical displacement relative to the resting surface $z = 0$. A slightly perverse approach would define it relative to *any* constant surface z_0, and then an additional term z_0, added to η, would appear. But the potential energy contribution arising from this displacement would not be *available* to generate kinetic energy. The definition of time-average $PE = g \langle \eta^2 \rangle /2h$ is the *available potential energy* (APE), or sometimes, the *gravitational available potential energy*.

Why would one be so perverse as to use any value other than $z = 0$ to define the potential energy? It is an unlikely choice in this situation, but in more complicated ones, which will be encountered later, defining a natural reference surface can be difficult and has led to a complex and opaque literature. *One* element of that complexity can be understood by supposing that the wave being considered is superimposed upon a second wave having a much smaller frequency, ε, and wavenumber, δ. Then the disturbed surface is

$$\eta = A \exp(ikx - i\sigma t) + B \exp(i\delta x - i\varepsilon t), \quad \delta \ll k, \ \varepsilon \ll \sigma,$$

and depending upon the magnitude of B, any average over a region that is small compared to $2\pi/\delta$ and over a time that is small compared to $2\pi/\varepsilon$, which is used to define the mean of η, can be different from the value calculated from $z = 0$. In that circumstance, the potential energy available to the wave with a high-frequency wavenumber for conversion into kinetic energy will appear different in the two calculations, and finding the available part becomes important.

REFLECTIONS AT BOUNDARIES

In the inviscid limit, the boundary condition at a vertical wall is $\mathbf{u} \cdot \hat{\mathbf{n}} = 0$, where $\hat{\mathbf{n}}$ is a unit normal vector. Let the wave be obliquely incident with wavenumber k, l onto a wall at $x = 0$. Elementary physics requires that the reflected wave at the wall must have the same frequency, σ, and along-the-wall wavenumber, l, as does the incident wave. These requirements lead to the classical result of Snell's Law, that the angle of incidence is equal to the angle of reflection.

With four walls, so that the waves exist within a bounded box, $0 \leq x \leq X, 0 \leq y \leq Y$, special solutions exist having specific wavenumbers, $k_n = n\pi/X$, $l_m = m\pi/Y$, where n, m are integers with special frequencies σ_{nm},

$$\sigma_{nm}^2 = g\sqrt{k_n^2 + l_m^2} \tanh\left(\sqrt{k_n^2 + l_m^2}\, h\right) \tag{4.30}$$

The waves then satisfy the boundary conditions, and they are the free-standing-wave solutions. The higher the x and y wavenumbers, the higher the associated frequency. At certain times their energy is all potential (when sea-surface displacement is a maximum) and sometimes all kinetic (when the sea surface is flat). All of the maximum potential energy, if defined about $\eta = 0$, is available for producing kinetic energy.

RADIATION CONDITIONS

In many wave problems, localized sources of motion exist, and if bounding walls are far away, it is required that the flow of energy be directed away from the source and not arrive from infinity. These requirements are generally called *radiation conditions*. In some problems such as the present gravity-wave ones, a wave of form, for example, $\exp(ikx - i\sigma t)$ is transmitting energy toward growing x—in its direction of phase propagation. For other waves of importance, when energy is not carried in the direction of the phase velocity, imposing radiation conditions becomes a more interesting problem, and it is easy to go astray (cf. Baines, 1971; Bühler and Holmes-Cerfon, 2011).

FORCED SOLUTIONS AND MODES

Suppose the system is forced with some spatial function $G(x, y) \exp(-i\sigma t)$ and, only to be specific, let the forcing occur in the hydrostatic equation in η (Eq. 4.17), so that

$$\frac{\partial^2 \eta}{\partial x^2} + \frac{\partial^2 \eta}{\partial y^2} + \frac{\sigma^2}{gh}\eta = G(x, y), \tag{4.31}$$

which is a forced Helmholtz equation, and that the motions are in an enclosed system (e.g., a large lake). Several ways exist to solve such problems. A useful one is to first find the free modes with their corresponding σ_{nm} and horizontal structures, $F_{nm}(x, y, \sigma_{nm})$, which satisfy the boundary conditions. Usually (not always), these are a complete set, meaning that an arbitrary function $G(x, y)$ can be expanded exactly:

$$G(x, y) = \sum_{n,m} B_{nm} F_{nm}(x, y, \sigma_{nm}), \tag{4.32}$$

where B_{nm} is obtained from the set of simultaneous equations

$$\iint_{area} G(x, y) F_{pq}(x, y, \sigma_{nm}) \, dxdy = \sum_{n,m} \iint_{area} B_{nm} F_{nm}(x, y, \sigma_{nm}) F_{pq}(x, y, \sigma_{nm}) \, dxdy, \tag{4.33}$$

and n, m are the integers. If the F_{nm} are assumed orthonormal (often possible), these last equations become

$$\iint_{area} G(x, y) F_{pq}(x, y, \sigma_{nm}) \, dxdy = B_{pq}, \tag{4.34}$$

with the integral described as the *projection* of G onto F_{pq}.

Expand the unknown solution in these same functions,

$$\eta(x, y, \sigma) = \sum_{n,m} A_{nm} F_{nm}(x, y, \sigma_{nm}), \tag{4.35}$$

and substitute into the governing equation,

$$\eta(x, y, t) = \sum_{n,m=0}^{\infty} \frac{B_{nm}}{\sigma^2 - \sigma_{nm}^2} F_{nm}(x, y, \sigma_{nm}) e^{-i\sigma t}. \tag{4.36}$$

The solution thus can be regarded as being made up of a sum of the free solutions (but having the frequency of the forcing). Each free solution (normal mode) is excited to an extent depending upon *both* the magnitude of its projection onto the forcing (Eq. 4.34), and the difference between the forcing frequency and that of the corresponding free mode. Whether this particular method of solution and corresponding representation (others exist) provides useful insight depends upon the problem and whether the projections are confined to a few or many modes. If the projection is finite, at $\sigma = \sigma_{pq}$, the response is resonant and formally infinite. Then the linearized inviscid perturbation assumption fails, and either dissipation and/or nonlinearity must enter to keep the result finite. Commonly, the prediction of a very large but finite response in the linear solution is found to be useful.

A closed basin is unnecessary for resonance. If a travelling disturbance has a frequency and wavenumber corresponding to a free wave in an entirely open system, a very large response will be generated from solutions similar to Equation (4.36).

THE WKBJ APPROXIMATION

The equation governing waves of many kinds can often be manipulated into the general form

$$\frac{d^2F}{dx^2} - r(x)F(x) = 0, \tag{4.37}$$

where $r(x)$ is slowly varying relative to the wavelength of the waves, for example, a bottom topography that changes slowly (note the minus sign) where "slowly" is measured in terms of wavelengths. Intuitively, a solution $F(x) = A(x)\exp\left(\pm\sqrt{r(x)}x\right)$ (sometimes called an *ansatz*—"a guess") might be expected to be approximately correct with a change in behavior as $r(x)$ passes from positive to negative. This approach, and its many generalizations and extensions, is named for Wentzel, Kramers, Brillouin, and Jeffreys (WKBJ), the first three of whom who exploited it in solutions of the Schrödinger equation (it should have been named for the much earlier Liouville and Green; see Olver and Wong, 2010, and the discussion of Stigler's Law in Appendix C). With some mild requirements on the behavior of $r(x)$, a procedure of substitution and ordering produces,

$$F(x) \sim r(x)^{-1/4}\exp\left\{\pm\int r(x)^{1/2}\,dx\right\}. \tag{4.38}$$

When $r(x)$ is negative, the behavior becomes oscillatory rather than exponential.

Many of the wave equations encountered in the ocean appear in the general form

$$\frac{d^2F}{dx^2} + p(x)\frac{dF}{dx} + q(x)F = 0.$$

By setting

$$F(x) = \exp\left(-\frac{1}{2}\int^x p(x')\,dx'\right)G(x), \tag{4.39}$$

then

$$\frac{d^2G}{dx^2} - \left(\frac{1}{4}p^2(x) + \frac{1}{2}\frac{dp(x)}{dx} - q(x)\right)G = 0, \tag{4.40}$$

and subject to proper behavior of the second term, it is a form to which the WKBJ approximation can be applied.

EFFECTS OF THE BOTTOM

In the shallow-water limit, with $kh \to 0$, and $l = 0$ (normal incidence) where the waves feel the bottom, several important properties can be deduced. If the topography is one-dimensional,

$h(x)$ and the wavelength is $2\pi/k >> h$, the motion is then hydrostatic. If h is permitted to vary (e.g., Lamb, 1932, p. 276), Equation (4.16) becomes, for fixed frequency,

$$\frac{\partial}{\partial x}\left(h\frac{\partial \eta}{\partial x}\right) + \frac{\sigma^2}{g}\eta = 0.$$

If the bottom depth is γx, increasing from $x = 0$, a possible solution is

$$\eta(x, t) = AJ_0\left(2\sigma\sqrt{x/(g\gamma)}\right)e^{-i\sigma t}, \tag{4.41}$$

J_0 being the Bessel function. This solution is a standing wave whose amplitude grows toward $x = 0$. When the water depth vanishes, the approximation, $\eta << h$, used in the linearizations breaks down. If a Hankel function solution, $\eta = H_0^{(2)}\left(2\sigma\sqrt{x/(g\gamma)}\right)$, were used instead, the wave is travelling from plus infinity toward $x = 0$, becoming arbitrarily large there with no reflection. The interpretation is of breakdown and dissipation, but its analysis requires physics going well beyond the present approximations. How much would be reflected in practice would depend upon the friction and nonlinearities such as the water shoals, and the more general solution would include both J_0 and $H_0^{(2)}$.

A two-dimensional wave, $A\exp(ikx + ily + i\sigma t)$, travelling obliquely toward diminishing x and impinging onto the slowly shallowing bottom $h = \gamma x$, can be guessed to be approximately of the form $\eta(x, y, t) = A(x)\exp(ik(x)x + ily - i\sigma t)$, and as x, h become small (but not zero), the local k increases, σ and l are preserved, and thus the wave turns ever more parallel to the shoreline in what is called *refraction*. Close to the origin, the nonlinearities and dissipational mechanisms already alluded to would be present. WKBJ methods can be used to find the changing amplitudes and energies.

SOME PARTING WORDS ON SURFACE GRAVITY WAVES

Surface gravity waves are a rich subject with a large and still-growing literature. Theory and observations can be extended to include full nonlinearity, surface tension and viscosity, examination of the implications of their stochastic nature, breaking, and the interesting subjects of their generation by wind and earthquakes.

4.3 CARTESIAN APPROXIMATIONS OF THE ROTATING SYSTEM

Let us return now to the more central subject of time-dependent motions in a rotating, and then stratified, ocean.

THE f- AND β-PLANE-UNBOUNDED OCEANS

For motions whose lateral scale is significantly smaller than the radius of the earth, a, but large enough and slow enough to be hydrostatic, of small amplitude, and in a fluid of constant density, and with the recognition that rotation might be important, the hydrostatic β-plane approximation version of Equations (3.30a–e) with ρ constant are

$$\frac{\partial u}{\partial t} - fv = -g\frac{\partial \eta}{\partial x} \tag{4.42a}$$

$$\frac{\partial v}{\partial t} + fu = -g\frac{\partial \eta}{\partial y} \tag{4.42b}$$

$$\frac{\partial \eta}{\partial t} + h\left(\frac{\partial(hu)}{\partial x} + \frac{\partial(hv)}{\partial y}\right) = 0, \tag{4.42c}$$

using $p = \bar{\rho}(g\eta - z)$. For motions of small meridional extent and/or small wavelength, $\beta = 0$ is a good approximation, with f treated as a constant in what is often called the *f-plane*. For even smaller-scale, higher-frequency, waves, the effects of f are negligible and the conventional theory of long gravity waves is recovered from these equations.

ROTATING CYLINDER AND BESSEL FUNCTIONS

In a rotating system, the cylinder provides the most accessible full solutions that exactly satisfy the governing Equation (4.43). When rewritten in the polar coordinates, $r = \sqrt{x^2 + y^2}$, $\tan \alpha = y/x$, the equation becomes

$$\frac{\partial^2 \eta}{\partial r^2} + \frac{1}{r}\frac{\partial \eta}{\partial r} + \frac{1}{r^2}\frac{\partial^2 \eta}{\partial \alpha^2} + \frac{\sigma^2 - f^2}{gh}\eta = 0,$$

and separating variables, $\eta(r, \alpha) = F(r)\exp(im\alpha)$, produces an ordinary differential equation,

$$\frac{d^2 F}{dr^2} + \frac{1}{r}\frac{dF}{dr} + \left(\frac{\sigma^2 - f^2}{gh} - \frac{m^2}{r^2}\right)F = 0,$$

known as a *Bessel equation*. Unless a barrier of some sort exists, m must be an integer to assure no discontinuities.

Putting $\lambda^2 = (\sigma^2 - f^2)/gh > 0$, the solutions are Bessel functions, $F(r) = J_m(\lambda r)$ or $F(r) = Y_m(\lambda r)$ which are the cylindrical coordinate analog of standing waves (sine and cosine) in r. $Y_m(\lambda r)$ are singular at $r = 0$ as a result of the coordinate convergence there. The two Hankel functions are the analog of travelling waves and are made from the two standing-wave solutions as $H_m^{(1)}(\lambda r) = J_m(\lambda r) + iY_m(\lambda r)$ and $H_m^{(2)}(\lambda r) = J_m(\lambda r) - iY_m(\lambda r)$, which for large r are proportional to $\exp(i\lambda r)$ (outgoing waves) and $\exp(-i\lambda r)$ (incoming waves), respectively. Should $\lambda^2 < 0$, the solutions become exponential rather than oscillatory and are called modified Bessel functions; these are usually written $I_M(\lambda r)$, $K_m(\lambda r)$. Numerous textbooks analyze Bessel function properties at length (e.g., Olver and Wong, 2010).

Consider free periodic motions, where $u, v, \eta \propto \exp(-i\sigma t)$. With f, h constant, the Helmholtz equation,

$$\frac{\partial^2 \xi}{\partial x^2} + \frac{\partial^2 \xi}{\partial y^2} + \frac{\sigma^2 - f^2}{gh}\xi = 0, \tag{4.43}$$

is satisfied by any of u, v, η. A solution in plane waves, $\xi = A \exp(ikx + ily)$, requires

$$\sigma^2 = f^2 + gh\left(k^2 + l^2\right), \tag{4.44}$$

the dispersion relation, which shows $\sigma > f$. Rotation increases the frequencies of the nonrotating system and is said to "stiffen" the fluid. Equation (4.44) is displayed in Figure 4.4 for two very different values of h. No plane wave motion exists below the Coriolis frequency, and its absence raises the question of what would happen if the system were *forced* with $\sigma < f$. (Solutions in cylindrical coordinates are often useful; see Lamb, 1932, Section 209, and the box "Rotating Cylinder and Bessel Functions.")

"CANAL THEORY," POINCARÉ, AND KELVIN WAVES

Discussions of f-plane solutions in "canals" and other bounded bodies of water exist (e.g., Lamb, 1932; Proudman, 1953; Defant, 1961), and so only one configuration is sketched. Consider a canal extending in the x-direction to $\pm\infty$. Walls exist at $y = \pm L$ where $v = 0$, and a wave of positive wavenumber k is propagating in the positive x-direction. Then choose $\xi = v$ in Equation (4.43) and the solutions at frequency σ require

$$l = \pm\sqrt{\frac{\sigma^2 - f^2}{gh} - k^2}. \tag{4.45}$$

Superimposing them and demanding that $v\left(x = \pm L\right) = 0$ results in

$$l = \frac{n\pi}{L}, \quad n = 1, 2, 3, \ldots, \tag{4.46}$$

$$v\left(x, y, t\right) = A \sin\left(\frac{n\pi y}{L}\right) \exp\left(ikx - i\sigma t\right),$$

where A is an arbitrary constant. But since Equation (4.45) is also required,

$$\sqrt{\frac{\sigma^2 - f^2}{gh} - k^2} = \frac{n\pi}{L} \tag{4.47}$$

or,

$$\sigma_n^2 = f^2 + gh\left(\left(\frac{n\pi}{L}\right)^2 + k^2\right), \quad \sigma > 0$$

which has a smallest value, $\sigma_{min} = \sqrt{f^2 + gh\left(\pi/L\right)^2} > f, k = 0$. If $\sigma_n < \sigma_{min}$ no wave can propagate.[4] Much of the interest in these *Poincaré waves* lies with their forcing by tides. Given v, u and η are determined from the original equations.

What happens if a wavemaker is placed in such a canal, for example, at $x = 0$, and oscillates with $\sigma < \sigma_{min}$? From Equation (4.44) and the requirement $l = n\pi/L$, any wave(s) excited must satisfy

$$k^2 = \frac{\sigma^2 - f^2}{gh} - \frac{n^2\pi^2}{L^2} < 0$$

[4] The low-frequency cutoff phenomenon is well known in the literatures on electromagnetic and optical waveguides.

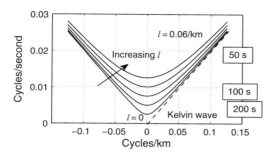
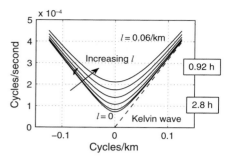

Figure 4.4: (a) The dispersion relationship for plane waves on an f-plane in dimensional form for $h = 4000$ m, $g = 10$ m/s, and with f appropriate to 30°N. For any fixed l, there is a minimum possible frequency, which for $l = 0$ is just f, which is very close to zero on this plot. Units are circular (not radian). (b) Same as (a) except the water depth is very small, $h = 1$ m, a value obtained using an equivalent depth. For the Kelvin wave, only the branch going in the positive x-direction is displayed. It is the only wave existing for $\sigma < f$, and only if a wall is present.

hence the response would be evanescent and trapped at the wavemaker, excluding the exponentially growing solution.

One important solution to Equations (4.42a–4.42c) has been missed completely. Suppose $v = 0$ everywhere at all times: then no equation in v could capture any solution. Does such a solution exist? With $v = 0$, the equations reduce to

$$\frac{\partial u}{\partial t} = -g\frac{\partial \eta}{\partial x}$$

$$-fu = -g\frac{\partial \eta}{\partial y}$$

$$\frac{\partial \eta}{\partial t} + h\frac{\partial u}{\partial x} = 0.$$

Assuming again that $u, \eta \propto \exp(-i\sigma t)$, and doing a bit of algebra, leads to

$$\eta(x,t) = A\exp\left(-\frac{f}{\sqrt{gh}}y\right)\exp(ikx - i\sigma t) + B\exp\left(\frac{f}{\sqrt{gh}}y\right)\exp(-ikx - i\sigma t), \quad (4.48a)$$

$$u(x,t) = \sqrt{\frac{g}{h}}\left[-A\exp\left(-\frac{f}{\sqrt{gh}}y\right)\exp(ikx - i\sigma t) + B\exp\left(\frac{f}{\sqrt{gh}}y\right)\exp(-ikx - i\sigma t)\right],$$

$$(4.48b)$$

$$\sigma/k = \pm\sqrt{gh}, \quad \sigma > 0 \qquad (4.48c)$$

with A, B being arbitrary constants. The dispersion relationship (4.48c) is plotted in Figure 4.4)

This solution is the *Kelvin wave* and contains two independent solutions—one is trapped against the wall at $y = -L$ and is travelling to the right (when $f > 0$); the second is trapped against the wall at $x = +L$ and is travelling to the left (they are evanescent in y, but propagating in x). The phase speed is $\sigma/k = \pm\sqrt{gh}$, here in the x-direction, which is the same phase speed as that of a long gravity wave in a nonrotating system. Being nondispersive, the group and phase velocities are equal. An important feature of the free Kelvin wave is that it exists at *any* frequency. Unlike the Poincaré waves, it is not restricted by f. Kelvin waves are very efficiently generated in real problems, have the fastest phase and group velocities, and are the most rapid transmitters of gravity-wave disturbance signals in a constant-density, rotating fluid.

FORCED FLOW IN A CYLINDER

As a prototypical forced problem in a closed system, consider the constant f-plane problem in a cylinder of depth h. Then the equation in η is

$$\nabla^2 \eta + \frac{\left(\sigma^2 - f^2\right)}{gh} \eta = \bar{\eta}, \qquad (4.48d)$$

where the Laplacian is in the cylindrical coordinates (r, α), and $\bar{\eta}(\sigma, r, \alpha)$ is a spatially arbitrary forcing function. α is the angle, positive clockwise, and σ is a fixed forcing frequency. Following Lamb (1932, Section 210) or the version in Wunsch and Stammer (1997, appendix), the free solutions are

$$\eta(r, \alpha) = A J_m(\lambda r) e^{im\alpha}, \quad m = 0, \pm 1, \pm 2, \dots, \quad \lambda = \frac{\sqrt{\sigma^2 - f^2}}{\sqrt{gh}}$$

and the $J_m(\lambda r)$ are the Bessel functions in the box "Rotating Cylinder and Bessel Functions." The boundary condition of no flow at $r = a$ is

$$\left. \frac{\partial J_m(\lambda r)}{\partial r} \right|_{r=a} - \frac{fm}{\sigma a} J_m(\lambda a) = \lambda J'_m(\lambda a) - \frac{fm}{\sigma a} J_m(\lambda a) = 0. \qquad (4.48e)$$

The transcendental equation (4.48e) has an infinite number of roots, $\lambda = \lambda_{mn}$, where n is again an integer. Each value of λ_{mn} defines a frequency $\sigma_{mn} = \left(\lambda_{mn}^2 gh + f^2\right)^{1/2}$ of the mode (see Table 4.1). One of those roots can be pure imaginary and corresponds to $\eta(r, \alpha) = A J_m(i|\lambda|r) e^{im\alpha} = A I_m(\lambda) e^{im\alpha}$—exponentially increasing towards the outer boundary, and is the cylindrical form of the Kelvin wave. It is the only free wave for which $\sigma < f$ is possible, and its radial velocity is small but nonzero. The free solutions are all of the form

$$\eta = A_{mn} J_m(\lambda_{mn} r) e^{im\alpha} \qquad (4.48f)$$

where A_{mn} is constant. If the forcing η is expanded in the free solutions,

$$\eta(r, \alpha) = \sum_{m=-\infty}^{\infty} \sum_{n=0}^{\infty} B_{mn} J_m(\lambda_{mn} r) e^{im\alpha} \qquad (4.48g)$$

and B_{mn} is now regarded as known. Substituting Equations (4.48f) and (4.48g) into Equation (4.48d) and noting that $\nabla^2 J_m(\lambda_{mn} r) e^{im\alpha} = -\left(\sigma_{mn}^2 - f^2\right)/gh J_m(\lambda_{mn} r)$,

$$\sum_{m=-\infty}^{\infty} \sum_{n=0}^{\infty} A_{mn} \left(\frac{\sigma^2 - \sigma_{mn}^2}{gh}\right) J_m(\lambda_{mn} r) e^{im\alpha} = \sum_{m=-\infty}^{\infty} \sum_{n=0}^{\infty} B_{mn} J_m(\lambda_{mn} r) e^{im\alpha}$$

continued

Table 4.1: a^2/R_D^2 adapted from Lamb, 1932, for the lowest roots, for $m=1$, of the free modes of the cylinder for two values of the ratio a/R_D. Negative value frequencies are larger in magnitude than the corresponding positive ones, because the motion is against the sense of rotation. For sufficiently small R_D relative to a, one of the roots corresponds to a Kelvin wave trapped at the outer boundary.

a^2/R_D^2	6	40
	−1.37	−1.1
	0.51	0.2
	−2.42	−1.32
	2.37	1.29

$$\eta\left(r,\alpha\right) = \sum_{m=-\infty}^{\infty} \sum_{n=0}^{\infty} \frac{ghB_{mn}}{\sigma^2 - \sigma_{mn}^2} J_m\left(\lambda_{mn}r\right) e^{im\alpha}. \tag{4.48h}$$

The Bessel functions are orthogonal over $0 \le r \le a$, in the sense that

$$\int_0^a rJ_m\left(\lambda_{mn}r\right) J_m\left(\lambda_{mn'}r\right) dr = 0, \quad n \ne n'$$

(Olver and Wong, 2010) and are also orthogonal in α for $m \ne m'$. Resonance is now possible if $\sigma = \sigma_{mn}$ and $B_{mn} \ne 0$. Forcing excites all of the free modes of the system, with the modes having an amplitude dependent upon both their spatial and temporal structure compared to that of the individual free modes. If the cylinder is unbounded (or the outer wall regarded as absorbing), the standing wave J_m is no longer appropriate, and it is replaced by the outward travelling wave $H_m^{(1)}\left(\lambda r\right)$, and no resonance of the form as in Equation (4.48h) is possible. (If the forcing coincides with the propagation structure of one of the Hankel functions, a travelling wave resonance can occur.)

If the Kelvin wave solution is applied to the ocean, with $h \approx 4000$ m at a latitude of 30°, and $f = 7.2 \times 10^{-5}/$s, the distance of exponential decay is $\sqrt{gh}/f \approx 2800$ km (the *barotropic Rossby radius of deformation*). This large value is inconsistent with the constant f, or local Cartesian, assumptions. If the period were 12 h, then the wavelength would be on the order of 8600 km—also too large for accuracy with the local Cartesian approximation. Similar objections apply to the Poincaré wave solutions. The utility of this whole development thus might be questioned. It would appear to be useful only for very shallow oceans, $h \to 0$, and to be of only marginal interest to a deep-water oceanographer. Surprisingly, this inference is not correct—stratification comes to the rescue, rendering solutions for very small h of great importance. If the solutions for large h are properly calculated in spherical geometry, the inaccurate f-plane results are seen to nonetheless reproduce many qualitative properties of the more accurate solution.

In the presence of rotation, analytical calculation of the normal modes of a rectangular box is extremely difficult. Although the algebra is a bit complicated, a Kelvin wave encountering a corner scatters energy not only into a transmitted Kelvin wave, but also into Poincaré waves.

Should the Kelvin wave frequency be less than the low-frequency cut-off, the Poincaré waves are trapped at the corners and cannot transmit energy; the outgoing Kelvin wave must then carry all of the energy of the incoming one. Taylor (1921) gave a series solution for Kelvin wave reflection from a wall at the end of a long canal.[5] Analytical solutions are known for the cylinder—no corners exist—(see the box "Forced Flow in a Cylinder") and can be extended to ellipses and other shapes.

THE β-PLANE WITH A RIGID LID

When the meridional extent of the motions is large enough to "feel" the change in f, some new physics enters. This physics is necessarily always present in the spherical approximation, but its consequences are, by experience, best understood first via the Cartesian β-plane. The further simplification that a rigid lid covers the ocean produces the most basic result. As discussed on p. 98, $w = 0$, identically. Absent a moving free surface, gravity has no role, and all ordinary surface gravity waves are suppressed—an important simplification. The continuity equation becomes

$$\frac{\partial u}{\partial x} + \frac{\partial v}{\partial y} = 0,$$

and is satisfied identically by using a stream function, $u = \partial\psi/\partial y$, $v = -\partial\psi/\partial x$. Substituting into Equations (4.42a–4.42c), produces

$$\frac{\partial}{\partial t}\left(\frac{\partial^2\psi}{\partial x^2} + \frac{\partial\psi}{\partial y^2}\right) + \beta\frac{\partial\psi}{\partial x} = 0.$$

Plane-wave solutions, $\psi = A\exp(ikx + ily - i\sigma t)$, require

$$\sigma = -\frac{\beta k}{k^2 + l^2}, \tag{4.49}$$

which is graphed in Figure 4.5. The dispersion relationship, Equation (4.49), has the strange property that the phase velocity $\sigma/k < 0$, that is, the waves can only travel westward. How could reflection take place? On the other hand, the group velocity (also shown in Fig. 4.5) can be in either direction, and since it is only energy propagation that has physical consequences in a resting fluid, no problems arise from the unidirectional phase velocity (which, again, is primarily kinematic in nature). Group velocities are westward for small k ($l^2 > 0$) and eastward for large k (small wavelengths). The solutions ψ satisfying Equation (4.49) are generally called *Rossby waves*, although they were known much earlier as solutions on a sphere (having a slightly more complicated, but nonetheless recognizable, dispersion relationship).

For $l \neq 0$, and with a north-south wall at, for example, $x = 0$, reflection of a wave with small k and having a westward-directed group velocity (and hence energy flux) requires that σ and l in the reflected wave must remain identical to that in the incoming wave, but k is greatly increased to produce an eastward energy flux. The angles of incidence and of reflection are not equal; Snell's Law is inapplicable.

In a closed basin, waves with the same frequency can be superposed to calculate the normal modes of a rectangular ocean of dimensions L_x by L_y. Longuet-Higgins (1964)

---

[5] See Buchwald, 1968, for a simplification. Modes of a rectangular box were calculated by Rao (1966).

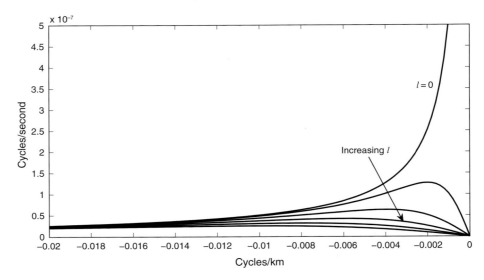

Figure 4.5: The dispersion relationship for the rigid-lid, midlatitude Rossby wave solutions for a reference latitude of 30°. Except for the $l = 0$ solution, all waves display a maximum frequency where $\partial\sigma/\partial k = 0$, with zero zonal group velocity. The $l = 0$ solution formally produces an infinite frequency as $k \to 0$, but the approximations involved fail long before the frequency even approaches f.

pointed out that for periodic solutions, $\psi(x, y, t) = A\exp(-i\sigma t)\hat{\psi}(x, y)$, the governing equation is

$$\frac{\partial^2 \hat{\psi}}{\partial x^2} + \frac{\partial \hat{\psi}}{\partial y^2} - \frac{\beta}{i\sigma}\frac{\partial \hat{\psi}}{\partial x} = 0,$$

and with the substitution,

$$\hat{\psi} = \exp\left(-i\beta x/2\sigma\right)\psi'(x, y),$$

(following Eq. 4.39) the new variable, ψ', satisfies the ordinary Helmholtz equation,

$$\frac{\partial^2 \psi'}{\partial x^2} + \frac{\partial \psi'}{\partial y^2} + \frac{\beta^2}{4\sigma^2}\psi' = 0. \tag{4.50}$$

Because the boundary condition of no flow into the sidewalls requires $\psi = \hat{\psi} = \psi' = $ constant, and the Helmholtz equation is known to give simple (separable) solutions in a variety of coordinate systems,[6] construction of the normal mode solutions is easy for basins natural to those coordinates. In a rectangular box,

$$\psi_{nm} = Ae^{-\frac{i\beta}{2\sigma}x}\sin\frac{n\pi x}{L_x}\sin\frac{m\pi y}{L_y}e^{-i\sigma t}, \tag{4.51a}$$

$$\sigma_{nm} = \frac{\beta}{2}\frac{1}{\sqrt{n^2\pi^2/L_x^2 + m^2\pi^2/L_y^2}} > 0, \quad n, m \text{ are integers},$$

where the arbitrary boundary constant was chosen as zero. The basin need not be oriented north-south, and solutions for circular, elliptic, and more exotic basins are available. These

---

[6] Morse and Feshbach, 1953, p. 656.

modes have another unusual property: that *the higher the wavenumber, the smaller the frequency*; they represent a westward-moving phase velocity superimposed upon a standing-wave pattern. In theory, if nothing intervenes, waves of arbitrarily small frequency are packed ever more tightly in wavenumber space as $n, m \to \infty$, and the packing eventually gets so tight that the mathematical idealization of a normal mode fails. In reality, friction and nonlinearity prevent what might be called an "infrared catastrophe" of an infinite number of modes with infinitesimal frequency. As will be discussed later, only the most tenuous evidence exists for the existence of basin-scale modes in the ocean, but the theory is elegant and simple, and indeed raises the ultimately important question of why such motions are not very apparent.

Reverting to the infinite domain, Equation (4.49) is the same as

$$k^2 + l^2 + \frac{\beta k}{\sigma} = 0,$$

which can be rewritten as

$$\left(k + \frac{\beta}{2\sigma}\right)^2 + l^2 = \left(\frac{\beta}{2\sigma}\right)^2, \tag{4.52}$$

which is a circle in k, l space of radius $\beta/2\sigma$ centered at $k = -\beta/2\sigma, l = 0$ (see Longuet-Higgins, 1964, which provides an elegant discussion). For fixed σ, the group-velocity vector points from the position k, l on the circle toward the center. Thus long waves have group velocities with a westward-directed component, and short waves have group velocities with an eastward one, which is consistent with Figure 4.5. As σ grows, the radii diminish, and so for any given value of β as a function of latitude, a maximum frequency, $\sigma_{max} = -\beta/2k$, exists, with $k = |l| \neq 0$ (see Fig. 4.5). When $\sigma = \sigma_{max}$, $\partial\sigma/\partial k = 0$, that is, the zonal component of the group velocity vanishes and no zonal energy propagation occurs (but $\partial\sigma/\partial l \neq 0$). See the paper by Durland and Farrar (2012) for discussion of the implications when one component of the group velocity vanishes. The frequencies and wavenumbers in Equation (4.51a) also satisfy Equation (4.52).

Gravity does not appear in the equations governing these rigid-lid Rossby waves. Thus, also unusually, they have no potential energy. All of their energy is kinetic, and so potential energy is not required for wave motion to exist. Movement through the background vorticity (the *effective* Earth rotation) gradient provides the restoring forces (as vorticity sources) that gives rise to the oscillatory motion.[7]

THE FREE SURFACE

The use of a rigid lid has suppressed the Poincaré and Kelvin waves, because their existence depends upon gravity. If the free surface is restored and the y-dependence in f is retained, the various fields, u, v, η, no longer satisfy the same equation. If the usual proportionality to $\exp(-i\sigma t)$ is assumed, then the equation in v proves to be the simplest one and is

$$\nabla^2 v - \frac{\beta}{i\sigma}\frac{\partial v}{\partial x} - \frac{f^2 - \sigma^2}{gh}v = 0, \tag{4.53}$$

with variable f.

---

[7] Longuet-Higgins (1965b) gave a physical interpretation of the westward movement in terms of the interacting relative vorticities, ζ_z, induced by displacement of neighboring fluid parcels north and south across a reference line. The analog is a row of line vortices of alternating sign in ordinary fluid dynamics and that self-propels.

Useful approximate solutions exist (having been checked against more complete solutions), in which f is treated as constant in Equation (4.53)—retaining its variation only where β appears explicitly. Then substituting plane waves, $\eta = \exp(ikx + ily)$, results in

$$\sigma = \frac{-\beta k}{k^2 + l^2 + f^2/gh} = \frac{-\beta k}{k^2 + l^2 + 1/R_D^2}, \tag{4.54}$$

which differs from Equation (4.49) only in having the extra term f^2/gh. The length scale, $R_D = \sqrt{gh}/f$ is the *Rossby radius of deformation* (it has already appeared in Eq. 4.48) and for an ocean of depth $h = 4000$ m has a value of about 3000 km at midlatitudes. Because other deformation radii appear later, the prefix *barotropic* is sometimes used—a code word for an unstratified solution. The dispersion relation is not unlike the one obtained for the rigid-lid approximation, except that the circle in Equation (4.52) is displaced away from the origin and for $l = 0$, σ no longer becomes arbitrarily large as $|k| \to 0$, but instead vanishes (an important dynamical consequence of including the free surface movement).

The zonal group velocity derived from Equation (4.54) is

$$v_{gx} = \frac{\partial \sigma}{\partial k} = \beta \frac{k^2 - l^2 - 1/R_D^2}{\left(k^2 + l^2 + 1/R_D^2\right)^2}, \quad v_{gy} = \beta \frac{2kl}{\left(k^2 + l^2 + 1/R_D^2\right)^2}, \tag{4.55}$$

whose direction, controlling the energy propagation vector, can vary by 360° depending upon k, l. v_{gx} has a maximum magnitude and value, $-\beta R_D^2$, when $k = l = 0$—the long-wavelength limit. The maximum frequency for a wave with fixed l follows by setting $v_{gx} = 0$ and

$$\sigma_{\max} = \frac{R_D \beta}{2(l^2 R_D^2 + 1)^{1/2}}, \tag{4.56}$$

whose own maximum occurs for $l = 0$ and is $\sigma_{\max} = R_D \beta/2$. At these frequencies, the zonal group velocity again vanishes. These various maxima and group-velocity variations become important in discussions of observed oceanic variability. See Hendershott, 1981, for a clear extended discussion of these waves.

THE EQUATORIAL β-PLANE

The f-plane, which treats f as a constant except when differentiated is a rough approximation, and it proves useful to understand some of its nature. The most elegant and useful approach is to allow f to vary with y, but to choose the y-origin at the equator, so that $f = \beta y$, resulting in the *equatorial β-plane*. The equation governing V is the simplest. Letting $v = A \exp(ikx) V(y)$, then from Equation (4.42a)

$$\frac{d^2 V}{dy^2} + \left(\frac{\sigma^2 - \beta^2 y^2}{gh} - k^2 - \frac{\beta k}{\sigma} \right) V(y) = 0, \tag{4.57}$$

an ordinary differential equation known as the *parabolic cylinder*, or *Weber*, equation, one of the classical equations of mathematical physics. Its solutions are sought by means discussed in detail in standard textbooks[8] and the oceanographic applications are analyzed by Gill

---

[8] Whittaker and Watson, 1996, or Morse and Feshbach, 1953.

(1982) and others. Solutions are sought that do not blow up at high latitudes (large $|y|$), and it is found that they exist only for special values of the frequency

$$\frac{\sigma_n^2}{gh} - \beta\frac{k}{\sigma_n} - k^2 = (2n+1)\frac{\beta}{\sqrt{gh}}, \quad n = 0, 1, 2, \ldots, \tag{4.58a}$$

$$V(y) = V_n(y) = A\exp(-\beta y^2/2gh)H_n\left(\left(\beta/(\sqrt{gh})\right)^{1/2}y\right), \tag{4.58b}$$

$$H_0(q) = 1, \quad H_1(q) = 2q, \quad H_2(q) = 4q^2 - 2, \ldots, \tag{4.58c}$$

where Equation (4.58a) is a "quantization condition" familiar from physics. With n an integer, the special parabolic cylinder functions are known as *Hermite functions* and are the product of the *Hermite polynomials*, $H_n(y)$, and a decaying exponential. Initially, the polynomial structure leads to growth poleward of the equator, but ultimately, the exponential factor overwhelms it, and the motions are thus oscillatory in latitude but trapped near the equator. Exponential decay means that the failure of the β-plane approximation for large y may be of no consequence. The trapping scale of these waves is the *equatorial deformation radius*, $L_E = \left(\sqrt{gh}/\beta\right)^{1/2}$ and again, for real values of h, L_E is so large that the equatorial β-plane approximation is too inaccurate. h must be small for the solutions to be useful except as a qualitative interpretation guide. (As before, many properties of the solutions are identifiable on a sphere; see, especially, Longuet-Higgins, 1965a).

Because the dispersion relationship (Eq. 4.58a) is now cubic in σ_n, up to three different free solutions with different horizontal wavenumbers can exist *for each frequency*, although not all are physically acceptable. Dispersion relations for the physically meaningful ones are displayed in Figure 4.6. (By using L_E and $1/\left(\beta\sqrt{gh}\right)^{1/2}$ to nondimensionalize length and time respectively, a universal plot can be generated; see Moore and Philander, 1977; Hendershott, 1981).

From a WKBJ point of view, a change in the behavior of the waves is expected when the quantity $\left(\sigma^2 - \beta^2 y^2\right)/gh - k^2 - \beta k/\sigma$ in Equation (4.57) changes sign—positive values correspond to waves oscillatory in y, and negative values to those exponential (and usually decaying) in y. For small k, the transitional latitude would be where $\sigma = \pm\beta y = f$. These solutions have a large literature (Moore and Philander, 1977; Gill, 1982). A branch of the solutions at high frequency is recognizable as Poincaré waves, trapped equatorward of the latitude where $\sigma \approx f$. Another branch with $\sigma \ll f$ is Rossby-wave-like, having a westward phase velocity only and being trapped equatorward of $|y| \approx L_E$. One new solution emerges, known as the mixed Rossby gravity wave[9]—it behaves like a Rossby wave when going westward but like a Poincaré (gravity) wave going eastward, and its dispersion relation is obtained by setting $n = 0$.

One important solution is entirely missed by this construct. As with Kelvin waves along a wall, a solution can be sought to Equations (4.42a–4.42c) with $v = 0$, identically. This approach produces a new solution, in which $\sigma/k = \sqrt{gh}$, and both phase and group velocities are eastward. Because of the analogy to the wave trapped by rotation along a wall, it is called the *equatorial Kelvin wave* and is once again the wave with the fastest eastward phase and group velocity. Crudely speaking, the equator acts as a boundary sustaining a Kelvin wave whose elevation, η, rises exponentially to the south of the eastward-going wave when north

[9] Sometimes known as a *Yanai wave*.

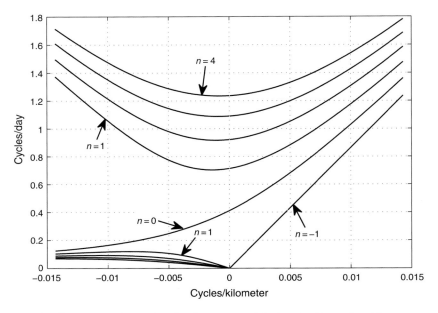

Figure 4.6: The dispersion relationship for the equatorial β-plane in dimensional form. $g = 10$ m/s, $h' = 0.1$ m and $\beta = 2.3 \times 10^{-11}$/m sec appropriate to the equator. A small value of h' is used to make the structure easier to perceive. Calculated from Equation (4.58a); see Wunsch and Gill, 1976, or Hendershott 1981. Labels $n = -1$, 0, 1, 4, are the meridional mode numbers, although identifications of $n = -1$, 0 for the equatorial Kelvin and mixed-mode wave is only formal.

of the equator and rises exponentially to the north when south of the equator. η matches smoothly at $y = 0$. Information and energy are carried very rapidly by this wave. The correct dispersion relationship is found from Equation (4.58a) by setting $n = -1$, but that is only a mnemonic, because the derivation of that equation does not apply to the Kelvin wave. (On a sphere, v is small but not zero.)

β is maximum on the equator and the stratification is strong; all of the time scales associated with motions at low latitudes are far faster than they are farther poleward. The equatorial region is thus much more volatile than is the ocean at other latitudes, with many important consequences.

THE REFLECTION PROBLEM

Reflection of waves on the midlatitude β-plane was touched on above. Because of the Hermite function structure of the solutions on the equatorial β-plane, reflection at meridional boundaries renders the algebra a good deal more complicated than for plane waves. If the equatorial region is included in the basin, the physics are interwoven with the understanding of El Niño. When the special properties of Rossby waves are combined with the anisotropic propagation of the Kelvin and mixed Rossby gravity waves, a very interesting mathematical problem arises in determining the hypothetical normal modes (Moore and Philander, 1977). Particular attention is called to the way in which an eastward-bound equatorial Kelvin wave reflects from a meridional boundary.

A classical wave problem (e.g., for electromagnetic and acoustic waves) concerns reflection from and penetration of waves through a wall with a narrow slit, or two nearby narrow slits. Although in the diffraction limit the resulting wave patterns can become complex, transmission of energy is, as expected, proportional to the area of the slit(s).[10] In contrast, for a Rossby wave, the transmission properties are quite different—a consequence of the need to conserve vorticity (the Kelvin Circulation theorem). The situation for a barotropic Rossby wave was analyzed by Pedlosky (2000), and some features also apply to the fully stratified fluid. Solutions are of particular interest because the complex geometry in the western Pacific presents a partially open barrier between the Pacific and Indian Oceans. Reflection from this boundary has been invoked as determining the time scales and nature of El Niño (the "delayed oscillator theory"), but its applicability is not discussed here.

4.3.1 Effects of Topography

The simplest case with topography is for a Cartesian ocean with a rigid lid but where the bottom is a function of position, $z = -h(x, y)$. With $\eta = 0$, Equation (4.42c) becomes

$$\frac{\partial (hu)}{\partial x} + \frac{\partial (hv)}{\partial y} = 0, \tag{4.59}$$

and the gravity effects have again disappeared. Equation (4.59) is satisfied identically by defining a *transport stream function*,

$$hu = \frac{\partial \psi}{\partial y}, \quad hv = -\frac{\partial \psi}{\partial x} \tag{4.60}$$

and then from Equations (4.42a–4.42c),

$$\frac{\partial}{\partial t}\left[\frac{\partial}{\partial x}\left(\frac{1}{h}\frac{\partial \psi}{\partial x}\right) + \frac{\partial}{\partial y}\left(\frac{1}{h}\frac{\partial \psi}{\partial y}\right)\right] + \frac{\partial}{\partial y}\left(\frac{f}{h}\frac{\partial \psi}{\partial x}\right) - \frac{\partial}{\partial x}\left(\frac{f}{h}\frac{\partial \psi}{\partial y}\right) = 0 \tag{4.61}$$

Consider the f-plane and suppose h is a function of y alone. By setting $\beta^* = f\partial(1/h)/\partial y$, an approximate mathematical analog of the β-plane equation (4.53) is found. Ibbetson and Phillips (1967) exploited this analog to produce a laboratory demonstration of a Rossby wave. In the more general case of Equation (4.61), the wave behavior is a function of $h(x, y)$, and a variety of phenomena can arise depending upon whether the topographic gradients are waveguide-like, how they are oriented, etc. An important inference is that wavelike motions can become focussed into relatively very small areas, rendering them uncorrelated over long distances and causing serious numerical and observational problems. In some cases, treating the bottom topography as random is the most useful approach and connects this problem to the interesting literature on waves in random media.

In a steady state,

$$\psi(x, y) = \psi(f/h), \tag{4.62}$$

with ψ arbitrary, which means the flow would follow the contours of f/h if physically possible; boundaries and depth discontinuities may intervene. f/h contours are shown in Figure 4.7

---

[10] The diffraction limit exists when obstacles such as edges or slits have spatial scales much smaller than the wavelengths present.

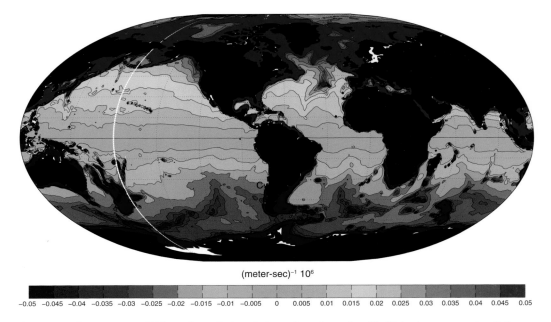

$(\text{meter-sec})^{-1} \ 10^6$

−0.05	−0.045	−0.04	−0.035	−0.03	−0.025	−0.02	−0.015	−0.01	−0.005	0	0.005	0.01	0.015	0.02	0.025	0.03	0.035	0.04	0.045	0.05

Figure 4.7: The contours of f/h at $1°$ resolution. Balanced motions that "feel" the bottom tend to follow this field, rather than h itself. The expected response can be very sensitive to the degree of smoothing imposed on the topography. Major deviations from zonality exist in certain areas. Large-scale regions of nearly uniform f/h have been observed with altimeters to be the locations of strong barotropic motions which can undergo a form of resonance there. Such known areas include the one west of the southern end of South America (e.g., Webb and de Cuevas, 2003) and the Argentine Basin (Tai and Fu, 2005).

4.4   A STRATIFIED OCEAN

Turn now from free, periodic disturbances to a resting, stratified fluid of buoyancy frequency $N(z)$ with water depth, h, still constant. What kinds of motions are possible? Seeking insight, rather than accuracy, start again by using local Cartesian coordinates, and set f to be constant, with no β-effect in Equations (3.30a–e) and (3.31).

Again let all variables be proportional to $\exp(-i\sigma t)$. The first three equations are just the linearized inviscid momentum balances. Equation (3.31) makes the important approximation that any changes in the density at a fixed position are due only to the vertical displacement of the background stratification, $\partial\rho_0/\partial z$, neglecting any lateral advection of density. (A climate theory must calculate the background stratification, but that goal is beyond a perturbation approach.)

As before, let $u \to u \exp(-i\sigma t)$, etc. the equations can be combined in various ways. For the pressure perturbation,

$$\left[\frac{N(z)^2 - \sigma^2}{\sigma^2 - f^2}\right]\left(\frac{\partial^2 p'}{\partial x^2} + \frac{\partial^2 p'}{\partial y^2}\right) - \frac{\partial^2 p'}{\partial z^2} = 0, \tag{4.63}$$

or defining,

$$c(z)^2 = \frac{\sigma^2 - f^2}{N^2(z) - \sigma^2}, \tag{4.64}$$

then

$$\frac{1}{c(z)^2}\left(\frac{\partial^2 p'}{\partial x^2} + \frac{\partial^2 p'}{\partial y^2}\right) - \frac{\partial^2 p'}{\partial z^2} = 0, \tag{4.65}$$

a deceptively simple equation usually named for Poincaré.

If the waves are hydrostatic, $N(z)^2 - \sigma^2$ is everywhere replaced by $N(z)^2$. Let N be a constant; then Equation (4.65) has constant coefficients. (In this case, it is readily confirmed that the partial differential equation for any of u, v, w, ρ', p' are all of the form of Eq. 4.65). Dropping the y-dependence, without loss of generality on an f-plane,

$$\frac{1}{c(z)^2}\frac{\partial^2 p'}{\partial x^2} - \frac{\partial^2 p'}{\partial z^2} = 0, \tag{4.66}$$

in a notation chosen to suggest that x is a time-like variable, $x \to t$. With time as a coordinate, Equation (4.66) is the same as one describing waves on a string travelling in the z-direction. Equation (4.66) has unusual properties because it is *hyperbolic*; such systems do not commonly occur in the spatial domain but are commonplace when describing waves evolving in time. Because t can only increase, but x is not so restricted, some mathematically interesting problems arise.[11]

If $N(z)$ and hence $c(z)$ are constant, the general solution to Equation (4.66) is

$$p'(x, z) = F(z - cx) + G(z + cx), \tag{4.67}$$

where F, G are arbitrary. $z \pm cx$ are the *characteristics* of the hyperbolic problem (confusingly sometimes, in this context, called "rays"). In the classical laboratory experiment, Mowbray and Rarity (1967) showed that by oscillating a rod in a tank with N taken to be approximately constant (Fig. 4.8), disturbances qualitatively similar to choosing $F = \delta(z - cx)$ and $G = \delta(z + cx)$ were excited and had a slope c was given with high accuracy by the theory.

The general solution is Equation (4.67) and the group velocity and energy flux are transmitted parallel to the characteristics (but in either direction). Conventional solutions by the method of characteristics need to carefully consider radiation conditions (Baines, 1971; Bühler and Holmes-Cerfon, 2011).

Restoring y and trying instead, a different-appearing, guessed solution to the same Equation (4.65),

$$p' = Ae^{ikx+ily+imz},$$

which on substitution gives

$$k^2 + l^2 = m^2 c^2.$$

and the horizontal axes can be oriented such that $l = 0$, again eliminating the y-dependence. Thus,

$$k^2 = m^2 c^2 = m^2\left(\frac{\sigma^2 - f^2}{N^2 - \sigma^2}\right) \tag{4.68}$$

If the waves are to be sinusoidal in the vertical and to travel horizontally, both k, m must be real, with both $k^2, m^2 > 0$ and hence $c^2 > 0$, or,

$$f < \sigma < N,$$

[11]Carrier (1949) and Greenspan (1963) discuss some analogous problems on a string. For the fluid, see also the papers by Bühler and Cerfon-Holmes (2011) and Maas (2011).

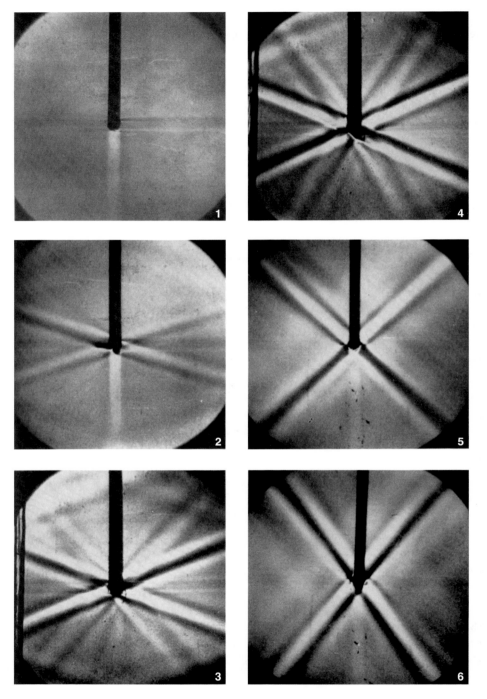

Figure 4.8: Disturbances emanating from a vertically oscillating rod in a stably stratified fluid. The four main characteristics emanating from the tip have an angle to the vertical determined by the frequency of oscillation which is different in each pane. Weaker "extra" structures are indicators of failure of the simplest possible theory. (Used with permission from D. E. Mowbray and B. S. H. Rarity [1967]. "A theoretical and experimental investigation of phase configuration of internal waves of small amplitude in a density stratified liquid." *J. Fluid Mech.* 28: 1–16.)

assuming $f < N$, which is almost universally true in the ocean. Waves can propagate only between these frequency limits. Should the inequality be reversed,

$$N < \sigma < f,$$

whose major application is with $N = 0$, a rotating homogeneous fluid. That limit makes it comparatively easy to construct laboratory analogs to internal wave propagation by using an unstratified, rotating fluid.[12]

With vector wavenumber $\kappa = (k, m)$, such that $k = |\kappa| \cos \theta$ and $m = |\kappa| \sin \theta$, travelling with constant phase speed,

$$s_p = \frac{\sigma}{|\kappa|}, \tag{4.69}$$

in the direction (relative to the horizontal) of $\tan^{-1}(m/k)$ (see Equation 4.10).

Equation (4.68) implies

$$\frac{k}{m} = \pm c = \pm \sqrt{\frac{\sigma^2 - f^2}{N^2 - \sigma^2}} = \cot \theta, \tag{4.70}$$

which also shows that *the angle, θ, to the horizontal is dependent only upon the frequency* and not upon the length of the wavenumber vector—another property not found in more familiar wave types and one that is consistent with the solution, Equation (4.67). Noting that as $\sigma \to N$ and $c \to \infty$, the orientation of the waves is such that $k/m \to \infty$. That is, from Equation (4.69), the phase velocity will, in that limit, be oriented in the x-direction with no vertical structure. In the opposite limit, as $\sigma \to f$ (from above), $k/m \to 0$, and thus inertial waves have no horizontal structure—in this approximation.

Solving Equation (4.70) for

$$\sigma = \sqrt{(k^2 N^2 + m^2 f^2)/(k^2 + m^2)} > 0 \tag{4.71}$$

The group velocity is

$$\mathbf{v}_g = \left(\frac{\partial \sigma}{\partial k}, \frac{\partial \sigma}{\partial m}\right) = \frac{(N^2 - f^2)}{(k^2 + m^2)^{\frac{3}{2}} \sqrt{k^2 N^2 + m^2 f^2}} (km^2, -k^2 m). \tag{4.72}$$

If m is positive, so that *the wave has a phase velocity in the upward direction, the vertical component of the group velocity is downward* and vice-versa. In the horizontal, the group velocity is directed with the sign of k. The dot product,

$$[k, m] \cdot \mathbf{v}_g = 0,$$

shows that *the phase and group velocities are at right angles*: the group velocity lies parallel to the crests and has a direction determined by the frequency alone. We have the interesting situation, not encountered in everyday life, of a wave apparently travelling in one direction but with its energy travelling at right angles to it. The reader might want to think about how a transient would appear.

If $\sigma = f$ and $k = 0$, then $\mathbf{v}_g = 0$, and any energy put in at that frequency cannot propagate, implying a buildup of energy locally. On the other hand, *any* extension of the theory that renders the group velocity finite would preclude the implied resonance, and f-plane response amplitudes are likely inaccurate.

[12] Waves satisfying the condition $0 < \sigma < f$ are sometimes known outside of oceanography as "inertial waves" (Chandrasekhar, 1968), but their oceanographic meaning will be defined differently below.

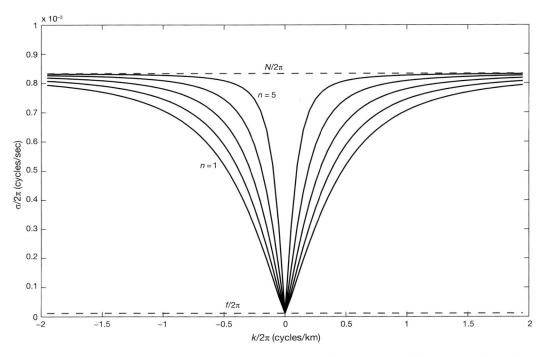

Figure 4.9: Dispersion curves for internal waves in an ocean of constant N, and f appropriate for $30°$ latitude and no lateral boundaries. A higher mode number, n, implies higher frequencies. Here $N = 2\pi/$ (20 min) $= 5.2 \times 10^{-3}/s$, $f = 7.3 \times 10^{-5}/s$ (at $30°N$). For a 12 h wave, $\sigma = 1.5 \times 10^{-4}/s$, $m = \pi/5000$ m (vertical mode $n = 1$ in the text below), $k = m\sqrt{(\sigma^2 - f^2)/(N^2 - \sigma^2)} = 1.5 \times 10^{-5}/m$, $2\pi/k = 155$ km. Without stratification, a long gravity wave at a period of 12 hours period has a wavelength of about 10,000 km, compared to about 100 km with stratification.

These solutions are appropriate in an infinite domain. Consider an ocean with a rigid bottom at $z = -h$, and top at $z = 0$ leading to boundary conditions $w(-h) = w(0) = 0$. Let

$$p' = Ae^{ikx+imz} + Be^{ikx-imz},$$

then the boundary conditions require

$$\sin mh = 0,$$

or, $m = n\pi/h$, where n are the integers, $1, 2, \ldots$, and,

$$p' = A\cos\left(\frac{n\pi}{h}z\right)e^{ikx-i\sigma t}, \qquad (4.73)$$

that is, a system of vertically standing waves. The dispersion relation becomes

$$\sigma_n^2 = \left(k^2 N^2 + \left(\frac{n\pi}{h}\right)^2 f^2\right) \Big/ \left(k^2 + \left(\frac{n\pi}{h}\right)^2\right), \qquad (4.74)$$

and is shown in Figure 4.9. A moving free surface introduces only slight shifts in the dispersion relationship and the vertical structure, and its consequences are discussed later.

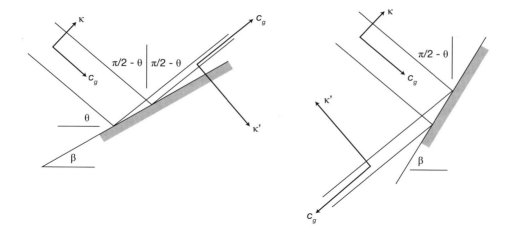

Figure 4.10: Internal waves in two dimensions reflecting from a sloping boundary. Because they have the unusual property of preserving their angle to the vertical, slopes can be *transmissive, or subcritical* (left panel), and *reflective, or supercritical* (right panel). In the transmissive case, characteristics along the slope are compressed. At the critical angle, when the slope is that of the characteristics, c, the motion is formally infinite. If the substitution $x \to t$ is made, the reflected characteristics formally correspond to backward running time—a hint of the sometimes interesting mathematical issues implicit here. (From Phillips, 1977.)

TOPOGRAPHY

The characteristic solution (Eq. 4.67) become of striking importance when the effects of topography are considered. Phillips (1977) and Eriksen (1982) considered the situation in which energy travelling along one such characteristic encounters a sloping boundary (Fig. 4.10). Because the reflected or transmitted energy can propagate only along the characteristic at the fixed angles for that frequency, reflection from the bottom again violates Snell's law. The situation is instead that shown in Figure 4.10, in which the energy can proceed either up the slope as shown, over a *subcritical*, or, better, *transmissive*, slope, or be back-reflected from a *supercritical*, or *reflective*, slope. If energy comes in along many parallel characteristics, the reflected or transmitted energy can become confined very close to the slope as shown; that is, when the bottom slope, $\gamma \approx c$, for a *critical* slope, the energy density is formally infinite. These effects are readily reproduced in the laboratory; see Figure 4.11. A similar, mode-like solution can also be constructed in the transmissive case, without using the characteristics, as

$$\psi = A\left[\exp\left(-iq\ln(cx-z)\right) - \exp(-iq\ln(cx+z))\right],$$
$$q = 2n\pi/\ln\left((c+\gamma)/(c-\gamma)\right), \quad \gamma < c, \quad n = 1, 2, \ldots$$

in a stream function, ψ.[13]

In a closed region with somewhat irregular reflecting surfaces, extremely complicated motions can be set up if multiple reflections of the characteristics are permitted without

---

[13] Cacchione and Wunsch, 1974; see also McKee, 1973. A WKBJ approach to these problems over varying terrain is described by Keller and Mow (1969).

Figure 4.11: Photographs from laboratory demonstration showing how a critical internal wave on a slope breaks up into a turbulent flow at the boundary. Panels are separated in time by 0.9, 0.3, 0.3 wave periods. Compare to Figure 7.13. (Used with permission from G. N. Ivey and R. I. Nokes [1989]. "Vertical mixing due to the breaking of critical internal waves on sloping boundaries." *J. Fluid Mech.* 204: 479–500.)

significant dissipation. Although these possibilities have long been known (e.g., Beardsley, 1970), the discussion has recently been recast into in the language of *attractors* and other dynamical system terminology (e.g., Maas, 2011). In a mathematically ideal setting with sloping walls, space can be filled with rays or characteristics undergoing an infinite number of reflections. This limit is not likely of oceanographic significance because nonlinearity, friction, and other loss mechanisms preclude observable multiple reflections.

Tracing the characteristics predicts a strong focussing of energy into regions such as ocean-margin canyons and V-shaped trenches, where the slopes are transmissive for rays arising from the mouth and/or from above. Much observational evidence exists for the existence of focussing in canyons (an example is Zhao et al., 2012).

Use of the thin-shell limit (leading to neglect of the component of the Earth rotation vector tangent to the seafloor) in what is known as the *traditional approximation* has been challenged in these and related situations by Gerkema and Shira (2005) and Winters et al. (2011). Retaining the extra terms leads to detailed modifications of the theoretical results, but because they have not been observed, the subject is not pursued here. (A possible exception in the very deep, weakly stratified basins of the Mediterranean is described by van Haren and Millot, 2004.)

4.4.1 Separation of Variables

Thus far, the dynamics of internal waves appear quite distinct from the discussion of the solutions for homogeneous fluids above, and nothing like a Kelvin or Rossby wave has appeared. Bringing the solutions together by assuming that everything is again proportional to $\exp(-i\sigma t)$, using the Equations (3.30a–e) and (3.31) and trying a *separation of variables*,

$$[u(x, y, z, t), \, v(x, y, z, t)] = [U(x, y, t), \, V(x, y, t)] \, F(z) \tag{4.75a}$$

$$w(x, y, z, t) = P(x, y, t)G(z)/\bar{\rho} \tag{4.75b}$$

$$p'(x, y, z, t) = P(x, y, t)F(z)/\bar{\rho} \tag{4.75c}$$

As worked out in the box "Separation of Variables", these assumptions do lead to equations in x and y solvable separately from those in z.

SEPARATION OF VARIABLES

Substituting Equations (4.75a–c) into Equations (3.30a–e) and (3.31),

$$F(z)\left[-i\sigma U(x, y) - f V(x, y)\right] = -\frac{\partial P(x, y)}{\partial x} F(z)$$

$$F(z)\left[-i\sigma V(x, y) + f U(x, y)\right] = -\frac{\partial P(x, y)}{\partial y} F(z)$$

$$0 = -P(x, y)\frac{dF}{dz} + \frac{N(z)^2}{i\sigma} P(x, y) G(z)$$

$$F(z)\left[\frac{\partial U(x, y)}{\partial x} + \frac{\partial V(x, y)}{\partial y}\right] + P(x, y)\frac{dG}{dz} = 0.$$

Dividing the first two of these equations by $F(z)$, the third equation by $P(x, y, \sigma)$, and the fourth one by $F(z) P(x, y, \sigma)$ gives

$$[-i\sigma U(x, y, \sigma) - f V(x, y)] = -\frac{\partial P(x, y, \sigma)}{\partial x} \tag{4.76a}$$

$$[-i\sigma V(x, y, \sigma) + f U(x, y)] = -\frac{\partial P(x, y, \sigma)}{\partial y} \tag{4.76b}$$

$$0 = -\frac{dF(z)}{dz} + \frac{N(z)^2}{i\sigma} G(z) \tag{4.76c}$$

$$\frac{1}{P(x, y, \sigma)}\left[\frac{\partial U(x, y, \sigma)}{\partial x} + \frac{\partial V(x, y, \sigma)}{\partial y}\right] = -\frac{1}{F(z)}\frac{dG(z)}{dz}. \tag{4.76d}$$

continued

The first two equations here depend only upon x, y, σ. In the last equation, the left-hand side is a function of x, y, σ alone, the right-hand side a function of z alone, and the only way equality is possible is if both are equal to a constant, or

$$\frac{1}{P(x,y,\sigma)}\left[\frac{\partial U(x,y,\sigma)}{\partial x}+\frac{\partial V(x,y,\sigma)}{\partial y}\right]=-\frac{dG(z)}{dz}\frac{1}{F(z)}=\text{constant}=i\sigma\gamma^2$$

(γ^2 is the separation constant; $i\sigma$ is introduced only for notational tidiness), or

$$\frac{dG(z)}{dz}=-i\sigma\gamma^2 F(z). \tag{4.77}$$

Equation (4.76c) is

$$\frac{dF(z)}{dz}=\frac{N(z)^2}{i\sigma}G(z). \tag{4.78}$$

Eliminating,

$$\frac{d^2G(z)}{dz^2}+N^2(z)\gamma^2 G(z)=0, \tag{4.79}$$

$$\frac{d}{dz}\left(\frac{1}{N^2(z)}\frac{dF(z)}{dz}\right)+\gamma^2 F(z)=0. \tag{4.80}$$

It suffices to solve either Equation (4.79) or (4.80), subject to the appropriate boundary conditions, because Equations (4.77) and (4.78) permit the deduction of either of F or G from the other. Thus the problem is reduced to equations in the horizontal and vertical coordinates, connected by the separation constant.

At the seafloor, $z=-h$ and $w(z=-h)=0$,

$$\left.\frac{\partial p'}{\partial z}\right|_{z=-h}=0,$$

requiring, from the box "Separation of Variables", either $G(z=-h)=0$, or $F'(z=-h)=0$. At the free surface, $z=\eta$, the dynamic boundary condition (Eq. 3.26) must be satisfied and which for F (in Eq. 4.80) is

$$F(0)+\frac{g}{N^2(0)}\frac{dF(0)}{dz}=0. \tag{4.81}$$

Although numerical solutions are not difficult to obtain, a few analytical solutions for specific choices of $N(z)$ are known, including when N is taken to be constant. $N=N_0 e^{bz}$ has solutions in Bessel functions (see the box "Forced Cylinder"). In general, finding γ, $F(z)$ is a *Sturm-Liouville problem* (see box "Sturm-Liouville Problems and Bessel Function Solutions") that can be solved analytically or numerically. A number of analytical profiles are discussed by Krauss (1966, p. 32). Dynamically, the influence of the free surface movement is essentially negligible. On the other hand, detection of these baroclinic motions from space by altimetry rests entirely on the comparatively slight induced motion.

STURM-LIOUVILLE PROBLEMS AND BESSEL FUNCTION SOLUTIONS

Sturm-Liouville problems are defined and analyzed at length, for example, by Morse and Feshbach (1953), and are essentially two-point eigenvalue problems for a restricted type of second-order differential system, of which the ocean linear eigenvalue problem is one example. Equations (4.88a,b) are solutions when N is taken to be constant.

A useful example of a more interesting analytical profile is the exponential, $N = N_0 \exp{(bz)}$, used by Garrett and Munk (1972) and many others as a way of accounting for intensified upper-ocean stratification. The governing equation produces

$$G(z) = A_1 J_0 \left(\frac{\gamma N_0}{b} e^{bz} \right) + A_2 Y_0 \left(\frac{\gamma N_0}{b} e^{bz} \right),$$

where J_0, Y_0 are the two lowest Bessel functions and γ is again the vertical eigenvalue. The bottom boundary condition of no normal flow is

$$A_1 J_0 \left(\frac{\gamma N_0}{b} e^{-bh} \right) + A_2 Y_0 \left(\frac{\gamma N_0}{b} e^{-bh} \right) = 0, \tag{4.82}$$

and the full free-surface boundary condition is

$$\frac{g\gamma}{N_0} \left[A_1 J_0 \left(\frac{\gamma N_0}{b} \right) + A_2 Y_0 \left(\frac{\gamma N_0}{b} \right) \right] - \left[A_1 J_0' \left(\frac{\gamma N_0}{b} \right) + A_2 Y_0' \left(\frac{\gamma N_0}{b} \right) \right] = 0. \tag{4.83}$$

Setting the coefficient determinant of Equations (4.82) and (4.83) to zero results in a transcendental equation for γ_n, which is readily solved numerically and which then relates A_1 and A_2.

The very slight failure of the rigid-lid approximation means

$$G(0) = A_1 J_0 \left(\frac{\gamma_n N_0}{b} \right) + A_2 Y_0 \left(\frac{\gamma_n N_0}{b} \right) \neq 0$$

for small n. For $n = 1$ and realistic stratification, a 10 m vertical displacement of an isotherm will move the sea surface by about 1 cm. The effect was used for tide-gauge data by Wunsch and Gill (1976) to detect internal gravity wave modes (see also Wunsch, 2013).

Solutions to Sturm-Liouville problems can be shown to produce a complete set, $\{F_n(z), \gamma_n\}$ or $\{G_n(z), \gamma_n\}$, satisfying the appropriate boundary conditions. The $G_n(z)$ are orthogonal, as are the $F_n(z)$ if weighted by $N(z)$. Then, for example, an arbitrary flow $u(z) = \sum_{n=0}^{\infty} \alpha_n F_n(z)$ can be expanded exactly in the F_n—whether or not the physics is appropriate. Tests of the physics can be made by examining the phase relations in the modal expansions among u, v, w if observations are available. Solutions to forced problems can be written in terms of the F_n (or G_n), and for sources extending over small vertical extents (e.g., mixed layer depths), $\alpha_n \approx 1/n$. That is, solutions tend to be dominated by the lowest modes, which have small n. When motions are observed to be dominated by high n modes, a reasonable inference is that they did not arise

continued

from direct forcing at the surface (of important application among other phenomena, to observed inertial motions in the ocean).

If the WKBJ theory described in the box "WKBJ Approximation" is applied to Equation (4.79), it produces for the hydrostatic limit ($\sigma \ll N$) (Kundu and Cohen, 2004, p. 624),

$$G(z) = \frac{B_1}{\sqrt{N(z)}} \exp\left(\gamma \int N(z)\, dz\right) + \frac{B_2}{\sqrt{N(z)}} \exp\left(-\gamma \int N(z)\, dz\right)$$

where A, B are determined by the boundary or radiation conditions. The corresponding solution in $F(z)$ is

$$F(z) = \frac{B_1 \sqrt{N(z)}}{-\sigma\gamma} \exp\left(\gamma \int \gamma N(z)\, dz\right) + \frac{B_2 \sqrt{N(z)}}{-\sigma\gamma} \exp\left(-\gamma \int N(z)\, dz\right), \quad \text{if } l = 0.$$

B_1, B_2, and γ are determined from the same pair of boundary conditions. Thus for the high modes dominating the internal wave field, the vertical structure of the kinetic energy is proportional to $N(z)$ and the potential energy to $N(z)^{-1}$, a conclusion generally well supported by observations. The vertical structure leads to the use, in analyzing data, of a "stretched" vertical coordinate, ζ, where $d\zeta = N(z)\, dz$ (Leaman, 1976).

If N is constant, then the solution to (4.80) is

$$F(z) = Ae^{iN\gamma z} + Be^{-iN\gamma z}. \tag{4.84}$$

Substituting into the two boundary conditions produces

$$Ae^{-iN\gamma h} - Be^{iN\gamma h} = 0, \quad z = -h \tag{4.85a}$$

$$A + B + \frac{g}{N^2}\left(AiN\gamma - BiN\gamma\right) = 0, \quad z = 0. \tag{4.85b}$$

Upon setting the coefficient determinant to zero,

$$\tan(N\gamma h) = \frac{N}{g\gamma}. \tag{4.86}$$

A graphical construct (Fig. 4.12) is useful for solving, and understanding, Equation (4.86). The point at which the hyperbola $N/g\gamma$ intersects the various branches of the tangent function defines the allowable values of γ.

The smallest value of γ satisfying Equation (4.86) produces

$$\gamma_0 \approx \frac{1}{\sqrt{gh}}, \tag{4.87}$$

almost exactly, which is the *barotropic* root. Notice γ is an inverse speed—here that of an ordinary long gravity wave without stratification. From Equation (4.85a), $B = A\exp(-2iN\gamma h) = A\exp(-2iNh/(\sqrt{gh})) \approx A$. Hence,

$$F_0(z) = A\left(e^{iNz/\sqrt{gh}} + e^{-iNz/\sqrt{gh}}\right) \approx 2A.$$

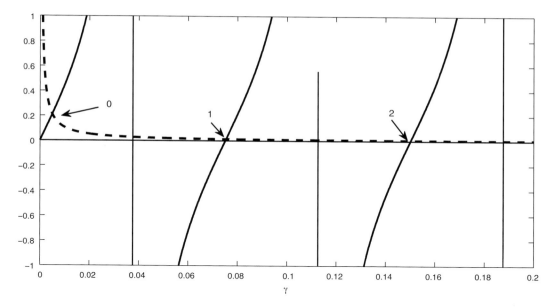

Figure 4.12: Intersections of the two curves, $\tan(N\gamma h)$ and $N/(g\gamma)$ that define the permissible values of γ. Intersection marked "0" is the barotropic root, $\gamma_0 \approx 1/\sqrt{gh}$, and "1" and "2" are the first two baroclinic roots, γ_2, γ_2, etc. As n increases, the roots asymptote to $\gamma_n = n\pi/Nh$, the values for a rigid-lid solution. Numerically, N corresponds to a 10 min buoyancy period, and $h = 4000$ m.

That is, the horizontal velocity is almost constant with depth for any physically reasonable values of N, h. The corresponding vertical velocity or displacement structure is obtained from Equation (4.77),

$$G_0(z) = -i\sigma 2A(z + h),$$

vanishing at $z = -h$ and growing linearly to the free surface. The vertical structure is thus that of the long gravity waves found previously for a homogeneous fluid.

For large γ, $\tan(N\gamma h) \approx 0$, $N\gamma h = n\pi$ is recovered, which is the same solution as if the sea surface had a rigid lid. With a rigid lid, the F, G solutions are simple sinusoids:

$$F_n(z) = A' \cos\left(\frac{n\pi z}{h}\right), \quad n = 0, 1, 2, \ldots \tag{4.88a}$$

$$G_n(z) = B' \sin\left(\frac{n\pi}{h}(z + h)\right) = B' \sin\left(\frac{n\pi z}{h}\right), \quad n = 1, 2, \ldots, \tag{4.88b}$$

that is, both the top and bottom are vertical displacement nodes corresponding to the solutions already discussed above (Eq. 4.73). Various ways exist to normalize these solutions: one possibility is $A' = B' = 1$: or the integral square can be set to unity, etc.

Here γ_n does not depend upon σ. If the motion is not hydrostatic, the typical case as $\sigma \to N$, determination of γ is dependent on the frequency. The WKBJ solutions for internal waves of high vertical wavenumber described in the box "WKBJ Approximation" are also the appropriate ones here for slowly varying $N(z)$.

The approximate solutions $\gamma_n \approx \pi n/Nh$ to Equation (4.86), are more accurately $\gamma_n = \pi n/Nh + N/g\pi n$, $n = 1, 2, \ldots$. The sea-surface movement no longer vanishes and is nearly

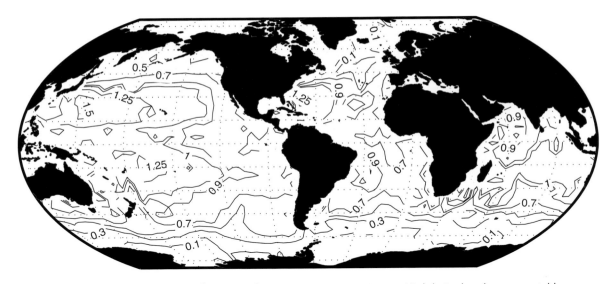

Figure 4.13: The equivalent depth h_1' (meters) from an ECCO state estimate. High-latitude values are notably very small compared to the subtropical and tropical ones.

$$G_n(0) = B'N^2 h/g\pi n, \quad n = 1, 2, \ldots, \tag{4.89}$$

diminishing with n.

With the γ_n known, the horizontal equations (4.75a–c) are,

$$-i\sigma U - fV = -\frac{\partial P}{\partial x} \tag{4.90}$$

$$-i\sigma V + fU = -\frac{\partial P}{\partial y} \tag{4.91}$$

$$\frac{\partial U}{\partial x} + \frac{\partial V}{\partial y} - i\sigma \gamma_n^2 P(x, y) = 0. \tag{4.92}$$

As with the homogeneous ocean, the equation in V is the simplest one,

$$\nabla_h^2 V - \frac{\beta}{i\sigma} \frac{\partial V}{\partial x} + \frac{(\sigma^2 - f^2)}{gh_n'} V = 0, \tag{4.93}$$

defining the *equivalent depth* as

$$h_n' = 1/\left(g\gamma_n^2\right) << h \tag{4.94}$$

for $n > 0$. Taking the usual $\exp(ikx)$ dependence, the y-dependence reduces to that in Equation (4.57), with h_n' replacing h. Figure 4.13 displays a global estimate of h_1'.[14]

Thus all of the theory for free long waves in a homogeneous fluid of uniform depth now immediately applies to the modes of a stratified fluid of depth h_n' on both f- and β-planes—a remarkable simplification. For typical values of N, h, and $h_n' \approx (1/n)$ m, for $n \geq 1$, decreasing

---

[14] Some authors prefer instead to define a *reduced gravity*, g_n', so that $g_n'h = gh_n'$.

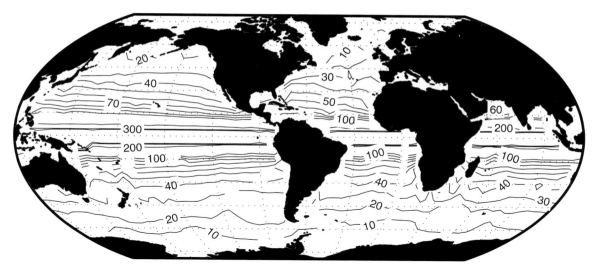

Figure 4.14: The first baroclinic Rossby radius, R_D, in kilometers from the same stratification as that used in Figure 4.13. Contouring near the equator is suppressed as f vanishes. Of greatest importance is the very wide range, from indefinitely large near the equator to below 10 km at high latitudes (and much smaller in shallow water). R_D is a major determinant of the spatial resolution required to reproduce observed fields. Diminished high-latitude values arise from the increasing value of f and the decreasing stratification there.

rapidly with n. The solutions already discussed are thus directly applicable to the baroclinic modes but operate as though the ocean depth were extremely small and justify the attention paid to solutions only accurate for very shallow oceans. All of the dispersion relations above remain valid with the replacement of h by h'_n. The corresponding Rossby radii become

$$R_{Dn} = \sqrt{\frac{gh'_n}{f^2}}$$

and for $n = 1$ have numerical values at midlatitudes on the order of 30 km (Fig. 4.14) instead of thousands of kilometers. The various Cartesian approximations are much more accurate for systems in which h'_n occurs rather than h. Similarly, the highest-frequency Rossby waves, Equation (4.56), have midlatitude periods of months to years rather than of days (Fig. 4.15). Vertical structures are, however, quite different from what is appropriate in a homogeneous ocean. The barotropic mode is also formally included, using $h'_0 = h$, although the f- and β-plane approximations are inaccurate. Figure 4.16 shows an estimate of the strongly latitude-dependent time for a first mode baroclinic Rossby wave to cross the Atlantic Ocean, determined by the changing values of β and R_{D1} in the zonal group velocity. With increasing n, h'_n, and R_{Dn} go rapidly to zero, and the corresponding vertical structures become ever more complex.

4.4.2 The Sphere

The same separation-of-variables approach works for the full spherical coordinate version of the inviscid dynamical equations (Eqs. 3.11a–d without viscosity and diffusion), and so *aquaplanet* solutions, some dating back to Laplace, are usable. If the bottom is flat, all of

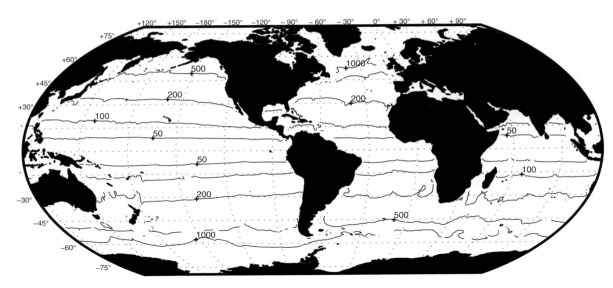

Figure 4.15: The *shortest* permissible period, in days, of a first-mode, linear baroclinic Rossby wave as computed from Equation (4.54). At high latitudes with decreasing β and $N(z)$, the permissible periods are so long that the linear time derivative terms become negligible compared to the nonlinear ones, and the applicability of the linear theory is increasingly doubtful.

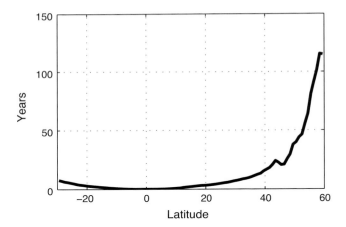

Figure 4.16: The time required for a long first-mode baroclinic Rossby wave signal to cross the North Atlantic Ocean. Computed as $L(y)/(\beta(y)R_d^2(y))$ in the nondispersive group-velocity limit, where $L(y)$ is the ocean width and $R_d(y)$ is the value of the zonal average first-baroclinic-mode deformation radius. Despite the poleward narrowing of the ocean, the reduction in both β and R_d greatly increases the signalling time with latitude. Equilibrium adjustment times would be far longer. (From Wunsch, 2011.)

the known barotropic solutions on a water-covered Earth become available for baroclinic motions of arbitrary vertical mode number. De Leon and Paldor (2009) demonstrated that the β-plane approximation can produce important quantitative errors relative to the full spherical equations. Normal modes of bounded basins on a sphere were discussed by Longuet-Higgins and Pond (1970), who resorted to numerical methods. Gill and Clarke (1974) summarized the history of the separation of variables method.

4.4.3 Inertial Waves

Using the separation of variables on a sphere, Longuet-Higgins (1968) derived an equation in P and analyzed it in the spherical geometry at the high frequency limit. His approximate governing equation was exploited by Munk and Phillips (1968) and Fu (1981) for the understanding of inertial waves when the f-plane restrictions were relaxed. Details of their solutions are omitted here, but the qualitative features can be understood from the equatorial β-plane equation (4.93), with the understanding that h' is the equivalent depth of O(1–2 m) or less.

If waves are traveling nearly north-south, so that $k \to 0$, the governing equation for V becomes

$$\frac{d^2 V}{dy^2} + \frac{\sigma^2 - \beta^2 y^2}{gh'} V = 0. \tag{4.95}$$

When $\beta y < \sigma$, a reasonable inference from the logic of the WKBJ approach is that the solutions are proportional to

$$\exp(ily) \approx \exp\left(\pm i\sqrt{(\sigma^2 - \beta^2 y^2)/gh'}\,y\right),$$

that is, travelling in the $\pm y$ direction, and that when $|\beta y| > \sigma$, they are proportional to

$$\exp(-ly) \approx \exp\left(-\sqrt{(\sigma^2 - \beta^2 y^2)/gh'}\,y\right),$$

that is, exponentially damped poleward of the latitude where $\sigma = |\beta y|$. This heuristic argument can be shown to be correct by examining the Hermite function solutions (4.58a–c) just equatorward and poleward of $|y| = \sigma/\beta$, and with more elaborate techniques applied to the full equations on a sphere.

In the transition region, across this latitude, called the *inertial latitude*, Munk and Phillips (1968) and Fu (1981), building on the results of Longuet-Higgins (1965a) and Dikii (1966), showed that a linear approximation in y (not the quadratic one in Eq. 4.95) is appropriate. Local solutions called *Airy functions* give a good representation that can be used to "patch" from the travelling to exponentially decaying forms, with the transition region describing inertial waves in an extension of the f-plane theory above. In the more general wave context, the inertial latitude is also a wave *turning latitude*, which produces a wave *caustic*.

Most of the oceanographic literature discusses inertial wave motion in the context of the f-plane, omitting the effects of β. For many purposes, that approximation is very useful. But it is an example where the approximations can be seen to break down in observations. The maximum response is often at $\sigma > f$ (a "blue-shift"), and the north-south structure is very short-scale rather than long (Munk and Phillips, 1968; Fu, 1981).

4.4.4 Vertical Propagation—Infinitely Deep Ocean

The ocean is a forced system, and various methods exist for understanding its response. For simplicity, consider an ocean in which the bottom is infinitely far away. No lower boundary condition is required, but the solution must be finite at infinite depth. In solving Equations (4.75a–c) with constant N, f above, the value of γ was determined from one of the vertical structure equations and its two boundary conditions. Without a lower boundary, Equation (4.85a) is no longer required and,

$$A + B + \frac{g}{N^2}\left(AiN\gamma - BiN\gamma\right) = 0, \quad z = 0, \tag{4.96}$$

and γ is unspecified. Turning then to Equation (4.93) and requiring

$$P = Ce^{ikx+ily-i\sigma t}$$

where σ, k, l are now fixed and arbitrary (that is, not satisfying any dispersion relationship), and substituting,

$$-\left(k^2 + l^2\right) + \frac{\beta k}{\sigma} + \gamma^2\left(\sigma^2 - f^2\right) = 0.$$

With k, l, σ imposed,

$$\gamma = \pm\sqrt{\frac{k^2 + l^2 - \beta k/\sigma}{\sigma^2 - f^2}}. \tag{4.97}$$

If $\sigma > f$, and setting $\beta = 0$, γ is real, and $F = A\exp\left(+iN\gamma z\right) + B\exp\left(-iN\gamma z\right)$, that is, sinusoidal in the vertical. The radiation condition, that energy not propagate from $z = -\infty$ requires $A = 0$ (recalling the group velocity sign). The vertical structure is now being set by the imposed horizontal wavenumber and frequency. Because energy escapes to infinity, this solution would have to be maintained by forcing.

On the other hand, if $\sigma < f$, $\beta = 0$, so that no free waves are possible without a wall boundary,

$$\gamma = \pm i\sqrt{\frac{k^2 + l^2}{f^2 - \sigma^2}}, \tag{4.98}$$

and the vertical structure is $F(z) = A\exp\left(N\gamma z\right)$, exponentially decaying downward. These are waves of arbitrary wavenumber and frequency $\sigma < f$ but are trapped against the free surface, unable to propagate vertically.

If β is restored, the nature of the solutions still depends upon the sign of γ^2, and with $\sigma \ll f$,

$$\gamma^2 = -\frac{\beta k}{f^2\sigma}. \tag{4.99}$$

Thus at low frequencies, $\sigma \ll f$, ignoring the first term, all forced eastward-bound motions, $k > 0$, must decay vertically away from the surface, while westward-bound ones will be wave-like radiating in the vertical. If $\gamma^2 < 0$, but small, the vertical decay scale will be very long, making the motion appear quasi-barotropic in nature. More generally if $k < 0$, then depending upon whether β is large enough to render $\gamma^2 > 0$ the solution can either have the same quasi-barotropic structure or become structured in the vertical, radiating energy downward. Because β grows toward the equator, it is anticipated that tropical disturbances at the sea surface will generate very different structures than ones at mid- to higher latitudes. Formally, the equivalent depth, Equation (4.94), is negative when $\gamma^2 < 0$. Such "negative depth" solutions (they are not modes in the conventional sense) appear only in forced problems. Solutions have been discussed generally by Lindzen (1967), Longuet-Higgins (1968b), Philander (1978), Frankignoul and Müller (1979), and Wunsch and Stammer (1998) and can be extended to the equatorial β-plane and the sphere.

As an example, suppose the ocean is forced at the annual period by a structure having a zonal wavelength of 5000 km, $l = 0$. Let the ocean have constant N with a buoyancy period of 1 h. If $k < 0$, then $\gamma^2 > 0$, $m = N\gamma$ is real, and a corresponding vertical wavelength is shown in Figure 4.17 as a function of latitude. The vertical wavelength at this frequency is so much larger than the ocean depth that the only response would be indistinguishable from a barotropic one, even accounting for the lower boundary condition, which we have omitted. (The corresponding density field disturbance is small.)

Taking the annual period as characteristic, Figure 4.17 shows how the vertical group velocity behaves as a function of latitude. In particular, as the latitude increases, the time to go 1000 m in the vertical becomes extremely small, as the penetration rate of the energy in large vertical wavelengths is very rapid. The zonal travel times are displayed in Figure 4.16.

When a finite depth-reflecting bottom is restored, and should k, l, σ correspond to the structure of one of the free modes, resonance is a possibility—a situation never observed. A forced solution with a flat bottom would correspond to the general response given in Equation (4.36)—a summation over the vertical modes satisfying the boundary conditions and formally producing unobserved resonances.

4.4.5 Nonlinearities

TRIAD INTERACTIONS

The linearization assumptions used here prove sufficiently accurate for order-of-magnitude descriptions of a very wide range of oceanic phenomena. Like all such approximations in fluid dynamics, they do fail, and the implications of failure must be understood. As described in Appendix A, under very general circumstances, nonlinear systems forced with one or more frequencies, $\sigma_1, \sigma_2, \ldots,$ will generate responses at frequencies $\sigma_i \pm \sigma_j, 2\sigma_i$, etc., that is, at all sum and difference frequencies including overtones and undertones. The same inference applies to wavenumbers, with outputs at $k_i \pm k_j$ etc.

Motions forced at sum and difference wavenumbers $\sigma_3 = \sigma_1 \pm \sigma_2$, with $k_3 = k_1 \pm k_2$ can, in some wave systems, correspond to a *free* solution to the linear wave equation. If so, that is the condition for resonance, and the sets $(\sigma_1, \sigma_2, \sigma_3), (k_1, k_2, k_3)$ are said to form a *resonant triad* (recall the denominator in Eq. 4.36). The third wave will grow rapidly out of the interaction of the first two waves. Resonant triad interactions exist in theories for internal waves and Rossby waves. They do *not* exist for surface gravity waves unless surface tension is included. But they do form *quartet* resonances, in which three waves can interact to produce a fourth wave that is a linear, resonant solution. Phillips (1977) summarized the general interactions. This type of theory is also applicable in two or three space dimensions with vector wavenumbers, $\mathbf{k}_i$.

If only one resonant triad can exist, nonlinearities and dissipation would produce a form of equilibrium in which the three waves coexisted, transferring energy among themselves. More generally, waves can form components of more than one triad, perhaps many, resulting in an exchange of energy across the entire frequency-wavenumber spectrum. If energy is fed

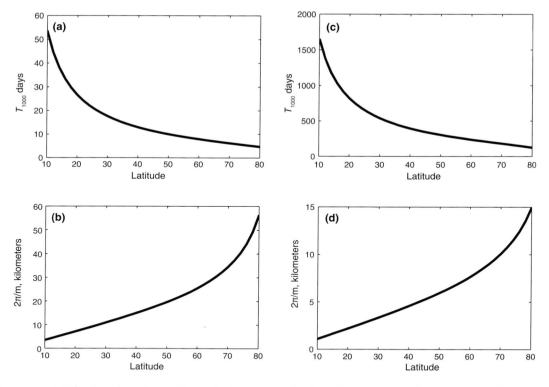

Figure 4.17: (a) The time, based upon the vertical group velocity, for a Rossby wave at the annual period having $k = -2\pi/5000$ km, $N = 2\pi/1$ h to travel a vertical distance of 1000 m. A very strong latitude dependence is apparent. (b) The vertical wavelength of the annual period wave in (a). At high latitudes, the wavelength greatly exceeds the water depth. (c) and (d) Same as (a) and (b), except the period has been increased to 10 years. Note the scale changes.

into a field of existing internal or baroclinic Rossby waves, it will be transferred at different rates to and from the differing frequencies and wavenumbers. When the energy source is maintained, a quasi-equilibrium distribution can be achieved in frequency and three-dimensional wavenumber space, producing a power density spectrum in frequency and three wavenumber dimensions, $\Phi(\sigma, \mathbf{k})$. See Lvov and Yokoyama, 2009, and Polzin and Lvov, 2011, for recent work and references; details of these processes are complicated and murky.

TURBULENT INTERACTIONS

Triad interactions, when they exist, are an efficient mechanism for transferring wave energy between disparate time and space scales. For example, if $\sigma_1 \approx \sigma_2$, $k_1 \approx k_2$, then the difference frequency and wavenumber can be very much smaller than those of the original waves. Another type of nonlinear interaction in fluids is that arising in classical turbulence theories in which energy is transferred to *neighbor* wavenumbers, $k \pm \Delta k$. The motion has no characteristic frequency, σ, because no dispersion relationship exists in turbulent flows. (In the presence of a large-scale advective current, U, an apparent frequency, Uk, appears.)

4.4.6 Low Frequency Limits

The wave-like motions described by the above systems of equations in homogeneous and stratified fluids run the gamut in time scales from seconds (surface gravity waves) to decades and longer (baroclinic Rossby waves). In the linear momentum equations as $\sigma \to 0$, the acceleration terms, $\partial u/\partial t$ and $\partial v/\partial t$, can be made arbitrarily small relative to the terms proportional to the pressure gradients and Coriolis forces. The Cartesian form of geostrophic balance (Eqs. 3.20a–b) is

$$-\rho f v \approx -\frac{\partial p}{\partial x}, \tag{4.100a}$$

$$\rho f u \approx -\frac{\partial p}{\partial y}. \tag{4.100b}$$

Substitution of numerical values for, for example, observed baroclinic Rossby waves, shows that this balance is very closely achieved. In practical terms, the omitted acceleration terms are very difficult to measure directly—because they are the small residuals of the differences of the near-balancing, much larger, terms in Equations (4.100a–b). On the other hand, the balance equations have no provision for any time evolution, as demanded in the wave solutions, and cannot account for injection or removal of energy from the system. Thus the small imbalances are critical to understanding the nature of the flow, where it came from, and how it is supported and dissipated. A major observational issue concerns the great difficulty in directly determining the magnitudes and phases of the vector components of the unbalancing terms including viscous and diffusive elements and nonlinearities.

4.5 INITIAL VALUE-ADJUSTMENT PROBLEMS

An important class of initial value problems exists in strongly rotating systems, commonly known as *adjustment*, or *Rossby adjustment*, problems and was reviewed by Blumen (1972). Consider here only a modification (Gill, 1982) of Rossby's original problem (1938)—because it contains the essence of the more complex situations that arise in nature.

On an f-plane, an initial height distribution is postulated,

$$\eta_0 = \left\{ \begin{array}{ll} \eta_I, & |y| \le y_1, \\ = 0, & |y| > y_1. \end{array} \right. \tag{4.101}$$

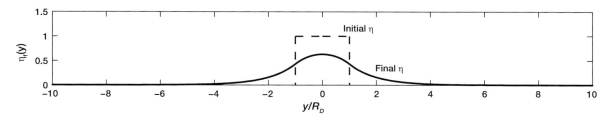

Figure 4.18: An initial elevation (dashed line), constant between $y = \pm 1$ (in units of the deformation radius) reaching a final equilibrium on an f-plane shown by the solid line, trapped by the rotation. This elevation is supported by flows into and out of the page depending upon the sign of f. "Final" applies only within the equations used—in practice, this solution would decay from friction or be disrupted by winds, eddies, etc.

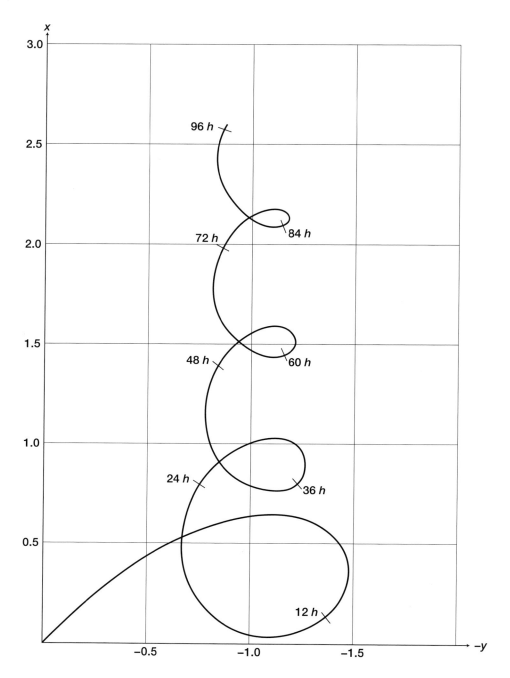

Figure 4.19: The theoretical trajectory of a fluid parcel during the transient stage of a Rossby adjustment problem in nondimensional units. Compare to Figure 3.11. The conspicuous, decaying inertial oscillations are visible. The particular frequency is a function of the latitude chosen. (From Cahn, 1945.)

that is, a confined "head" of water symmetric about $y = 0$, and which is "let go" from rest (Fig. 4.18). It is not difficult to show that a geostrophic steady-state solution is eventually achieved. From conservation of potential vorticity (see Eq. 3.19), the final elevation must satisfy

$$\frac{d^2\eta_F}{dy^2} - \frac{1}{R_D^2}\eta_F = \frac{\eta_I}{2R_D^2}\left[\text{sgn}\left(y + y_1\right) - \text{sgn}\left(y - y_1\right)\right], \quad R_D = \frac{\sqrt{gh}}{f} \tag{4.102}$$

(sgn is the sign function), whose solution is

$$\eta_F\left(y\right) = \eta_I\left(1 - \exp(-y_1/R_D)\cosh(y/R_D)\right), \quad 0 \le y \le y_1 \tag{4.103}$$
$$= \eta_I \exp\left(-y/R_D\right)\sinh\left(y_1/R_D\right), \quad y_1 < y$$

and is symmetric about $y = 0$, as shown in Figure 4.18. This elevation, and its corresponding hydrostatic pressure field, will nominally remain in this form forever, being a fully balanced, geostrophic flow. (If nothing else intervenes, friction leads to decay.)

The transition from Equation (4.101) to the final steady-state Equation (4.103) is an interesting one, in which long waves (in this limit) are radiated away as the initially discontinuous elevation collapses. One form of the transition was worked out by Cahn (1945), for a slightly different set of initial condition, and he found that the transient motions were dominated by f-plane inertial oscillations long after the initial "dam break" had taken place (Fig. 4.19).

The adjustment problem is prototypical of an entire suite of situations in which a flow that is initially unbalanced approaches one that is balanced, with the final equilibrium sometimes taking extremely long times to achieve. Stratification, the β-effect, topography, nonlinearity, viscosity, spherical geometry, lateral boundaries, the presence of preexisting time-mean flows, etc., can be included. In essentially all cases, an initial generation of wavelike motions occurs, which carry away much of the initial energy (be it kinetic or potential or both), leaving behind a remnant flow field that is asymptotically balanced.

CHAPTER 5

Observations of Internal and Inertial Waves

5.1 INTRODUCTION

Theory shows that internal waves in the ocean should be found in the band of frequencies $f(\phi) \leq \sigma \leq N(\lambda, \phi, z)$, that is, between periods nominally approaching the infinite near the equator and a few minutes in regions where N is very large. The terminology "nominal" is used because these bounds are not absolute. The review by Munk (1981) remains probably the best available overview of the main elements of the subject.

That a background "noisiness" existed in oceans and lakes, and that it was likely owing to internal wave motions, has been known for a long time, with the basic theory for layered media having been constructed in the nineteenth century. Spectra such as in Figure 3.10 show that energy exists at the expected frequencies everywhere in the ocean. At a period of 5 h, in an ocean 4000 m deep with a buoyancy period of 1 h, the horizontal wavelength of the first vertical mode at midlatitudes is about 40 km and the second mode about 20 km (Eq. 4.74). Until about 1970, however, most papers about observed internal waves consisted largely of the publication of wiggly lines. Some authors even questioned whether internal waves existed at all, choosing to interpret the temperature variations as a form of conventional turbulence of an unstratified or weakly stratified fluid. As described by Munk (1981), the major issue was that measurements made close together in the horizontal, much closer than 40 km, bore little resemblance to each other—an observation mystifyingly counter to the expectations from the long wavelengths predicted by the theory and the experience with surface gravity waves (Fig. 5.1).

From the linear theory discussed in Section 4.4, Fofonoff (1969) and Lien and Müller (1992) developed consistency tests for moored current meter and temperature records. These tests involve ratios of power densities of velocity components and temperatures as a function of the frequency at single points. Although not proving the results to be internal waves, the test outcomes convinced most oceanographers that this explanation of what was recorded was the correct one.

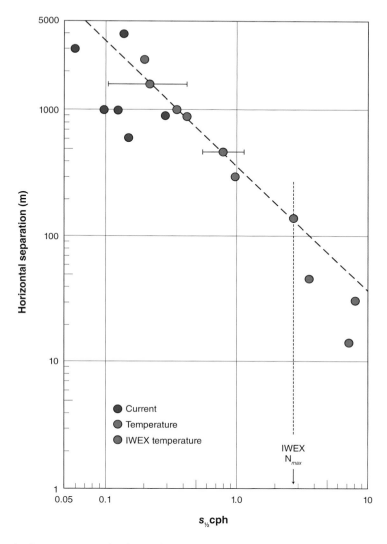

Figure 5.1: The frequency in cycles/hour where the coherence magnitude (see Eq. A.72) between two horizontally separated instruments (measuring either temperature or velocity) falls to a value of 1/2. Thus, at a 1000 m (1 km) horizontal separation, fluctuations at 10 h periods and shorter will show little or no coherence. At zero separation, the coherence magnitude is always 1. Recall the large wavelengths of the waves at these periods. (See Briscoe, 1975, for identification of the cited sources.) Levine (2002) discusses extensions of such calculations using his version of the model spectrum. (After Briscoe, 1975, from the Internal Wave Experiment, IWEX.)

5.2 THE SURPRISING GARRETT AND MUNK RESULT

The paper of Garrett and Munk (1972) revolutionized the study of internal waves, and also provided a prototype for the handling of other observed stochastic fields. It has permitted the calculation of the effects of internal waves on myriad phenomena ranging from acoustic propagation to oceanic mixing. A number of unexpected conclusions were drawn about the waves that are in striking contrast to expectations arising from the more familiar surface or acoustic or other cases. Each member of this list is accompanied by the words "almost everywhere," because none of the statements is strictly or universally true. They do have a remarkable qualitative near universality, when measured over "long-enough" time intervals and sufficiently far from lateral boundaries. The weasel words "long enough" are included because in any stochastic field, short-duration measurements can differ greatly from their time-average properties.

1. Top and bottom boundaries appear irrelevant in observations made more than about 100 m from those boundaries, as though the ocean extended to infinity in both directions.

2. The energy levels of internal waves, when divided by $N(z)$ are constant within a factor of about 2. When so normalized, the RMS motion, as measured either in velocity or vertical displacement, is almost universal in both time and space.

3. As much energy is being transmitted upward as downward.

4. The horizontal transmission of energy is isotropic.

5. Two internal wave records, be they of temperature or velocity variation, show no visible similarity if they are separated in the horizontal by more than about 1 km or in the vertical by about 100 m (see Eq. A.72).

In marked contrast to the experience with ordinary oceanic surface waves, internal waves normally exhibit neither calms nor storms, and they cannot be perceived as propagating preferentially from a region of generation to a region of decay.

Garrett and Munk attempted to synthesize into a frequency-wavenumber spectrum as much as they could of the relevant observations available to them. The synthesis has been polished over the years by a number of workers, and so variant forms exist. The Garrett-Munk spectrum (hereafter GM[1]) in one form (Munk, 1981) is, assuming $N(z) = N_0 e^{bz}$ with N_0, b chosen for any locale and using the internal-wave solutions of Chapter 4, the vertical displacement, and the horizontal kinetic energy density spectra are

$$\Phi_{VD}(\sigma, n) = b^2 \frac{N_0}{N(z)} \frac{\sigma^2 - f^2}{\sigma^2} E(\sigma, n), \tag{5.1a}$$

$$\Phi_{KE}(\sigma, n) = b^2 N_0 N(z) \frac{(\sigma^2 + f^2)}{\sigma^2} E(\sigma, n), \tag{5.1b}$$

where

$$E(\sigma, n) = \frac{2}{\pi} \frac{f}{\sigma} \frac{E_0}{(\sigma^2 - f^2)^{1/2}} \left[\frac{(n^2 + n_*^2)^{-1}}{\sum_1^\infty (n^2 + n_*^2)^{-1}} \right], \quad E_0 \approx 6.3 \times 10^{-5}, \quad n_* \approx 3, \tag{5.2}$$

and n is an integer index determining the vertical wavelengths (normal modes are not implied, because the uncorrelated upgoing and downgoing components do not have the phase locking

---

[1] Not to be confused with the later Gent-McWilliams eddy flux parameterization.

required to make a mode). Vertical displacements are about 7 m RMS, and the velocities about 7 cm/s RMS in the upper ocean (values from Munk, 1981). Other representations (e.g., Olbers, 1983; Levine, 2002) can be more convenient or more accurate.

This model is from a curve fit to results such as those in Figure 5.1 and is not based upon any physics other than the dispersion relationship, and the implied relationships between, for example, u, v and ρ' in Equations (3.30a–e). The qualitative applicability of this spectrum over most of the ocean has withstood four decades of observational tests, and while it is definitely inadequate in detail, its general form is surely qualitatively correct.

Many attempts have been made to derive and justify the GM spectrum, and the most plausible explanations of the general structure are based upon the nonlinear resonant interaction theories sketched in Chapter 4. These theories, sometimes described as "wave turbulence," have been used to show that the GM spectrum *may* be a near equilibrium shape. Because they generally rest upon wavenumber interactions, testing them in the field is very difficult. At the present time, it is perhaps fair to say that the *shape* of the frequency-wavenumber spectrum can be understood at least qualitatively from these ideas (Müller et al., 1986). But controversy lingers even over the basic assumptions of these theories (see Holloway, 1980, and Lvov et al., 2012, and p. 151 above).

5.3 SUBELEMENTS OF THE SPECTRUM

5.3.1 *Inertial Waves*

The large peak in energy in the current meter spectral density occurs at the lower end of the internal wave band near, but not usually exactly at, $\sigma = f$. The physics here is sufficiently different from that of the rest of the internal wave band that these inertial motions are usefully considered as distinctive.[2] Equation (5.2) has a factor $\left(\sigma^2 - f^2\right)^{-1/2}$, which is a place holder, roughly accounting for this excess. If a current meter record is run through a narrow-bandpass filter to isolate these motions, the result (Fig. 3.11) is seen to be a clockwise (in the northern hemisphere) narrow-band random process (that is, the amplitude and phase vary slowly but randomly over the course of the record). Separation into clockwise and counterclockwise motions can be achieved through data-processing methods adapted from optics (Mooers, 1973; Calman, 1978). In Figure 5.2, the dominant visual structure arises from the inertial wave propagation.

If a forcing term is put into the f-plane equations such as (4.65), an apparent resonant response appears at $\sigma = f$ from components having a clockwise component, a result consistent with the vanishing group velocity (Eq. 4.72). Energy accumulates because the group velocity cannot carry it away, eventually building up to near infinity. The resonance, however, is only apparent, because in this limit, $k \to 0$, and the assumption that f is taken to be constant fails. Theory extended to account for the variation of f with latitude and of small background shears produces an amplified, but finite, response near, but not at, $\sigma = f$.

At least three separate proposals have been made for the energy producing the inertial peak. (I will not continue to write "near-inertial peak.") The first is direct, local, wind generation at the sea surface. Any change in the wind field, either direction or magnitude or both, occurring over a time interval short compared to the inertial period, is particularly efficient at exciting inertial oscillations (See Leaman, 1976; D'Asaro, 1985, 1995). Theory suggests that the initial response would be in low modes/long vertical wavelengths.

---

[2] See Garrett, 2001.

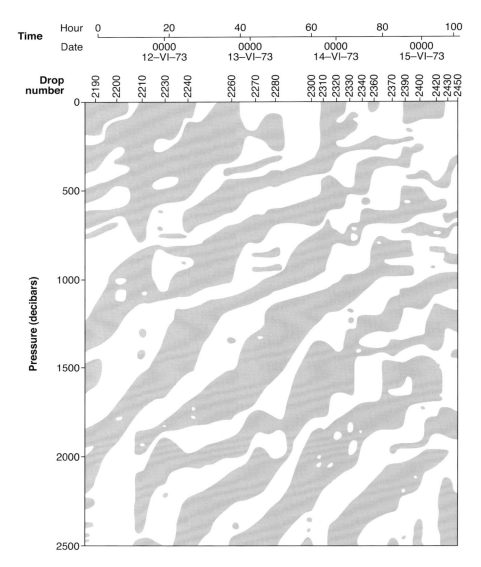

Figure 5.2: A time-depth diagram of the east component of velocity as seen by a vertical profiling instrument. Filled areas mark negative velocities. The upward-bound phase velocity is consistent with a *downward-bound* group velocity and hence energy input at the sea surface (recall Eq. 4.72). (After Leaman, 1976.)

The near resonance-like behavior has the consequence that any impulsive change in a flow, whatever the cause, tends to preferentially excite near-inertial motions. An example is in the Rossby adjustment problem—where the transient motions resulting from an initial-value problem have a strong inertial component (Fig. 4.19).

A second mechanism relies upon the wind field, but only indirectly. Through a process of frontogenesis extracting energy from the winds, the near-surface fronts become unstable, radiating inertial waves as part of the instability process. A few observations have only recently become available (Alford et al., 2013), and it is premature to say anything about the importance of this mechanism. Ultimately, the combination of fronts, submesocale motions generally, and mixed-layer dynamics must all interact to both produce and, perhaps, absorb radiated inertial waves.

The third mechanism relies upon the ambient background internal wave field. Fu (1981) provided a review of inertial wave theory and the data to that time for the open-ocean problem including, for example, the loss of coherence and the energy levels in the near-inertial band. Among his conclusions are that the inertial peak is less universal than is the remainder of the spectrum, but that over flat bottom topography and below about the top 100 m, a nearly universal character exists. Consider any latitude, ϕ_0, with inertial frequency $\sigma_{f_0} = f(\phi_0)$. Then equatorward of that latitude, motions are described by the GM spectrum, with approximately half the energy flux directed poleward. Internal wave energy at σ_{f_0} is freely propagating poleward until it encounters latitude ϕ_0—where it can progress no further, undergoing a dynamical reflection manifested as a wave caustic (see Eq. 4.57).

The theory involves the Airy functions, as local approximations to the equations on a sphere or β-plane near $\sigma = f$. Fu (1981) showed that this theory predicts, among other features, that the inertial peak should be centered at a slightly higher frequency than the local f (a "blue-shift") and that the bandwidth (giving an apparent Q; see the box "Quality Factor: Q") of the peak could be realistically calculated. (See Fig. 5.3.)

Yet further possibilities exist, including transient flow interaction with topography and failures of balance within the geostrophic eddy field (discussed later). Probably all of these generation mechanisms are operating somewhere some of the time, but no global summary statement appears possible now.

The existence of inertial motions depends directly upon the background planetary vorticity, represented by f, and modified by its nonzero gradient, β. Unsurprisingly, relative vorticity, ζ, owing to intense background flows can modify the apparent value of f, qualitatively changing the properties of the generation and, particularly, the propagation of inertial waves. (See Kunze and Boss, 1998, or Chavanne et al., 2012.)

On an f-plane, $w(\sigma = f) = 0$, and indeed spectra of temperature (e.g., Fig. A.10) show only a small excess over the background values. Even in the β-plane or spherical geometry, the relationship between the horizontal velocity components and the vertical velocity, and hence the induced temperature variations, is still predicted to produce a weak temperature signal at the inertial frequency. In practice, the observed excess energy is likely partially a consequence of either or both the advective effects of strong lateral currents acting on weak horizontal temperature gradients, or the vertical movement of the mooring under the induced drag producing spurious apparent energy.

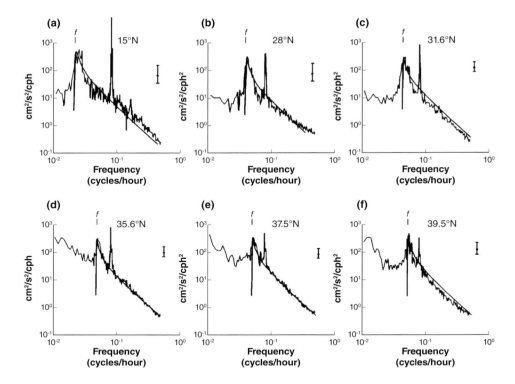

Figure 5.3: Various estimated kinetic energy spectra from 4000 m deep current meters and Fu's calculation of the implied Garrett and Munk (1971) spectrum at the inertial-frequency turning latitude. The M_2 tide peak is prominent. (From Fu, 1981.)

5.3.2 Internal Tides

Many, if not most, spectra of velocity or temperature in the ocean display an excess of energy in the band of frequencies surrounding the principle tidal lines (Figs. 3.10 and 5.3). A major contribution to these local peaks is from internal waves of tidal period (Section 7.6). The broadband, $f \leq \sigma \leq N$, internal wave component usually, however, contains far more energy than is found in tidal period temperature fluctuations. Velocity can contain a significant, even dominant, component from the ordinary (barotropic) tide. As with inertial motions, spurious temperature signals at tidal periods can arise from vertical motions of the instruments induced by drag forces on the moorings at tidal periods.

5.3.3 Vortical Modes

Small-scale, so-called, vortical modes, not obeying internal wave dynamics but being in cyclostrophic balance (meaning that pressure forces are balanced by local centrifugal rather than Coriolis forces), have been postulated by Müller et al. (1988) as existing in data. D'Asaro

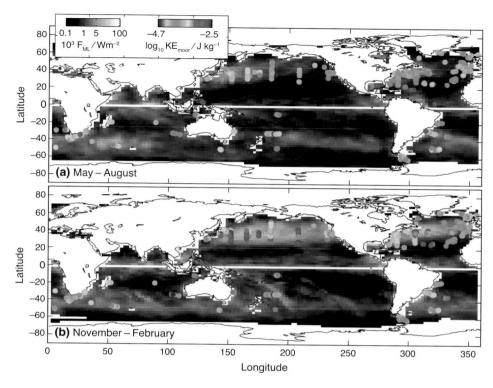

Figure 5.4: Estimated depth-averaged inertial frequency kinetic energy on moorings (colored dots), and the mean energy flux from the wind to mixed-layer inertial motions, F_{ML} (grayscale) from a model using reanalysis winds, for (a) May–August and (b) November–February. (From Alford and Whitmont, 2007.)

and Morehead (1991) described the appearance of such motions as finestructure in the Arctic Sea. These motions, with a very different time scale, are best described as balanced motions, rather than as gravity waves, and appear as a noise component in internal wave data.

5.3.4 Deviations from the GM Spectrum and the Energy-Source Problem

WHERE DO INTERNAL WAVES COME FROM?

Attempts have been made to calculate global energy distributions and rates of energy input to inertial motions. An example is from Alford and Whitmont (2007; see Fig. 5.4), who used data from moored current meters. Perhaps the most significant of their results is that at midlatitudes, the energy input is enough to replenish the entire internal wave band about every 10 d (longer at low latitudes) and, as expected, a strong seasonal cycle exists.

Munk (1981), Thorpe (2005), and others have more generally discussed the energy sources that could sustain a temporally constant internal wave field. The discussions are bound up in the question of how rapidly, and how, the field dissipates. If dissipation is strong, the field would rapidly diminish spatially and temporally as sources became more distant in space and time. Thus from the observed approximate uniformity of energy, a plausible inference is that sources must themselves be largely space and time independent. On the other hand, if dissipation is slight, the internal wave field can propagate long distances away from the initiators, and the near-constant energy level would be a consequence not of the local source levels, but of internal, nonlinear equilibration among the triad resonances. No generally accepted description yet exists and both processes are probably important.

Sources of internal wave motions include

1. the wind field, both directly and indirectly near the surface through, for example, turbulence in the mixed-layer disturbing its lower boundary, including the strong inertial-wave inputs;

2. the tides;

3. the breakdown of the geostrophic (balanced) motions in the eddy fields, particularly over topography (Molemaker et al., 2005; Nikurashin and Ferrari, 2010), where their physics overlaps the generation of so-called lee waves.[3] This mechanism implies a coupling of balanced and ageostrophic motions.

No particular reason exists to believe that a single dominant energy source exists, although in particular areas at particular times, one might. Some hints about the different sources can be obtained from observed deviations from the standard GM spectrum.

VARIATIONS FROM THE STANDARD MODEL

The GM spectrum is a remarkably good semiquantitative estimate of internal waves over the great bulk of the ocean, but important deviations are observed. Because $\partial \rho / \partial z$ is affected by the waves themselves, as well as by other phenomena present, and because a derivative is involved, the measured $N(z)$ appearing in Equations (5.1a–b) is itself commonly uncertain to within about a factor of two. Its variations likely account for some of the observed temporal variability in energy levels.

Perhaps most important, however, is the basic inference that a universal, horizontally and vertically isotropic spectrum of propagation is physically impossible—no provision exists for maintaining the spectrum against dissipation. Deviations from the universal form *must* occur, and large energy-level differences have been detected near topography, within canyons, very near the surface and bottom, etc. Convincing, durable, directional deviations from horizontal isotropy in the open ocean have not been observed except in the components identified with the internal tides. Vertical propagation, particularly in the inertial wave band is widely apparent (Fig. 5.2). Blumenthal and Briscoe (1995) showed that decoupling of the upward- and downward-bound motions was only an approximation—significant indications of vertical modes could be detected in their records—implying that the motions were "aware" of both the top and bottom boundary conditions, and presumably the possible energy sources there.

Levine et al. (1985) compared spectral-energy levels in a variety of locations. A major problem with the results was, as in many oceanographic measurements, the duration of the time series, which varied in length from one(!) day to several months. Proximity to boundaries causes major changes in energy, both increases, where the topography is focussing, as in

---

[3] Because the wind field is the primary energy source for the "mean" flow and the thus of the eddies that arise from it, the wind is ultimately the source of internal waves generated through geostrophic breakdown. This somewhat pedantic distinction will not be insisted upon. Apart from geothermal effects and lunar tides, the Sun is the ultimate source of oceanic motions.

canyons, and reductions, from reflections generating velocity zeros near wall-like features. A main result was that the energy levels appear lower generally in the ice-covered Arctic, and ice cover is believed to diminish the effects of wind stress (see also Rainville and Woodgate, 2009). Tides are also weaker in the, mostly isolated, Arctic Sea. Surprisingly, no definitive measurements seem to have been reported from the deep Mediterranean Sea, which has "normal" winds but weak tides (e.g., Perkins, 1972; van Haren and Millot, 2004). Forty years after the first discussion of the GM spectrum, understanding of its origins remains incomplete.

Velocity anisotropy is conspicuous near topographic features at the low end of the internal wave band, where flows necessarily become parallel to the obstructions. The canyons incised into continental margins and with intense, focussed internal waves (and often internal tides), have a specialized literature of their own. Monterey Canyon on the U.S. west coast has been the subject of particularly elaborate studies of the phenomenon (e.g., Kunze et al., 2002; Zhao et al., 2012). More recently, observations showing relative internal wave amplification in the canyons of the mid-ocean ridges have become available (e.g., Thurnherr et al., 2005).

5.3.5 Instabilities and Breaking of Internal Waves

OPEN OCEAN

Whitecapping and breaking of surface waves are familiar phenomena. Analogous, but more complex, mechanisms govern the breakdown of the internal wave field in the open ocean. This breakdown is of intense interest for at least two reasons: (1) if the resulting energy removal rate were known, it becomes a check on calculations of the rates with which energy must be supplied; (2) water parcels overturning within the water column can stir the fluid, enhancing mixing, and thus acting to reduce its stratification. This latter process—breaking of internal waves—is today believed to be the major way in which the ocean is stirred and then mixed.

A concrete indication of the action of the process goes back to the now-classical in situ photographic observations by Woods (1968), shown in Figure 5.5, depicting overturning of the water column by internal wave breakdown in the Mediterranean Sea. Two dominant instability mechanisms are believed to be acting on the open-ocean internal wave field: an overturning instability, somewhat like that occurring when an ordinary surface wave breaks on a beach by toppling over on itself, and a version of the Kelvin-Helmholtz instability deriving its kinetic energy from the shear in the internal waves (and thought to be dominant in Fig. 5.5). Application of these ideas to internal waves must account for the fact that the waves are a stochastic continuum in frequency and wavenumber. Calculations of stability parameters become essentially ones of probability and statistics. The situation can be different on the continental margins, where strong unidirectional internal tides may well dominate (e.g., Moum et al., 2007).

Extended discussions exist concerning the functioning of the Kelvin-Helmholtz instability (Turner, 1973; Thorpe, 2005) and, more or less, all reduce to showing a dependence upon the nondimensional *gradient Richardson number*,

$$Ri = \frac{N(z)}{dU(z)/dz}.$$

When this number is sufficiently *small*, the tendency of the stratification, $N(z)$, to stabilize the fluid, can be overcome by the kinetic energy contained in the flow, U, where U is an RMS value. Depending upon the details of the assumptions, the necessary and sufficient conditions

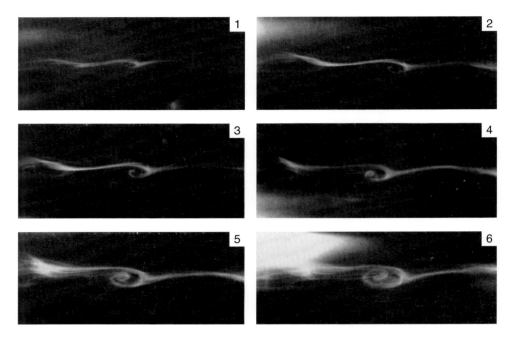

Figure 5.5: Demonstration of an apparently unstable sheared interface at about 40 m deep in the ocean (Mediterranean). The physics are thought to reflect that of Kelvin-Helmholtz instability. Wavelengths vary, but are on the order of 1 m. (Used with permission from J. Woods, [1968]. "Wave-induced shear instability in the summer thermocline." *J. Fluid Mech.* 32: 791–800.)

for instability require the critical values $Ri = 1$, and $Ri = 1/4$. The inertial peak dominates the calculation of shear, and its intermittency is a major element in controlling Ri. Because the inertial peak broadens and breaks up into multiple structures as the equator is approached, a loss of shear, and hence greater stability, is expected and observed (Gregg et al., 2003).

Eriksen (1978) measured the stratification and shear on vertical scales of centimeters in the main thermocline near Bermuda and found that the calculated Richardson number almost never became less than 1 (Fig. 5.6)—which result was interpreted to mean that the wave field breaks whenever that floor was penetrated (but see Gregg et al., 1993). Such measurements are indicative, rather than definitive, because, among other problems, the vertical scales over which N and dU/dz should be calculated are not predicted by the simplest theory.

Internal wave motions along slopes and along the sloping floors of the continental-margin canyons are unstable in several different ways, some of which have been tested in the laboratory (e.g., Cacchione and Wunsch, 1974; Ivey et al., 2000; see Figs. 4.11 and 7.13) and in lakes and in the ocean (Petruncio et al., 1998; Moum et al., 2002).

As noted already, the overturning, or *convective*, instability is roughly analogous to what one sees with toppling breakers in the surf zone—the waves becoming too steep to sustain themselves. Thorpe (2010) discussed the analogies between surface- and internal-wave breaking and concluded that shear instabilities, dominated by the near-inertial periods are likely the more important. He did emphasize how few relevant measurements exist.

Some instantaneous profiles of measured density appear to be unstable on small scales. If the data are reordered to stabilize the profile and the RMS of the required displacements are

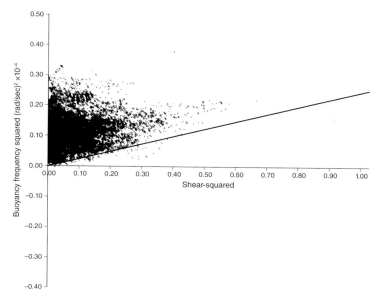

Figure 5.6: Measurements by Eriksen (1978) of shear versus stratification, and the near absence of values below the corresponding $Ri = 1/4$, denoted by the straight line. Shear was measured over 6 m and the stratification over 7 m during a 78 h period.

computed, one has the definition of the *Thorpe scale*. With enough assumptions, that in turn can be connected to the rate of mixing in the ocean (e.g., Smyth et al., 2001; Stansfield et al., 2001; Lozovatsky et al., 2008). Because these estimates have essentially all been made from instruments launched or lowered from ships, the existing sampling is spatially and temporally very thin, and no basin or global-scale inferences about averages are possible. As the data base of autonomous profiles with adequate vertical resolution (e.g., Argo) grows, large-scale estimates should become possible.

BREAKDOWN ON SLOPING TOPOGRAPHY

Theory and laboratory results, such as depicted in Figure 4.11, show the very strong influence of sloping topography on internal wave instability, leading to the expectation of comparatively violent interaction with continental margins and other topographic features (Fig. 3.5). These expectations are met in a very wide range of circumstances from open-ocean continental slopes (e.g., Nash et al., 2004) to the canyon-like features, which focus energy.

INTERNAL WAVE STOKES DRIFT

In a form analogous to the phenomenon in surface gravity waves, the vertical inhomogeneity of internal wave normal modes gives rise to a Stokes drift (e.g., Fig. 5.7). The time-mean Eulerian flow vanishes to a high approximation so that $\langle u_L \rangle = \langle u_S \rangle$. Consider, however, that somewhere in the direction of propagation, the isopycnals must intersect a solid boundary, vertical or sloping. Because the particles cannot penetrate the boundary nor can they, in the absence of diffusion, leave the isopycnals some phenomenon must reduce the mean Lagrangian velocity (the mean Eulerian plus the Stokes drift; Eq. 4.25) to zero. Theory says

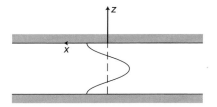

Figure 5.7: The Stokes drift in a first-mode internal wave travelling from left to right. Top surface is rigid. In the presence of a stochastic internal wave field, the Stokes drift also becomes stochastic and its time and space structure are random fields (see Dewar, 1980). (From Wunsch, 1971.)

that a time-mean Eulerian flow, equal and opposite to the Stokes drift, is established by the nonlinear terms generated by the linear wave train, and it exactly cancels the particle velocity at the solid boundary. Such effects, although probably real, have not been seen in the ocean because the boundary regions are so noisy. For internal waves obliquely incident on a sloping boundary, significant long-shore flows can in principle be established, ones that would be influenced by rotation (Hogg, 1971; Thorpe, 2000). Again, however, observations are lacking because so many competing phenomena exist in coastal regions. Of course, if the waves break, as they would in a wedge-shaped boundary or canyon, then the Lagrangian flow can be finite.

As in the discussion of surface gravity waves, the existence of rotation introduces complicating factors: a "time-mean" Stokes and Lagrangian drift persisting for a significant fraction of $1/f$ in an ocean of infinite lateral extent feels Coriolis forces, and these must be supported in a steady state (Eqs. 3.20a–d). Absent walls, no pressure gradient can be created, and the existence of second-order mean Eulerian effects, $\langle u_E \rangle$, such that $\langle u_L \rangle = \langle u_E \rangle + \langle u_S \rangle = 0$, can be inferred. The consequences of this effect for internal inertial waves were established by Hasselmann (1970; Cf. Wunsch, 1971), and must exist in nature. Observed persistent long trains of low-mode internal tides would be prime candidates for generating these second-order flows, but since they have never been observed in the noisy ocean, the subject is not pursued here.

STRESSES AND MOMENTUM

Perturbation disturbances to a resting fluid generate average second-order product quantities in the Eulerian representation such as $\langle u'v' \rangle$. The spatial dependencies in these products exert forces on the fluid that contain time dependencies at much lower frequencies, usually interpreted as the time-mean flow. In a wave context, these are often called *radiation stresses* and include those that can bring a mean Lagrangian flow to zero, as above, if required. In a rotating stratified fluid, various mathematical manipulations prove helpful in the interpretation of the net effects on the mean. Thus a field of deterministic or random internal waves will drive flows manifested at larger scales and lower frequencies. A special vocabulary surrounds the general study of second-order effects in such fluids, including the notions of *pseudo-momentum*, *Eliassen-Palm fluxes*, etc. The reader is referred to McIntyre, 1981, and Bühler, 2009.

In the ocean, observations of such effects arising from internal waves are very difficult because of the need to suppress the more energetic wave and other motions so as to perceive the second-order effects. A possible example is that of Pinkel et al. (2012), who make plausible the detection of internal tide-induced stresses.

CHAPTER 6

The Tide Disturbing Potential and the Milankovitch Forcing

6.1 ORIGIN

Measurements in the open ocean of changes in sea surface height, η, are dominated by the tides, which contribute on the order of 1 m in the root mean square (RMS). The residual variations, representing all other ocean variability in elevation, are on the order of 0.1 m RMS. Unlike the other forces acting on the ocean, which are primarily meteorological in origin, the structure and nature of the tidal forcing is known almost exactly, thus removing a major source of uncertainty in understanding the oceanic response to specific forces. Because the central discussion of the oceanic response to tidal forcing relies most directly upon the comparatively simple long wave theories, and because of the intense interest in the theory of tidal response that began with Isaac Newton, study of the tides was a central theme of nineteenth century physicists and mathematicians. Although the problems of open-ocean tide measurements were never overcome, and technical difficulties were encountered in obtaining analytical solutions to the equations of motion in bounded basins, the fundamental physics of the problem nonetheless was well understood by the early twentieth century. Tidal prediction in most places was so routinely accurate that it became scientifically boring. Few, if any, measurements were available except in coastal ports and at a handful of open-ocean islands. Consequently, the study of ocean tides became peripheral to physical oceanography as a whole, generally fading from the oceanographer's consciousness. It appears only in passing mention in many textbooks.

This situation changed abruptly about 1992, with the arrival of data from the TOPEX/POSEIDON altimeter mission,[1] which produced high-accuracy measurements almost everywhere of the tidal contribution to η. Initially, the main issue was the removal of strong tidal signals, which otherwise obscured all other physics, but the construction of globally consistent

---

[1] Predecessor missions, such as GEOSAT and SEASAT, had much larger error budgets and some unfortunate tidal aliases.

dynamical tidal models revived interest in the underlying mechanisms, and specifically, in the understanding of tidal dissipation. The real excitement was generated, however, by the realization that tidal-period internal waves were globally visible in the altimetric data. These observations opened a new window on the physics connecting the tides to the unending quest to understand the mechanisms of ocean mixing as well as such interesting problems as the evolution of the lunar orbit.[2] Because of this recent revival of interest in tides, this chapter sketches the origin of the tide-producing forces and the nature of the oceanic response. Cartwright's book (1999) is an interesting and useful overview of tidal studies from antiquity to the recent past.

Another, perhaps unexpected, reason exists for taking the trouble to outline the origin and structure of tidal forcing. One of the central elements of all discussions of low-frequency climate change is the hypothesis that changes in the Earth-Sun orbital configuration, which influences the solar radiation incident on the Earth, have important consequences. This hypothesis has become known as the Milankovitch theory, and its history is described by Imbrie and Imbrie (1986). Tidal and Milankovitch forcings are both the result of configuration changes among the Earth, Moon, Sun, and planets and of observers on the spinning Earth, and they are very closely related but not identical. Discussing them together is helpful, even though effects on the time scale of climate are not the subject of this book. Tides are a mature and comparatively well-understood phenomenon that have been studied since antiquity. In contrast, serious interest in insolation forcing arose only in the past few decades. The number of useful tidal records is orders of magnitude larger than those representing the insolation response. On the other hand, Milankovitch forcing is, in some ways, simpler than the tidal one, because it involves only the relative positions of the Sun and Earth, whereas the tides also involve the Moon—whose orbital motion is extremely complicated.

6.1.1 Solar Gravitational Tides

The framework for discussing both tides and solar radiation is the movement of a nearly spherical Earth in a near-elliptical orbit about the Sun. It is often simpler, and entirely accurate, to take a Ptolemaic view and regard the Sun as in orbit about the Earth, and that vocabulary is sometimes used here. This discussion begins with solar tides, assuming the Moon to be absent. Straightforward derivations of the tidal potential are given by Lamb (1932) and in the appendix to Munk and Cartwright's paper (1966). The fundamental positional astronomy is described by Smart (1962), Green (1985), and Seidelmann (2013). A notationally complete and consistent derivation is given by Lambeck (1980). Astronomers use many different and confusing coordinate systems, time measurements, and notations (see Bowditch, 2002).

Consider the geometry of Figure 6.1 showing an observer at point P with longitude λ, colatitude θ, and distance r from the center of the Earth (most often the Earth's radius, $r = a$). The colatitude is being used rather than the latitude, $\phi = \pi/2 - \theta$, because it is conventional for these problems. The Sun is at S, at a distance R_S from the center of the Earth and a distance ρ_S (not to be confused with density) from the observer. The solar mass is M_S. For the observer at P, the disturbing *gravitational potential* of the Sun is

$$V_S = \frac{GM_S}{\rho_S},$$

(6.1)

[2] Tidal forces are also very important in the modern understanding of other planets and their satellites. See Murray and Dermott, 1999.

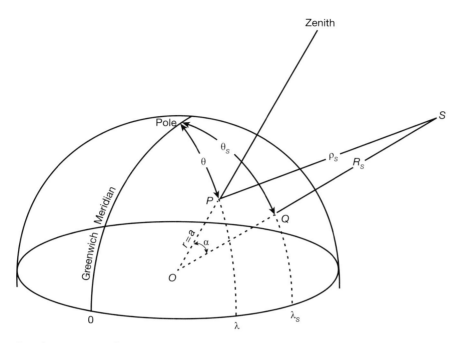

Figure 6.1: The geometry of the subsolar point Q relative to that of an observer at P. Only the angle α relative to the center of the earth, and the distance ρ_S are required to calculate the gravitational forcing or insolation at P. Here S is the solar position. (After Munk and Cartwright, 1966.)

where G is the gravitational constant. Let α be the angle, shown between the observer's vertical (the *zenith*) and the line connecting the solar position to the center of the Earth. Then,

$$V_S = \frac{GM_S}{\rho_S} = \frac{GM_S}{R_S \left(1 - 2\left(r/R_S\right) \cos \alpha + \left(r/R_S\right)^2\right)^{1/2}}. \tag{6.2}$$

The ratio, r/R_S, $r \le a$, is the *solar parallax* and is a very small number (the angle subtended by the Earth's disk of radius r as seen from the center of the Sun). Expanding Equation (6.2) in the parallax,

$$V_S = \frac{GM_S}{R_S} \sum_{n=0}^{\infty} \left(\frac{r}{R_S}\right)^n P_n \left(\cos \alpha\right) = \frac{GM_S}{R_S} \left(1 + \frac{r}{R_S} \cos \alpha + \left(\frac{r}{R_S}\right)^2 \left[\frac{3}{2} \cos^2 \alpha - \frac{1}{2}\right] + \ldots\right), \tag{6.3}$$

where the $P_n \left(\cos \alpha\right)$ are the *Legendre polynomials* listed in the box "Spherical Harmonics." Values for the astronomical constants can be found in Appendix D.

The gradient operator in a spherical polar coordinate system $(r, \alpha, \gamma$, with the coordinate pole at the subsolar point, and γ the longitude-like coordinate about the pole) is

$$\nabla = \mathbf{e}_r \frac{\partial}{\partial r} + \mathbf{e}_\alpha \frac{1}{r} \frac{\partial}{\partial \alpha} + \mathbf{e}_\gamma \frac{1}{r \sin \alpha} \frac{\partial}{\partial \gamma},$$

where the $\mathbf{e}_p$ are unit vectors pointing in the direction of increase of each of the three coordinates. The first term in Equation (6.3) is constant—and so produces no force; it can

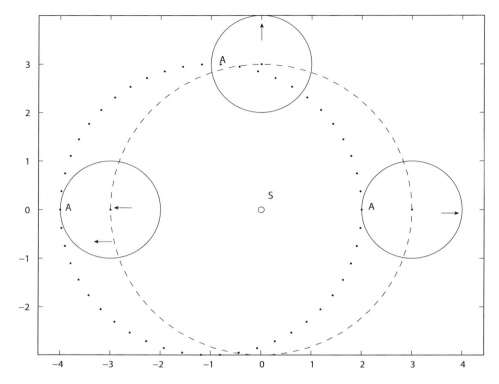

Figure 6.2: A schematic (after George Darwin) of orbital geometry. For a nonspinning earth in orbit about the sun at S, all points in and on the earth, for example, A, move in circles of a common diameter, albeit about different centers. Thus all points on the earth experience an identical centrifugal force (arrows). The forces owing to the spin motion—once per day—are simply superimposed.

be safely dropped from V_S. No γ dependence appears in V_S, and the gradient of the second term produces

$$\frac{GM_S}{R_S} \nabla \left(\frac{r}{R_S} \cos \alpha \right) = \frac{GM_S}{R_S} \left(\cos \alpha, \ -\sin \alpha, \ 0 \right), \tag{6.4}$$

which has a constant magnitude, GM_S/R_S, directed toward the Sun. When multiplied by the Earth's density, $\rho(\alpha, \gamma, r)$, and integrated over the sphere ($0 \le r \le a, 0 \le \alpha \le \pi, 0 \le \gamma \le 2\pi$), the magnitude is found to be $GM_E M_S/R_S^2$, where M_E is the Earth mass—and it is the net attractive force between the Sun and Earth. The Earth does not fall into the Sun, and this force is necessarily balanced by the centrifugal motion of the Earth about the Sun. Figure 6.2 shows that the centrifugal force acting at any point on or in the Earth is a constant, and its integral over the planet is equal and opposite to the integral of Equation (6.4). (For this purpose, think of the Earth as not spinning.)

The magnitude of Earth's gravity at the surface (with a small correction for the centrifugal force of *rotation*) is

$$g = \frac{GM_E}{a^2}, \tag{6.5}$$

where M_E is the Earth's mass, and a is its radius (in spherical approximation). So, $G = a^2 g / M_E$, and

$$\frac{V_S(r, \alpha)}{g} = \frac{a^2}{R_S} \frac{M_S}{M_E} \left[\left(\frac{r}{R_S} \right)^2 \left(\frac{3}{2} \cos^2 \alpha - \frac{1}{2} \right) + \ldots \right]. \tag{6.6}$$

Setting $r = a$,

$$\frac{V_S(r, a)}{g} = a \frac{M_S}{M_E} \left[\left(\frac{a}{R_S} \right)^3 \left(\frac{3}{2} \cos^2 \alpha - \frac{1}{2} \right) + \left(\frac{a}{R_S} \right)^4 P_3 (\cos \alpha) + \ldots \right], \tag{6.7}$$

at the Earth's surface. Surfaces $V_S(r, \alpha)/g =$ constant are rotationally symmetric ellipsoidal bulges pointing at the subsolar and antipodal points, where $\alpha = 0, \pi$. All forces derived from the various terms in the potential Equation (6.7) vanish when integrated over the Earth's surface. A physical interpretation is that the gravitational attraction of the Sun is larger than the outward centrifugal force on the side of the Earth nearest the Sun with a net attraction toward the Sun. On the far side, the gravitational force is weaker than the centrifugal one, and the apparent gravity is directed away from the Sun. An observer on the spinning Earth moves in and out of a bulge in the disturbing potential twice per day, and unless the Sun is over the equator or the observer is there, one potential bulge will appear larger than the other (Fig. 6.3). A simple way to represent a twice-daily sinusoidal variation with one sinusoid bigger than the other is to write it as the sum of a twice-per-day (semidiurnal) sinusoid and a once-per-day (diurnal) sinusoid, for example, $\sin(2\pi t / T_1) + b \sin(\pi t / T_1 - \phi)$. This geometric effect is the origin of the dominant semidiurnal and diurnal component tides. Because the Sun is over the equator twice per year (at the vernal and autumnal equinoxes), the diurnal component, and hence its gradient, disappears at those times.

The coefficient of the leading term, $H_S = a M_S / M_E (a/R_S)^3 \approx 0.16$ m from the values in Appendix D. The magnitude of the tide-disturbing effect is thus proportional to the ratio of the solar mass to that of the Earth, and to the cube of the solar parallax, a/R_S. The coefficient of the next term, P_3, is smaller by a factor of a/R_S.

Equation (6.7) is still written in a coordinate system based upon the subsolar point, and it is needed in ordinary latitude and longitude. Consider Figure 6.4, which shows the position of the subsolar point in Greenwich coordinates. θ_S and λ_S are the colatitude and longitude of the instantaneous subsolar point. λ_S is measured not from the Greenwich meridian, but from the point (direction) Υ, which is defined below. From the spherical trigonometry in Figure 6.1,

$$\cos \alpha = \cos \theta_S \cos \theta + \sin \theta_S \sin \theta \cos (\lambda_S - \lambda), \tag{6.8}$$

Substituting (6.8) into (6.7) and doing a bit of manipulation,

$$\frac{V_S}{g} = \underbrace{\frac{3}{2} H_S \left(\cos^2 \theta_S - \frac{1}{3} \right) \left(\cos^2 \theta - \frac{1}{3} \right)}_{1} + \underbrace{\frac{1}{2} H_S \sin(2\theta_S) \sin(2\theta) \cos(\lambda_S - \lambda)}_{2} \tag{6.9}$$

$$+ \underbrace{\frac{1}{2} H_S \sin^2 \theta_S \sin^2 \theta \cos 2(\lambda_S - \lambda)}_{3} + \underbrace{O\left((a/R_S)^4 \right)}_{4},$$

$$H_S = \frac{M_S}{M_E} \left(\frac{a}{R_S} \right)^3 \approx 0.16 \text{ m}$$

Figure A.16 displays the basic spatial structure of the three leading terms. Equation (6.9) is now in a coordinate system useful to an observer at any location on the Earth's surface—if

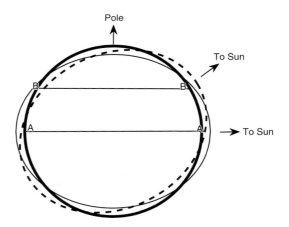

Figure 6.3: Geometry of the net effect of the orbital centrifugal force and the gravity disturbance of the Sun. The thick solid line is the Earth's surface. With the Sun over the equator, as occurs at the two equinoxes, an observer on the rotating Earth at latitude B sees two equal tidal potential maxima per day (thin solid line). With the Sun at a nonzero latitude (dashed line), the observer at B sees that one maximum is higher than the other, giving rise to a diurnal variation. In contrast, an observer on the equator at A never sees a diurnal component. For the Moon, a similar construct applies except that the Moon is over the equator twice per month.

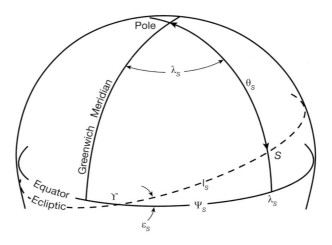

Figure 6.4: The latitude and longitude of the subsolar point in Greenwich coordinates (λ_s, θ_s). Angles l_s, ψ_s are measured from the vernal equinox Υ. (Adapted from Munk and Cartwright, 1966.)

the subsolar point λ_S, θ_S is known as a function of time. Variations in the spatial gradients of Equation (6.9) describe the time-varying forces an observer on the spinning Earth, under a moving Sun, would see. The dominant near-24-hour variation arises because the observer is spinning, and the other temporal variations occur because the Sun is moving in its orbit about the Earth (in the Ptolemaic view).

Equation (6.9) provides the basic structure of the most important tidal lines (some higher-order terms, represented by the ellipsis points, are usually carried along in standard representations, but they are significantly weaker). Term 1 has no dependence upon λ_S. Thus the time dependence in term 1 is controlled only by yearly changes in the solar colatitude, θ_S, and in the solar radius, R_S (contained in H). Because $\cos^2 \theta_S = (1/2) (1 + \cos 2\theta_S)$, and θ_S varies between $\pm 23.5°$ annually, the trigonometric term has a dominant 6-month periodicity. With the Sun moving in an ellipse, R_S will also vary with an annual periodicity; its reciprocal will have an annual period plus all of its harmonics, as is readily seen by expanding

$$1/R_S = \left(\bar{R}_S^2 + \varepsilon^2\right)^{-1/2} = \bar{R}_S \left(1 + \varepsilon^2/\bar{R}_S^2\right)^{-1/2} = \bar{R}_S \left(1 - 1/2 \left(\varepsilon^2/\bar{R}_S^2\right) + 1/4 \left(\varepsilon^2/\bar{R}_S^2\right)^2 + \ldots\right),$$

where $\bar{R}_S$ is the mean radius, and ε is the small deviations from it. The expansion varies with a period of a year and all of its harmonics, producing in term 1 of Equation (6.9) the solar *long period tides*.

Term 2 in Equation (6.9) is proportional to $\cos(\lambda_S - \lambda)$. Because λ_S changes by $360°$ in one day, its dominant periodicity will be 24 hours, representing the solar *diurnal tides*. As anticipated, it vanishes for $\theta = \pi/2$; that is, for an observer on the equator, no *local* diurnal forcing exists. The diurnal term is also modulated (split) by changes in $\sin(2\theta_S)$ and by any variation in R_S throughout the year. Term 3 is dominated by the time change in $\cos 2(\lambda_S - \lambda)$ and will have a dominant near-12 hour period; this term represents the solar *semidiurnal tides*. Splitting arises from multiplication of two or more sinusoidal functions (see p. 385) and leads to the presence in V of changes at all combinations of the frequencies and of their sum and difference frequencies. A detailed calculation is required to find the magnitude and phase of the forcing. The individual terms are known as tidal *constituents*. Some terms of (6.9) are *time-independent*, producing a permanent tidal deformation of the Earth, which is important in geodesy.

Conventional Fourier analysis shows that two nearby frequencies in a record of length T can be distinguished only if their frequency difference, $\Delta s > 1/T$ (Rayleigh's criterion; assuming super-resolution is not attempted). Thus the solar diurnal constituent is split by the annual cycle into two constituents differing by one cycle in a year. Unless the length of a record exceeds a year, they will appear to be a single frequency, whose amplitude and phase will depend upon the particular time interval of measurement, rather than as a "doublet."

6.1.2 Solar Motion

It is useful to sketch the ways astronomers specify R_S, θ_S, λ_S. Much of what follows is no longer done in practice, as the advent of computers has removed the need for expansions and approximations that were essential in the days of hand computation. But many of those approximations were based upon physical insights that are helpful to recall, particularly in the context of the much less familiar Milankovitch forcing.

The basic geometric considerations are in Figures 6.5 and 6.6. Consider, by way of example, the geocentric picture, regarding the Sun as in motion in an elliptical orbit about the Earth. In

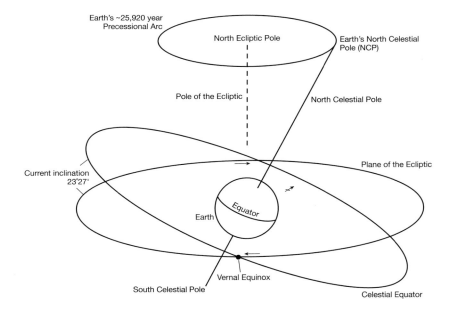

Figure 6.5: The geometry and notational definitions for discussion of the precession of the equinoxes. The rotation axis, which is perpendicular to the Earth equatorial plane, is tilted relative to the perpendicular to the plane of the ecliptic on which the Sun moves. The intersection of the equatorial and ecliptic planes defines a line that is called the vernal equinox, or the First point of Aries. Because (primarily) of the torque exerted by the Moon on the Earth's equatorial bulge, the vernal equinox precesses westward along the ecliptic with a period of about 26,000 years, too long a period to be of interest in modern tidal dynamics.

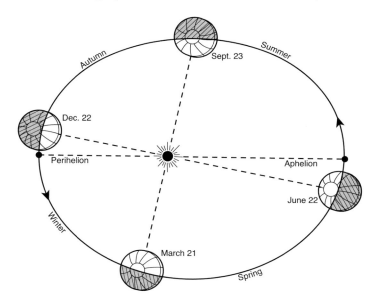

Figure 6.6: The seasonal configuration of the Earth's orbit. In the present epoch, the date of closest approach (perihelion) differs only slightly from the date when the noon sun is lowest in the sky (northern hemisphere winter solstice). In the past and future these dates differ by up to six months. (From Baker, 1955.)

the formulas above, the angle λ_S does not change uniformly with time, because the angular movement of a body in an ellipse is not uniform. Methods exist for generating more convenient variables (see box "The Kepler Equation").

Astronomical references tabulate the fundamental parameters as functions of time in Taylor series in terms of *Julian centuries*, T_J, of 36525 days. So, for example, the motion of the mean sun, represented by the mean anomaly (defined in the box) is

$$M_a = 358°28'33.00'' + 129596579.10''T_J - 0.54''T_J^2 + \ldots \tag{6.10}$$

relative to Υ on 1 January 1950 (Anonymous, 1961, p. 98). The mean anomaly is *not* exactly a uniformly increasing angle with time—but the correction is very small and can be ignored in ordinary tidal studies. When T_J becomes very large—as it does on Milankovitch time scales—the higher order terms are important.

THE KEPLER EQUATION

Consider Figure 6.7. The position of the Sun in the ellipse of eccentricity e, and semimajor axis a_S is given by its radius, R_S, and the angle v, called the *true anomaly*. Kepler's second law says that in an elliptical orbit, "equal areas are swept out in equal time," implying that the rate of change of v varies with time; its use is inconvenient. Instead, define the angle E, the eccentric anomaly measured from the origin of a circumscribing circle, which can then in turn be related to another angle, the *mean anomaly*,

$$M_a = \frac{2\pi}{1 \text{ tropical yr}} (t - \tau), \tag{6.11}$$

which has a uniform rate of change and which defines the concept of the mean Sun with a corresponding mean year and mean solar day (MSD) of 24.0000 mean solar hours. Then with some elementary trigonometry and ellipse geometry, Smart (1962, p. 111) showed that

$$\tan\frac{v}{2} = \frac{(1+e)^{1/2}}{(1-e)^{1/2}} \tan\left(\frac{E}{2}\right). \tag{6.12}$$

The radius vector is,

$$R_S = a_S (1 - e\cos E). \tag{6.13}$$

Some more elementary geometry and Kepler's law shows that the relationship between E and M is given by the *Kepler equation*,

$$E - e\sin E = M_a = \frac{2\pi}{1 \text{ tropical yr}} (t - \tau). \tag{6.14}$$

For small e (typically about 0.02 for the Sun), the Kepler equation can be solved iteratively for E in terms of t, or M_a. Then v is known from Equation (6.12), and R_S from (6.13). Ignoring all of the details, it is apparent that v and R_S will carry periodicities at 1 year, 6 months, and all of the higher harmonics of the year, and all of these frequencies will further split the basic tidal lines in Equation (6.9).

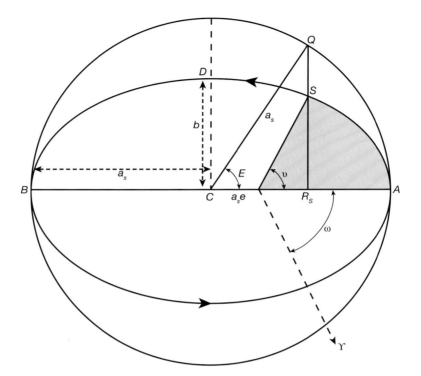

Figure 6.7: The geometry of the sun, at S, moving in an exaggerated ellipse about the earth at a focus. v is the true anomaly of the body (the Sun), and E, measured relative to the center of the ellipse, is the eccentric anomaly. ω defines the angle between the vernal equinox Υ and perihelion. a_S, b are the semimajor and semiminor axes of the ellipse, and a_S is also the radius of the bounding circle. v, E, M_a (the solar anomalies) are all measured from perigee. M_a is the angle of the position of the fictitious mean sun moving uniformly along the circle. (Adapted from Smart, 1962.)

The point Υ in Figure 6.7 lies along the line of intersection of the plane of the Earth's equator and the plane of the ecliptic (the two make an angle of about 23°). The Sun crosses the equator twice per year at the equinoxes with the *spring*, or *vernal, equinox* defined as the time when the Sun moves into the northern hemisphere, today around 21 March. The *line of nodes* points in space toward Υ, the *First Point of Aries*.[3] This intersection of two planes is used as a coordinate origin for defining the positions of various astronomical bodies within the planes. Because the Earth rotation pole precesses in space, with a period of about 26,000 years, the line of intersection moves westward (clockwise), completing a 360° rotation in that period of time. For conventional astronomical work over short intervals, a fixed line—for example, its position at noon on 1 January 1850 or 2000—called the *epoch* is used. For tidal work in the modern era, the choice of epoch is a detail; but in a climate theory, because it defines the onset of astronomical springtime, accounting for its moving position is essential. Over long enough periods of time, and if the year origin of 1 January is kept fixed, Υ will occur in all months, eventually running through the entire calendar.

[3] In ancient times, Υ lay in the direction of the constellation Aries.

For solar tides, the only remaining complication is the nonclosure of the orbital ellipse. The point of closest approach of the Earth to the Sun (perihelion), or perigee if the Sun is orbiting the Earth, moves counterclockwise in Figure 6.7, in the direction of the solar motion. In an absolute frame, perigee would take about 110,000 years to go through 360° (*perigee frequency of* $1/1.1 \times 10^5$ *years*). But the vernal equinox is moving clockwise, *toward* *perigee*, and thus relative to Υ the period of ellipse closure is only about 21,000 years. In the current configuration (Fig. 6.6) in which we live, the time of perihelion is near 3 January, coincidentally close to the time of winter solstice, now about 21 December, when the noon Sun reaches its lowest northern hemisphere latitude. These dates will separate as precession carries Υ clockwise around the orbit and perihelion moves counterclockwise toward it.

To obtain a complete description of the solar tide disturbing potential, $V_S/g, \theta_S, \lambda_S, R_S$ are expressed in terms of the mean anomaly, M_a. Everything is periodic in three frequencies—1 cycle/mean solar day, $1/T_1$; 1 cycle per year, $1/T_3$; and 1 cycle per 21,000 years, $1/T_6$—and all of their harmonics. The mean solar day is necessarily slightly longer than the Earth rotation period, $2\pi/\Omega$, because the orbital motion of the Sun is in the same direction as the Earth rotation, and the observer has to "catch-up" to it. The algebra of V_S shows that *all* of the frequencies present can be represented as integer multiples of the reciprocal of the mean solar day (12.000 hours); of the year defined, for example, as the interval of the passage of the Sun through the vernal equinox Υ, the so-called *tropical year*; and of the time required for the orbital ellipse to close. By inspection,

$$\frac{V_S}{g} = \sum_{n_1,n_3,n_6=-\infty}^{\infty} A\left(n_1, n_3, n_6, \theta, \lambda\right) \left\{ \begin{array}{l} \cos\left[2\pi\left(\frac{n_1}{T_1}t + \frac{n_3}{T_3}t + \frac{n_6}{T_6}t\right)\right] \\ \sin\left[2\pi\left(\frac{n_1}{T_1}t + \frac{n_3}{T_3}t + \frac{n_6}{T_6}t\right)\right] \end{array} \right\}. \tag{6.15}$$

Either the sine or cosine is chosen in a particular term depending upon whether $m + n$ in the corresponding spherical harmonic is even or odd. In practice, only small integers, n_i, are required for very high accuracy (never more than ±5). Skipped indices are reserved for the Moon. Terms with $n_1 = 0$ are termed long-period tides, and the form of their spatial structure is $A\left(0, n_3, n_6, \theta, \lambda\right) = B\left(0, n_3, n_6\right) P_2\left(\cos\theta\right)$, with no longitude dependence; $n_1 = 1$ terms are the diurnal tides, and $A\left(1, n_3, n_6, \theta, \lambda\right) = B\left(1, n_3, n_6\right) P_2^1\left(\cos\theta\right)\left(\cos\lambda, \sin\lambda\right)$ are the diurnal components; and for $n_1 = 2$, $A\left(2, n_3, n_6, \theta, \lambda\right) = B\left(2, n_3, n_6\right) P_2^2\left(\cos\theta\right)\left(\cos 2\lambda, \sin 2\lambda\right)$ are the semidiurnals. $P_l^m\left(\cos\theta\right)$ are the *associated Legendre polynomials* of order l and degree m (see box "Spherical Harmonies"). Other tides exist: $n_1 = 3$ are the ter-diurnals and are readily observed in the ocean, but are significantly smaller than the semidiurnals and diurnals.

Although the three fixed frequencies used to describe the solar tidal disturbance are a very accurate approximation for tens of thousands of years, on geological time scales, the disturbing potential is dependent upon the slow changes in the planetary configuration of the solar and Earth-Moon system, which significantly modify these frequencies. For example, the changing length of day becomes very important after hundreds of millions of years—it was probably less than about 20 h, 1 billion years in the past (Williams, 2000), but the changes are negligible in any instrumental record. (Modern length-of-day changes are measurable with appropriate instruments,[4] but such changes are far too small to affect tidal studies.)

---

[4] Munk and Macdonald, 1960; Lambeck,1980.

6.1.3 Lunar Tides

The underlying principle of the representation of lunar tides is identical: a local *instantaneous* elliptical approximation to the lunar orbit is used, and the lunar disturbing potential, V_L/g, is identical to (6.9), with the lunar mass M_L and orbital radius R_L replacing those for the Sun, and similarly for the various relative position angles, λ_L, θ_L. Amplitudes are proportional now to $H_L = M_L/M_E \, (a/R_L)^3 \approx 0.36$ m. Instantaneously, the Moon and Earth are orbiting each other just as the Sun and Earth do. The lunar mass is much less than the solar mass, but because the disturbing potentials depend upon the cube of the parallaxes, the close lunar distance "wins" ($H_L > H_S$). But the lunar orbital motion is much more complicated than that of the Sun, because the position of the Moon depends upon the relative location of the Sun, but the opposite is not true. Terms must be included representing the distortion of the lunar orbital ellipse as the relative positions of the Sun and Moon change (e.g., when aligned, the lunar radius increases slightly).[5] The lunar orbit is also tilted with respect to both the equator and the plane of the ecliptic, so that even more angles appear. The plane of the lunar orbit (or equivalently, the normal to its plane) precesses with a period of about $T_5 = 18.6$ years, and the perigee point rotates through $360°$ relative to Υ in a period of about $T_4 = 8.9$ years. With the addition of these three periods (1 cycle/month $= 1/T_2$, 1 cycle/8.9 years, and 1 cycle/18.6 years), an expression essentially identical to (6.15) is obtained,

$$\frac{V_L(\theta, \lambda, t)}{g} = \sum_{n_i = -\infty}^{\infty} A_L(n_1, n_2, n_3, n_4, n_5, n_6, \theta, \lambda) \begin{Bmatrix} \cos(2\pi s_{\mathbf{n}} t) \\ \sin(2\pi s_{\mathbf{n}} t) \end{Bmatrix}, \quad (6.16)$$

$$s_{\mathbf{n}} = \mathbf{s} \cdot \mathbf{n}, \quad \mathbf{s} = [s_1, s_2, \dots s_6], \mathbf{n} = [n_1, n_2, \dots n_6] \quad (6.17)$$

with the solar frequencies appearing because they influence the lunar position. Once again, long-period, near-diurnal, and near-semidiurnal groupings are present. The fundamental long period is a month, rather than a year, and periodicities from splitting by the additional frequencies, such as a cycle of 18.6 years, appear.

6.1.4 Combined Tides

Customarily, the lunar and solar tides are combined as

$$V_{tide}/g = V_L/g + V_S/g,$$

tabulating them generally in terms of the six numbers n_i. The arguments of the cosines and sines are determined by the 6-tuple, $\mathbf{n}$, called the Doodson number (5 was added to n_2 to n_6, in many tabulations, so as to produce positive numbers and save printing space (as in Table 6.2); context reminds the reader to remove it). Cartwright and Edden's (1973) is the most recent version. Because the lunar tides are stronger than the solar ones, compilers of the tidal expansions often use a *mean lunar day* for the frequency s_1 based upon a fictitious

---

[5] The orientation of the earth in space, and hence its rotation axis, does depend upon the lunar position, and hence the relative separation of the subsolar point and the observer has lunar terms in it. Thus for example, Earth's obliquity is forced at the 18.6-year lunar period that introduces a periodicity into the insolation function. The relative position problem of the Earth, Moon, and Sun constitutes a *full three-body problem*, which is still not an easy calculation, particularly when the other planets are also taken into account.

Table 6.1: The six frequencies and corresponding periods used in ordinary tidal work (periods less than 21,000 years) during the current epoch and using the lunar day for s_1 but with the frequencies in mean solar days. Equally valid use can be made either of the solar day or of any of the other definitions of a month and year. The time origin is generally measured from the vernal equinox at a fixed date. Perigee period is the interval for the ellipse to close on itself, and the lunar nodal period is that of the precession of the lunar orbital plane. Values are known to considerably greater accuracy than given here. This table can be extended to lower frequencies to encompass the Milankovitch periodicities below 1 cycle/solar perigee period. The tropical year and the mean solar day are defined in the text. (Adapted from Doodson, 1921.)

Label	Name	Frequency (°/mean solar day)	Period (tropical years)
s_1	1 cycle/lunar day	347.81	2.83×10^{-3}
s_2	1 cycle/tropical month	13.18	7.47×10^{-2}
s_3	1 cycle/tropical year	0.986	1.0
s_4	1 cycle/lunar perigee period	0.1114	8.98
s_5	1 cycle/lunar nodal period	0.0529	18.63
s_6	1 cycle/solar perigee period	4.71×10^{-5}	20,940

mean Moon, rather than on the mean solar day. Table 6.1 gives the nominal frequencies and periods used—in mean *solar* days, but where s_1 corresponds to the period of the lunar day.

At least two different days, three different years and four different months are in use (see Appendix D). The frequency corresponding to the solar day is $s_1 + s_2$, describing the shorter time interval between zenith (overhead) passages of the mean Sun compared to those of the mean Moon. The tropical year was used in the table; the anomalistic year is the time for successive passages through perigee, and because it is moving in the direction of the Sun, the anomalistic year, with frequency $s_3 - s_6$, is slightly longer than the tropical one. Similar relations occur for the differing month definitions (see Anonymous, 1961, or Smart, 1962). Different definitions of year length are important in climate studies but are incidental for gravitational tides.

The discussion given thus far is adequate for all ordinary modern tidal analyses, and indeed the dependence upon s_6 is often ignored (no instrumentally measured sea level records approach such long periods). s_6 appears as a splitting frequency of the much-higher-frequency terms—no significant forcing amplitude occurs at s_6 itself.

At much longer periods (which are of interest in the insolation problem), lower frequencies present in the solar system must be accounted for. For example, the Earth eccentricity varies, and can go to zero.

The frequencies s_i are irrational multiples of each other. So by choosing the n_i, possibly using very high absolute values, it is possible to produce with arbitrary accuracy *any* frequency, $s_\mathbf{n} = n_1 s_1 + n_2 s_2 + \ldots + n_6 s_6$, and this possibility has led to a numerological literature in which claims are made for important tidal forcing at a variety of periods. But if the forcing amplitudes at that frequency are calculated, they turn out to be minuscule (Munk et al., 2002).

Table 6.2 lists the most important of the tidal frequencies. The names and symbols are conventional, and date back to Kelvin and G. Darwin.

Table 6.2: The main tidal constituents. The argument is the Doodson number, which is the vector **n** in which 5 has been added to n_2, n_3, ... to avoid the presence of minus signs. Note that in Doodson's treatment, **n** is based upon a lunar, rather than a solar day, and which differ by one cycle/lunar month. Cartwright and Edden (1973) has a listing for hundreds of frequencies. The names and Darwin symbols are historical in origin. (Adapted and corrected from Defant, 1961, p. 267.)

Name of Partial Tide	Darwin Symbol	Doodson Argument	Period in Mean Solar Hours	Relative Amplitude
Principal lunar	M_2	25555	12.42	1
Principal solar	S_2	27355	12.00	0.47
Larger lunar elliptic	N_2	24565	12.66	0.19
Luni-solar semidiurnal	K_2	27555	11.97	0.13
Larger lunar evectional	ν_2	24745	12.63	0.04
Variational	μ_2	23755	12.87	0.03
Luni-solar diurnal	K_1	16555	23.93	0.58
Principal lunar diurnal	O_1	14555	25.82	0.42
Principal solar diurnal	P_1	16355	24.07	0.19
Larger lunar elliptic	Q_1	13565	26.87	0.08
Lunar fortnightly	Mf	07555	327.86	0.09
Lunar monthly	Mm	06545	661.3	0.05
Solar semiannual	Ssa	05755	4382.86	0.04

6.2 POLE TIDE

A scientifically interesting, exotic, gravitational tide exists—one that does not arise from the presence of either the Sun or Moon. Called the *pole tide* by Munk and Macdonald (1960), it occurs because the Earth spin axis does not coincide with a fixed geometrical position but, rather, orbits around it (Fig. 6.8). This *Eulerian free wobble* exists because solid body, spun about an axis not coinciding with a principle moment of inertia will respond to the resulting torques by conserving angular momentum through movement of the spin axis. In a rigid Earth, the period ought to be about 9 months but was found by a nineteenth-century amateur astronomer named Chandler to be at about 14 months. The explanation of the difference lies with the finite elasticity of the Earth, which can then be estimated from the period shift. The fluid envelope gives a similar, smaller, contribution.

From the point of view of an observer at a fixed location, her apparent latitude, and hence the centrifugal contribution to the gravitational equipotential, shifts in a quasi-periodic way about every 14 months. The result is a gravitational tide forcing uniquely having the character of a narrow-band random process. The equilibrium amplitude of this motion is a few millimeters, and so it is of little oceanographic concern. It remains of considerable geophysical interest because the motion can be used to make inferences about dissipation rates and mechanisms in the solid Earth. Those rates are sufficiently large that the motion, if not constantly reenergized, should have died away after a few thousand years; see Munk and Macdonald, 1960, or Lambeck, 1980. Time-varying ocean seafloor pressure variations are now considered to be a significant element in the excitation maintaining the motions (Gross et al., 2003). A *forced* wobble exists at the annual period owing to the movement, primarily, of atmospheric air masses.

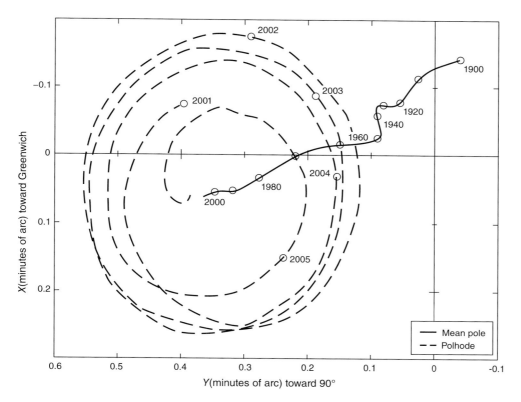

Figure 6.8: The observed motion of the Earth's rotation pole about an arbitrary starting position. The motion is known as the *Chandler Wobble* and causes the latitude of any observer on the Earth to change with a period of approximately 14 months. The wobble exists independent of the Moon and Sun and is excited (energized against damping) by a combination of atmospheric, oceanic, and seismic disturbances. Polar motion also includes a mean drift, as indicated, plus an annual band and a general spectral continuum. The *polhode* is a technical term for the path of the pole on the ellipse defined by the moments of inertia (Goldstein, 1980). The annual mean position over many years is also shown, depicting its slow drift. (From the website of the Jet Propulsion Laboratory.)

6.3 THERMAL TIDES

Some of the sea level variability observed at semidiurnal and diurnal periods is due to meteorological phenomena, and much more obviously so at annual and semiannual ones (Munk and Cartwright, 1966). Daily heating and cooling of the ocean surface causes the water column to expand and contract. In some places, sea breezes influence the sea-surface elevation. To remove these effects, which otherwise contaminate the analysis of the ocean or Earth response to the gravitational disturbances, they defined a thermal, or *radiational*, tide-forcing function. It is at this stage that tidal theory encounters insolation and becomes the Milankovitch problem, albeit greatly simplified because the duration over which the representation has to be accurate is extremely short compared to the full span of climate concerns ranging out to billions of years. They wrote a *radiation function* (here called the *insolation function*),

$$\mathcal{R} = S_0 \left(\frac{\bar{R}_S}{\rho_S} \right) \cos \alpha$$

$$= S_0 \frac{\bar{R}_S \cos \alpha}{R_S \left(1 - 2 \left(a/R_S \right) \cos \alpha + \left(a/R_S \right)^2 \right)^{1/2}}, \quad \cos \alpha > 0 \qquad (6.18)$$

$$= 0, \ \cos \alpha < 0,$$

where α is the same angle appearing in the tidal potential, S_0 is the solar constant and $\bar{R}_S$ is the time-mean solar distance. The basic geometry is identical to that of the gravitational tides. The denominator of Equation (6.18) is again expanded,

$$\mathcal{R} = S_0 \frac{\bar{R}_S \cos \alpha}{R_S} \left(1 + \frac{a}{R_S} P_1 \left(\cos \alpha \right) + \left(\frac{a}{R_S} \right)^2 P_2 \left(\cos \alpha \right) + ... \right), \quad \cos \alpha > 0 \qquad (6.19)$$

$$= 0, \quad \cos \alpha < 0.$$

The behavior of $\mathcal{R}$ differs from that of V in three ways: (1) It is physically sensible only on $r = a$. (2) The restriction $\cos \alpha > 0$ means that the Earth is opaque to radiation, and hence only the illuminated hemisphere sees a nonzero $\mathcal{R}$. (3) The $P_1 \left(\cos \alpha \right) = \cos \alpha$ term must be retained, because no analog to the centrifugal force arises to balance it. Equation (6.19) is expanded in Legendre functions, taking account of the vanishing for $\cos \alpha < 0$. The result (Munk and Cartwright, 1966) is

$$\frac{\mathcal{R}}{S_0} = \frac{\bar{R}_S}{R_S} \left\{ \frac{\cos \alpha}{2} + \sum_{n=2,4,6,...} \frac{2n+1}{2} \left[\frac{1(-1) ... (3-n)}{2(4) ... (2+n)} \right] P_n \left(\cos \alpha \right) \right\}, \quad 0 \leq \alpha \leq \pi, \quad (6.20)$$

n even, with a similar expression for n odd, and then expression (6.8) is used to calculate $\mathcal{R}/S_0$ in terms of the solar position. Evidently, semidiurnal, diurnal, annual, and perigee frequencies, all of their harmonics, and all integer multiple sum and difference frequencies will be included. As with the gravitational tides, the amplitudes at any given frequency must be calculated to see if the forcing is significant. In this restricted definition, $\mathcal{R}$ includes several approximations: for example, when the solar center is at the horizon with a setting or rising Sun, illumination exists in both hemispheres (the abrupt cutoff with $\alpha = 0$ is not physical). Insolation over the ocean would be influenced by path length changes through the atmosphere at large α, among several other issues, including the long twilight associated with the latitudes of polar night.

6.4 SOLID-EARTH TIDES, SELF-ATTRACTION AND LOAD, AND ATMOSPHERIC TIDES

Every part of the Earth is subject to tidal forcing, including the atmosphere and solid Earth. Understanding their responses are subjects in their own right. Tides of the solid Earth (Harrison, 1985) are of special concern to the ocean problem because tide gauges measure η relative to the moving seafloor whereas altimeters measure η relative to the center of the Earth, $r = 0$. In addition, the deformation of the solid Earth produces another gravity disturbance, which has also to be accounted for in calculating the water tide.

If the Earth were perfectly rigid, no solid-Earth tide would occur. With finite elastic rigidity, μ, the Earth does deform, producing two important corrections to the interpretation of tide gauge records: (1) The deforming Earth generates a significant change in the disturbing gravity field, which in turn affects the fluid ocean. (2) Tide gauges attached to the solid Earth move vertically with the tide, producing values of η differing from those measured from space. (Nineteenth-century determinations of μ were obtained by comparing hypothetical tides with infinite μ to those observed.)

Consider an elastic Earth that is deformed by the leading tidal term of Equation (6.7),

$$V/g = H\left(3/2\cos^2\alpha - 1/2\right) = HP_2\left(\cos\alpha\right). \tag{6.21}$$

Then in a static response, the Earth will deform proportionally to the disturbance as

$$h_L H\left(3/2\cos^2\alpha - 1/2\right),$$

where h_L is a constant.[6] If the Earth were infinitely rigid, $h_L = 0$. If the Earth were a fluid, $h_L = 1$, and if equilibrium were achieved, the fluid will move to render the displaced equipotential identical to the disturbed sea surface, η. An observer at the surface would then perceive a vertical motion, if she had a very good accelerometer, but no change in the gravitational equipotential. But the shifting mass modifies (increases) the tide disturbing potential, V at fixed height, also by an amount proportional to V/g, so that the apparent disturbing potential is $(1 + k_L) V/g$. Then if a thin ocean lies on top of the Earth, and the motion were at equilibrium, the *apparent* static deformation would become

$$\bar{\eta} = (1 + k_L - h_L) H\left(\frac{3}{2}\cos^2\alpha - \frac{1}{2}\right).$$

k_L, h_L are known as Love numbers, after Love (1911), and are defined from the elastic response to a disturbing potential and clearly depend upon the physical properties of the solid Earth. Textbooks on geophysics all discuss their evaluation,[7] which today is done with numerical models of the Earth's interior properties. $h_L \approx 0.6$, $k_L \approx 0.3$ and hence the decrease in the apparent or effective tidal potential for a tide gauge attached to the solid Earth is about 30%.

Disturbance in gravity by the moving solid Earth has an analog in the movement of the water—that, too, disturbs the gravitational field, with an additional effect on the water movement. The phenomenon is called *self-attraction*. With finite rigidity, μ, the weight of the water displaced by the tides presses down (or relieves the pressure) on the seafloor deflecting it by discernible amounts, particularly on the continental shelves, where strong ocean tides occur. By deforming the continental shelves, the tide gauges attached to the continents are deflected, and another gravity disturbance occurs owing also to the solid-Earth displacement. This effect is the *load*. It is commonly treated using another set of *load Love numbers*, k'_L, h'_L. The tidal disturbing potential, V, is augmented by the Earth deformation as $(1 + k'_L) V$, and the actual movement of the surface is $h'_L V$. Load Love numbers are numerically different from k_L, h_L.

---

[6] The subscript on k_L, h_L is not conventional but is used because of the multiple use of k, h in the overlapping fields discussed here.

[7] Munk and Macdonald, 1960; Lambeck, 1980; Stacey, 2008; Cartwright, 1999, p. 140.

Loads from moving water will not in general correspond to $P_2(\cos\alpha)$ but will have contributions from arbitrarily high order values of Y_n^m, $n \geq 0$, $m \leq n$, and one thus defines k'_{Lnm}, h'_{Lnm}, which will differ because the solid Earth responds differently to short-scale disturbances than it does to long-scale ones. Tidal models described later account for these load numbers (Farrell, 1972).

The fluid dynamical behavior of the ocean tides (taken up in Section 6.5) is complex because the ocean has "free," or normal, modes with periods comparable to the semidiurnal, diurnal, and (perhaps) long-period tides. In contrast, the normal modes of the solid Earth have periods generally shorter than 1 h (e.g., Jeffreys, 1976; Stacey, 2008). Thus the latter response is nearly a static one, and the "equilibrium theory" of a deformed elastic solid is nearly directly applicable. (The interior of the Earth is neither strictly solid nor dissipationless, and some complications arise from dynamical modes in the fluid core with periods comparable to a day; Lambeck, 1988).

6.4.1 The Atmosphere

Atmospheric tides are of interest in a book on the oceans for at least two reasons: the ocean responds to the atmospheric tides, and they are a high-frequency analog of the atmospheric response to the Milankovitch/insolation forcing. Although the gravitational tides are measurable in the atmosphere, it was recognized long ago that barometric measurements at the surface were dominated by the solar tide constituents, S_2, S_1, and were thus surely primarily of insolation rather than gravitational origin (Chapman and Lindzen, 1970; Cartwright, 1999).

Several generations of scientists were perplexed that the surface-pressure signal is strongest in S_2, instead of S_1 where the insolation function forcing is much stronger. An attractive explanation was the postulated existence of a resonant mode of the atmosphere much closer to 12 than 24 h. Almost nothing was known of the vertical structure of the atmosphere, particularly at very high altitudes, until balloons could reach there and sounding rockets were developed after World War II. The Krakatoa explosion in 1883 had generated a pressure pulse observed around the world on recording barometers (barographs), and the inverse problem of determining the atmospheric structure from its arrival patterns was studied by Taylor (1929) and others to determine if there was such a resonant mode. The now-accepted explanation is that the observed surface S_2, S_1 are thermally forced by ozone heating in the upper atmosphere and water vapor heating in the lower, resulting in dynamical solutions that are amplified at the ground for S_2, but not for S_1. A fuller explanation requires some dynamical theory (see the box "Atmospheric Tide Basics") and that is also of oceanographic utility. Ray and Ponte (2003) describe the structure of observed atmospheric tides at the surface.

6.5 TIDAL ANALYSIS

Serious tidal analysis in modern form began in the nineteenth century and is associated with the names of Whewell, Kelvin, Darwin, Proudman, Rossiter, etc. (Cartwright, 1999). The relevance for the Milankovitch problem is that tidal analysts long ago confronted the problem of understanding forced, nearly strictly periodic, motions in the presence of other motions usually best described as stochastic. The experience and tools developed are thus directly relevant for understanding forced insolation changes, up to the differences in the nature of the forcing and the gradual failure of the periodic approximations over extended time scales. Tidal signals exist in almost every geophysical time series from the ionosphere to temperature changes at the bottom of the ocean to the deformation of the Earth, including its core. In many of these records, the tidal motions are relatively weak, and extracting them may be onerous. For sea level, often 99% of the variance lies in the tidal bands, depending upon the record lengths, and one has an unusual and (for the tidal analyst) happy situation of a very large signal-to-noise ratio, but an unwelcome situation for anyone interested in the nontidal motions.

Two basic approaches exist to understanding tidal signals in records of any sort, no matter what the signal-to-noise ratio: (1) in the frequency domain, and (2) in the time domain. These approaches represent the two flavors of time series analysis described in Appendix A. The main differences are those of convenience, in the implementation of numerical analysis algorithms, in the ease with which various a priori physical constraints can be imposed on the analysis, and in the ability to interpret the result.

6.5.1 Frequency Domain Analysis

Historically, frequency domain analysis of tides emerged first in scientific use (associated largely with Kelvin), and it remains today the most common, and more readily physically interpretable approach.[8] The physical interpretation is accessible because ocean physics is different in the various frequency bands. Beyond that, the assumption of linearity of a physical system says that if it is forced at a frequency s, then the response will occur only at that same frequency, with at most an amplitude and phase shift relative to the forcing. For ocean tides, this first approximation is an excellent one; it does fail impressively in some places and is strictly correct nowhere.

The tidal potential, Equation (6.9), is a linear sum of pure frequencies, s_n given by the Doodson numbers and the calculated amplitudes, $H(s_n)$, and phases, $\alpha(s_n)$ relative to a fixed time origin (usually the Greenwich meridian). Linearity implies that the ocean response will be at this same frequency, with amplitude $A(s_n)$ and phase $\gamma(s_n)$, with these values depending upon the physical variable being analyzed (surface elevation, water velocity, temperature, pressure, etc.). This response defines the transfer function, defined immediately below.

Computers rapidly calculate A, γ from long records. The process is related to, but somewhat different from, ordinary Fourier analysis, because the tidal frequencies are not integer multiples of each other; ordinary least-squares works, as does interpolation from conventional Fourier series. Complications arise in a number of ways. An important one is accounting for the contributions in A, γ of motions also present in the ocean but unrelated to the tides themselves.

Because the tidal forcing contains nearby frequencies, they are often not resolved. Unseparated lines will have an apparent amplitude and phase dependent upon the relative amplitudes and phases of the unseparated frequencies, and a correction is made for the effect. Consider the sum of two nearby sinusoids,

---

[8] Before Kelvin, private commercial tide predictions were based upon empirical time-lag formulas. See Cartwright, 1999.

$$H_1 \sin(2\pi s_1 t) + H_2 \sin(2\pi (s_1 + \Delta s) t - \eta), \quad \Delta s/s_1 \ll 1$$
$$= A \sin(2\pi s_1 t - \beta), \tag{6.22}$$
$$A = \left[H_1^2 + H_2^2 + 2H_1 H_2 \cos(2\pi \Delta s t - \eta) \right]^{1/2},$$
$$\beta = -\arctan\left[H_2 \sin(2\pi \Delta s t - \eta) / (H_1 + H_2 \cos(2\pi \Delta s t - \eta)) \right]$$

that is, with a slowly (depending upon Δs) changing amplitude and phase. Older standard tables left the 18.6-year lunar terms unresolved, and tabulated the shifting A, β as the apparent amplitude and phase (using so-called f, u factors—not to be confused with the same symbols used with the fluid equations of motion) of the unresolved lines (Schureman, 1958). This approach works very well, although the physics can conceivably change in the small frequency interval Δs, for example, owing to a near resonance (these exist in the real ocean in the semidiurnal and diurnal bands).

A slightly more general notation is useful. Let $\zeta(t)$ be any physical variable thought to exhibit linear tidal responses (velocity, temperature, elevation, etc.) and let its Fourier transform be $\hat{\zeta}(s_n)$, where s_n is the frequency. Let $q(t)$ be the tidal forcing. Then the hypothesis of a linear response, in a noise-free system is, from the convolution theorem, equivalent to

$$\hat{\zeta}(s_n) = \hat{h}(s_n) \hat{q}(s_n) \tag{6.23}$$

so that the response at frequency s_n multiplies the forcing by an amplitude $|\hat{h}(s_n)|$ and shifts it by a phase $\beta(s_n) = \tan^{-1}\left[\text{Im}(\hat{h}(s_n)) / \text{Re}(\hat{h}(s_n)) \right]$, where Im, Re denote the imaginary and real parts of $\hat{h}(s)$. No other relationship is consistent with the linearity hypothesis. $\hat{h}(s_n)$ is the *transfer function*. Determining and explaining the transfer functions and how and why they vary with frequency and location, leads into the fluid physics. Understanding the relationship between tidal forcing and the system response can be viewed as an attempt to find the transfer function. A limitation is that $\hat{q}(s_n)$ has nonzero amplitude only in clusters lying in the narrow frequency bands surrounding the basic astronomical frequencies. Attempts to find the transfer function can be seen in Figures 6.9a and 6.9b, and in the paper by Garrett and Munk (1971). Wunsch (1972a) and Webb (1974; see Fig. 7.4) emphasized the problem of determining the tidal Q. However, the measure of response is only a local one: the ratio of local water response to the local forcing function. But oceanic tidal motions are waves, and the response at any point is the summation of wave disturbances coming from the entire ocean, meaning that $\hat{\zeta}(s_n)$ has to be known over large areas, not just single points. Point values, however, can be a good indicator of the presence of near-resonant or null locations.

6.5.2 Hydrodynamic Nonlinearities

The multiple frequencies, s_n, of the tidal forcing arise from nonlinearities in the Newton-Kepler equations of motion that produces many sum and difference frequencies of the basic orbital ones. If the hydrodynamic equations governing the oceanic response were truly linear, the tidal motions would exist only at the s_n. But the fluid equations are also nonlinear, and so the set of frequencies s_n feeds into a second nonlinear system, which in ways is similar to those sketched for the tidal potential, generating harmonics and all of the sum and difference frequencies of the s_n. Some frequencies, such as that corresponding to the 2-week period of the lunar fortnightly tide, Mf, almost coincides with the difference frequency of the two much more powerful lunar and solar semidiurnal tides, M_2, S_2, giving rise to their difference frequency, $s = 1/12$ h $- 1/12.42$ h $= 1/354.9$ h. The result is a *compound* or *nonlinear* tide, usually labelled MSf. *Overtides* are those particular ones corresponding to simple harmonics (e.g., two times the frequency of M_2, called M_4), or sum frequencies of two separate constituents (frequency of M_2 plus that of S_2, giving rise to MS_4), and so forth. Compound tide frequencies usually have linear counterparts occurring in the tidal potential at s_n, and the linear response *could* dominate. In many cases (such as MSf), the corresponding linear tide is extremely weak, and the compound tides provide a way of studying hydrodynamic nonlinearities. But in some places (notably Southampton in the United Kingdom) and in fjords, the nonlinearities can become so strong that the ordinary analyses focussed on the s_n are not very useful. Special methods are used there (Gallagher and Munk, 1971).

6.5.3 Time Domain Analysis

In most data analysis with computers, functions are uniformly sampled at intervals Δt, and the convolution relationship (Section A.3.3) for a linear system forced by a sequence $q(j\Delta t)$ is[9]

$$\zeta(n\Delta t) = \Delta t \sum_{m=0}^{\infty} h(m\Delta t) q(n\Delta t - m\Delta t), \tag{6.24}$$

or more concisely and setting $\Delta t = 1$,

$$\zeta_n = \sum_{m=0}^{\infty} h_m q_{n-m}. \tag{6.25}$$

Here, h_m is the *Green function* of the linear system (Eq. A.34). Sums cannot run to infinity, and they do not need to. The problem of understanding the behavior of a linear system then becomes that of finding approximating values h_m, because knowledge of the Green function of a linear system completely specifies it. The sums start at zero to reflect system *causality*—that is, that it does not respond before being disturbed.

Several ways exist to determine the h_m in Equation (6.25) using finite upper and lower limits (which must be determined). The most straightforward way is through ordinary least-squares, finding the minimum over h_m of

$$\delta^2 = \sum_{n=-N_1}^{N_2} \left(\zeta_n - \sum_{m=-N_1}^{N_2} h_m q_{n-m} \right)^2. \tag{6.26}$$

The standard reference on this subject is by Munk and Cartwright (1966) who discuss choices of Δt, N_1, N_2, etc. Causality is abandoned because it requires knowledge of the Green function at *all* frequencies, not just those associated with tides.

As a summary statement, Figures 6.9a and 6.9b shows an analysis by Munk and Cartwright (1966) in both frequency and time domains of the tide gauge record at Hawaii. In these figures, the second and third panels from the top show the *coherence*-squared between the forcing (V/g) and the tide gauge record (Appendix A). The black bar shows the coherent part (the fraction of the variance explained by the forcing in that frequency band). Two plots are required for each figure because a logarithmic display cannot show powers that are additive, $\log(a) + \log(b) \neq \log(a + b)$. The lowest panel in each shows the estimated transfer function,

---

[9] Various dimensions (units) must be tracked in moving from an integral to a sum. Often $\Delta t = 1$ is set, suppressing the units, as is done here, but they come back to haunt one eventually.

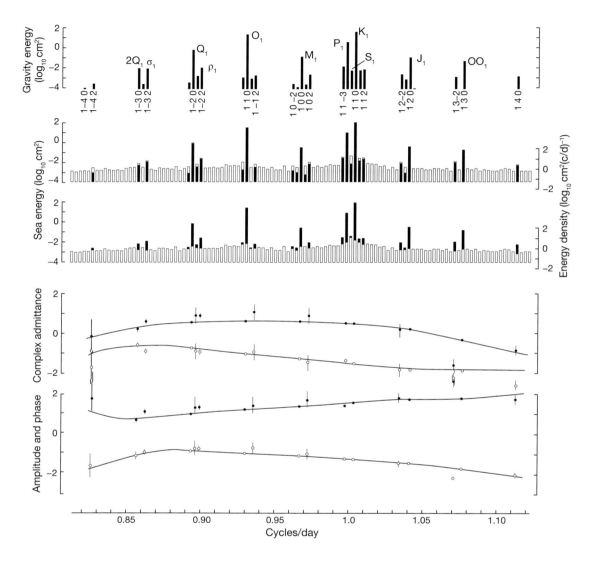

Figure 6.9a: The diurnal band tidal frequencies (*constituents*) separable at 1 cycle/year as well as an analysis of the behavior of the tides at Honolulu (see the text and Munk and Cartwright, 1966). None of the frequencies lies very far from 1 cycle/lunar or solar day. The uppermost panel shows the energy in the diurnal band and which is proportional to the spherical harmonic $Y_2^S(\theta, \lambda) \propto P_2^1(\cos\theta)\exp(\pm i\lambda)$. Numbers are abbreviated Doodson values. Note the clustering of the constituent lines (the clusters are separated by multiples of 1 cycle/month. Within the clusters, lines are separated by multiples of 1 cycle/tropical year and the large frequency gap between the diurnals and semidiurnals —the latter shown in Figure 6.9b. Lines separated by multiples of 1 cycle/8.9 years or 18.6 years are unresolved here. In the second and third panels, black lines indicate the energy coherent with the tide, repeated to show the smaller contribution on a logarithmic scale. The lowest two panels show the real and complex parts, and equivalently, the amplitude and phase of the local complex admittance function (Eq. 6.23).

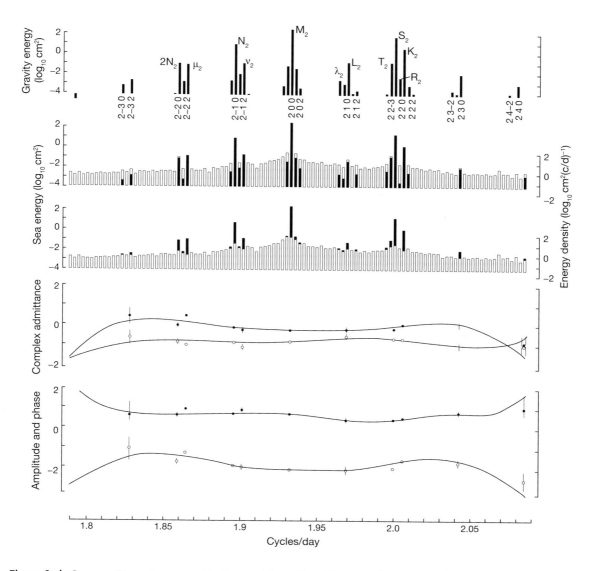

Figure 6.9b: Same as Figure 6.9a except in the semidiurnal band (note the frequency scales in the two figures) where the relevant spherical harmonic is $Y_2^{\pm 2}(\theta, \lambda) \propto P_2^2(\cos\theta)\exp(\pm 2i\lambda)$.

$\left|\hat{h}\left(s_{\mathbf{n}}\right)\right|, \beta\left(s_{\mathbf{n}}\right)$, which is the magnitude and phase of the Fourier transform of the Green function at Honolulu. Points indicate the frequencies, $s_{\mathbf{n}}$, of actual tidal energy. The smooth curves are obtained by making assumptions about the behavior of h_m. Generally speaking, the frequency domain representation is more enlightening, even when the time domain method is used to find it.

If the least-squares problem of minimizing δ^2 is solved, the coefficients so estimated, $\tilde{h}_m$, tend to be sensitive functions of N_1, N_2 although their Fourier transforms are more stable and interpretable. Groves and Reynolds (1975) suggested working with an orthogonalized set of h_m, later explored by Alcock and Cartwright (1978).

SPHERICAL TRIANGLES

Figure 6.10 shows a spherical triangle measured on the surface of a unit sphere. Note that there are two types of angles in a triangle on a sphere: those measured along the arc (a, b, c) and that correspond to angles measured with respect to the center of the sphere; the other angles, C, are interior angles to the triangle.

Spherical triangles satisfy a spherical law of cosines:

$$\cos c = \cos a \cos b + \sin a \sin b \cos C.$$

The identity may be derived by considering the triangles formed by the tangent lines to the spherical triangle subtending angle C and using the plane law of cosines (Passano, 1918.) Moreover, it reduces to the plane law in the small area limit. They also satisfy an analog of the law of sines,

$$\frac{\sin a}{\sin A} = \frac{\sin b}{\sin B} = \frac{\sin c}{\sin C}$$

where the capital letter angles are the interior angles opposite to the small-letter arcs.

6.6 THE MILANKOVITCH PROBLEM[10]

Orbital insolation forcing and the climate system's response to it on very long time scales can be skipped by anyone interested only in modern physical oceanography. The Milankovitch problem, however, a direct analog of the tidal problem, is of widespread scientific interest, and for that reason, a brief digression is made into the insolation or thermal forcing.

The main issues are finding accurate expressions for $\cos \alpha$ and R_S (Eq. 6.18) that are valid over hundreds of thousands and millions of years. This task is a problem in positional astronomy that at long periods must account for the changing disturbances of the other planets. Over sufficiently long periods (beyond about 20 million years), the system appears to become chaotic in the relative angular positions (not in the radii!).[11] The tidal disturbing potential would be even more complex from the need to calculate the lunar position.

---

[10] I had assistance from P. Huybers in writing this section.
[11] See Laskar et al., 2004. On extremely long time scales (billions of years), radii may also become unstable.

Major new elements are the variations in the Earth's obliquity (the angle ε_S in Fig. 6.4) and in the eccentricity e. Two phenomena dominate the former: the angle of the rotation axis of the Earth varies relative to the normal to the ecliptic (equivalently, the angle between the equator and the plane of the ecliptic), which today is about $23° \; 27' \; 8.26'' - 46.84'' \; T_J + \ldots$, where T_J is in Julian centuries; that is, the angle is decreasing.[12] It varies between about $22°$ and $25°$, and the plane of the ecliptic itself undergoes a forced motion relative to the so-called invariable plane. The latter is defined as being perpendicular to the net angular momentum axis of the entire solar system.

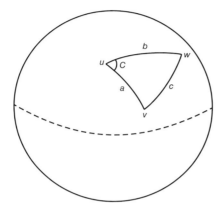

Figure 6.10: Angles of triangles on a sphere are of two types: those measured at the surface (C), and those measured as arcs from the center of the sphere (c).

Starting with an expression identical to Equation (6.19), Rubincam (1994) omitted all terms dependent upon λ, thus suppressing daily and higher frequency variations. To the lowest order (following a correction, Rubincam, 2004),

$$
\mathcal{R} = S_o \left\{
\begin{aligned}
&\tfrac{1}{4} P_0 \left(\cos \theta\right) \left[1 + \tfrac{e^2}{2} + 2e \cos M_a + \tfrac{5}{2} e^2 \cos 2M_a \right] + \\
&\tfrac{1}{2} P_1 \left(\cos \theta\right) \left[
\begin{array}{l}
\left(1 - \tfrac{e^2}{2}\right) \sin \varepsilon_S \sin \left(\omega + M_a\right) + 2e \sin \varepsilon_S \sin \left(\omega + 2M_a\right) + \\
\tfrac{e^2}{8} \sin \varepsilon_S \sin \left(\omega - M_a\right) + \tfrac{27}{8} e^2 \sin \varepsilon_S \sin \left(\omega + 3M_a\right)
\end{array}
\right] + \\
&+ \tfrac{5}{16} P_2 \left(\cos \theta\right) \left[
\begin{array}{l}
\left(1 + \tfrac{e^2}{2}\right) \left(\tfrac{3}{4} \sin^2 \varepsilon_S - \tfrac{1}{2}\right) - \tfrac{3}{4} \left(1 - \tfrac{7}{2} e^2\right) \sin^2 \varepsilon_S \cos \left(2 \left(\omega + M_a\right)\right) \\
+ 2e \left(\tfrac{3}{4} \sin^2 \varepsilon_S - \tfrac{1}{2}\right) \cos M_a + \tfrac{3}{4} e \sin^2 \varepsilon_S \cos \left(2\omega + M_a\right) \\
- \tfrac{9}{4} e \sin^2 \varepsilon_S \cos \left(2\omega + 3M_a\right) + \tfrac{5}{2} e^2 \left(\tfrac{3}{4} \sin^2 \varepsilon_S - \tfrac{1}{2}\right) \cos 2M_a \\
- \tfrac{39}{8} e^2 \sin^2 \varepsilon_S \cos \left(2\omega + 4M_a\right) + \ldots
\end{array}
\right]
\end{aligned}
\right.
$$

$$\tag{6.27}$$

Here, ω is the angle between perihelion and the equinox, Υ. S_o is the solar constant—the intensity of the Sun's radiation at a distance equal to the semi-major axis of the Earth eccentric orbit. M_a is, again, the mean anomaly. Recall that $P_0 = 1$. A representation of the frequency content (power spectrum) of $\mathcal{R}$ is displayed in Figure 12.1.

---

[12] Anonymous, 1961; Seidelmann, 2006. Precise numbers will differ with the time origin being used (the epoch) and the date of determination.

Equation 6.27 can be usefully considered under various limiting scenarios. If the global average is taken, only the zeroth order Legendre polynomial, P_0, remains, and no dependence upon ε_S survives. Changing the tilt of the spin axis only redistributes the total insolation across the face of the Earth, as indicated by the higher-order Legendre polynomials. Global average insolation can also be computed directly by noting that the radius from the Earth to the Sun is given by, $\rho = \bar{R}_S \left(1 - e^2\right) / \left(1 + e \cos v\right)$, and that global average insolation is $\mathcal{R} = S_o \left(\bar{R}_S / \rho\right)^2$, where $\bar{R}_s$ is the length of the semimajor axis of Earth orbit. For a relatively large eccentricity of 0.05, Earth receives approximately 20% more insolation at perihelion than at aphelion.

Long-term changes in insolation can be obtained, to good approximation, by eliminating all terms dependent upon M_a in Equation (6.27), thus isolating the insolation changes at periods longer than a seasonal cycle, and yielding,

$$\mathcal{R} = S_o \left\{ \frac{1}{4}\left(1 + \frac{e^2}{2}\right) + \frac{5}{16}P_2\left(\cos\theta\right)\left(1 + \frac{e^2}{2}\right)\left(\frac{3}{4}\sin^2\varepsilon_S - \frac{1}{2}\right)\right\}. \tag{6.28}$$

ω never appears independent of M_a in Equation (6.27) and, thus, longer-than-seasonal effects only enter from the variations of eccentricity, e, and obliquity, ε_S. The influence of ω—the precession of the equinoxes—is expected to average close to zero for any linear process that has a long-adjustment time scale relative to the seasonal forcing. To the extent it appears conspicuously in climate records, the climate system is inferred to be strongly nonlinear. e is very small, never exceeding 0.06 over the last five million years, so that the dominant nonseasonal forcing arises from ε_S. Much more can be said about the interesting problem of the response to such slowly varying forcing function, but that discussion is left for another time and place.

CHAPTER 7

Observations of Tides and Related Phenomena

7.1 TIDAL DYNAMICS

STATICS

The vertical component of gravity in the tide disturbance is overwhelmed by the local vertical (by definition) component, g, of the Earth's static field. g has no horizontal component, and the tidal disturbance necessarily dominates, and it is this "sideways" forcing that moves the water.

The undisturbed local gravitational potential is $V_p = -gz + V_0$, where V_0 is a constant evaluated at $z = 0$. From Chapter 6, the disturbing tidal potential is V_{tide}, and

$$V_{total}\left(\mathbf{r}, z, t\right) = -gz + V_0 + V_{tide}\left(\mathbf{r}, t\right), \tag{7.1}$$

where $\mathbf{r}$ is a horizontal position vector in any coordinate system. The magnitude of the vertical derivatives of V_{tide} being very small relative to V_0, the new height, $\bar{\eta}$, where the equipotential has the value V_0, can be obtained as

$$-(z = 0 + \bar{\eta})g + V_0 + V_{tide} = V_0,$$

or

$$\bar{\eta} = \frac{V_{tide}\left(\mathbf{r}, t\right)}{g}. \tag{7.2}$$

In the static situation in Figure 7.1 the resting sea surface—normal to local gravity—is disturbed by a mass, m_P, at the bottom of an ocean of depth h. Then the gravitational equipotential that was at $z = 0$ before the mass was added is moved vertically a distance, $\bar{\eta}\left(x, y\right)$, such that

$$-g\bar{\eta} + V_0 + \frac{ga^2 m_P}{m_E}\frac{1}{\sqrt{x^2 + y^2 + h^2}} = V_0,$$

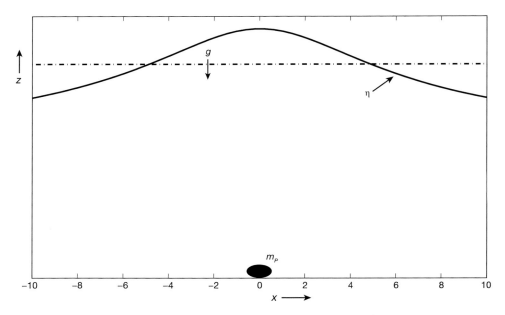

Figure 7.1: Schematic of the displaced free surfacem η, in the presence of a disturbing mass m_P. The dashed line is the nominal position of an undisturbed local sea surface $\eta = 0$. The gravity disturbance of a point mass, m_P, pulls fluid toward itself. By definition, the horizontal component of its attraction dominates that from ambient gravity, g, whose horizontal component is zero. An integral over a finite-area basin must vanish to conserve mass, and the effective potential, and hence of $\eta = \bar{\eta}$, is defined only up to a constant determined by mass conservation. The line $\eta = 0$ is drawn here as though mass is conserved over the volume shown, so that the spatial average vanishes. Units are arbitrary.

or,

$$\bar{\eta} = \frac{a^2 m_P}{m_E} \frac{1}{\sqrt{x^2 + y^2 + h^2}},$$

which is a maximum when $x = y = 0$ as expected from the horizontal forces.

If the sea surface should coincide so that $\eta = \bar{\eta}$, an *equilibrium tide* is the result. The misnamed *equilibrium theory of the tides* is a description of the movement of the gravitational equipotentials under the disturbance of the moving Sun and Moon. Whether the ocean *could* be in static equilibrium with the tidal disturbance is answered through analysis of the fluid dynamics.

If fluid equilibrium should occur or is assumed to do so, a sometimes important adjustment is required in the presence of land—one that accounts for mass conservation. Let A be the area of any finite ocean basin. Then in general,

$$\iint_A \bar{\eta}(\theta, \lambda, t) \, dA \neq 0.$$

But because the gravitational potential is undefined up to a spatial constant, $C(t)$,

$$\iint_A [\bar{\eta}(\theta, \lambda, t) + C(t)] \, dA = 0,$$

or

$$C(t) = -\frac{1}{A} \iint\limits_A \bar{\eta}(\theta, \lambda, t) \, dA, \tag{7.3}$$

and the result is known as a *corrected equilibrium tide*.[1]

7.2 TIDES OF A FLAT-BOTTOM OCEAN ON A SPHERE 1

Dynamical study of the tides began with Newton, but it was Laplace (1839)[2] who, about 1774, produced the first dynamically consistent theories. He focussed on the tides of water-covered globes (see Lamb, 1932, Chap. 6). His equation set is called the Laplace Tidal Equations (LTE). On the full sphere, $\bar{\eta}$, the tidal potential is expressed in spherical coordinates. Let the motion be periodic and proportional to $\exp(-i\sigma t)$, $\sigma = 2\pi s$. From Equations (3.11) and (3.12), replacing latitude ϕ, with colatitude θ, and omitting the friction terms,[3] the result is

$$-i\sigma u - 2\Omega v \cos\theta = -\frac{g}{a \sin\theta} \frac{\partial(\eta - \bar{\eta})}{\partial\lambda} \tag{7.4}$$

$$-i\sigma v + 2\Omega u \cos\theta = -\frac{g}{a} \frac{\partial(\eta - \bar{\eta})}{\partial\theta} \tag{7.5}$$

$$-i\sigma\eta + \frac{h}{a \sin\theta} \left\{ \frac{\partial u}{\partial\lambda} + \frac{\partial(v \sin\theta)}{\partial\theta} \right\} = 0. \tag{7.6}$$

If Equations (7.4) and (7.5) are multiplied through by a constant density, $\bar{\rho}$, they show that the mass times acceleration is equal to the horizontal gradient of the difference between η and $\bar{\eta}$—that is, the force is given by the horizontal derivatives of the tidal potential $V = g\bar{\rho}\bar{\eta}$. Should $\eta = \bar{\eta}$, the tide would be equilibrium. Such a response is not formally a solution to the equations, because the two momentum equations then have only the solution $u = v = 0$. Equation (7.6), however, would require $-i\sigma\eta = 0$, which contradicts the assumption that $\eta = \bar{\eta}$, unless $\sigma = 0$. Equilibrium can sometimes be obtained as a limiting behavior, for example, when h becomes very large or at low frequencies in the presence of friction.

For a zonally travelling disturbance, as in the tides, that is proportional to $e^{im\lambda}$, a single equation in η can be found:

---

[1] The standard reference is by Thomson and Tait (1912), who also discuss many other aspects of tidal theory.

[2] The 1839 date is that of a translation by Nathaniel Bowditch of Laplace's *Mécanique Celeste* (circa 1775). (Bowditch, an autodidact, produced the handbook of navigation used by the US Navy and most US civilian ships. The Navy's handbook, much updated, is still known almost everywhere simply as "Bowditch.") This translation is particularly useful and interesting, because Laplace was (in)famous for asserting that something was "obvious," and thereby omitting all of the intermediate steps in his mathematics. Bowditch says that upon coming upon such a place, he knew he was in for days of labor to fill in the gaps. Thus the translation often has a few lines of Laplace, with the whole rest of the page being Bowditch's explanation.

[3] The notation is essentially that of Longuet-Higgins (1965a), except he writes λ for longitude and puts $a = 1$. v is in the direction of *decreasing* colatitude θ. In Lamb's (1932) notation, u is positive to the south, v positive to the east and the time dependence is proportional to $\exp(+i\sigma t)$.

$$\frac{\partial}{\sin\theta\,\partial\theta}\left\{\frac{h\sin\theta}{(\sigma/2\Omega)^2-\cos^2\theta}\left(\frac{\partial\eta'}{\partial\theta}-\frac{m}{(\sigma/2\Omega)}\cot\theta\eta'\right)\right\} \tag{7.7}$$

$$-\frac{h}{(\sigma/2\Omega)^2-\cos^2\theta}\left(\frac{-m}{(\sigma/2\Omega)}\cot\theta\frac{\partial\eta'}{\partial\theta}+m^2\csc^2\theta\eta'\right)+\frac{4\Omega^2a^2}{g}\eta'$$

$$=-\frac{4\Omega^2a^2}{g}\bar{\eta},\quad \eta'=\eta-\bar{\eta}$$

Note the appearance of the forcing on the right-hand side and also its implicit presence in η'. The apparent singularities at $\sigma=\pm2\Omega\cos\theta$ are related to inertial waves, but no infinities appear in the solutions to the equation when forced by tides.

A number of serious attempts were made to solve the LTE in geometries with horizontal boundaries. Apart from the solutions in canals discussed by Airy in the nineteenth century and sketched below, not much progress was achieved because the boundary conditions of no flow into walls in the presence of rotation are mathematically awkward, involving both the normal and tangential derivatives of η' at a wall.

7.3 CARTESIAN APPROXIMATIONS

For motions whose lateral scale is significantly smaller than the radius of the Earth, a, but large enough to be hydrostatic, and if the forcing is not too strong, a formal approximation by the local Cartesian coordinate, β-plane version of Equations (7.4–7.6) can be derived:

$$\frac{\partial u}{\partial t}-fv=-g\frac{\partial\eta'}{\partial x} \tag{7.8a}$$

$$\frac{\partial v}{\partial t}+fu=-g\frac{\partial\eta'}{\partial y} \tag{7.8b}$$

$$\frac{\partial\eta}{\partial t}+h\left(\frac{\partial u}{\partial x}+\frac{\partial v}{\partial y}\right)=0,\quad \eta'=\eta-\bar{\eta}. \tag{7.8c}$$

v is *positive* to the north. As discussed in Chapter 6, the f-plane is useful for motions of small meridional extent.

"CANAL THEORY"

The free solutions in a canal described in Chapter 4 produced the Poincaré and Kelvin waves. Consider now a simple "tide-like" forced motion,

$$\bar{\eta}(x,y,t)=A_0\exp(ik_0x-i\sigma_0t)\,Y(y),$$

ignoring rotation for the moment. An equation in η' is

$$\frac{\partial^2\eta'}{\partial x^2}+\frac{\partial^2\eta'}{\partial y^2}+\frac{\sigma_0^2}{gh}\eta=0, \tag{7.9}$$

which involves a mixture of terms in η' and in η. Subtract $(\sigma_0^2/gh)\,\bar{\eta}$ from both sides to produce a forced equation in η', and the substitution $\eta' = A_0 \exp(ik_0 x)\,\hat{\eta}(y)$ requires

$$\frac{d^2\hat{\eta}(y)}{dy^2} + \left(\frac{\sigma_0^2}{gh} - k_0^2\right)\hat{\eta}(y) = -\frac{\sigma_0^2}{gh}A_0 Y(y). \tag{7.10}$$

As in all such problems, it can be solved by first finding a particular solution, and then by using the free solutions to satisfy the boundary conditions (the procedure is not unique, although the result is). Suppose $Y(y) = 1$ so that the north-south scale of the "tide" is much larger than the canal's width. Then a particular solution is

$$\hat{\eta}_p(y) = \frac{-A_0 \sigma_0^2/gh}{\sigma_0^2/gh - k_0^2},$$

with complete solution,

$$\eta'(x, y, t) = \frac{-A_0 \sigma_0^2/gh}{\sigma_0^2/gh - k_0^2}e^{ik_0 x - i\sigma_0 t} + \eta'_h(x, y, t), \tag{7.11}$$

where the homogeneous part η'_h is any solution to Equation (7.9) with $\bar{\eta} = 0$. These solutions are used to satisfy the boundary conditions, $v = 0$ on $y = -L, L$ or,

$$\left.\frac{\partial\eta'}{\partial y}\right|_{y=-L,L} = 0.$$

Because $Y(y) = 1$, it is automatically satisfied and $\eta'_h = 0$, a special case.
   Then from η',

$$\eta(x, y, t) = \frac{-A_0 \sigma_0^2/gh}{\sigma_0^2/gh - k_0^2}e^{ik_0 x - i\sigma_0 t} + A_0 e^{ik_0 x - i\sigma_0 t}.$$

As $h \to \infty$, $\eta \to \bar{\eta}$, in an equilibrium response. Equilibrium is possible in this limit because the horizontal divergence required at any depth to supply or remove the fluid sufficient to the vertical displacement η becomes vanishingly small as the water column's length becomes infinitely large. At the other extreme, the possibility of a very large response exists if $(\sigma_0/k_0)^2 = gh$, with the forced wave coinciding with both the frequency and wavenumber of a free wave in the canal. In this *travelling wave resonance*, very large amplitudes imply the breakdown of various assumptions including linearity and no friction. The sign of the difference between η and $\bar{\eta}$ changes across the resonance—the tide goes from *direct* (when both are positive) to *indirect* (when the response is 180° out of phase with the forcing).
   The situation with the rotation restored is not very different. Now the free motions have frequency,

$$\sigma = \sqrt{f^2 + gh\left(\frac{n\pi}{L}\right)^2}, \quad n = 1, 2, \ldots \tag{7.12}$$

which, as discussed in Chapter 4, has a smallest value, $\sigma_{min} = \sqrt{f^2 + gh\,(\pi/L)^2} > f$. As before, however, Kelvin wave solutions can exist at any frequency. Taylor's (1921) theory of a forced wave in a canal encountering an end wall showed that if the forcing occurred at $\sigma < f$, a Kelvin wave would be generated and travel toward the end wall, where trapped

Poincaré waves would be produced. These would sort themselves out so that only an outward bound Kelvin wave would emerge from the canal on the opposite wall. To the extent that the returning Kelvin wave has a smaller amplitude than the inbound one, the dissipation within the system can be calculated from far-field measurements alone. This elegant result was exploited to both describe the tides of the Gulf of California (Sea of Cortes) and infer the energy loss (Hendershott, 1981). Rao (1966) computed the normal modes of a rotating f-plane rectangle.

7.3.1 Atmospheric Tides

As previously noted, the main reason for including a bit of discussion of atmospheric tides here is that they provide a high-frequency analog to the Milankovitch forcing of the climate system, being predominantly a thermally forced response. Atmospheric gravitational tides are observed and reasonably well understood, but they are comparatively weak.

A standard reference is Chapman and Lindzen (1970). Holton (1975, p. 53) provides a succinct derivation of the governing equations that most readily map onto the analysis done here (and see the box "Atmospheric Tide Basics"). For a resting stratified atmosphere, he shows that the same separation of variables can be used as was derived above for the ocean. The major distinction is that the upper-boundary condition is one of boundedness or of outgoing radiation, rather than that of a rigid or moving surface. Under these open conditions, the atmosphere produces free oscillations with only a single equivalent depth h' from γ in Equations (4.97) or (4.98) for a given wavenumber and frequency. The central message for climate dynamics is that even on diurnal time scales, the response to thermal forcing is already global and the behavior of the thermal response could never be understood as a local one.

7.4 TIDES OF A FLAT-BOTTOM OCEAN ON A SPHERE 2

Consider once again the equations on a water-covered sphere (Eqs. 7.4–7.6). Some exact solutions of these equations are known, and what follows here is taken almost directly from Lamb (1932), with only minor notational changes.

FREE MOTIONS

Setting $\bar{\eta} = 0$, the normal modes are defined as the solutions to the above equations, subject to the requirement of finiteness on the sphere (which takes the place of the boundary conditions and is required because of the meridian convergence at the two poles). Suppose, first, $\sigma = 0$ (steady motion). Then,

$$u = -\frac{g}{2\Omega a \cos\theta}\frac{\partial\eta}{\partial\theta}, \quad v = \frac{g}{2\Omega a \sin\theta \cos\theta}\frac{\partial\eta}{\partial\lambda}, \tag{7.13}$$

which describes a *geostrophic motion* in which the flow field is produced exactly by a balance between the pressure gradient and the Coriolis forces (recall that v is positive to the north here). Substituting into Equation (7.6) produces an equation that can be written

$$\frac{\partial(h/\cos\theta, \eta)}{\partial(\theta, \lambda)} = 0,$$

that is, that the Jacobian of η and $h/\cos\theta$ must vanish, implying that $\eta = Y(h/\cos\theta)$, where Y is an arbitrary function of $h/\cos\theta$. If h is a constant, this flow is strictly zonal and in this inviscid, linear limit could persist forever. More generally, a strong tendency will exist to follow the contours displayed in Figure 4.7. The result is the extension of Equation (4.62) to spherical coordinates.

When the frequency is nonzero, and h is constant, calculation is carried out for a fixed zonal dependence $e^{im\lambda}$, where m is an integer, leaving a differential equation (see Eq. 7.7) in θ alone. By expanding the solution in the spherical harmonics, $Y_n^m(\theta,\lambda)$ (not his notation), Hough (1897) in a remarkable accomplishment, calculated the free modes of a rotating spherical ocean, which Lamb summarized. Unsurprisingly, they tend to divide into a gravity-wave set and a set corresponding to Rossby waves—which in the limit of zero rotation become steady zonal flows. Longuet-Higgins (1964, 1968b) reexamined this general problem and in particular looked at the asymptotic solutions in the limit as $h \to 0$. In that limit, equatorial Kelvin waves, Poincaré waves, etc., all can be identified in the spherical solutions.

FORCED MOTIONS

Long-period-tide forcing has no zonal dependence, and absent meridional boundaries, this feature permits the suppression of the λ derivatives in the equations of motion, simplifying the calculations, although a series expansion in $\cos\theta$ is still necessary. The best-known solution is that of Kelvin, who used his calculation of the theoretical amplitude of the fortnightly tide, Mf, compared to that of the ostensibly observed Mf, to estimate the yielding of the seafloor, and hence to estimate the rigidity, μ, of the Earth. Owing to noise, his estimates of the amplitude and phase of Mf were inaccurate. The result is also dependent upon the assumptions that the dynamical response was independent of λ and that friction could be ignored.

When $m = -1$ or -2, which is appropriate for the diurnal and semidiurnal tides respectively, series solutions are generally required. But the special case $m = -1$, $\sigma = \Omega$, which is a weak constituent, is analytically tractable. As Laplace found, with

$$\bar{\eta} = A\sin\theta\cos\theta\exp(-i\sigma t - i\lambda),$$

(note the westward-going phase velocity characterizing all semidiurnal and diurnal tides), an exact solution to Equations (7.4–7.6) is

$$\eta = 0, \quad \eta' = -\bar{\eta},$$

$$v = i\frac{gA}{\Omega a}\exp(-i\sigma t - i\lambda)$$

$$u = -\frac{gA}{\Omega a}\cos\theta\exp(-i\sigma t - i\lambda)$$

that is, *the tide is a pure current system with no vertical displacement.* Tidal responses are always a combination of an elevation change and a flow field; Laplace's solution is an extreme case in which the elevation change vanishes, producing only a tidal current. Laplace apparently regarded this solution as an explanation of why observed oceanic diurnal tide elevations were so weak. He only knew about the North Atlantic Ocean ones; later on it was found that in the Pacific Ocean they were quite strong. No lateral boundary conditions can be satisfied with such a solution while retaining $\eta = 0$.

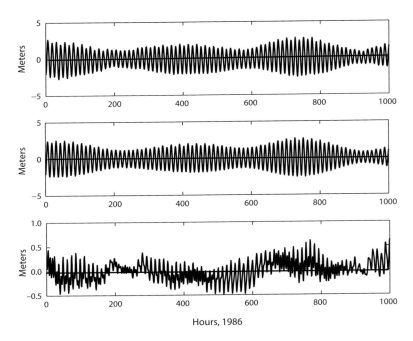

Figure 7.2: The top panel shows the tide gauge record at Newlyn, England, in 1986 displaying the strong semidiurnal tide and the fortnightly beating by the M_2 and S_2 constituents. The middle panel is from a least-squares fit to the first six most important tidal frequencies, and the lower panel is the residual of that fit. Six tidal components are only a fraction of the tidal and other variance. Note the scale change.

THE STRATIFIED OCEAN

The separation of variables applied in Chapter 3 to find the equations governing a horizontally uniform stratified fluid of constant depth on a β-plane can also be used on the sphere as well, putting $[u(z, \theta, \lambda, \sigma), v(z, \theta, \lambda, \sigma)] = F(z)[U(\theta, \lambda), V(\theta, \lambda)] \exp(-i\sigma t)$, etc., and going through the same separation exercise. Then h is everywhere replaced by h'_n and the asymptotics as the apparent water depth goes to zero become appropriate. But the vertical structure of the tidal forcing is uniform, the horizontal forcing wavelengths are also very long compared to those of baroclinic waves. Thus the projection of the forcing onto these baroclinic motions essentially vanishes. It is only with the introduction of bottom topography that *internal tides* can be generated (see Section 7.6).

7.5 BASIC TIDAL OBSERVATIONS

Before discussion of the science, a brief digression into the nature of the raw records is useful. Historical tide gauge records as obtained from any one of the various technologies in use (Chap. 2) are point observations of sea level and, as with all observations, have various errors and uncertainties. Figure 7.2 shows a somewhat typical one from Newlyn, England, that was sampled every hour. Visually, the record is dominated by a kind of quasi-periodicity that is reasonably inferred to be tidal, with nothing much else very obvious.

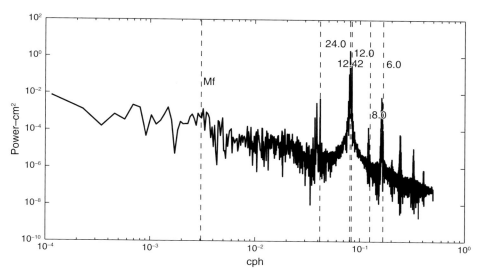

Figure 7.3: A periodogram (square of the Fourier coefficients, $a_n^2 + b_n^2$) for the year 1987 at Newlyn, UK, using hourly values. A few periods are marked, as is the location of the Mf constituent. Note the log-log scales. The result is a mixture of deterministic sinusoids and a stochastic background continuum. In the continuum, the variance is proportional to the square of the estimated periodogram value. At the deterministic lines, the uncertainty arises from the background continuum and the statistics can be inferred from the behavior of a sinusoid plus white noise (see Middleton, 1960).

Some perspective on the problem is given by Figure 7.3, the periodogram for the year 1987 in the Newlyn record. Many prominent peaks occur, particularly at high frequencies as overtones and at the sum and difference frequencies of the major tidal constituents, a consequence of the slightly nonlinear tidal hydrodynamics that is known to be present at Newlyn.[4] The presence, however, of nonzero values at non-tidal frequencies shows that the tidal response of the ocean is not the only physics acting. In particular, the Mf constituent might be present, but if so, it does not stand above the background power continuum. (Routine production of a figure such as 7.3 was computationally impossible for tidal analysts prior to the availability of cheap, fast computers.)

Computer codes exist for calculating the tidal disturbing potential, or the equivalent elevation, $V_{tide}/g = \bar{\eta}(\theta, \lambda, t)$, from expressions such as Equation (6.9) at any location and time (e.g., Pawlowicz, et al., 2002). By analyzing $\bar{\eta}$ with the same representation and transfer function method as in Equation (6.23), equilibrium tidal amplitudes, A_E, B_E, can be found, and hence the transfer function at $s_{\mathbf{n}} = n_1 s_1 + n_2 s_2 + \ldots$, as in the example in Figure 7.4.

For several reasons, including those connected with the history of oceanic mixing and the evolution of the lunar orbit, the extent to which parts of the ocean are at or near resonant conditions has been of considerable interest. Results are usually defined in terms of the local value of Q as estimated from the local transfer function. The result in Figure 7.4 has

[4] Munk and Cartwright, 1966.

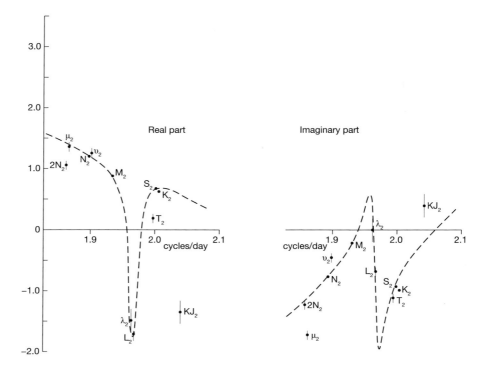

Figure 7.4: The real and imaginary parts of the complex transfer function at Cairns, Australia. The rapid transition with frequency is an indicator of resonance, as is the amplification near the L_2 component (a lunar component lying between M_2 and S_2 in Table 6.1 with Doodson number 265455). The width of the apparent resonance is directly related to the estimated Q of the system, as described in the box "Quality Factor: Q." The estimated decay time was about 18 d. The corresponding Q is about 225, a remarkably high value for any oceanic fluid phenomenon. (From Webb, 1974.)

an exceptionally high value for Q. More common regional semidiurnal tidal values for Q lie between about 3 and 30. The larger value implies an energy decay time of about 2–3 d, although Q values are usually lower bounds (see Garrett and Munk, 1971; Wunsch, 1972a; Arbic et al., 2007).

7.5.1 Distribution

Before the altimetric satellites, ocean tides were determined by analyzing ordinary tide gauge data for amplitudes and phases relative to the tidal potential for particular constituents (e.g., the principal lunar tide, M_2). Armed with these results, an analyst would contour the amplitudes and phases as a function of position over entire ocean basins, sometimes using some reduced dynamics for extrapolation. The central difficulty was that, with the exception of gauges on a handful of open-ocean islands (Hawaii, Tristan da Cunha, Bermuda, etc.), the measurements were all at the oceanic edges, sometimes shoreward of wide continental shelves and often inside harbors with complex regional physics. One example of such a construction is shown in Figure 7.5. The charts do show the characteristic rotation of phase lines about

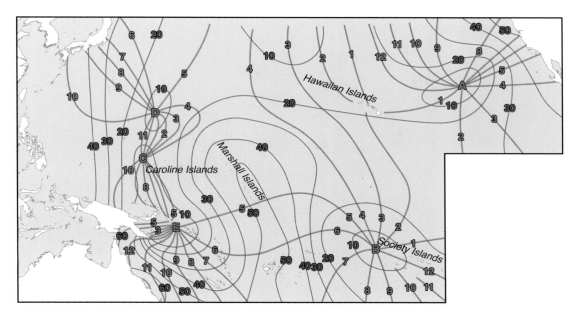

Figure 7.5: A *cotidal* chart for M_2 tide in part of the Pacific Ocean—one having more than the usual number of oceanic islands per unit area. The amplitudes (red lines) are in centimeters, and the phases (green lines) are in hours delay relative to equilibrium high tide at Greenwich, UK. (From Luther and Wunsch, 1975.)

interior positions of zero amplitude. (Called *amphidromic points* or *amphridromes*. Multiple phases can coexist only at zero amplitude.) The phases are with respect to Greenwich time. These results are now primarily of historical interest.

Munk et al. (1970) fit the California coastal tide gauge and some offshore pressure-recorder data to a Kelvin wave and a sum of Poincaré waves as shown in Figure 7.6—reproducing the amphidrome visible in Figure 7.5, but giving it a dynamical description. Such syntheses provide reassurance, should any be needed, that the basic physics of long hydrostatic gravity waves are appropriate to real tides.

High-precision satellite altimetry completely changed the global tide-observation problem. Coverage by TOPEX/POSEIDON and Jason spacecraft is shown in Fig. 2.18 and is nearly global (the high-northern-latitude ocean is not covered), and the measurements are dense along the tracks (with measurements roughly every 7 km). Some wide separations occur between the tracks in the "diamond-shaped" areas, but fortunately the expected spatial scale of the open-ocean tides is generally much larger than these intertrack distances (see Fig. 7.5). One novel problem arises: the satellite returns to any given location only every 10 d, and the resulting alias produces a signal not at the actual M_2 period of 12.42 h, but at 62.1 d. Because the signal is so strong, it stands out very clearly at the aliased period and the analysis can proceed.[5] Table 7.1 displays the various alias periods, depending upon different satellite sampling intervals.

[5]One of the major design problems for the orbit choice of TOPEX/POSEIDON was the requirement that *none* of the many tidal frequencies should alias into the annual or semiannual cycles or the time mean. (See Parke et al., 1987.) The effort succeeded well, although problems remained with the diurnal tides and their alias near the semiannual period.

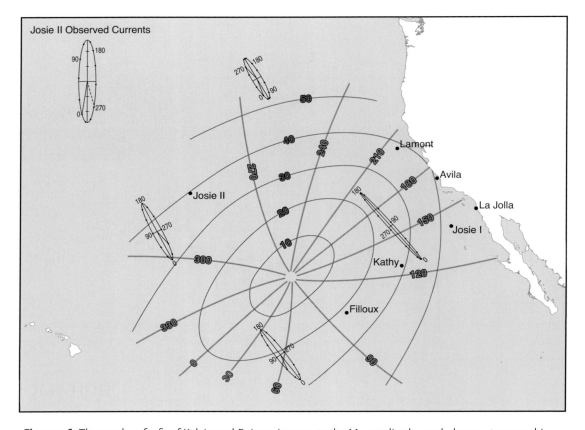

Figure 7.6: The results of a fit of Kelvin and Poincaré waves to the M_2 amplitudes and phases at an amphidrome offshore of the California coast that also appears in Figure 7.5. The amplitudes are in centimeters, and the phases in degrees relative to Greenwich mean time. Measured tidal-current ellipses are also shown, with tick marks at 1 cm/s intervals. Labels such as "Josie," are instrument names—some after the partners of the investigators. (From Munk et al., 1970.)

The amplitude and phase of each constituent, derived for each aliased constituent at every sampling point can be determined and then represented classically in cotidal charts. For the global estimates now most widely used, a somewhat more sophisticated procedure is implemented. Motions are *assumed* to satisfy the Laplace Tidal Equations written in discrete (numerical form), and which are then least-squares fit to the data. Parameters that are adjustable to produce the fit (known as *controls*) can include the local dissipation parameterization, the bottom topography shape, rigidity of the seafloor, load and self-attraction, etc., with different investigators making different assumptions about which are subject to adjustment and by how much. See Appendix B for the methodology.

Table 7.1: The periods of the principal tidal lines and their apparent periods as aliased by three altimetric satellites with different repeat periods. (From LeProvost, 2001.)

Tides	Tidal Period (hours)	Aliased periods (days)		
		TOPEX/POSEIDON 10-d repeat orbit	ERS-1 35-d repeat orbit	GEOSAT 17-d repeat orbit
M_2	12.42	62.11	94.49	317.11
S_2	12.00	58.74	∞	168.82
N_2	12.66	49.53	97.39	52.07
K_2	11.98	86.60	182.62	87.72
K_1	23.93	173.19	365.24	175.45
O_1	25.82	45.71	75.07	112.95
P_1	24.07	88.89	365.24	4466.67
ε_2	13.13	77.31	3166.10	98.75
$2N_2$	12.91	22.54	392.55	58.52
μ_2	12.87	20.32	135.06	81.76
ν_2	12.63	65.22	74.37	41.56
λ_2	12.22	21.04	129.53	35.36
L_2	12.19	20.64	349.25	39.20
T_2	12.02	50.60	365.26	313.89
S_1	24.00	117.48	∞	337.63
Q_1	26.87	69.36	132.81	74.05
OO_1	22.31	29.92	102.31	49.38
J_1	23.10	32.77	95.62	60.03
$2Q_1$	28.01	19.94	5250.89	43.88
σ_1	27.85	21.81	214.30	55.77
ρ_1	26.72	104.61	80.73	54.46
χ_1	24.71	26.88	7974.60	38.86
π_1	24.13	71.49	182.63	397.79
ϕ_1	23.80	3354.43	121.75	89.48
θ_1	23.21	38.97	178.53	46.48
M_1	24.83	23.77	200.71	35.64

7.5.2 The Many Tidal Constituents

With more than 20 years of high accuracy altimetry, marginal frequency separation is possible in the Rayleigh criterion for lines separated by the lunar-orbit precession frequency of 1/18.6 years. Figure 7.7 shows the estimated amplitude and phase of Mf, K_1, S_1, M_2, and M_4 (Lyard et al., 2006). M_4 is the *overtide* (Also see Weis et al., 2008, although these results are based upon shorter record lengths.)

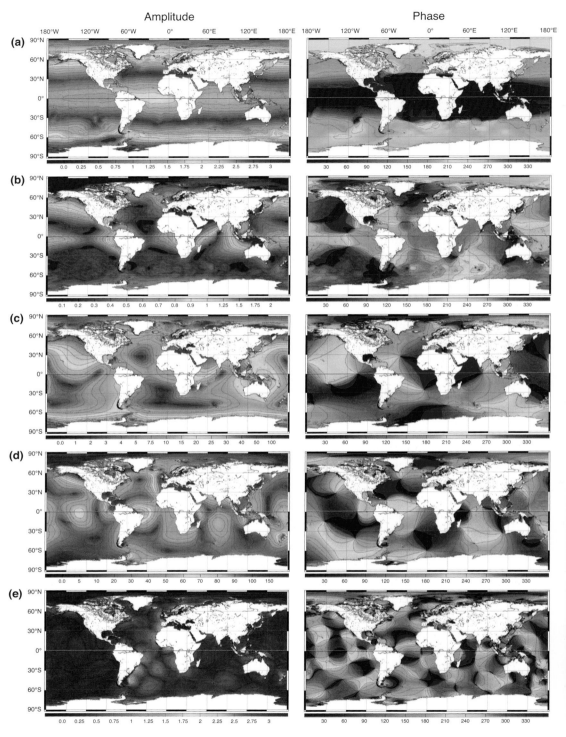

Figure 7.7: The estimated amplitudes (in meters, left) and phases (in degrees, right) of Mf, K_1, S_1, M_2, and M_4 in descending order from a dynamical model and altimetric and tide gauge data. The model accounted for the self-attraction and load effects, as well as viscous dissipation. (From Lyard et al., 2006, and pers. comm., 2014.)

The synthesized results have high accuracy and are consistent with the data on the order of 1 cm in elevation in deep water (roughly deeper than about 1000 m). Results are less satisfactory in shallow seas and over continental margins generally. Altimetric data tend to be degraded near continents for a number of reasons, including difficulties with atmospheric water vapor corrections. With altimetric arcs starting over land, the instrument takes a finite distance to achieve "lock" on the sea surface, and thus a data loss occurs near shore. Tides in shallow water can be large and strongly nonlinear, and the tidal models are much less skillful there than in deep water. See Vignudelli et al. (2011) for the flavor of coastal altimetry generally.

MODAL SYNTHESIS

As in any linear forced problem, the tidal response at any frequency can be represented as a summation over the normal modes of the global ocean. Platzman et al. (1981) made a notable attempt at computing the modes of the real oceans in the period range between 8 and 80 h.[6] Because geostrophically balanced modes can become locally trapped and amplified over small-amplitude topographic features, the computation was defeated by the numerical difficulties in going to arbitrarily long periods. Nonetheless, using the modes they were able to determine, Platzman (1984) synthesized the then available data for eight constituents (using periods to 96 h), recalling the forced solution in Equation (4.36). The far-denser altimetric era data has not apparently been so synthesized, but it is now possible.

The major difficulty, apart from the numerics at low frequencies, is that normal modes of rotating, realistic ocean basins do not always have simple intuitive interpretations. It thus is not so clear that much is learned from the modal representation.

VELOCITY FIELDS

Periodograms, or spectral density estimates from current meter moorings, typically display prominent peaks at tidal frequencies. One example is shown in Figure 7.8, with the spectrum of the east component (u) displayed in Figure 7.9. A peak at the M_2 frequency is marked, as is its first overtone, M_4. In contrast to the periodograms or spectral estimates of η, the tidal velocity components are a very small fraction of the record variance. The problem is particularly acute for diurnal tides measured at midlatitudes and where the relevant frequency band is usually swamped by the inertial motions near f. Dynamical tidal models for η also all predict the corresponding flow fields, u, v. Comparisons, however, with direct velocity measurements prove difficult, both because the signal-to-noise-ratio in the records is typically very small and the predicted velocities are usually for the barotropic component alone. In many places, the baroclinic (internal wave) tide can be very strong in the velocity field, while remaining small (but detectable) in η. Consequently, extraction of the barotropic velocity requires coverage of the whole water column—a rarity in mooring data. Luyten and Stommel (1991) attempted a pre-altimetric-era comparison and found rough agreement between a then-state-of-the-art barotropic model and Atlantic current meter records.

LONG PERIOD TIDES

Semidiurnal and diurnal tide contributions to sea-surface motion dominate the records through their excitation of the nearby gravity modes, sometimes producing nearly resonant

[6] See also Müller, 2007; Sanchez, 2008.

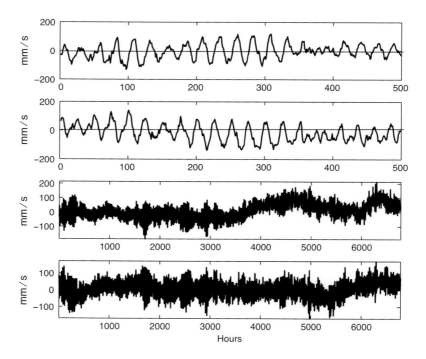

Figure 7.8: East and north components of a North Atlantic current meter record at 55°W, 28°N, at 498 m. $\Delta t = 1$ h. Records are visibly dominated by a near-24-hour inertial motion, *not* the tide. The top two panels show the first 500 h, the lower two, the entire record.

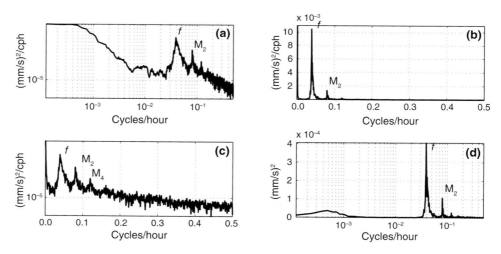

Figure 7.9: Spectral estimates, $\tilde{\Phi}(s)$, of the east component of the record in Figure 7.8 (using the entirety of the records) plotted in different ways: (a) log-log, (b) linear-linear, (c) log-linear, and (d) linear-log as area preserving $s\tilde{\Phi}(s)$. Each gives a different impression of the energy content.

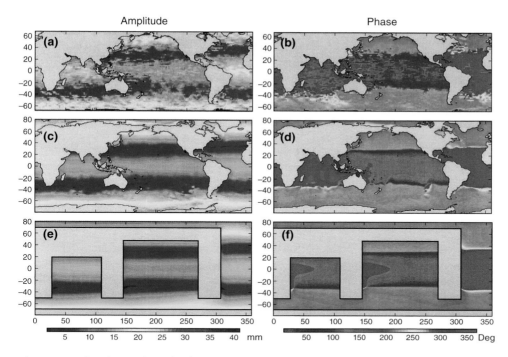

Amplitude   Phase

Figure 7.10: The observed amplitude (a) and phase (b) of the Mf tide. Panels (c) and (d) show a calculation of the fields in panels (a) and (b) using a tidal model. Panels (e) and (f) show the amplitude and phase from an extremely simplified, schematic ocean. Compare to the top panels in Figure 7.7. (From Egbert and Ray, 2003a.)

responses. In contrast, the far weaker long-period gravitational tides, principally the lunar fortnightly, Mf, and lunar monthly, Mm, proved very difficult to detect in tide gauge records but did finally emerge out of the background continuum in the altimetric records when they became sufficiently long. The solar gravitational long-period tides at 6 months and 1 year are masked by the stronger insolation effects.

The chief interest in Mf arises from its ability to directly force geostrophically balanced flows. Altimetric observations were discussed by Egbert and Ray (2003a) and are displayed in Figure 7.10. The small-scale structures, which are visible in the amplitude and phase, arise from wave modes required to satisfy no-flow boundary conditions at the continental margins. Their spatial derivatives, through geostrophic balance, generate significant long-period tidal velocities.

A small literature (Wunsch et al., 1997; Kantha et al., 1998; Egbert and Ray 2003a) exists on the physics of long-period tides as seen in the altimetry. As expected, the observed motions can be represented as a summation of heavily damped Rossby waves and basin-scale modes. As with the semidiurnal and diurnal tides, much of the interest lies in estimates of damping rates, which at these low frequencies are probably dominated by topographic scattering and trapping, not friction.

A brief summary of the pole tide results is worthwhile, if only as a cautionary tale. Despite published claims to the contrary, no convincing evidence exists for a non-equilibrium oceanic pole tide, anywhere. What had appeared to be a remarkably anomalous nonequilibrium pole

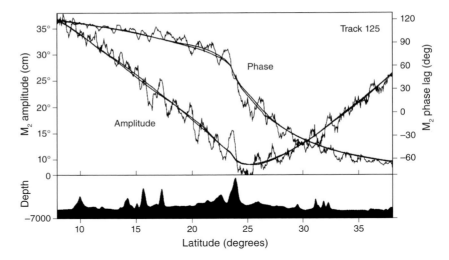

Figure 7.11: Altimetric measurements in which the short-wavelength first-baroclinic-mode internal tide is obvious and is superimposed upon the much-longer-wavelength barotropic tide (M_2). (Adapted from Ray and Mitchum, 1996.)

tide in the North Sea and Baltic Sea (Haubrich and Munk, 1959; Miller and Wunsch, 1973) now appears to have been the fortuitous result of an atmosphere temporarily displaying excess energy in the Chandler-wobble frequency band—a random occurrence (Tsimplis et al., 1994; Wunsch et al. 1997; O'Connor et al., 2000). A spectral peak was observed—but one which has been diminishing ever since, as the records lengthened. The primary message is that spurious spectral peaks are expected—and are seen.

7.6 INTERNAL TIDES

Tidal velocities extracted from the current meter record are, with good reason, believed to have major contributions from internal waves of tidal period, or *internal tides*. Altimetry made them visible globally in a way that the sparse coverage by current meters could not and is largely responsible for the existing intense interest in their behavior.

The visibility of internal waves in altimetric data is a consequence of the free surface boundary condition discussed in box "Sturm-Liouville and Bessel Function Solutions." Because modern altimeters have measurement accuracies close to 1 cm, the lowest-mode internal tide motion at the sea surface is readily measured with sufficiently long records. The internal tide in the first vertical mode was clearly detected by Ray and Mitchum (1996; see Fig. 7.11) in the altimeter data. Independent tomographic confirmation was offered by Dushaw et al. (1997). It is important in mixing the abyssal (below about 2 km) ocean, and the study of its generation and propagation has recently become the focus of considerable attention.

Internal tides are not directly generated by the tidal potential but instead are produced when external, or barotropic, tidal waves encounter topographic obstacles such that the isopycnals are disturbed as the fluid flows over and around the feature in the motions described in Section 4.4. The pioneering theoretical papers here include those of Cox and Sandstrom (1962), Baines (1973, 1982), and Bell (1975a,b). Until the advent of the altimeters, the ubiquity

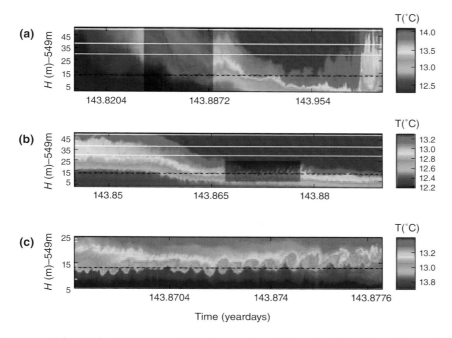

Figure 7.12: Kelvin-Helmholtz billows in the temperature field on the side of Great Meteor Seamount. Depth above the bottom is shown. Horizontal white lines show missing sensors. Their inference is that motions are the result of a tidal-boundary-layer instability in the stratified fluid. Compare to Figure 4.11. (From van Haren and Gostiaux, 2010.)

and strength of the internal tide was hardly appreciated. Motivated by the modern observations, the earlier theories have been extended, and a major in situ experiment to better understand the generation and propagation was conducted near Hawaii in the 2000s. A useful overall review was written by Garrett and Kunze (2007), and the field has been very active subsequently.

The theoretical discussion above showed that internal wave solutions can be written either in terms of modal summations or of characteristics. A characteristic can be written as a sum of modes (extending to very high mode numbers if the disturbance is narrow), and a mode can in turn be written as a summation over the characteristics, again involving a large number of them. The choice of representation in any particular situation lies between efficiency and convenience. In practice, characteristic-like disturbances at tidal frequencies are seen (e.g., Pingree and New, 1991; Cole et al., 2009; Holbrook et al., 2009), but diffusely so. In many other cases, generally those far from boundaries, a few low modes are far more efficiently representative of the observations (Hendry, 1977).

Ocean topography, having strong three-dimensional components, does not readily cooperate in producing situations in which the two-dimensional mathematical representations can be used accurately. Figure 7.13 shows an estimate of the radiation of the first-mode baroclinic tide in the vicinity of the Hawaiian arc (Zhao et al. 2010). The narrow regions of high flux are apparent. They correspond to the characteristics in direction but might be best described as narrowly focussed, in the horizontal, low modes.

The generation theories predict narrow, intense beams in the vertical plane in the vicinity of critical slopes (when the bottom slope $\gamma \approx c$, where c is the characteristic slope; see Figs. 4.10 and 7.2). Critical slopes can act both as effective generators of internal tides when disturbed by the barotropic components and, conversely, as regions of intense dissipation of these same motions (Figs. 4.11 and 7.12) with consequent strong mixing. In a dissipation-free ocean, noncritical characteristics would reflect from the top and bottom and travel across the ocean basins. Such behavior is almost never seen: ordinary dissipative processes as well as nonlinear interactions and scattering effects tend to rapidly degrade the high modes relative to the low ones, and thus the low modes dominate the vast bulk of the open ocean. Cacchione et al. (2002) have suggested that the slopes are sculpted so as to become critical for the tidal frequencies, but a dynamical mechanism for doing so has not been provided.

The scattering of the barotropic tide by topography into baroclinic flows represents an irreversible loss of energy from the barotropic tide, and thus from the orbital energies of the Sun and Moon and from the spinning Earth. At the present time, estimates of the fraction of the total ocean tidal dissipation (3.2 TW; Egbert and Ray, 2003b) owing to the baroclinic conversion process range from 10 to 50%, with the remainder still primarily attributed to shallow-water turbulence.

THE LUNAR NODAL TIDE

The normal vector to the plane of the lunar orbit precesses about the normal to the Earth's equatorial plane with a period of 18.6 years and corresponds to Doodson number n_5. Linear forcing corresponding to the period of 18.6 years produces an equilibrium variation of about 1 cm and because of the omnipresent background noise has never been convincingly seen in any tide gauge record.

On the other hand, the powerful nodally split semidiurnal and diurnal tides go in an out of phase over 18.6 years. In an influential paper Loder and Garrett (1978) suggested that in

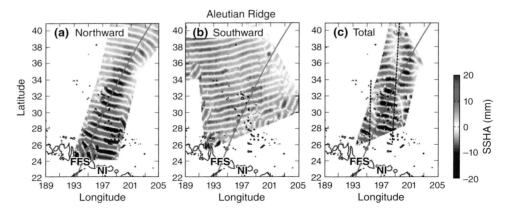

Figure 7.13: The internal tide as inferred from altimetry radiating northward from Hawaii (left panel), southward from the Aleutians (middle panel), and their sum (right panel). SSHA is the altimetric measurement of the height η. Solid lines are a typical altimetric track and dots are mooring positions. (From Zhao et al., 2010.)

shallow water, however, the 18.6-year envelope of the beating tidal frequencies could give rise to a measurable change in the intensity of tidal currents and thus to the generation of baroclinic motions leading to mixing of the oceanic water column. Ku et al. (1985) showed an approximate 3% effect on the semidiurnal amplitudes in the Bay of Fundy. For the open ocean, however, the evidence for any effect remains weak despite some enthusiastic reports. (See Munk and Bills, 2007, for a general overview of tide-modulated mixing.)

7.6.1 Changing Tidal Constituents

Gravitational tides are embedded in a moving, changing fluid, and so fluctuations and apparent trends are expected in the amplitudes and phases of the tides. Ray (2009) has discussed small observed changes in the western North Atlantic, and a number of analyses of long records (e.g., Colosi and Munk, 2006; Muller et al., 2011; Fig. 12.16) have shown usually slight, slow, changes in the constituent amplitudes and phases. Manifold reasons for such shifts exist, including changing water depths, shorelines, stratification that modifies the internal tide contributions, observing technologies, measurement sites, atmospheric forcing, general circulation, etc.) but the wider significance of these constituent disturbances remains unclear. During the last glacial period, sea level was lower than today by 120–130 m, a change big enough to shift parts of the ocean in and out of near resonances (see Egbert et al., 2004, for a modeling study). On geological time-scales, major continental and topographic rearrangement occurs, leading to expectations of qualitative shifts in tidal character.

7.7 DISSIPATION: TIDAL FRICTION

Discussion of tidal friction is usually traced to Immanuel Kant (Cartwright, 1999; Williams, 2000) and had a somewhat delicate history because it implied an imperfection in the supposedly perfect celestial mechanics of Newton, Laplace, et al. That dissipation had to be important was recognized in the nineteenth century, when lunar observations produced position deviations incompatible with a nondissipative theory. Laser reflectors left on the Moon by US astronauts have been used ever-since then to make highly accurate measurements of the lunar orbital radius, which is increasing by 3.82 ± 0.07 cm/year (Dickey et al., 1994; Bills and Ray, 1999). Because a receding Moon is gaining orbital angular momentum (from Keplerian physics), and because the Earth-Moon system conserves angular momentum, the spinning Earth must be losing an equal and opposite amount. (Solar-orbit effects are very small even though the solar-tide dissipation is important.)

Schematically, the sketch in Figure 7.14 is helpful. A tidal bulge, as shown there, mimics the equilibrium form. The spinning Earth, which, in the absence of friction, would leave the bulge aligned with the Earth-Moon line, drags the bulge "forward" through an angle δ in the direction of the spin. The resulting mass displacement exerts a torque on the Moon, accelerating it in its orbit and leading to the observed recession. A corresponding negative torque from the frictional effects brakes the Earth's spin. The distribution of tides in the oceans looks very unlike the equilibrium bulge, but if the actual distribution is expanded into spherical harmonics, one term in the expansion is proportional to the $P_2 (\cos \theta)$ of the equilibrium form. Angles $\delta = 0$, or $\delta = 90°$ in the figure would produce no torque and hence correspond to a frictionless tide.

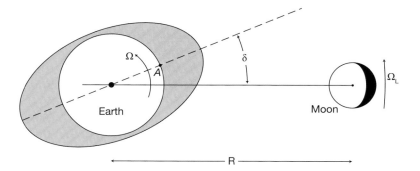

Figure 7.14: The net effect of tidal friction is equivalent to a tidal bulge carried *forward* of the Earth-Moon line and thus accelerating the Moon in its orbit and increasing its angular momentum, but at the expense of Earth's rotational energy and a decrease in its angular momentum. The ocean tidal distribution looks nothing like this simple bulge, but it is the projection onto the bulge (Legendre function, $P_2(\cos\alpha)$) that provides the torque. The calculated angle, δ, is about $3°$ and $\tan\delta = 1/Q$. (From Stacey, 1992, after a sketch of G. Darwin.)

Although the Moon is gaining orbital angular momentum as it recedes, it is losing total energy. Knowledge of the spin rate of the Earth and the orbital characteristics of the Moon again permit, via Keplerian mechanics, a calculation of the total energy (potential plus kinetic) in the orbit (the change in spin energy of the Earth can be neglected). Computation leads to the estimate of about 3.7 TW with an uncertainty on the order of 0.05 TW as the rate of lunar plus solar tidal dissipation (Munk, 1997).

Tidal motions are observed everywhere in the Earth including the atmosphere and solid earth, in lakes, wells, the ionosphere, etc., as well as on the Moon itself. They are inferred to occur in the liquid outer and solid inner cores of the Earth. Although the study of general tidal responses is fascinating in its own right, note here only that these nonoceanic motions are believed to account for about 0.2 TW (Munk, 1997), leaving about 3.5 TW to be dissipated in the ocean.

Modern understanding began with Taylor (1919), who showed that he could obtain a near balance between the turbulent dissipation by tides in the bottom boundary layer of the Irish Sea (Fig. 7.15), and the difference between the net tidal energy entering and leaving. (The calculation was corrected many years later by Garrett, 1975.) Taylor's calculation, with Garrett's correction, outlines the elements of the tidal dissipation problem. Taylor assumed that the turbulent dissipation within the sea would be of the form

$$\mathcal{D} = C_D \rho |\mathbf{u}| \mathbf{u} \cdot \mathbf{u}_{tide}, \tag{7.14}$$

where C_D is an empirical *drag coefficient*, and where $\mathbf{u}$ would be the ambient flow, including, but not limited to, the tidal velocity, $\mathbf{u}_{tide}$. He noted that when integrated, $\mathcal{D}$ should equal the difference of the flux of energy into the sea minus the amount exiting plus any work done by the moon on the water within the sea. An energy principle can be obtained for the tidal equations (e.g., in the Cartesian form) by first modifying the momentum equations to have

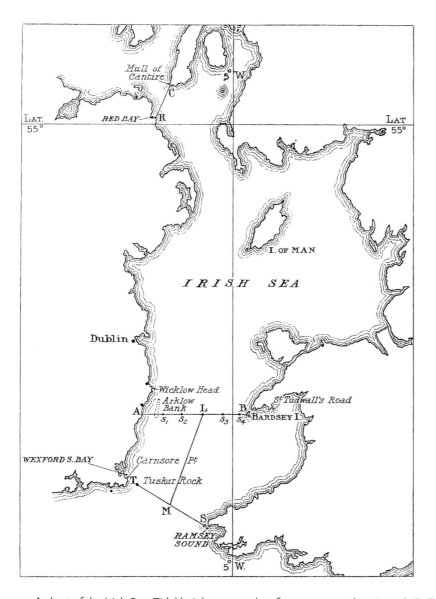

Figure 7.15: A chart of the Irish Sea. Tidal heights were taken from gauges at locations A, B, T, S, C, R. Velocities were calculated from the height (pressure) differences, the assumption of uniform inflow-outflow, and the dynamical equations. Energy fluxes in and out were computed across the three lines shown and compared to integrals of the drag law over the area of the Sea. Account was also taken of the small amount of direct work by the moon on the elevation within the sea. (From G. I. Taylor [1919], "Tidal Friction of the Irish Sea." *Phil. Trans. R. Soc.*, London.)

generic dissipation terms, F_x, F_y so that they become (from Eqs. 7.8a–7.8c),

$$\frac{\partial u}{\partial t} - fv = -g\frac{\partial \eta'}{\partial x} - F_x \tag{7.15}$$

$$\frac{\partial v}{\partial t} + fu = -g\frac{\partial \eta'}{\partial y} - F_y \tag{7.16}$$

$$\frac{\partial \eta}{\partial t} + h\left(\frac{\partial u}{\partial x} + \frac{\partial v}{\partial y}\right) = 0, \quad \eta' = \eta - \bar{\eta}. \tag{7.17}$$

Following the same logic as used in obtaining Equation (4.27),

$$\frac{\partial \left(\left(u^2 + v^2\right)/2 + g\eta^2/2h\right)}{\partial t} =$$

$$-\frac{\partial}{\partial x}(ug\eta) - \frac{\partial}{\partial y}(vg\eta) + \frac{\partial}{\partial x}(ug\bar{\eta}) + \frac{\partial}{\partial y}(vg\bar{\eta}) + \frac{g}{h}\bar{\eta}\frac{\partial \eta}{\partial t} - F_x u - F_y v.$$

In a coordinate free form:

$$\frac{\partial \left(\left(u^2 + v^2\right)/2 + g\eta^2/2h\right)}{\partial t} =$$

$$-\nabla \cdot (g\eta \mathbf{u}) + \nabla \cdot (g\bar{\eta}\mathbf{u}) + \frac{g}{h}\bar{\eta}\frac{\partial \eta}{\partial t} - \mathbf{u} \cdot \mathbf{F} \tag{7.18}$$

Integrating over the area of the sea and invoking the divergence (Stokes) theorem:

$$\frac{\partial}{\partial t} \iint\limits_{Area} \left[\left(u^2 + v^2\right)/2 + g\eta^2/2h\right] dA =$$

$$- gu\eta|_{bdy} + gu\bar{\eta}|_{bdy} + g \iint\limits_{Area} \frac{1}{h}\bar{\eta}\frac{\partial \eta}{\partial t} dA - \iint\limits_{Area} \mathbf{u} \cdot \mathbf{F} dA$$

Denoting an average over one tidal cycle as a bracket, $\langle . \rangle$, and making the good assumption that the total tidal energy is time invariant, produces

$$\iint\limits_{Area} \langle \mathbf{u} \cdot \mathbf{F} \rangle dA = -\left\langle gu\eta|_{bdy}\right\rangle + \left\langle gu\bar{\eta}|_{bdy}\right\rangle + g \iint\limits_{Area} \frac{1}{h}\left\langle \bar{\eta}\frac{\partial \eta}{\partial t}\right\rangle dA.$$

Taylor showed the approximate equality of the two sides of this equation from observed data and concluded that he had a reasonable budget for the Irish Sea by itself accounting for about 5% of the global total. Evidently, the shallow areas of the world ocean could account for all of the required tidal dissipation.

Jeffreys (1920) and Heiskanen (1921) then almost immediately constructed global budgets and showed that the shallow seas of the world could accommodate the required total (although the estimated total in 1920 was about two-thirds of the present revised value; Munk and Macdonald, 1960). Dependence on the cubed velocity in $\langle \mathbf{u} \cdot \mathbf{F} \rangle$ in the drag law means that the shallow seas must dominate (e.g., Thorpe, 2005, Chap. 10, or Eqs. 7.14 and 11.6 below). Jeffreys's calculation was troubling because of the apparent dominance of the Bering Sea. Miller (1966; see Fig. 7.16) redid the calculation and found only a comparatively minor contribution there,

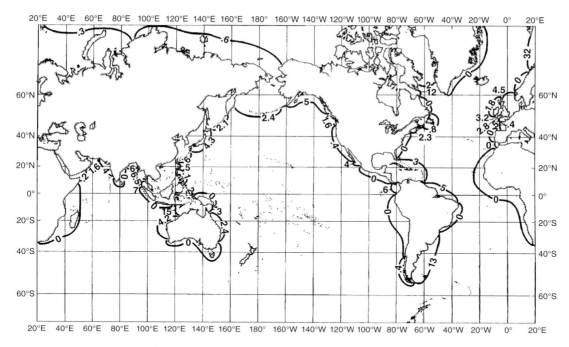

Figure 7.16: An estimate of the regions of total lunar semidiurnal tidal dissipation, 10^{10} W assuming it all occurs in shallow seas. Largely based upon use of the cubic-velocity dissipation rule. The numbers are the net flux into the bracketed shallow areas. (Used with permission from G. R. Miller [1966], "The flux of the tidal energy out of the deep oceans." *J. Geophys. Res.* 71: 2485–2489.)

but compensatingly larger values elsewhere. Calculations using the cubes of observed velocities are subject to very large errors. Munk (1997) reviewed this subject, and Egbert and Ray (2003b) in turn considerably modified Miller's estimate of the spatial distribution (Fig. 7.17).

From the model used in Figure 7.17, Egbert and Ray (2003) estimated that the tides had a total energy of 4.6×10^{17} J, of which 1.9×10^{17} J was potential and 2.7×10^{17} J was kinetic energy. In their estimate, M_2 alone had 1.3×10^{17} J and 1.8×10^{17} J, respectively, for a total of 3.1×10^{17} J and that its dissipation was 2.4 TW (of their total dissipation of 3.5 TW, which was close to Munk's 1997 estimate). Thus dividing the total energy by the rate of loss produces a time of 1.5 d and the Q of M_2 can be estimated as about 18. So about every three cycles the M_2 tide is fully dissipated, and the system is a heavily damped one.

The only minor caveat to the conclusion that shallow seas control tidal dissipation arose from a few observations showing that strong internal waves also existed at tidal periods and that they might account possibly for anywhere between 10 and 50% of the total tidal dissipation.[7] Little attention was paid to this subject because the internal wave data were very sparse and uncertain, and ordinary tidal dynamics were so well understood.[8]

The greatest importance of the conversion into internal tides lies with the implications for oceanic stirring and mixing. A recent review is by Garrett and Kunze (2007). Maas (2011)

---

[7] Wunsch, 1975.

[8] This situation is an example of that alluded to in the preface of how conventional wisdom depends directly on what is measurable.

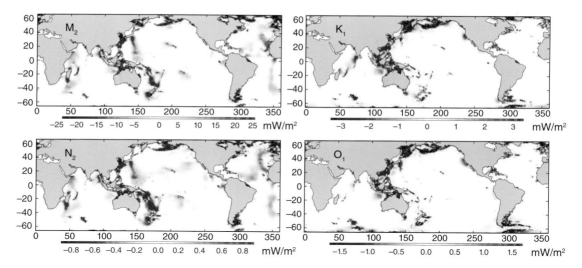

Figure 7.17: The dissipation regions of the tides for four different constituents, as computed by Egbert and Ray (2003b) from an inverse calculation. About one-third of the total was calculated to occur in the open ocean topographically rough areas—a change from the older inference that only the shallow seas were of importance. Compare to the older Figure 7.16.

produced an interesting theoretical exception.[9] A number of elaborate field programs in recent years have been directed at studying the generation, propagation, and breakdown of the internal tide, particularly in the vicinity of the Hawaii Ocean Mixing Experiment (HOME, Carter et al., 2008; see Fig. 7.12 and the later discussion of ocean mixing).

7.7.1 Paleotides and the History of Earth Rotation

Until the discovery of the strong and ubiquitous internal tides, most of the interest in tidal dissipation arose from recognition of the dramatic problem of the lunar orbit. The connection with modern oceanography is indirect, through the problem of understanding the nature and rates of tidal friction in the ocean and the analysis of how they might have been different in the past. If the orbital equations are integrated backward in time (Bills and Ray, 1999, is a survey of this problem) and the present rate of energy dissipation is extrapolated, the Moon would have been catastrophically close to the Earth only 1–1.5 GY ago. Munk (1968) labeled this approach the *Gerstenkorn event*, after the scientist who first pointed it out. Kilometers-high boiling ocean tides, a melted Earth and lunar crust, and a day length of only a very few hours would all have occurred. One interpretation was that the lunar orbit itself would have flipped over the pole (running in the time-forward direction) during this interval, and the scenario was interpreted as a capture of the Moon in an initially retrograde orbit. But the Gerstenkorn event never occurred: among other evidence, terrestrial and lunar rocks are known not to have melted so recently, leading to the inference that tidal dissipation must

---

[9] Another peculiarity of the governing Poincaré equation is the theoretical existence in two dimensions of solutions for topography in which complete destructive interference of the scattered internal wave field occurs. At specific frequencies, the baroclinic disturbance vanishes identically except over the obstacle itself, there being no far-field effect.

have been much lower in the past. Figure 7.18 shows a somewhat simplified calculation of the time to catastrophic approach as a function of a fixed dissipation rate.[10]

For the climate hundreds of millions of years in the past, the actual lunar orbit has non-trivial consequences. For example, when the length of a day was significantly shorter than it is today, the value of the Coriolis parameter f in all of the theoretical calculations would have been much larger, and of course, forcing by the lunar tides would have been amplified, and having a higher absolute frequency.

Reconstructing the history of Earth rotation and the lunar orbit is a fascinating problem, involving such studies as determining the observed positions of solar eclipses in ancient Babylonia and China and the sex lives of marine animals and their relationship to the tidal cycles.[11] "Since the study of ancient observations requires competency in both astronomy and antiquities, the field has never been over-crowded" (Munk and Macdonald, 1960, p. 187).

Stephenson and Morrison's paper (1995) is a good starting point for a description of the generic problems of determining the past length of a day and the corresponding rates of tidal friction. The problem generally divides into the "historical" interval, when written records first occurred in Babylon and China (circa 700 BCE), and the much longer time before that when only natural paleontology records are available. In the historical interval, reports of total eclipses of the Sun are the most useful because they do not require precise timing (mechanical clocks did not exist until the seventeenth century). Consider that if the Earth's rate of rotation were constant when going backwards in time, a total eclipse could be predicted to have occurred, for example, in Constantinople. But no such report exists there, and in contrast, scribes in Paris hypothetically reported that they observed a total eclipse (farther west than predicted). Assuming the particular eclipse can be identified, which can be a robust result because they are sufficiently infrequent to sometimes permit a unique determination, Earth rotation must have slowed in the intervening time (Fig. 7.19). This bland statement disguises a host of interesting practical questions arising from the interpretation of ancient archives, which are often in archaic languages, and for example, did the scribe actually observe the eclipse, or did he merely report one from years before on the basis of hearsay? See Stephenson, 2008. Figure 7.19 shows the estimated change in the length of day since 500 BCE. The angular position useful for eclipse calculations requires integrating this curve backward in time.

At one time (Wells, 1963), cyclic-appearing growth layers in various near-shore fossil organisms were regarded as the most promising method for obtaining length-of-day estimates in the remote past. Growth would be expected to occur primarily in the coincidences of the solar day (controlling light) and times of high or low water. The number of lunar tidal cycles per lunar month (in its various definitions) are fixed, and thus with the lunar-day periodicities beating against the solar-day periodicities producing spring neap cycles, and the whole system having apparent seasonal cycles, orbital mechanics permits determination of both the length of day and the length of month.

---

[10] Current estimates of the age of the Moon are about 4.5GY from the oldest lunar rocks and showing no Gerstenkorn event. The present favored origin hypothesis is that it originated in a collision between the Earth and a "Mars-sized body." For an interesting animation of the collision hypothesis for the origin of the Moon see http://www.youtube.com/watch?v=m8P5ujNwEwM (accessed 22 July 2014). History suggests that the origin debate is unfinished.

[11] Wells, 1963; Rosenberg and Runcorn, 1975; Stephenson and Morrison, 1995.

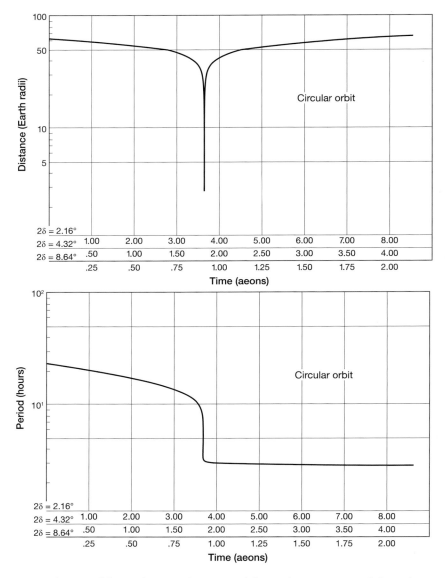

Figure 7.18: A history of the Earth-Moon distance and the Earth rotation period through time using a simplified physics (circular lunar orbit) with time scales set by different assumed uniform dissipation rates, here assigned by choosing different values of δ in Figure 7.14. Dissipation-rate assumptions are reflected in the different time scales along the x-axis. The interval of the jump in parameters would correspond to the Gerstenkorn event—if it ever occurred. To the left of that event, the Moon would be moving in an orbit counter to the Earth's direction of spin, spiraling ever closer to the Earth. Having moved through this near singularity, to the right, it enters an orbit moving in the same direction as Earth's spin and spirals out. Other parameters also vary, including the Earth's obliquity. The Gerstenkorn event singularity was originally interpreted as a capture event, but that is not now believed to be a credible theory of lunar origin. (From Macdonald, 1964.) See Macdonald's paper (1964) for the many specific details assumed in the calculations.

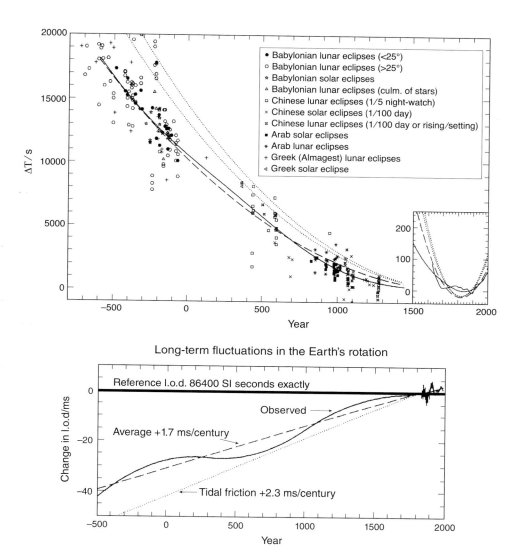

Figure 7.19: (top) The estimated time shift from various eclipse sources as compiled by Stephenson and Morrison (1995), who list the particular references. The inset shows estimates from the period 1620–1990 on an enlarged scale. (bottom) The estimated rate of change in the length of day (l.o.d.) from astronomical phenomena since 500 BCE. A smooth curve was fit to the raw data points from the top and then differentiated. (From Stephenson and Morrison, 1995.)

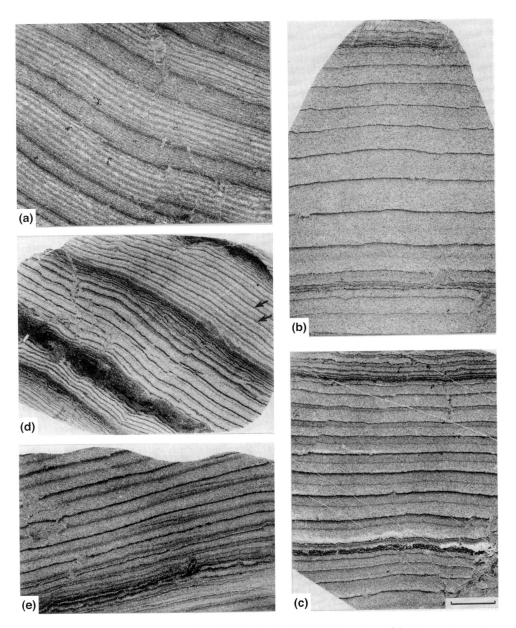

Figure 7.20: Thin sections of rocks from the banded geological formation of the Neoproterozoic in South Australia interpreted as "rhythmites." Bar at lower right in (c) is 1 cm. The primary banding is postulated to be diurnal and semidiurnal. (Used with permission from G.E. Williams [2000], "Geological constraints on the Precambrian history of Earth's rotation and the Moon's orbit." *Revs. Geophys.* 38: 37–59.)

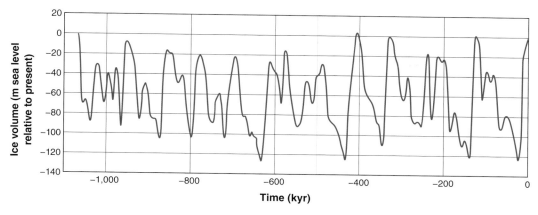

Figure 7.21: The estimated shifts in sea level over the last 425,000 years. The accuracy of these estimates is not clear. With continental margins fixed, the main influence on the tide would be the changing water depths over the shelf areas. But these will be strongly influenced by the weight of the continental glaciers during the low stands of sea level, and an actual reconstruction of water depths is a challenging problem (Peltier, 1994). On much longer time scales, that is, hundreds of millions to billions of years, the movement and size of the continents will dominate the tidal response. (After Waelbrock et al., 2002.)

Unhappily, as various fossil shells were studied (e.g., Pannella, 1972), it became apparent that what had appeared to be the comparatively straightforward business of counting layers and the length of their modulation, was actually a morass of incomplete data and complex growth behavior. Consequently, this approach has fallen from favor.

Williams (2000) reviewed the purely geological (nonbiological) evidence, and in particular, discussed so-called rhythmites in Australian Neoproterozoic (−1 billion to −500 million years) rocks (see Fig. 7.20). He estimated that at about −620 million years, 400 ± 7 days occurred in each solar year, that the day length was 21.9 ± 0.4 h (modern hours), and that the lunar recession rate was a little over one-half of the present-day value. These inferences remain susceptible to the possibility of major dating errors.

What accounts for a reduced tidal dissipation rate over geological time—an inference that is far more robust than any of the detailed determinations? This question becomes one of oceanic dynamics in the presence of changing continental configurations and mean sea level shifts (Fig. 7.21). In the last million years, the movement of continents has been sufficiently small that the modern continental configuration is an accurate depiction, with the major element being the fall and rise of the glacial-interglacial sea level. Because much of the dissipation occurs on the shallow continental shelves with the velocities generally proportional to $1/h$, the cubic dependence on the flow renders reconstruction of the low-stand ocean topography of first-order importance. Unfortunately, doing so is not simple, because the continents are strongly deformed by the weight of the continental glaciers (Laurentide and Fenno-Scandian in the most recent glacial period) as they exchange mass with the ocean (e.g., Peltier, 1994). The scattering of significant energy into internal waves is dependent upon the stratification ($N(z)$) in the abyss, about which almost nothing is known for the past. Direct calculation of the rate of conversion into internal waves remains uncertain even in the modern ocean.

Over most of Earth's history major glaciers did not exist, and so continental movement would have dominated the changing tides over tens of millions of years (Kagan and Sündermann, 1996). During one long interval, the Neoproterezoic, the Earth may have been ice-covered ("snowball Earth", Hoffman and Schrag, 2002). Dissipation is sensitive to proximity to resonance (e.g., Munk, 1968; Egbert et al., 2004). The best inference is that over most of the last 1 GY, the ocean basins have been configured in such a way that on average, the dissipation has been reduced compared to that of the last several thousand years (see Bills and Ray, 1999).

ATMOSPHERIC-TIDE BASICS

Atmospheric tides can be understood as similar to that of an infinitely deep ocean, except it is assumed that the system is turned upside down, so that the "ocean" floor is at $z = 0$ and extends to positive infinity upward, with an exponentially decaying density, $N = N_0 e^{-bz}$. γ is set by the horizontal wavenumbers and the frequency in Equation (4.97). From the assumed $N(z)$, $G(z) = AH_0^{(1)}\left(\gamma N_0 e^{-bz}\right) + BH_0^{(2)}\left(\gamma N_0 e^{-bz}\right)$, where the $H_0^{(1,2)}$ are Hankel functions of the first and second kind on the order of 0 (see the box "Rotating Cylinder and Bessel Functions"). Assume that the lower boundary is rigid, so that for an assumed solution in which dF/dz vanishes at $z = 0$, and $B = -AH_0^{(1)}(\gamma N_0)/H_0^{(2)}(\gamma N_0)$. Now $H_0^{(1)}\left(\gamma N_0 e^{-bz}\right) e^{-i\sigma t}$ represents a wave motion with downward group velocity, and unless there is an energy source at infinity, $A = 0$, and hence that $B = 0$ and no free motions possible. But forced motions can exist.

Example: Insert a density source (perhaps a heating-cooling one) centered at height $z = z_s$ having the form $R(z)\exp(ik_0 x - i\sigma_0 t)$ (that is, separable in x, z) into Equation (3.30c) so that it becomes, in the hydrostatic limit,

$$0 = -\frac{\partial p'}{\partial z} - g\frac{\rho'}{\bar{\rho}} + R(z - z_s)e^{ik_0 x},$$

or with $\rho' = (w/-i\sigma) = \bar{\rho}N^2/g$,

$$0 = -\frac{\partial p'}{\partial z} + \frac{N(z)^2}{g}w - \frac{g}{i\sigma}R(z - z_s)e^{ik_0 x}.$$

The equations are still separable, but the vertical structure equation is now

$$\frac{d^2 G(z)}{dz^2} + \gamma^2 N^2(z)G(z) = g\gamma^2 R(z - z_s), \tag{7.19}$$

subject to boundary conditions, and a similar one exists for $F(z)$. Let $G^{(p)}(z)$ be any particular solution of Equation (7.19). In general it cannot be expected to satisfy the rigid-floor boundary condition, and so without actually finding it, define the vertical velocity it implies at $z = 0$, as w_p. To obtain a solution that does satisfy the boundary condition, add to it any solution(s) of the free equation

$$\frac{d^2 G}{dz^2} + \gamma^2 N^2(z)G(z) = 0, \tag{7.20}$$

but now with the inhomogeneous boundary condition $w(z=0) = -w_p$. Evidently, any such "free" solution must be proportional to $\exp(ik_0 x - i\sigma_0 t)$ and thus the eigenvalue γ is known from the horizontal velocity equation, which remains unchanged. For example, ignoring β, it must be true that $\gamma^2 = k_0^2/(\sigma_0^2 - f^2)$ which is positive only if $\sigma_0 > f$. For $\gamma^2 > 0$, the solutions to Equation (7.20) are wavelike and can travel unattenuated away from the disturbance level, z_s. However, if $\gamma^2 < 0$, the disturbances are exponential and will be greatest in the vicinity of $z = z_s$, corresponding to the negative-depth eigenvalues previously discussed. Given the homogeneous solutions (sines/cosines or sinh/cosh for constant $N(z)$, or Bessel functions for the exponential version) particular solutions can be found by variation of parameters methods.

This development is only a sketch. The vertical structure equation uses an $N(z)$ computed from the density of a resting gas, and the source term is assumed, in part, to arise from solar heating of atmospheric ozone. Many complications, including the influence of strong background shear flows, can be considered.

CHAPTER 8

Balanced Motions

8.1 THE NATURE OF THE VARIABILITY

Oceanic motions on time scales exceeding about $2\pi/f$ and occupying spatial scales exceeding about 10 km, are well described through a near balance of the Coriolis and pressure forces and are thus generically referred to as *balanced*, or *geostrophic*, *motions*. Completely steady motions, as well as slowly time-varying ones, can be balanced to a high approximation. Balance says nothing about their generation, dissipation, or temporal evolution. These latter elements can take on a great many forms, that involve both direct and indirect forcing and dissipation of many types, including topographic interactions, linear and nonlinear frictional effects, and mechanisms that also arise as products of various instabilities. As expected from this large number of possibilities and combinations, several different types of balanced motions exist in the ocean, and making useful generalizations about their distribution and physics is not easy, but a few overall characteristics can be found.

As described in earlier chapters, the observational technologies available to oceanographers until the 1970s only occasionally permitted the acquisition of time series of any kind, the lack of which led to the depiction of the ocean circulation as essentially quasi-steady on basin scales. Although it was known that important variations on small scales existed (e.g., Maury, 1855, Helland-Hansen and Nansen, 1920), little or nothing could be said about their physics or their role, and they came to be regarded as a nuisance—primarily a noise in the larger-scale data.

With the advent of self-contained moorings and subsurface drifters in the 1970s, and particularly as record lengths grew, delineating the nature of this apparent noise became possible. Out of those observations and the growth in computer power came the field of geostrophic (or mesoscale) eddy research (e.g., MODE Group, 1978; Robinson, 1983; McWilliams, 2008), and how to represent eddies in inadequately resolved models is a central focus of research.

8.1.1 The Forcing

Without a continual input of energy, the ocean would come to rest on a time scale on the order of of a few thousand years, although major subelements would be "spun down" far faster—in a few days, years, decades, or centuries. (Ocean energetics will be discussed later.) For purposes of understanding oceanic variability, it largely suffices to separate the nominal time-mean forcing fields, such as heating through the seafloor (geothermal), the time-average winds, etc., and focus primarily on those elements whose temporal variations are significant over the duration of the instrumental records. In all cases, time averages depend upon the record length, sometimes very sensitively so, and even the geothermal fluxes do vary with time.

Tidal forcing of the ocean is a major influence, but apart from the long-period tides, the motions generated primarily involve gravity wave physics. Most important for balanced motions are the direct meteorological drivers, and only secondarily, freshwater input extraction from land runoff and storage and ice melt and formation. At the present time, the most widely used estimates of large-scale atmospheric forcing of the ocean are derived from meteorological *reanalyses*. These are the product of what meteorologists call *data assimilation*, as discussed in Appendix B. While they are useful approximations, they are not "truth," differing significantly among themselves and suffering from a variety of errors, some of great importance to ocean variability.

Figure 3.35 depicts the dominant time scales of the wind field, which are derived from a reanalysis and which range from on the order of 10 d in the tropics to 2–3 d at high latitudes. Luther (1980) and Chave et al. (1991) provided comparatively rare discussions of the atmospheric frequency-wavenumber structure, and those discussions are restricted primarily to periods of days.

8.2 IN SITU OBSERVATIONS

8.2.1 Moorings

Apart from a handful of island tide gauge data, observations of low-frequency variability[1] began with time series on moored instrumentation and from the various float types. The physics of open-ocean variability divide usefully, if not completely, by time scale. Thus referring back to the time series and frequency spectra shown in Chapter 3, the internal wave band, $f \leq \sigma \leq N(z)$, and the band inferred to be in balance, $\sigma < f$, were found to be almost ubiquitous (the equatorial region is excluded); see Figure 8.1. It was inferred, primarily on the basis of theory rather than on direct observations, that a third, roughly three-dimensionally isotropic band existed, $N(z) < \sigma$, having a high-frequency cutoff thought to be at the molecular scale (e.g., Thorpe, 2005, Chap 6).

Stream function maps (calculated pressure fields at a fixed depth in a geostrophic balance) from coherent arrays and float clusters, such as those from the MODE Group (1978), produce a good idea of the physical scales in the balanced motions for durations of a few months (Fig. 8.2). Some crude velocity- and temperature-covariance estimates emerged, but little insight into frequency-wavenumber characteristics was possible until the advent of accurate satellite altimetry in 1992. Maps such as that in Figure 8.3 show how sparse the in situ Eulerian coverage of the global ocean continues to be.

---

[1] Here "low-frequency" is by comparison with inertial-internal wave motions.

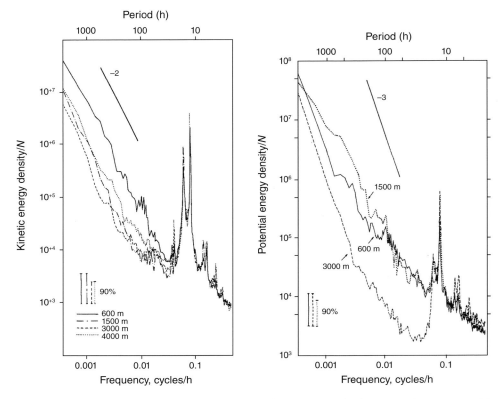

Figure 8.1: Typical midlatitude current meter spectral estimates in the frequency domain (cm/s)^2. The internal wave, balanced bands, and inertial peak are apparent. The measurement region is the Porcupine Abyssal Plane in the northeastern Atlantic Ocean, with the power laws and different confidence intervals shown. (From Mercier and de Verdiere, 1985.)

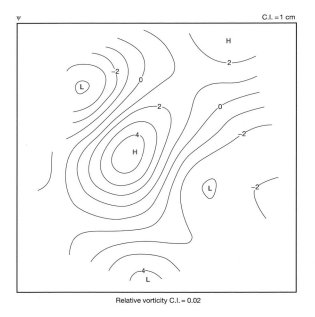

Relative vorticity C.I. = 0.02

Figure 8.2: Streamfunction as estimated on particular day during the Mid-Ocean Dynamics Experiment (MODE). 750 m is given as elevation $p' = \rho g \psi$. The region is about 500 km on a side. Compare to Figure 3.31. (After McWilliams, 1976.)

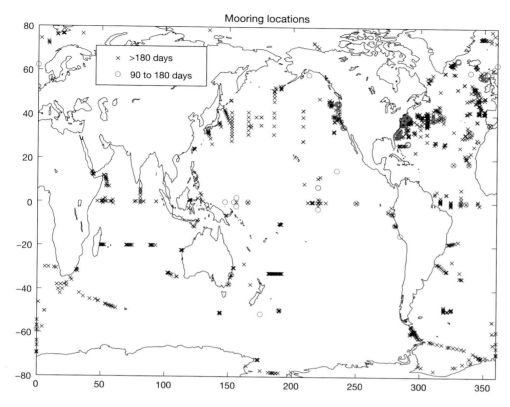

Figure 8.3: Positions of current meter records available for analysis circa 2008. Their vertical extent through the water column is very variable, with many more records near the seafloor than anywhere else in the water column. *x* denotes records at least 180 days long, and *o*, those between 90 and 180 days. Almost no durations exceed two years. Compare these durations to those seen in Figure 11.15. (From Scott et al., 2010.)

Almost from the first measurements, it was noticed (e.g., Freeland et al., 1975; MODE Group, 1978) that the phase lines of oceanic disturbances had a strong tendency to move westward, and this feature was taken as a demonstration of the existence of Rossby-wave-like dynamics. Such westward tendencies are made obvious by plotting longitude-time diagrams for almost any observable property, such as temperature or color.

Following MODE and similar field programs, a number of exploratory, isolated, current meter and temperature-measuring moorings were deployed around the world. Because these were not coherent arrays, the results were confined primarily to determinations of the kinetic energy with depth and some notion of the shape of the frequency spectrum (Schmitz and Luyten, 1991). Occasionally, moorings were densely enough instrumented to provide an indication of the vertical mode structure.[2]

In the decades since the technology became robust, thousands of moorings have been set in the ocean—with most time durations ranging from days to about two years, with a few isolated longer exceptions. Figure 8.3 shows the positions of many of the records available in the early 2000s. A major problem was that the moorings, installed for other purposes, were

[2] Wunsch, 1997.

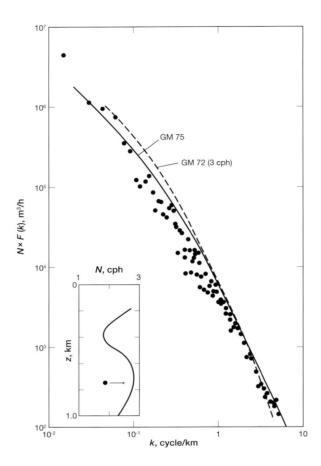

Figure 8.4: A wavenumber power density spectral estimate for the vertical displacement of the 12° C isotherm, converted to potential energy, south of Bermuda. The "red" character is apparent. The comparison made is to the Garrett-Munk internal wave spectrum, with the inference that it can account for the entire variability. GM 72 and GM 75 denote two different forms of the spectrum using $N_0 = (3\text{cph})2\pi$. $N(z)$ shown in the inset. (From Katz, 1975.)

either instrumented only near the seafloor, or only very thinly in the vertical. Few records are of sufficient duration to produce stable statistics for periods of months to years, much less for decades. Without measured horizontal coherence, the direct determination of lateral scales is impossible.

8.2.2 Shipborne Instruments

Isolated moored devices produce records permitting frequency-domain separation of physical phenomena. Determination of lateral scales has relied heavily on the ship-borne towing of instruments through the ocean. Results such as that shown in Figures 8.4 and 8.12 have been the basis for extracting spatial information.

Figure 8.4 shows a wavenumber spectral estimate compared to the Garrett-Munk internal wave spectrum. A great complication, however, is that theories of the balanced motion, discussed below, produce significant amounts of energy on these same scales, down to about

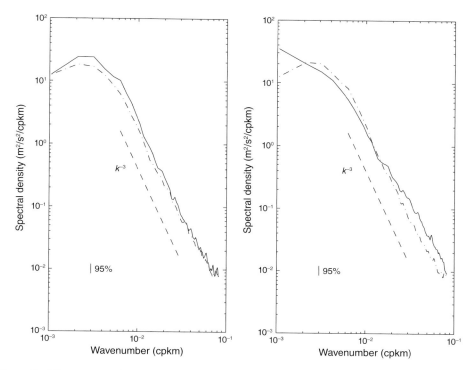

Figure 8.5: The wavenumber spectral estimates for kinetic (left panel) and potential energy (right panel) in the Gulf Stream region estimated by Wang et al. (2010). The solid line shows the zonal values, the dash-dot line, the meridional. An approximate 95% confidence interval is indicated, as is a k^{-3} power law. Here, in contrast to Figure 8.4, the authors interpret the results as arising from balanced motions.

10 km wavelengths and below (see Fig. 8.5). Flow fields in the very near surface layer must also include a strong contribution from directly wind-driven frontogenesis and turbulence on all space and time scales. Separating the contribution from these distinct physical regimes requires both temporal and spatial resolution—still a very rare commodity in oceanic data.

8.2.3 Float Observations

An ideal, neutrally buoyant float is a Lagrangian tracer moving in lockstep with the surrounding water parcel. Deviations from perfection inevitably occur. Many floats, including the vast majority of those deployed thus far, have been isobaric, tending to stay at a fixed pressure-depth despite vertical flows past them. Some floats with active power sources do attempt to track particular isotherms or isopycnals, rendering them closer to ideal behavior.

Ignoring nonideal characteristics, it has been observed that clusters of floats undergo both a shift in their center of mass, as well as a spatial dispersion over time. That dispersion reflects both temporal and spatial scales and has been of intense interest in the study of oceanic mixing processes. LaCasce's paper (2008) is a good starting place for understanding the data. Results, unfortunately, have remained primarily regional in nature, and the center-of-mass movement of clusters reduces the coverage time in any particular place. Most of the time, the Argo floats are sufficiently distant to represent an incoherent sample.

8.3 ALTIMETRIC DATA:
GLOBAL CHARACTERISTICS

By the late 1970s, it had become evident that many decades would be required if moorings or floats were used to characterize the intense oceanic variability on time scales of approximately a year. A major feature of altimetry is that, to a good approximation, it detects only the geostrophic portion of the flow field. Although this character may seem like a serious limitation—surface flows contain a variety of physical components—it proves to be advantageous. Only the geostrophic portion directly mirrors the physics deep into the interior. Near-surface ageostrophic components, which may well dominate the kinetic energy, are thereby weakened, and it is this simplifying physics of approximately "seeing" only the geostrophic part that has led to the revolutionary impact of altimetry on the study of balanced motions. (The observed, aliased, internal tide shows a violation of this approximation.)

Figure 2.18 displays the basic TOPEX/POSEIDON-Jason ground track, which has been maintained for 20 years, plus "infill" from other satellites. The latter data are commonly more dense in space, but at the price of longer repeat intervals (the basic trade-off as dictated by orbital dynamics). Subject to various residual errors, the altimetric data are often gridded uniformly at time intervals, Δt, of 7–10 days. Because several different altimetric satellites have flown simultaneously, having differing repeat patterns and hence ground tracks, no obvious choice of a global value of Δt exists after their data are combined, although the 10 d coverage by the TOPEX/POSEIDON-Jason series is a plausible one, meaning that altimeters are capable of depicting oceanic variability with frequencies less than 1/20 d (the Nyquist frequency). The situation is, however, not quite that simple owing to the aliasing of higher-frequency oceanic variability. In particular, the largest time variable signals in altimetric data are the tides, with the principal lunar tide, M_2, appearing clearly at a frequency of about 1/61 d and the other tides being visible as spectral lines distributed over the entire frequency band between 1/record length and the Nyquist frequency (see Table 7.1). Tides are effectively removed from the raw data because they are both predictable and have relatively very high signal-to-noise ratios. The possibility of any further aliasing from the background continuum is then usually ignored.

8.3.1 What Does the Altimeter See?

Assuming that the altimeter records the elevation, η, then in a hydrostatic system $\rho g \eta$ is the surface pressure distribution. That in turn is a dynamical quantity related in different ways to the motions below, and the essence of the science lies with understanding possible inferences about the deep interior.

A snapshot of η relative to an 18-year mean is displayed in Figure 3.31, which shows the complex patterns and geographical inhomogeneities of surface pressure anomalies. With the horizontal gradients at the surface known, then under the geostrophic assumption, the temporal kinetic energy in the variability can be computed from the multiyear altimetric record, resulting in Figure 3.32. It shows one of the most important, zero-order, descriptive features of the ocean circulation—the extreme geographically inhomogeneous character of the variability at the sea surface. Western boundary current regions, the Circumpolar Current, and the equatorially enhanced variations among other regions are visible. This complexity suggests that no universal theory of surface elevation change is likely. Differing physics varying with space and time, geographical position, and in some cases, season, must be considered.

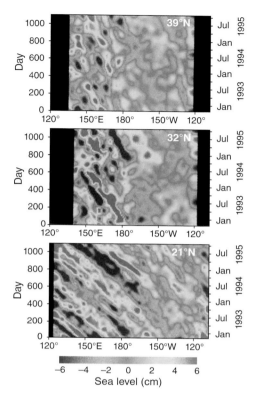

Figure 8.6: A time-longitude diagram for three latitudes in the Pacific Ocean showing the pronounced westward propagation of identifiable phase lines. (From Chelton and Schlax, 1996.)

8.3.2 Time-Domain Representations

Animations of the nearly 20 years (as of this writing) of altimetric measurements are available on a number of websites. Away from the equator and the Southern Ocean, and as seen in the Mid-Ocean Dynamics Experiment, a visibly striking, dominantly westward, propagation of features in ocean circulation is apparent. A time-longitude diagram, Figure 8.6, shows the conspicuous westward propagation seen in altimetric data over much of the ocean. Surface properties, such as ocean color (Cipollini et al., 2001), have similar features. As in many aspects, the Southern Ocean is distinct in showing dominantly eastward movement.

Chelton and Schlax (1996) noted that the dominant phase speed visible in Figure 8.6 at midlatitudes was systematically faster than that predicted by the basic Rossby wave dispersion equation (4.54). This inference struck a chord in the theoretical community and led to a literature attempting rationalization. That westward propagating motions with a definable scale are important in the surface ocean is apparent; what they imply about the oceanic interior, in terms of the vertical modes or structure, is less obvious.

8.3.3 Frequencies and Wave Numbers

Visual inspection of Figures 3.31 and 8.6 and the animations now available are strongly suggestive of a very diverse set of phenomena, all of which are present simultaneously. To

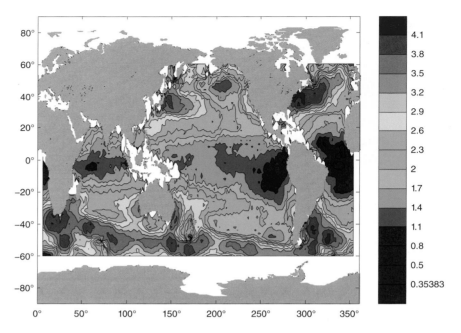

Figure 8.7: Estimated regional variations in the power-law behavior of the one-dimensional spectra of η on wavelengths of 7–250 km. A multiplicity of regimes is apparent. Noise problems arise at the high wavenumber end (see Fig. A.13). Compare to the spatial energy changes in Figure 3.32. (From Xu and Fu, 2011.)

isolate different dynamical regimes, a frequency-wavenumber spectral depiction, $\Phi(k, l, \sigma)$ of the motions is helpful. The natural representation of the spatial structure of wave-like motions on a sphere is in terms of spherical harmonics, but their straightforward interpretation for the oceans is precluded by the complicated ocean geometry.[3] The conceptually simpler, Cartesian zonal-wavenumber–frequency problem, is used here, while acknowledging its limitations.

Stammer (1997) and others have attempted to describe regional variations underlying the global average values most commonly by integrating out two of the variables, k, l, σ, to produce one-dimensional wavenumber or frequency spectral estimates, or by reducing $\tilde\Phi$ to the scalar wavenumber $\kappa = \sqrt{k^2 + l^2}$. Most recently Xu and Fu (2012) summarized the wavenumber, κ, characteristics as a function of region (Fig. 8.7). The range of wavenumber power laws is roughly from -1 to -4, being "flattest" in the low-energy regions. Hughes and Williams (2010) mapped the frequency power law variations as shown in Figure 8.8, which shows a very crude similarity to the wavenumber power law variations.

Scharffenberg and Stammer (2011)[4] exploited the so-called tandem mission when TOPEX/POSEIDON and its successor Jason-1 were flying for nine months simultaneously in horizontally displaced orbits. They were able to discuss wavenumber spectral shapes for zonal and

---

[3] As described by Wunsch and Stammer (1995), the expansion of quantities such as $\eta(\theta, \lambda, t)$ into spherical harmonics can be used to calculate equivalent Cartesian spectra $\tilde\Phi(k, l, \sigma)$. Because η is not defined over the continents, either the nonorthonormal expansion over the oceans must be done by least-squares, or some plausible extrapolation over the continents (such as $\eta = 0$) is required.

[4] See the corrections to these results discussed by Wortham et al. (2014).

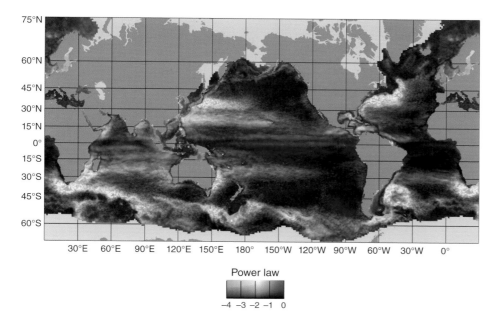

Figure 8.8: The estimated power law of the *frequency* spectrum of altimetric variability. The range is from 4 (red) to zero (blue). Compare to the wavenumber power laws in Figure 8.7. (From Hughes and Williams, 2010.)

meridional velocity (u, v) separately. It is worth emphasizing that in terms of the horizontal vector wavenumber (k, l) that even in an isotropic velocity field, one-dimensional spectra $\Phi_u(k) \neq \Phi_v(k)$. That result is an immediate consequence of quasi two-dimensional mass conservation,

$$\frac{\partial u}{\partial x} + \frac{\partial v}{\partial y} \approx 0, \text{ or}$$
$$k\hat{u}(k, l) + l\hat{v}(k, l) \approx 0,$$

the carats denoting the Fourier transforms.

Figure 8.9 displays for η an estimated frequency zonal-wavenumber $(s, k'$, both circular) power density spectrum along a midlatitude line in the eastern Pacific. The intense motions along the *nondispersive* line is apparent, as is the presence of energy at all frequencies and wavenumbers.

A number of proposals have been made for approximate analytical expressions describing the frequency and horizontal wavenumber power densities. These have been based upon altimetric and various compilations of in situ observations, but without any physics beyond the requirement of the simplest reproduction of observed spectra. Recognizing the inevitable large spatial variations, these spectra contain slowly varying parameters that are adjusted to particular areas. A recent such form is[5]

[5] Wortham and Wunsch, 2014.

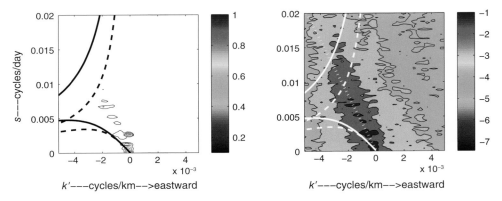

Figure 8.9: A frequency, s, circular zonal-wavenumber, k', spectral density estimate (linear amplitude scale on the left, logarithmic amplitude scale on the right) computed from altimetric data along a zonal line at $27°$N in the North Pacific. Recalling the discussion of Figure 4.1, the assumption is made that the meridional wavenumber, $l \approx 0$. Solid lines are the flat-bottom resting-ocean dispersion relationship for the barotropic (upper) and first baroclinic modes, while the dashed lines are the corresponding curves when $l = k$. The dotted line is the *nondispersive line*, which produces the dominant westward phase velocity. Note, however, that energy exists at all frequencies and wavenumbers. (From Wunsch, 2010.)

$$\Phi_\psi\left(k',l',s,n,\mathbf{r}\right) = \tag{8.1}$$

$$E(n)\,I(\mathbf{r})\left[\underbrace{\frac{1}{\left(k'^2 L_x^2 + l'^2 L_y^2 + 1\right)^\alpha\left(s^2 T_s + 1\right)}}_{\text{"turbulence" + waves + instabilities}} + \underbrace{\exp\left[-k'^2 L_x^2 - l'^2 L_y^2 - T_s^2\left(k' c_x + l' c_y - s\right)^2\right]}_{\text{non-dispersive waves}}\right]$$

Here, k', l' are circular wavenumbers (cycles/km), and s is the circular frequency (cycles/d). L_x, L_y, T_s are spatial and temporal parameters taken to be slowly varying functions of geographical position, as is the exponent, α. c_x, c_y control the slope of the *nondispersive* line. Φ_ψ is intended to represent the spectral density for the geostrophic stream function, for vertical mode, n. That is, all vertical modes are depicted as having the same distributions of frequency and horizontal wavenumber. The first term in Equation (8.1) represents a general background continuum. The second, exponential, term arises from the nondispersive line behavior seen at most latitudes outside the Southern Ocean and the northern extremes. $I(\mathbf{r})$ is a scale factor dependent upon the horizontal position $\mathbf{r}$ and $E(n)$ is the relative contribution from mode n. $c_x \approx -\beta R_D^2$—the long-wavelength/low-frequency limit of the first-baroclinic-mode dispersion relation, which is shown in Figure 8.9. This term diminishes rapidly with latitude (both β and the Rossby radius, R_d become very small), so that at high northern latitudes, even a 20-year record is not long enough for the nondispersive line to be visible (if it exists there).

Corresponding spectra of all fields, $u, v, w, p, \ldots$, can be derived from Φ_ψ. One and two-dimensional frequency, or wavenumber, or frequency-wavenumber spectra are also implied. Wortham and Wunsch (2014) propose strawman values globally and provide charts of estimated values of c_x, etc.

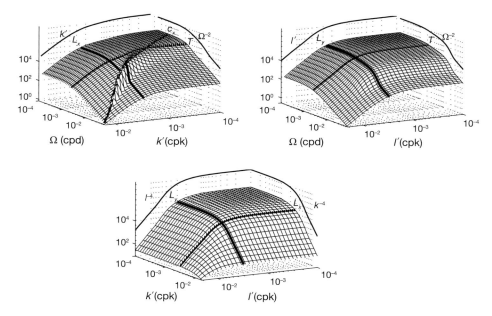

Figure 8.10: The three two-dimensional projections of the spectrum of surface elevation η near 30°N, 190°E in the Pacific. The top left is for frequency, s, and zonal wavenumber k'; the top right, for s and l'; and the bottom for k', l'. The nondispersive line is most conspicuous in the s, k' form. (From Wortham and Wunsch, 2013.)

The spectrum of surface elevation, η, can be found and is displayed in Figure 8.10 as three two-dimensional forms. Equivalent results can be calculated for the one-dimensional spectra and for the corresponding values for all physical parameters of the dynamical equations. These can be compared in detail with the underlying spectra from individual instruments.

It is important to keep in mind, however, that the high-wavenumber behavior, beyond about 1/200 km, of the altimetric data is essentially unknown. The internal wave spectrum of Garrett and Munk (1972) contains energy at 100 km and shorter, and as yet there is no practical observational method for separating internal wave energy from that of the geostrophically balanced flows[6] in the region of overlap (see Fig. 8.4). The conspicuous appearance of internal tides in altimetric data shows emphatically that internal waves of large vertical scale (including low modes) will be present there. That altimeters respond only to balanced motions is a useful, but incomplete, approximation. A full oceanic frequency-wavenumber spectrum must ultimately reflect the contribution from balanced motions, internal waves, and other ageostrophic energy.

8.4 VERTICAL STRUCTURE

Manipulation of the geostrophic equations shows that balanced disturbances of horizontal scale L, seen at the sea surface, must reflect motions of the same horizontal scale into the interior. A strong tendency exists (not universally so) for the motions to have vertical scale, d,

[6]See Callies and Ferrari, 2013.

so that $Nd \approx fL$. The main issue is whether the physics are barotropic (in the mode-definition sense), or baroclinic. If the former, penetration is unattenuated to the seafloor. If baroclinic, penetration to the seafloor can still occur, but with a complex vertical structure.

The modal picture of time-dependent motions is at best a starting point—based as it is upon the theory describing a resting, horizontally stratified ocean without topography. Various studies show, nonetheless, that a comparatively small number of these modes does a quantitatively good job of reproducing what measurements do exist. This theory produces a complete set of functions, implying that for any single field (e.g., temperature), an arbitrary vertical structure can be reproduced exactly, if enough modes are included. That result is useful, however, only if the number of functions required is small. Furthermore, no guarantee is available that the same modes and their amplitudes and phases would be consistent with related fields. Thus if particular modes $F_n(z)$ describe the velocity fluctuations, the corresponding $G_n(z)$ will not necessarily describe the temperature changes.

The MODE Group (1978) had already shown an apparent dominance by the barotropic and first baroclinic modes in temperature and velocity data from moored devices (generally periods less than about 2 years). A very crude summary of the global situation would be that the variance at these periods contributing to the total balanced kinetic energy of the *water column* on a global basis is partitioned roughly as 40% in barotropic motions, 40% in the first baroclinic mode, and the remainder in higher modes and otherwise unexplained "noise."

These values do not describe the energy partition at the sea surface because the baroclinic modes are amplified there, roughly in accord with the expected behavior from the WKBJ approximation. $N(z)$ tends to grow near the surface (but is often overlain with a mixed layer, which is nearly invisible to low modes; see the box "Sturm-Liouville and Bessel Function Solutions"). Roughly 5–10% of the *surface* kinetic energy is in the barotropic mode, with perhaps 70% in the first baroclinic one (Fig. 8.11). The most energetic barotropic motions have a much larger lateral scale than do the baroclinic ones, and thus a correspondingly reduced velocity. Higher baroclinic modes produce only a small surface deflection—making them undetectable to the altimeter in the presence of noise.

OTHER INTERPRETATIONS

In recent years, alternative interpretations of the surface balanced motions as manifestations of the interior geostrophic modes have been offered. Chief among these is the so-called surface quasi-geostrophic (SQG) theory, an approach borrowed from studies of the lowest part of the atmosphere. One motivation for this view is to recognize that large-scale temperature (or density) fields visible at the sea surface undergo motions potentially controlled by the underlying geostrophic motions. In the standard theory, these motions would be regarded as essentially passive responses to the interior modes and not having any effect on their physics. In the SQG theory, the presence of surface buoyancy features is regarded as central to the motions and is introduced by changing the surface boundary condition in Equation (4.81) to one with a given density perturbation, ρ'. Some model results seem to support this interpretation, and various extensions have been made to the basic theory.[7] Interaction within and below the mixed layer of the strongly ageostrophic flows with the assumed balanced flow is not entirely clear (see the cartoon in Fig. 11.1). In summer at least, the temperature at the

---

[7] Klein and Lapeyre, 2009; LaCasce, 2012.

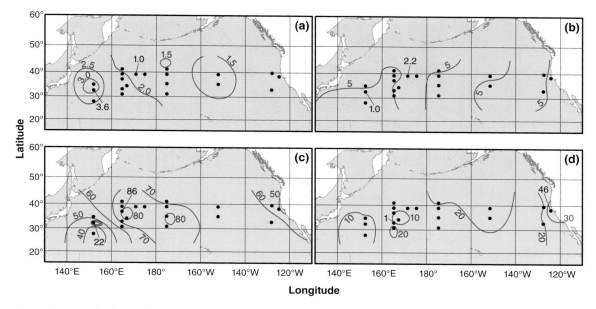

Figure 8.11: (a) The $\log_{10}$ of the surface kinetic energy in moored data. (b) The percentage estimated to be in the baroropic mode, (c) the percentage in the first baroclinic mode, and (d) the percentage in the second baroclinic mode. Relative contributions to η are very different. (From Wunsch, 1997.)

sea surface may have little or no connection with that at the base of the mixed layer, which is invisible from space (see Fig. 8.12).

A further complication comes from the ubiquitous presence of wind forcing. Wind-forced motions will drive not only direct changes in the sea surface temperatures, but also excite the conventional positive and negative depth modes, with Ekman pumping and suction involved (see Eq. 10.13). Distinguishing the negative or positive depth solutions from those of the SQG models has never been done convincingly in any mooring data available thus far.[8]

In a more general sense, growing attention has been paid to what are sometimes called "submesoscale" motions (below about 10 km horizontal scale), particularly as they are manifested in the surface boundary layer, and often connected with the formation of front-like features.[9] Some of the growing understanding of the observed structures can be found in the paper by Ferrari and Rudnick (2000) and in the subsequent literature.

8.5 SPECTRAL INTERPRETATION

Observed frequency, wavenumber (including the vertical), and frequency-wavenumber spectra have complex forms. Rationalizing these forms, with most attention being directed at the apparent power laws appearing in the various one-dimensional projections, is ongoing. A great difficulty in this endeavor is the clear presence at any location in the ocean of several competing phenomena, including various baroclinic and barotropic instabilities, wind-forced

---

[8] See Callies and Ferrari, 2013; Wang et al., 2013.
[9] The 10 km transition is somewhat arbitrary. Sometimes criteria such as requiring that the Rossby number exceed one are used. Ubiquitous weaker motions on these small scales also exist.

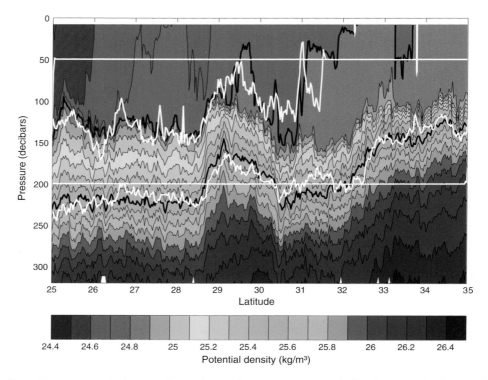

Figure 8.12: Isopycnals along 140°W in the Pacific Ocean. The jagged white lines are the depths of a towed vehicle. The horizontal gradients within and at the surface of the mixed-layer are clear, but their relationship to those at the mixedlayer base is not a simple one. (From Ferrari and Rudnick, 2000.)

motions, linear Rossby waves, coherent structures, and vortical modes. The hope, not often realized, is that clear dominance by one of these physical systems in some region can be shown, so that a single physical process can be demonstrated to explain the observations. Whether such regions exist is not obvious.

In any case, much theoretical endeavor has been directed at the wavenumber power laws, what they predict, and the degree to which they describe the data, and we now turn to an overview.

8.5.1 Wavenumber Power Laws

Theories and numerical experiments support the suggestion that the balanced turbulence of a rotating, stratified ocean follows certain simple power laws. What follows is based loosely on the discussions by Vallis (2006, Chap. 8 and 9) and Ferrari and Wunsch (2009), which should be consulted for more detail. In the "conventional" theory of homogeneous, isotropic turbulence (no rotation, stratification, or boundaries), a simple and elegant theory owing to Kolmogoroff predicts that the distribution of kinetic energy in scalar wavenumbers, $\kappa = \left(k^2 + l^2\right)^{1/2}$, should be proportional to $\kappa^{-5/3}$ and extend to the very small scales (large κ) where laminar friction dissipates energy. The basic inference is that nonlinearities in the large-scale, essentially, inviscid, components of the flow interact with their neighboring wavenumbers so as to drive energy toward the small, dissipating scales. That is, there exists a *direct cascade* of energy from large to small scales. The Kolmogoroff theory is well supported by laboratory, numerical, and some field experiments.

Discussion of such processes in the ocean started with Charney's (1971) calculations for two-dimensional, balanced barotropic turbulence, instead of the three-dimensions of the nonrotating Kolmogoroff theory. As in many fluid-dynamical situations, two-dimensional flows are strikingly different from three-dimensional ones. Arguments show that if energy is put in at a particular wavenumber, κ_I, that energy should flow toward *larger* scales. Thus, in such systems, the powerful constraint of two-dimensionality *reverses* the direction of the energy cascade—it becomes *upscale*—to larger scales, where the energy dissipation must ultimately take place. This *inverse cascade* runs counter to the conventional assumption that energy is dissipated in a turbulent fluid by transformation to ever-smaller scales. Some other process must act to remove energy at the large scale—it is a serious modelling problem that has in recent years focussed particular attention on bottom and sidewall topography.

In contrast, the *enstrophy*, or vorticity squared of the fluid, which is defined in two dimensions as

$$Z = \iint dA \left| \nabla_h^2 \psi \right|^2 , \tag{8.2}$$

ψ being the stream function, has a direct cascade. Vorticity, and its most visibly prominent structures, congregate at the smallest scale.

For wavenumbers less than κ_I, upscale two-dimensional turbulence is predicted to have the same kinetic-energy wavenumber power law as in Kolmogoroff turbulence, $\kappa^{-5/3}$, although the physics is different.[10]

If the ocean were of constant density, unforced, and with a flat bottom, the two-dimensional turbulence argument could be applicable, insofar as homogeneous rotating fluids tend to be at the Taylor-Proudman limit. Possibly some of the behavior of the demonstrably barotropic components of oceanic variability is determined by this physics. Applying this much-simplified theory to the ocean requires adding numerous complicating factors. The ocean is stratified, and basic observations suggest a mixture of barotropic-like and baroclinic-mode-like behavior, as already described. Purely barotropic motions aside, two-dimensional motions do not exist, although quasi-geostrophic-balanced motions share many of those properties because the vertical velocities are very small compared to the horizontal ones. Simple scaling laws show that the β-effect cannot be ignored entirely, that topographic gradients are very strong in many places, that the wind field injects energy into the ocean over a broad, rather than an identifiable single scale, and that instability theories of the large-scale flow apply only to the situation where no other disturbances are already present.

Rhines (1977) showed that baroclinic energy tended to become more barotropic through nonlinear processes. The baroclinic modes have a downscale cascade, but as the energy flows into the barotropic mode, it moves to larger scales. Eventually it has wavenumbers sufficiently small that $\kappa \approx \sqrt{U/\beta}$, where U is an RMS flow strength. At that point the cascade stops—with the motion becoming wavelike. But given the strong wavelike motions on the nondispersive line, β must be important, at least at low and midlatitudes. Wavelike motions, including baroclinic Rossby waves, interact through triad interactions in frequency and wave-number, not nearest-neighbor ones (Longuet-Higgins and Gill, 1967), and are unstable,

---

[10] The coincidence of the predicted and observed power laws is evidently a necessary, but not sufficient, demonstration of the correct physics being used.

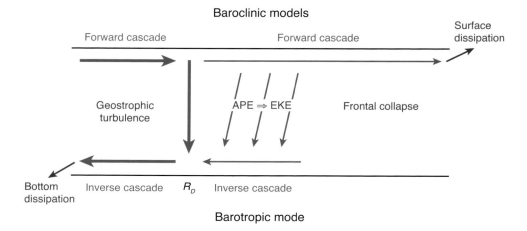

Figure 8.13: A schematic of the coupled baroclinic and barotropic energy cascades. At scales shorter than the baroclinic deformation radius, R_D, eddy available potential energy (APE) is converted efficiently into eddy kinetic energy (EKE) some of which is converted to barotropic motions, and some of which continues the forward cascade to the submesoscale and fronto-genesis. (From Ferrari and Wunsch, 2009.)

in some conditions, to other disturbances. Figure 8.13 is a schematic of the processes, with the barotropic and baroclinic energy cascades being in different directions, but coupled. This cascade story is an intricate one and its details are being brushed aside here (for which, see the references).

Xu and Fu (2012) have mapped the power laws as seen in the altimetric data (Fig. 8.7). These do not correspond in any simple way to the two-dimensional theories. In some limited regions near the western boundary currents, the power laws are consistent with the SQG theory, albeit the high wavenumbers of greatest significance in that theory lie beyond the existing altimetric resolution. As already noted, similarity of power laws is a necessary but insufficient condition for inferring the proper physics.

8.5.2 Frequency Spectra

Turbulence theories generally predict only wavenumber partitioning, there being no equivalent of the dispersion relationship for waves. Many more frequency than wavenumber spectral estimates from observations exist, but relations to the theoretical predictions have been elusive. Some calculations rest upon the Taylor hypothesis, that the turbulence is carried along by a large-scale advective flow, U, with the pattern "frozen" as it is carried past a fixed sensor, so that

$$\sigma = Uk, \tag{8.3}$$

in a form of Doppler-shifting. (U is not the same variable as in the Rhines theory.) Such a hypothesis is difficult to use in the ocean (Fig. 8.10) because no obvious scale separation exists in the spectral continuum in either frequency or wavenumber that would permit a choice of U that would not itself be effectively a random variable.

Regions of intense time-mean currents may be exceptional in showing Taylor advection (although spatial isotropy and homogeneity assumptions implicit in the theory become suspect). Even then, to the degree that the motions are "wavelike" in having a dispersion relation, the Taylor hypothesis would fail unless the phase velocity were much smaller than U—and again, Figure 3.17 shows that it is difficult to identify such regions (the Antarctic Circumpolar Current being a possible exception; see Wortham, 2012). In the open ocean, Doppler effects such as those in Equation (8.3) must be operating, but only in a statistical sense in which U is part of the spectrum. Mutual advective effects interact over all wavenumbers, so as to produce the observed frequency-wavenumber interaction. Arbic et al. (2012) made a brave attempt to diagnose frequency power spectra. No theory is available to describe all existing results.

8.5.3 Wave-Like Features

Chelton and Schlax (1996) had pointed out the dominance of the near-linear ridge in σ, k in Figures 8.9 and 8.10, being referred to here as the nondispersive line. Motions along that line are wavelike in that they obey the relation,

$$\sigma/k \approx -\beta R_D^2,$$

where R_D is the first baroclinic Rossby deformation radius. This phase velocity is appropriate to the first baroclinic mode only in the low-frequency limit (Eq. 4.54) and departs from it at higher frequencies and wavenumbers. (As already noted, McWilliams and Flierl, 1976, and the MODE Group, 1978, had already concluded that the "MODE eddy" was inconsistent with the basic linear theory.) That, in general, phase velocities are not quantitatively in accord with the perturbation theory for a resting, uniformly stratified, flat-bottom ocean is unsurprising. That the phase velocities were systematically too fast is more difficult to explain (see Killworth and Blundell, 2003). Wortham (2012), invoking the weak-bottom interaction topographic theory of Bobrovich and Reznik (1999) and related work, showed that the systematic effects of topography are to couple the barotropic and first baroclinic modes so as to minimize the bottom velocity and hence the topographic interaction. The coupled modes have phase velocities that are faster than that of first baroclinic modes alone, but slower than that of the barotropic one, and at least roughly consistent with the non-dispersive line. Little is directly known of the behavior of higher modes, and in general, details of topography must be very important.

In the interim, Chelton et al. (2011) argued that the dominant physics of the nondispersive-line disturbances should not be regarded as linear waves at all, but are much more closely akin to a nonlinear, coherent vortex motion. Invoking the Dirac maxim that if a phenomenon is physically possible then it does exist, it is still a challenge to make a quantitative partition into these different types of motion.

8.5.4 Balanced Barotropic Basin Modes

Despite its importance to η, oceanic barotropic variability has received far less attention than the baroclinic components, because it does not usually contribute any significant signal to temperature or salinity measurements. The altimetric data show, however, that at least

50% of the η variance is in components that appear barotropic. With their large scale and deformation radii, the handy Cartesian simplifications (f- and β-planes) cannot be used quantitatively. Finally, because of their large scales, such motions contribute only weakly to local geostrophic flow fields.

Study of barotropic motions involves basin-scale boundary phenomena, which are not included in spectral models such as Equation (8.1). Longuet-Higgins and Pond (1970) calculated the normal modes of a bounded hemispherical ocean. Platzman et al. (1981) in turn determined the normal modes of realistic ocean basins including variable topography. Great difficulty is encountered in computing such modes at low frequencies, where they represent balanced motions with a powerful tendency to follow localized topographic features (f/h contours), sometimes becoming trapped or confined to small areas. Such effects are obvious in the observations in some oceanic regions (e.g., Tai and Fu, 2005, for the Argentine Basin; Stammer et al., 2000, for the southeast Pacific Ocean; Warren et al., 2002, for the Mascarene Basin).

Do basin modes, either in the balanced or gravity field limits, exist outside of textbooks? Luther (1982) discussed evidence in tide gauge records for a weak excitation by the atmosphere of a barotropic balanced mode in the range of 4–6 d in the Pacific Ocean. He inferred that the mode was heavily damped, with an e-folding time for energy of less than 3 d, estimating $Q \approx 4$. Woodworth et al. (1995) suggested that a weak, sea level feature analogous to that of Luther (1982) was present in the South Atlantic. Figure 8.14 shows the power spectra and coherence observed at the Port Stanley, Falkland Islands, tide gauge. The broad coherence near five days is the main evidence, and the excited mode is not visible in the power density spectrum—being lost in the background noise. Periods on the order of 5 d are aliased in altimetric data and so are inaccessible in those measurements except by dealing with the energy in the aliased bands. (A 5 d oscillation with a 10 d sampling will alias into and near the sample time mean.) Most authors have inferred that the damping mechanism was dominated by small-scale topographic trapping of energy and some form of topographic scattering.

Wunsch et al. (1997) estimated a $Q \approx 5$ for the balanced barotropic fortnightly tide (Mf). In the gravity wave branch, Rabinovich et al. (2013) made an estimate of tsunami dissipation rates in the Pacific corresponding to a $Q \approx 4.5$.[11] Semidiurnal tide values of Q were described on p. 203.

Q in the oceans is very low for barotropic motions of any sort, owing to the complex geometry of the lateral boundaries, and in particular to the powerful scattering and trapping effects of the multiscale bottom topography. In some theories of El Niño (not discussed in this book) reliance is placed upon finite reflectivity for *baroclinic* motions at the western Pacific coastline. Theory suggests a very complex interaction with walls (see Moore, 1968; Lighthill, 1969; Pedlosky, 2000; Clarke, 2008; Zhai et al., 2010), and some published interpretations of the observed data are contradictory.

A reasonable summary is that if basin modes exist in the real ocean, their amplitudes are small; they are heavily damped by topographic scattering, trapping, and bottom friction; and they are unlikely to be very important. The picture emerges of the open ocean as acting qualitatively as though distant-enough boundaries were in practice infinitely remote, for both barotropic and baroclinic dynamical motions alike.

---

[11]Interpreting the natural period as $2 \times$ (width of the Pacific) $/\sqrt{gh} \approx 28$ h. Their results depend somewhat on location and the particular tsunami, which is made up of a superposition of much-higher-frequency gravity waves.

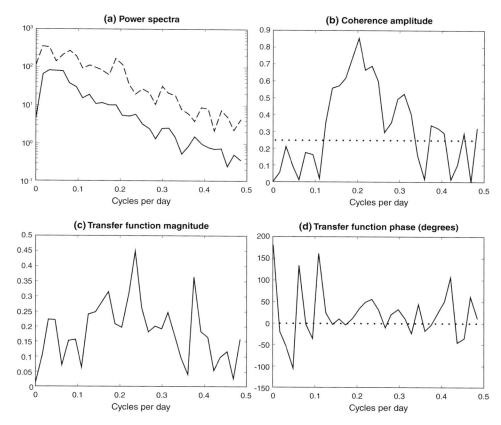

Figure 8.14: (a) Power density of sea-level (dashed line) and atmospheric pressure (solid line) at Port Stanley, Falkland Islands. (b) Coherence amplitude between atmospheric pressure and sea level at Port Stanley. The broad maximum around 5 d is interpreted by Woodworth et al. (1995) as showing excitation of a 5 d mode of the South Atlantic by the atmosphere. Despite the coherence, any excess energy in the power density at around 5 d periods does not appear as even a local peak—implying that the coherent modal energy is lost in the background noise. The magnitude of the transfer function (Eq. A.46) and its phase are shown in panels (c) and (d). (From Woodworth et al., 1995.)

8.5.5 Other Contributions: Vortical Modes

These relatively small, vertical-scale motions were briefly mentioned in the discussion of internal waves, as their signature is often found embedded in vertical profiles used to study those waves. They are meant to be in geostrophic balance on their somewhat larger scales, and in cyclostrophic balance at the smallest ones (see, e.g., Polzin et al., 2003). Vortical modes do not propagate like internal waves and their overall significance remains unknown.

SUBMESOCALE MOTIONS

The contribution to surface manifestations of velocity arising from the submesoscale remains an open question. Much of the discussion concerns the interaction with surface density fronts and a generalization at this time is premature. The most conspicuous submesoscale features in the open ocean have been the so-called *Meddies*, formed from the high-salinity Mediterranean outflow water, far below the sea surface, and travelling as nearly closed circulations for thousands of kilometers over many years (see, e.g., Shoosmith et al., 2005).

8.6 GENERATION OF BALANCED VARIABILITY

Large-scale oceanic variability arises from several distinct energy sources including (1) direct forcing by the time-dependent atmosphere and the gravitational tides; and (2) through instabilities of large-scale quasi-steady flows. Total energy in the balanced motions globally has been crudely estimated as 10 EJ (Ferrari and Wunsch, 2009).

Direct generation by the wind field has been estimated, also very crudely, as about 0.05 TW (Frankignoul and Müller, 1979), which would produce a renewal time of 10^{19} J$/0.05 \times 10^{12}$ W or about 7 years. On the other hand, the dissipation rate is thought to be about 0.5 TW (with likely a factor of 2 uncertainty), which is much too large for sustenance by direct wind driving. In contrast, rates of generation by various instability processes have been calculated also to be roughly 0.5 TW (Huang and Wang, 2003; Ferrari and Wunsch, 2009; Xu et al., 2011). At the present time, rough agreement exists, with the major production apparently coming from a mixture of baroclinic- and barotropic- and mixed-instabilities. Very large regional variations in these numbers are expected (see Fig. 11.6).

8.7 DISSIPATION OF BALANCED VARIABILITY

GENERALITIES

As a thought experiment, consider the fate of a dynamical oceanic disturbance, for example, in the flow field, with characteristic space scale L and that is "let go" to freely evolve. After a sufficiently long time, this initial disturbance will become arbitrarily weak. The routes to disappearance are varied and are here simply lumped together as "dissipation." Keeping in mind the immense range of space and time scales occupied by balanced oceanic variability, from the global to the submesocale, several forms of dissipation can be identified:

1. Friction. Kinetic energy loss to molecular friction is the ultimate dissipation, but it is not directly visible because of the very small scales involved. Nonetheless, it is inferred to equal the entire energy input to the ocean. More visible is the loss to mesoscale and other turbulence, whose own dissipation must then be considered. In models, and not necessarily the ocean, the loss to turbulent dissipation appears in several different forms, commonly represented at least schematically as horizontal and vertical friction in the Reynolds analogy as $A_H \nabla^2 u$, $A_V \partial^2 u / \partial z^2$, etc. within the water column. For orders of magnitude, $A_H \approx 10^2 - 10^4$ m^2/s, $A_V \approx 10^{-1}$ (see the later discussion of time scales). Upscale energy transfers lead to the invocation of bottom friction and topographic interactions as intermediaries able to reduce the spatial scales to ones where the Reynolds-like terms can operate.

2. The bottom frictional layer. Dating back at least to Taylor's (1919) discussion of tidal flows, the bottom stress is assumed to be of the form

$$\boldsymbol{\tau}_B = C_D \rho |\mathbf{v}| \mathbf{v}, \tag{8.4}$$

(Eq. 7.14) where **v** is the *total* bottom velocity, and C_D is again an empirical drag coefficient, the analog of Equation (11.3) for the air-sea interface. τ_B acts on all scales, and indeed couples them through its nonlinearity.

3. Scattering. Large-scale flows encountering topography will generate smaller-scale flows through irreversible processes here lumped as "scattering," but also including topographic trapping, and thought to be important in suppressing basin modes. A high-frequency example has already been encountered as the generation of internal tides by the barotropic component flowing over smaller-scale topographic features. More generally, large-scale, low-frequency flows will tend to generate so-called *lee waves*, which can occur at all frequencies from internal waves and the balanced motion band. Energy extracted from the background flow is never returned to it, and it is effectively dissipated. In the case of balanced motions, the precise nature of this scattering loss is much less clear, ranging from the generation of eddy-like flows to "loss of balance." Direct generation of nongeostrophic flows becomes of central interest in oceanic dissipation. For general discussions see Molemaker et al., 2010; Nikurashin and Ferrari, 2011; or Vanneste, 2013. The trapping mechanism refers to the tendency for near-bottom flows to follow lines of constant f/h, and hence to potentially produce smaller horizontal scales of flow than found in the incident wave.

4. Coastal absorption. A balanced motion encountering the continental margins (Fig. 3.5) will undergo enhanced scattering and frictional processes, but with the scattering also generating trapped wavelike disturbances capable of propagating along the continental margin. The result would be a reflection coefficient for energy that could be very small. (See Zhai et al., 2010; Szuts et al., 2012; Marshall and Johnson, 2013.)

5. Forced destruction. Large-scale time-variable flow phenomena, having been established by an atmospheric disturbance, can be then destroyed by later such disturbances. Visibly compelling examples are seen in altimetric animations of the tropical Pacific Ocean surface elevation, in which westerly wind bursts give rise to equatorial Kelvin waves and which on their way east across the equatorial Pacific Ocean are annihilated by easterly winds before arrival at South America, thus aborting incipient El Niños[12]. The only quantitative estimates pertain to the destruction rate of mesoscale eddies through the wind-stress interactions described later.

6. "Negative viscosity" phenomena. They are the opposite sign from conventional viscous effects and do not represent dissipation but acceleration. The most common discussions of instability describe the production of smaller spatial scales from larger-scale established flows. Smaller-scale flows, however, can interact with larger-scale ones so that energy and/or momentum can be fed *into* the larger scale. Much of this discussion pivots around *critical layer* phenomena, in which a *phase velocity*, c_p, of a time-variable motion coincides with a local flow velocity $U(\mathbf{r},t)$ of the larger scale. Owing to the very great difficulty of determining the structure of $c_p - U(\mathbf{r}, t)$ at depth, little is known of this process except for some model estimates, particularly in the Southern Ocean.

---

[12] Perhaps more properly *Los Niños*.

CHAPTER 9

The Time-Mean Ocean Circulation

The earliest physical oceanographers recognized the presence of basin- to global-scale contourable features in temperature and salinity, and eventually in oxygen and in various nutrients such as silica, phosphate and nitrate. Measurements could be combined into the striking pictures of the sort shown in Chapter 3 and now in online atlases. A steady-flow field was hypothesized whose spatial patterns were commensurate with the observed hydrographic patterns.

Beginning in the late 1940s, as ocean circulation theory gradually emerged along with geophysical fluid dynamics, the focus was very much on producing theories of the inferred quasi-steady, large-scale, fluid flow. The impressive and very interesting theoretical structure that emerged is the subject of the several textbooks already mentioned.

Most inferences about the ocean circulation were made using shipborne hydrographic measurements under the assumption that the data described some time average, usually only over a comparatively small region, such as an ocean basin or gyre. With the World Ocean Circulation Experiment (WOCE) and its associated CTD, tracer, satellite, float, etc., measurements in the early 1990s, it was possible for the first time to form estimates having the character of a true average over intervals of a few years and approaching a global scope. That possibility, however, opens the Pandora's box of determining the reliability of time averages over any finite interval of M−years, that is, whether they represent true long-term means. At the time this chapter is being written, about 20 years of high-accuracy altimeter data exist, with upper-ocean Argo float data having become quasi-global about 2005.

Ocean general circulation models (GCMs) also improved rapidly during that period so that it made sense to combine their information content about the physics with that contained within the data in a process of *state estimation*, as discussed in Appendix B, and which, in practice, is usually a form of least-squares fitting in a problem of immense dimension. Hypothetical proper use of a fitted, adequate GCM solves several problems: (1) All of the relevant physics (and chemistry and biology) can be included, and known equations are being satisfied; (2) Diverse measurement types can jointly address estimates of important elements; (3) Temporal variability existing in data sets spanning years and decades can be accounted for; (4) Explicit, quantitative descriptions can be used for the inevitable observation and model errors; and (5) Estimates can be made of important but not-directly observed variables.

The first part of the present chapter is, however, devoted to the results of the historical methodologies that formed the backbone of the subject until recently.

9.1 GEOSTROPHY AND THE DYNAMIC METHOD

The central feature of almost all observations of the ocean circulation and of their explanatory power again lies with the accuracy and utility of the balanced geostrophic, hydrostatic equations. In local Cartesian approximation the appropriate equations, repeated here, are

$$-f\rho v = -\frac{\partial p}{\partial x}, \tag{9.1a}$$

$$f\rho u = -\frac{\partial p}{\partial y} \tag{9.1b}$$

$$0 = -\frac{\partial p}{\partial z} - g\rho. \tag{9.1c}$$

For many purposes below, it is adequate to replace ρ by $\bar{\rho}$, except where multiplied by g. As balances, no provision exists for time evolution or dissipation or sources. The pressure field does not "drive" the flow, nor does the flow field "drive" the pressure gradient.[1] Geostrophic/ hydrostatic balance is a *description*, but not an *explanation*, of the ocean circulation. This descriptive power of geostrophy is quite remarkable—and it has been a huge asset to the entire subject. Simultaneously however, its accuracy is a major challenge to both observation and theory. Most of the physics useful for understanding the circulation lies in the small deviations from this balance: the slight discrepancies that govern the generation, dissipation, and maintenance of the circulation. Determining where, when, and how these equations fail to describe observations can be extremely difficult.

The application of geostrophy in oceanography is inextricably entwined with its use at sea, and the so-called dynamic method is thus briefly recapitulated. As outlined in Chapter 2, measurements at sea of temperature, salinity, depth, and geographic location were available by the middle of the nineteenth-Century. These remained the dominant form of physical measurement until near the end of the twentieth-Century. Laboratory measurements gave an equation-of-state for density, $\tilde{\rho} = \rho\,(T, S, p)$, where a tilde, ˜, is used to indicate an estimate, rather than an exact value. In what follows, *all* variables (including x, y) should carry tildes, as none is known exactly, but in the interests of notational tidiness, they are usually suppressed below.

Eliminating pressure by cross-differentiation among Equations (9.1a–9.1c) gives rise to the *thermal wind shear*:

$$f\frac{\partial \rho v}{\partial z} = -g\frac{\partial \rho}{\partial x} \tag{9.2a}$$

$$f\frac{\partial \rho u}{\partial z} = g\frac{\partial \rho}{\partial y}, \tag{9.2b}$$

all of whose terms could be estimated from measurements at sea—by replacing the derivatives with finite differences. Integrating in z (summing in practice) produces the *thermal wind*:

$$\rho\,(x, y, z)\, v\,(x, y, z) = -\frac{g}{f} \int_{z_0(x,y)}^{z} \frac{\partial \rho}{\partial x} dz + \rho\,(x, y, z)\, b\,(x, y) \tag{9.3a}$$

---

[1]See Appendix C.

$$\rho\,(x, y, z)\,u\,(x, y, z) = \frac{g}{f} \int_{z_0(x,y)}^{z} \frac{\partial \rho}{\partial y}\,dz + \rho\,(x, y, z)\,c\,(x, y),\qquad(9.3b)$$

where b, c are the v, u velocities at the arbitrary depth z_0 and which we will call the *reference-level velocities*.

What is usually regarded as the definitive test of Equations (9.2a, 9.2b) was made by Wüst (1924). He showed that direct velocity measurements in the Florida Straits by Pillsbury in 1891 produced shears consistent with the equations, and he went on to infer the integration constants $b\,(x, y)$ for the meridional flow. (Wüst's figures are reproduced, e.g., by Sverdrup et al., 1942; Wunsch, 1996; and in other textbooks.) Warren (2006), however, showed that with hindsight, the comparison is unconvincing—having involved many assumptions and coincidences.

Nonetheless, direct efforts with more modern data (Bryden, 1977; Swallow, 1977; Horton and Sturges, 1979) have failed to detect any failure of the thermal wind shear predicted by geostrophic balance. Terms on both sides of these equations are noisy, and so although perfect geostrophic balance is impossible, noise in the observations usually precludes the demonstration of that failure using local, short-term measurements of the two sides of Equations (9.2a, 9.2b).[2]

But in a much-told story, the great controversy surrounding the use of geostrophy pertained to the second step, Equations (9.3a and 9.3b), and the seemingly innocuous problem of determining the integration constants. This discussion occupies large sections of textbooks (e.g., Sverdrup et al., 1942; Defant, 1961; Wunsch, 1996, 2006a) and a summary will suffice. b, c are the geostrophic velocities at the depth, z_0, where the integration in Equations (9.3a and 9.3b) begins—an arbitrary depth that can be any place between the sea surface and the seafloor. Unable to measure that velocity, most investigators chose z_0 to be somewhere in the middle of the water column and *assumed* that the corresponding constants $b, c = 0$—thus supposing the existence of a *level-of-no-horizontal-motion*. (The word "horizontal" is commonly omitted in these phrases.)

This assumption was plausible, given the inference that the circulation was largely wind-controlled, and hence diminishing with depth. Papers on the ocean circulation written after about 1900 almost universally made some assumption about a simple form for $z_0\,(x, y)$. Among the consequences of the gradual conversion of the assumption of its existence into a fact—were a greatly diminished interest in making difficult, expensive, hydrographic observations at great depths below z_0, and it led to the conclusion that the circulation was "slablike."

The unsatisfactory nature of this situation emerged as a crisis in the work of Worthington (1976). A prominent hydrographer at the Woods Hole Oceanographic Institution, he concluded that geostrophic balance had to fail at the lowest order in the North Atlantic Ocean if mass and salt were to be conserved (this inference is, however, stated only in a small footnote

[2]Geostrophic balance is expected, a priori, to fail on and near the equator; in locations where strong currents undergo major directional shifts; and in the Southern Ocean in the range of latitudes of the Drake Passage where no continental margin supports a zonal pressure gradient. Even in these places, extensions (not pursued here) can be made preserving much of the balance form: Eriksen (1982) describes a form of "equatorial geostrophy"; *cyclostrophic* flows, where local centrifugal forces are accounted for, occur in particularly intense eddies; and in the Southern Ocean, surface outcrops of density surfaces can support some pressure gradients (Mazloff et al., 2013).

in his book). His conclusion was that major flows existed in that ocean *for which no balancing pressure gradient existed*, and he published corresponding flow diagrams. But his inferred flows produce Coriolis forces. If they are not balanced by pressure forces, either the circulation could not be in a steady state (Worthington assumed it was steady), or some other, suppressed, term in the equations of motion would need to balance them; otherwise Newton's laws of motion were being violated. But neither viscous nor nonlinear terms appeared to be large enough, and a paradox had appeared.

9.1.1 Resolving the Paradox: The Box Inverse Method

The conventional dynamic method makes no explicit use of the perfect fluid mass conservation equation,

$$\frac{\partial(\rho u)}{\partial x} + \frac{\partial(\rho v)}{\partial y} + \frac{\partial(\rho w)}{\partial z} = 0, \tag{9.4}$$

nor that of effective incompressibility,

$$\frac{\partial u}{\partial x} + \frac{\partial v}{\partial y} + \frac{\partial w}{\partial z} = 0. \tag{9.5}$$

Equation (9.4) can be generalized for the conservation of any scalar property, C, per unit mass, so that

$$\frac{\partial(\rho u C)}{\partial x} + \frac{\partial(\rho v C)}{\partial y} + \frac{\partial(\rho w C)}{\partial z} = 0. \tag{9.6}$$

Combining Equations (9.4) and (9.5) produces

$$u\frac{\partial \rho}{\partial x} + v\frac{\partial \rho}{\partial y} + w\frac{\partial \rho}{\partial z} = 0. \tag{9.7}$$

Consider a region of ocean bounded by hydrographic stations and land (see Figure 9.1). Redefine v, for the moment, as the component normal to the section. From each pair of stations, call them pair i, v can be computed from Equations (9.3a and b). Label the result as $v_i(z_j)$, where z_j are the midpoints of layers bounded by density surfaces. Corresponding to each such v_i is the thermal wind component, v_{Ri}, which is computed directly from the density field and the unknown, b_i, for that pair. In a steady state, the mass of fluid within the box should be conserved, a requirement written in discrete form as

$$\sum_i \sum_j \rho(i,j) A_{ij} n_i \left(v_{Ri}(z_j) + b_i\right) = 0, \tag{9.8}$$

where the sum is taken over both defining sections, accounting for the fact that a positive v on a southern boundary is a flow into the box, and on a northern boundary it is out of the box (by assigning $n_i = \pm 1$, as appropriate). A_{ij} is the areal cross-section of layer j in station pair i.

This last equation is a numerical rendering of Equation (9.4) integrated from bottom to top. Furthermore, in a steady state, the total amount of salt entering the box must also be leaving it, or

$$\sum_i \sum_j S(i,j) \rho(i,j) A_{ij} n_i \left(v_{Ri}(z_j) + b_i\right) = 0, \tag{9.9}$$

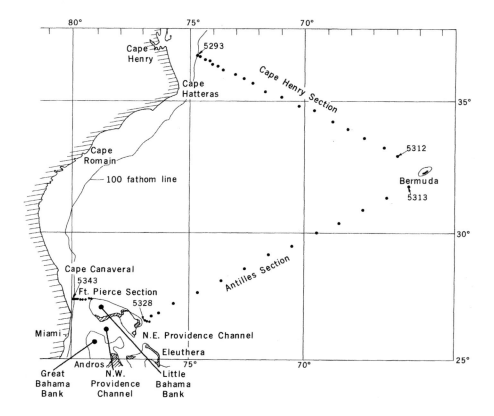

Figure 9.1: The roughly triangular region of the North Atlantic used to formulate the box inverse method. Because storage terms are known to be small, fluid flowing in must be nearly equal in amount to the fluid flowing out. The Florida Current (the Gulf Stream) enters the region confined between the land of Florida and the Bahamas, making it relatively straightforward to measure its transport properties. Worthington (pers. comm., 1975) had suggested that this region would provide a crucial test, and he provided the hydrographic data to carry it out. Because of the presence of the Bahama island chain, three separate sections are involved, one confined to the Florida Straits. v_i in Equation (9.8) is the normal component. (From Wunsch, 1977.)

vertically integrating (9.6). Any number of of observed fields are nearly conserved within the volume, for example, silica, phosphate or even, under some circumstances, potential vorticity. Define the concentration per unit mass of any one of them as $C_p(i, j)$. With $C_p = 1$, mass is included, too. Then,

$$\sum_i \sum_j C_p(i, j) \rho(i, j) A_{ij} n_i \left(v_{Ri}(z_j) + b_i \right) = 0, \quad p = 1, 2, \ldots. \tag{9.10}$$

is a generic conservation statement.

The collection of equations (9.8–9.10) is a set of M-simultaneous equations for the L-unknown b_j. Rearrange these equations to put all the known terms on the right,

$$\sum_i \left(\sum_j A_{ij} n_i C_p(i, j) \right) b_i = - \sum_i \sum_j C_p(i, j) \rho(i, j) A_{ij} n_i v_{Ri}(z_j), \tag{9.11}$$

or

$$\mathbf{Ex} = \mathbf{y}, \tag{9.12}$$

$$E_{ij} = A_{ij}n_iC(i,j), \quad y_i = -\sum_j C(i,j)\rho(i,j)A_{ij}n_iv_{Ri}(z_j).$$

Ignoring errors, (9.12) is a set of linear simultaneous equations for the unknown reference level velocities. For the bounded triangular region of Figure 9.1, an infinite number of acceptable solutions is easily found because $M < L$ and none of the equations is formally contradictory.[3]

The equations should never be regarded as exact—noise always exists in the measurements, and the assumptions of steady states, perfect conservation, and geostrophy are never fully accurate either. Thus the M-equations in L-unknowns are rewritten as[4]

$$\mathbf{Ex} + \boldsymbol{\varepsilon} = \mathbf{y}, \tag{9.13}$$

where $\boldsymbol{\varepsilon}$, a vector of length M, represents, approximately, all of these errors. Equation (9.13) is *always* underdetermined: $\boldsymbol{\varepsilon}$ is part of the solution, and it contains an element for every equation. Determining both $\mathbf{x}$, $\boldsymbol{\varepsilon}$ means finding $M+L$ unknowns from M equations. So-called inverse methods, many based on ordinary least-squares, are described in Appendix B. Most scientists seek problems that are *formally overdetermined*, meaning $M > L$, but in the presence of noise, truly overdetermined problems are rare, if they ever exist at all; inverse methods exist to cope.

Applications of inverse methods applied to ocean boxes have been made with modern transoceanic hydrographic sections, often quasi-zonal ones. Such sections typically have about 100 station pairs. A nearly closed box can be formed either with two such sections taken at different latitudes, or with one section along with the assumption that the ocean is closed to the north or has known inflows/outflows there. With conservation requirements applied to the resulting box the set of equations (9.13) contains about 200 formal unknowns, b_i. Even adding conservation of other properties, such as silicate, potential vorticity, etc., leaves the system underdetermined. Conservation within individual layers does add more information, but at the expense of additional unknowns representing transfers between layers (and to the atmosphere if the surface layer is involved). Inadequate information exists to find the individual b_i, but a great deal of information *does* exist about them in various linear combinations, as well as about the structure of all possible solutions.

Worthington's (1976) conclusion was equivalent to the inference that the underdetermined system in Equation (9.13) had *no* solution, instead of an infinite number of them. Following historical practice, he wanted a level-of-no-horizontal-motion that was a simple function of the lateral coordinate for example, that the total flow had to vanish at the 3° C isotherm, equivalent to tightly constraining the b_i in such a way that $v_{Ri}(z_{j_0}) + b_i = 0$, for some particular known or guessed z_{j_0}. Because these requirements would be an additional constraint on every one of the b_i, it is unsurprising that the system (if Worthington had

---

[3] By "formally" is meant that equations such as $x_1 + x_2 = 1$, $x_1 + 1.00001x_2 = 2$ are *not* a mathematical contradiction and do have a finite solution, but one that is unlikely to be physically acceptable. Commonly, understanding of the accuracy with which the numbers have been obtained would lead a scientist to infer that they were contradictory in practice.

[4] For the matrix/vector notation, see the box on "Notation and Basic Machinery."

written it down) became both formally overdetermined and inconsistent (contradictory) if the ε were assumed to be zero. The remedy was simple: allow the b_i to be whatever they needed to be to satisfy the equations and permit small imbalances within estimated noise levels. No theoretical or observational basis exists for the assumption of a spatially or temporally simple level-of-no-horizontal-motion. Out of the infinity of possible solutions, some solution or classes of solution are readily chosen by a good inverse method—exploiting any ancillary information, such as expected spatial covariances, or requiring minimal deviations from a theoretical form. Alternatively, the explicit structure of the indeterminate components can be obtained.

Many extensions of these equations are possible and have been used. For any scalar property, C, modifying, for example, Equation (9.6) to

$$\frac{\partial (\rho u C)}{\partial x} + \frac{\partial (\rho v C)}{\partial y} + \frac{\partial (\rho w C)}{\partial z} - \nabla \cdot (\rho \mathbf{K}_C \nabla (C)) = 0. \tag{9.14}$$

Then for any circulation field, consisting of flow $\mathbf{v}$, and (as a generic place-holder) diffusive parameters contained in tensor, $\mathbf{K}_C$, net transport of that property is

$$M_C = \rho C \mathbf{v} + \rho (\mathbf{K}_C \cdot \nabla C). \tag{9.15}$$

The transport of C across, for example, a latitude line with unit normal $\hat{\mathbf{n}}$ is,

$$M_{C_{net}} (y) = \int_{x=x_1}^{x=x_2} dx \int_{z=-h(x)}^{0} \hat{\mathbf{n}} \cdot (\rho C \mathbf{v} + \rho (\mathbf{K}_C \cdot \nabla C)) \, dz, \tag{9.16}$$

ignoring the elevation contribution, η, and the integrals again evaluated in practice as numerical sums and differences. Parameters such as $\mathbf{K}_C$ can become part of the set of formal unknowns. *Any* relationship that can be written as an algebraic or other equation on a computer can be used, as can variant coordinate systems. Many methods are known for obtaining solutions for simultaneous equation sets whether linear or nonlinear. In particular, the errors in $\mathbf{E}$, which do render the problem nonlinear, can be handled.

All available information should be used. The triangular region of Figure 9.1 includes the Gulf Stream near Florida transporting about 31 Sv, and it is an easy matter to write another equation involving the b_i within that current, setting the total inflow to about 31 Sv and weighting it appropriately with its estimated error. Use of inequality and other relationships is also possible.

9.1.2 The β-Spiral

Like the box inverse method, the β-spiral method in its basic form combines the density (or any other) conservation equation with those used in the conventional dynamic method. Equation (9.7) can be written as

$$w \frac{\partial \rho}{\partial z} = -\left(u \frac{\partial \rho}{\partial x} + v \frac{\partial \rho}{\partial y} \right) = \frac{uf}{g} \frac{\partial \rho v}{\partial z} - \frac{vf}{g} \frac{\partial \rho u}{\partial z}, \tag{9.17}$$

where the density gradients have been replaced by the corresponding thermal wind shears, and Equation (9.5) was used. Then following Bryden (1977), $(u, v) = V(z)(\cos \alpha (z), \sin \alpha (z))$ and substitute it into Equation (9.17):

$$\frac{d\alpha (z)}{dz} = \frac{\rho}{\bar{\rho}^2} \frac{gw}{f V(z)^2} \frac{\partial \rho}{\partial z} \approx \frac{gw}{\bar{\rho} f V(z)^2} \frac{\partial \rho}{\partial z} \tag{9.18}$$

The geostrophic vorticity equation, found by eliminating the pressure by cross-differentiation of the β-plane geostrophic momentum equations (Equations 9.1a–c) is,

$$\beta v = f \frac{\partial w}{\partial z},$$ (9.19)

so if v is nonzero, w must also be nonzero, and from Equation (9.18), the horizontal velocity in a stratified fluid *must* turn with depth—that is, a *spiral*. (Even when $\beta = 0$, if $w \neq 0$ for *any* reason, wind forcing for example, then Equation (9.18) still requires a spiral.)

Stommel and Schott (1977) defined the depth of any isopycnal, ρ, as $h'(x, y)$ and rewrote Equation (9.7) as

$$u \frac{\partial h'}{\partial x} + v \frac{\partial h'}{\partial y} = w,$$ (9.20)

employing

$$\frac{\partial \rho}{\partial x} = -\frac{\partial \rho}{\partial z} \frac{\partial h'}{\partial x}, \quad \frac{\partial \rho}{\partial y} = -\frac{\partial \rho}{\partial z} \frac{\partial h'}{\partial y}.$$

Taking the z-derivative of Equation (9.20), w can be eliminated via Equation (9.19). Terms involving the vertical shears, $\partial u / \partial z, \partial v / \partial z$, can then be replaced by their corresponding horizontal density derivatives written in terms of h. Then substituting $u(x, y, z) = u_R(x, y, z) + c(x, y)$, $v(x, y, z) = v_R(x, y, z) + b(x, y)$ as before,

$$c(x, y) \frac{\partial^2 h'}{\partial x \partial z} + b(x, y) \left(\frac{\partial^2 h'}{\partial y \partial z} - \frac{\beta}{f} \right) + u_R(x, y, z) \frac{\partial^2 h'}{\partial x \partial z} + v_R(x, y, z) \left(\frac{\partial^2 h'}{\partial y \partial z} - \frac{\beta}{f} \right) = 0$$ (9.21)

If $b(x, y), c(x, y)$ were known, Equation (9.21) would be a partial differential equation for h' subject to boundary conditions, defining a forward problem for h'.[5] Instead, the derivatives of h' are here treated as known from observations, and b, c are sought such that Equation (9.21) is satisfied as best possible at each point x, y, z. If discretized in z, the result is again a set of linear simultaneous equations for $b(x, y), c(x, y)$ written for different values of z_j. Let $(x, y) = (x_i, y_i)$ be fixed and apply Equation (9.21) at a series of depths $z_i, i = 1, 2, \ldots, M$. Then M-equations in the two unknowns $b(x_i, y_i), c(x_i, y_i)$ result, and a solution and its error components can be found easily, for example, using the procedures for coping with noisy simultaneous equations. The neutral surfaces defined by McDougall (1987) can be used instead of ordinary isopycnal depths, as can spherical coordinates, and the system can be rewritten in terms of ρ without introducing h'.

As a point balance, Equation (9.21) requires both x and y derivatives of the density, and so shipboard observations need to be specially arranged to permit their estimate. Armi and Stommel (1983) organized four expeditions to an open-ocean region (see Figs. 9.2 and 9.3) and were able to demonstrate the existence of a spiral in each survey. The strong time dependence is apparent and leads to attempts to time-average the resulting inferred flows.[6] When applied independently at a series of adjacent points, the β-spirals and derived horizontal velocity fields need not satisfy any large-scale conservation relations; some authors (e.g., Fukumori, 1991) have therefore added the box inverse constraints—which do impose such requirements—in effect, combining the two methods.

---

[5] See Appendix B for the definitions of forward and inverse models.
[6] Time averages of box inversions are by Tziperman and Hecht (1988) and Katsumata et al. (2013).

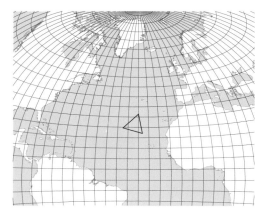

Figure 9.2: Triangular region whose boundary was revisited four times by Armi and Stommel (1983) so as to implement the β-spiral calculation with derivatives of the hydrography in both latitude and longitude.

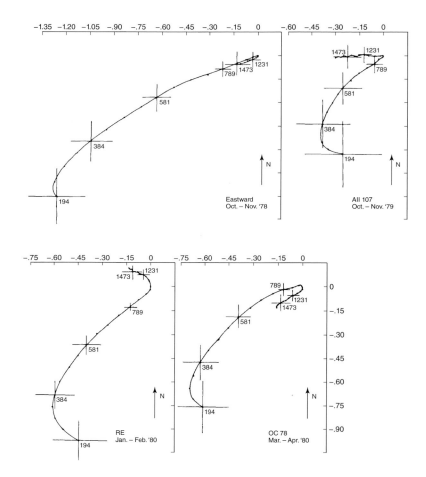

Figure 9.3: Four upper-ocean geostrophic velocity spirals determined by Armi and Stommel (1983) using the β-spiral equation. Quasi-equilibrium spirals exist for all four cruises, but the temporal changes among them are obvious. Depths in decibars are labelled, and the cruise designations are also shown. Units are cm/s.

EXTENSIONS

Many modifications of these basic ideas have been discussed, including the addition of time-dependent terms, mixing coefficients, etc. The so-called Bernoulli method of Welander (1983), Killworth (1986), and Chu (2006) has received particular attention. Zika et al. (2010) developed a method directed specifically at determining mixing coefficients from time-averaged hydrographic data. But because none of these variants has seen widespread practical use, no more will be said about them here.

Numerical finite difference representations are also box models, sometimes with more complex connections than with only nearest-neighbor boxes. If written out, whether in steady state or otherwise and with noise elements added to each grid box, they are seen to be once again a set of simultaneous equations—analogous to those used in the box inverse or β-spiral systems.[7]

9.1.3 Needler's Formula

Some further perspective on the resolution of Worthington's paradox can be obtained by a later theoretical development. Needler (1985) manipulated the collection of Equations (9.2a, 9.4–9.5) into the expression

$$\rho\,(x,y,z)\,[u,v,w] \tag{9.22}$$

$$= g\,\frac{\hat{\mathbf{k}}\cdot(\nabla\rho\times\nabla q)}{\nabla(f\partial q/\partial z)\cdot(\nabla\rho\times\nabla q)}\,[\nabla\rho\times\nabla q],\quad q = f(y)\,\frac{\partial\rho}{\partial z}$$

$$\nabla = \left[\frac{\partial}{\partial x},\frac{\partial}{\partial y},\frac{\partial}{\partial z}\right]$$

for the absolute, *three-components*, of velocity. Obtaining this expression is an exercise in vector calculus and versions of it can be seen in the books by Pedlosky (1996) or Wunsch (2006a). Derivation makes explicit use of the y- dependence of f and thus applies to the β-plane or sphere. Needler's formula is not a practical relationship for use with shipboard data—because it involves the third derivatives of the density field, as well as derivatives in both x, y directions. What the formula does show formally is that enough information exists in a steady-state density field to infer the absolute velocity under geostrophic, hydrostatic balance on a β-plane or sphere. Equation (9.22), along with the discussion of Davis (1978), provides a unifying theoretical principle.

9.1.4 Qualitative Circulation Estimates

The dynamic method, relying upon arbitrarily chosen levels-of-no-motion, gave rise to a very long series of depictions of the inferred ocean circulation that were mostly confined to the upper levels. Many of them can be seen displayed in Warren, 1981, Reid, 1981, and Richardson, 2008. Peterson et al. (1996) provided a historical overview of attempts to describe the global circulation from antiquity to the present. Level-of-no-motion assumptions were able to define many of the most important qualitative elements of the circulation, including

---

[7] Such finite-difference inverse models have been used by Martel and Wunsch (1993), Schlitzer (2007), and Gebbie and Huybers (2011) among others.

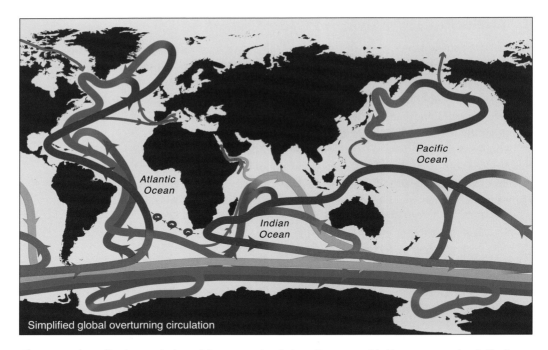

Figure 9.4: A qualitative rendering of the ocean circulation. Compare with Figures 9.5 and 9.6. Notice the absence here of numerical values and the generally greater complexity compared to earlier attempts, as more observations became available. The colors are intended to indicate the dominant named water masses. Representing such flows in two dimensions (zonal-integrals) can defeat understanding of the flow physics. (From L. D. Talley [2013] "Closure of the global overturning circulation through the Indian, Pacific, and Southern Oceans: schematics and transports."*Oceanog.* 26: 80–97.)

the western boundary currents, the deep western boundary currents (DWBCs), the much more sluggish return flows in the interior, and the kind of schematics of eastern boundary and other currents of the sort appearing in most textbooks. The synthesis by Talley (2013), using various estimates by J. Reid, is perhaps the culmination of this method. (See Figure 9.4, and Talley, 2003.)

Circulation diagrams are interesting, and a goad to theorists. But normally they remain qualitative—without numerical estimates of volume or other property transports or any estimate of their representativeness of the long-term average—and potentially misleading as representations of a time-dependent, three-dimensional turbulent flow. For many scientists, the chief importance of ocean circulation lies not so much in the detailed pattern of flow at any given instant, or even when averaged over arbitrarily long times, but rather, in the divergences of properties carried by the circulation including enthalpy (heat), fresh water, carbon, oxygen, nutrients, or even petroleum. Detailed streamline or vector plots showing sometimes complex, even tortuous, pathways are of interest to the fluid dynamicist who would like to understand them. Whether they have some larger significance, biological, chemical, or climatological, must be determined in each case.

9.1.5 Circulation Estimates: Inverse Methods

Because of the stable, large-scale temperature and other hydrographic properties, a widespread intuitive expectation is that the time dependences on global and basin scales are negligible. Much of the present difficulty lies, however, with determining, for example the relationship between the flow and properties across a trans-Atlantic section made during a month in 2002 with the flow and properties averaged over a decade or more. The flows are directly dependent not on the properties themselves but upon their much less stable spatial derivatives. This discussion involves not only the real physical changes in the ocean, but also accounting for shifts in sampling rates and locations and for changes in measurement technologies. Transport estimates made at two different times always differ to some extent, but how different they will be, and why, is a central question. Thus the ability to detect expected weak trends in properties such as the meridional temperature transport in the ocean is hostage to considerable uncertainty in the accuracy of existing results.

GLOBAL INVERSE ESTIMATES

Several attempts at producing near-global, steady-state estimates of ocean circulation exist, based upon the long hydrographic sections.[8] All are comparatively recent, because their use of the box inverse method relied heavily upon the ability to write total-mass and tracer-conservation constraints, which require hydrographic sections extending all the way to the seafloor and from continent to continent.[9] Apart from the R/V *Meteor* sections obtained in the South Atlantic in the 1920s and the Atlantic International Geophysical Year (IGY) sections of 1957–1959, top-to-bottom, continent-to-continent data were extremely rare until the 1980s. They did not become relatively commonplace until WOCE—beginning about 1990—and its successor programs.

How best to render the movement of a complicated three-dimensional flow field on a sphere has no ready answer. Macdonald's (1998) estimates for the horizontal circulation divided into two layers by the 3.5° C isotherm is shown in Figure 9.5 and that of Lumpkin and Speer (2007) is in Figure 9.6. Both are quantitative and are provided with useful estimates of their uncertainties.

Existing estimates of a time-mean global ocean circulation are descriptive approximations: either (1) they are based upon individual hydrographic surveys whose relationship to a true time average is not known; or (2) they are based upon time-averaged hydrographic data (the climatologies) for which the governing equations are unknown—because the space or space-time averages of the nonlinear terms in the equations of motion have not been included. Thus estimates that have been made from combining individual hydrographic sections bear some relationship to the average flow or flow properties, but as with any instantaneous value of any field, whether the result is close to or far from the true time average depends upon the probability distributions of the flow. Accurate analysis of time-varying fields requires appropriate time-varying estimates, which can then be averaged to form a mean.

[8]Macdonald 1998; Ganachaud, 2003b; Lumpkin and Speer, 2007.
[9]The Schlitzer (2007) and Gebbie and Huybers (2011) representations are in terms of comparatively small boxes more closely resembling the finite difference form and rely upon climatologies rather than averaged hydrographic sections.

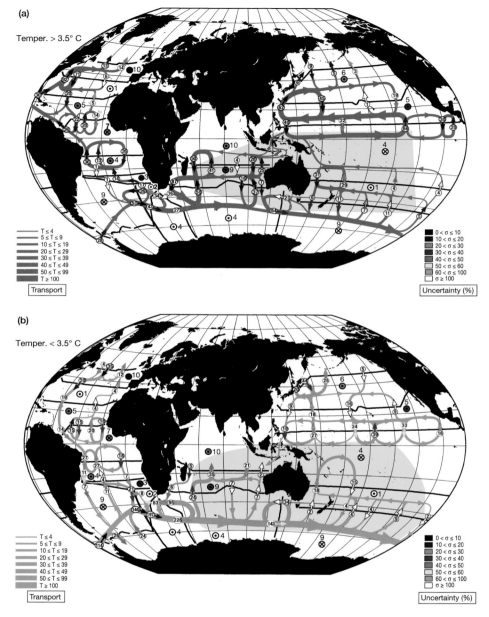

Figure 9.5: An estimate of the horizontal circulation pathways in the first global inverse calculation. Note that many of the upper-level features are perceptible in Figure 3.29, keeping in mind however, that flows are related to the spatial derivatives of the pressure field and that the eye is a poor instrument for quantitative estimates. Values in Sv. (Used with permission from Macdonald A. M. (1998). "The global ocean circulation: a hydrographic estimate and regional analysis." *Prog. Oceanog.* 41: 281–382)

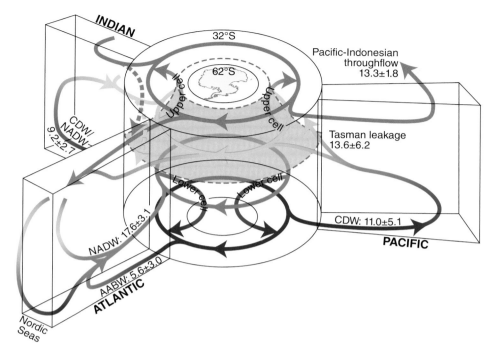

Figure 9.6: Schematic of the global circulation (Sv) from an inverse method calculation of Lumpkin and Speer (2007)—after an earlier version of W. J. Schmitz. The presence of error estimates is very important. Colors indicate different density ranges. This representation emphasizes the role of the Southern Ocean. It is probably true, however, although unproven, that fluid ultimately passes through all points in all oceans. CDW (Circumpolar Deep Water); NADW (North Atlantic Deep Water). Colors indicate approximate water densities. (R. Lumpkin and K. Speer [2007]. "Global ocean meridional overturning." *J. Phys. Oc.* 37: 2550–2562. © American Meteorological Society. Used with permission.)

As the database has grown, the putative time-mean circulation diagrams have become ever more complex, and at some indeterminate stage, the fact of the complexity becomes more significant than the specific details of the flows. In a fluid like the ocean, a reasonable, if unproven, inference is that fluid parcels located at any point **r** will ultimately, after sufficient time, pass arbitrarily close to *all* other points **r′** in the fluid, in a type of ergodic assumption. Mixing renders fluid parcel identification ultimately impossible; nonetheless, an ever-growing tangle of pathways in diagrams such as Figures 9.4–9.6 is expected as more details of the circulation become documented. For this reason, their import grows increasingly obscure; focus shifts to the integrated transport properties for mass or heat or carbon and their time variations. These integrals have directly observable consequences.

9.2 GLOBAL PROPERTY-WEIGHTED TRANSPORTS

The movement of mass (or volume, within the Boussinesq approximation) is the most fundamental element in the science of the ocean circulation. For many applications however, their resulting property transports are of greatest interest. M_C and its integrals in Equations 9.15 and 9.16 represent for any scalar field, C, a C-weighted version of the mass or volume transports. Properties such as temperature weight the integrals heavily toward the upper ocean, and others, such as silicate, weight it toward the seafloor. Sensitivities of the results then tend to be strong functions of that weighting; thus temperature (heat) transports will have a strong tendency to depend much more directly on upper-ocean flows and diffusion, while silicate estimates will be much more dependent upon the details of the abyssal flow field. (These statements are not absolute because cancellations of major transport components can occur, leaving the integrals primarily dependent upon the more weakly weighted flows.)

In discussions of differently weighted transports, the vertical integral in Equation (9.16) is commonly broken into a series of layers as already described for the box inversions (Equations 9.10, 9.11). Alternatively, and sometimes more informatively, any monotonic coordinate system can be used. If $C = T$, the temperature, then a temperature coordinate system is attractive (assuming that temperature is monotonic in depth in the domain under consideration; see Bocaletti et al., 2005).

The various integrals are often broken into two pieces,

$$M_{C_{adv}} = \underbrace{\iint\limits_{\text{section}} \rho C \mathbf{v} \cdot \hat{\mathbf{n}} dA}_{} = \underbrace{\iint\limits_{\text{section}} \rho \bar{C} \bar{\mathbf{v}} \cdot \hat{\mathbf{n}} dA}_{} + \underbrace{\iint\limits_{\text{section}} \overline{\rho C' \mathbf{v}'} \cdot \hat{\mathbf{n}} dA}_{}, \qquad (9.23)$$

where the overbar denotes the spatial *averaging* operator. In a simple world, $M_{C_{adv}}$ would be dominated by the first term, the product of the means, and could be computed accurately by multiplying the section average velocity by the section average concentration. But it is easy to see from real data that this product is commonly overwhelmed by the spatial covariance terms, $\overline{C' \mathbf{v}'}$, including the contributions from the ambient eddy field, and by the structures in the strong boundary currents.

A large literature describes property transports, particularly for temperature, in terms of various representations of the flow field (upper versus lower ocean, gyres versus zonally integrated overturning, etc.; see, for example, Figure 9.7). Because all elements of the ocean circulation are connected, via global mass, salt, vorticity, etc., conservation requirements, these representations of the integrals of $\overline{C\mathbf{v}}$ are, mainly, useful descriptions if one element dominates. Discussion of component breakdowns of the various integrals are provided by Bryden and Imawaki (2001), Talley (2003), Boccaletti et al. (2005), and others. If one volume transport component—for example, the abyssal flow, or the DWBC, or the western boundary current (WBC)— should change, then the variety of conservation rules require shifts in some or all of the others. This comment is simply another way of saying that (1) steady-balanced flows are difficult to use for understanding causal relationships and, (2) it is the changes, real and potential, that are a central concern in understanding ocean circulation.

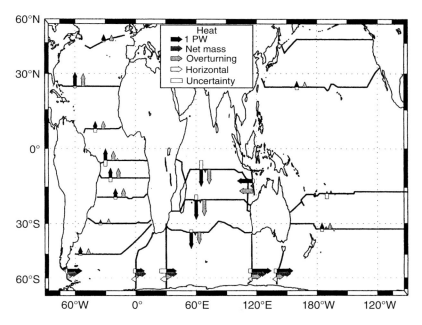

Figure 9.7: The estimated meridional enthalpy transports across the available zonal sections. The black arrow is the total, and the white tail is an indication of its uncertainty. Gray and white arrows are, respectively, a decomposition of the total into nominal overturning and "horizontal" components, here using neutral density coordinates (a decomposition in pressure is very similar). Most notable is the equatorward transport in the South Atlantic Ocean. The zonal integrals where common latitudes extend around the world are shown in Figure 9.8. A 1 PW scaled arrow is shown in the small box. (From Ganachaud and Wunsch, 2003.)

9.2.1 Global Heat, Freshwater, Nutrient and Oxygen Transports

HEAT

Property-weighted global transport estimates have been made for temperature, salinity, and some of the nutrient and oxygen fields. Enthalpy (Warren, 1999) transport by the oceans is a component of the Earth energy balance and so must be described and understood as part of the basic workings of the climate system. This meridional transport component permits the Earth to maintain a near-constant average temperature while being heated in the tropics and radiating back to space at high latitudes. Sverdrup et al. (1942) had concluded from the very limited oceanic data then available that the ocean could not be contributing more than about 10% of the total movement of heat from equator to pole.

That inference, along with the almost complete absence of three-dimensional global ocean data, meant that little further attention was paid to the subject. Eventually, Oort and vonder

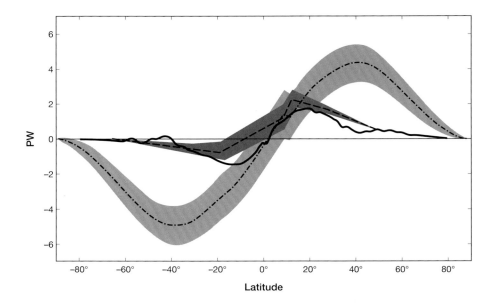

Figure 9.8: Estimates of meridional global heat transport. The dashed line shows the estimate from Ganachaud and Wunsch (2002) as linearly interpolated between latitudes where complete zonal sections were available, along with a one-standard deviation-error band. (Compare to Figure 9.15.) The dash-dot line shows the estimated atmospheric component of heat transport computed as a residual of the total—as estimated by satellites, subtracting the oceanic component (see Wunsch, 2005). The error bars are large, and likely optimistic, as all these results are susceptible to systematic errors. The thick solid line is an estimate of the zonally integrated meridional heat transport by the ocean as obtained from a state estimate (Wunsch and Heimbach, 2007). The apparent oceanic transport extreme near 15°S was missed in the WOCE hydrographic survey, thus exaggerating the apparent hemispheric asymmetry.

Haar (1976), using satellite data, inferred from estimated atmospheric residuals that the oceanic component could be much larger than the earlier estimates had shown—a conclusion that generated renewed interest in determining the transports. Figure 9.7 displays an estimate of the meridional heat transport in the ocean by basin, and Figure 9.8, the estimated global integrals for the ocean and atmosphere.

The results in Figure 9.8 have several important features. At low latitudes, particularly in the northern hemisphere, the estimated oceanic meridional transport exceeds that of the atmosphere. This oceanic dominance is the result of the intense Ekman flux at elevated temperatures, as f goes to zero near the equator. At higher latitudes, the atmospheric contribution grows through mechanisms associated with the generation of the synoptic weather systems. At the highest latitudes, the area of ocean rapidly diminishes (vanishing altogether in the south at about 65°S) almost all of the oceanic heat having been released by then to the atmosphere. (See Jayne and Marotzke, 2001.)

Most of the ocean's volume is in the southern hemisphere, and yet the poleward transports by the ocean are more nearly antisymmetric than the area ratios would suggest. To the contrary, once the northward transport of heat by the South Atlantic was recognized (Figure 9.7), it

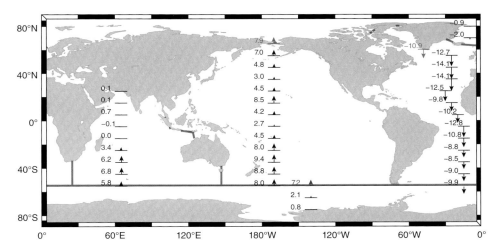

Figure 9.9: Estimated meridional transport of freshwater by the oceans in tenths of Sverdrups, with positive values to the north. Only the divergences are physically significant, with the transport value set arbitrarily at the Bering Strait. The authors suggest an error of roughly 10% for the numbers shown, but that is likely to be optimistic. Compare to Figure 3.38. Values in red involve externally prescribed injection; see the reference. (From Schanze et al., 2010.)

was expected that the magnitudes of southern-hemisphere net oceanic heat transport would be much below that of the northern. In practice, however, the flows in the huge areas and the volume transports of the Pacific and Indian Oceans nearly compensate for that northward component. The WOCE hydrographic survey did not obtain a section at the latitude of the estimated southern-hemisphere maximum heat transport visible in the figure, leading to the mistaken impression of a greater hemispheric asymmetry than appears correct.

The structure and magnitudes of the meridional transport of heat by the ocean and atmosphere (the "why of it") has been the subject of many recent papers. The ocean and atmosphere combine to maintain the overall net heat balance of the Earth, with an accuracy of better than $1 \, W/m^2$ over hundreds of years. The residual imbalance, the very small difference of large numbers, is generally attributed to the ongoing greenhouse warming of the Earth. Stone (1978) made an interesting attempt at explaining the meridional distribution of the combined ocean-atmosphere transports, an explanation challenged more recently by Enderton and Marshall (2009) from their model results. The near antisymmetry in the total, which is apparent in Figure 9.8, does call out for a simple explanation, given the markedly different distributions of land and sea in the two hemispheres.[10]

The enthalpy transport by the ocean and atmosphere and its variability can be regarded as a coupled mode of the combined system: the two together prevent the growth of any large global imbalances. Much of the atmospheric transport is in the form of latent heat (moisture); but the meridional moisture transport must be balanced by the oceanic return flow of freshwater (e.g., Figure 9.9).[11]

Because of the need to maintain the overall heat balance of the Earth, a fluctuation in either atmosphere or ocean contributions must be balanced by the rest of the system.

[10] Discussed by Wunsch (2005).
[11] See Bryden and Imawaki, 2001.

Bjerknes (1964) speculated about compensation mechanisms on comparatively short time scales. Arguments that fluctuations in the climate system as a whole are controlled for example, by the North Atlantic oceanic heat transport, are not easy to accept without an accompanying understanding of how the coupled-system global balance is maintained. Within the time span of the instrumental record, no indication exists of any persistent large-scale changes in ocean-atmosphere partitioning, although compensating regional fluctuations can exist for many years.

FRESHWATER AND SALT

Changes in oceanic salinity arise from internal redistributions (advection and mixing) and from the addition or removal of freshwater across the air-sea interface. The atmosphere carries no significant amount of salt.[12] Thus net changes in the salt content in ocean basins can be converted into estimates of the net air-sea exchange of freshwater (evaporation minus precipitation) between the two systems. An estimate of the salt transport from the global inversion was produced by Ganachaud and Wunsch (2003) and its divergence is readily converted into an estimate of the necessary freshwater flux into and out of the ocean basins. Other estimates, using combinations of section transports and the bulk formula calculation with satellite data, have been produced. Figure 9.9 is an example of such a calculation.

NUTRIENTS AND OXYGEN

Nutrient and oxygen distributions are known with varying degrees of coverage and accuracy from shipboard sampling. Sarmiento and Gruber (2004) and Williams and Follows (2011) provide extensive discussions of these fields. Figure 9.10 is representative of the understanding of the transports of silicate, nitrate, and oxygen and their divergences as determined from box inversions.

Interpretation of these fields in terms of the combined chemistry and biology with which they interact would take us too far afield, and the reader should consult the references for further information.

9.3 REGIONAL ESTIMATES

A large number of regional estimates have exploited the box inverse and β-spiral-like techniques. As regions diminish in area, temporal variations become of increasing concern in two ways: (1) the relative values of temporary storage of properties can become very large; and (2) the magnitudes of temporally varying flows become proportionally larger compared those of to any quasi-steady structure. Boxes formed from hydrographic lines crossing in the ocean interior can incur very large inconsistencies from their nonsynoptic nature. So for example, a survey of the time-shifting Gulf Stream in a hydrographic box could suffer the misfortune of having observed the stream entering the box but not its departure, had it moved in the interim.

---

[12] For numerical convenience, modelers most often replace freshwater budgeting by salt budgeting, making the useful fiction that the atmosphere does transport salt. The surface source/sink terms thus appear in the salinity conservation equation rather than in the one for mass conservation, leading to questions about the impact on the resulting fluid circulation. See Huang and Schmitt, 1993.

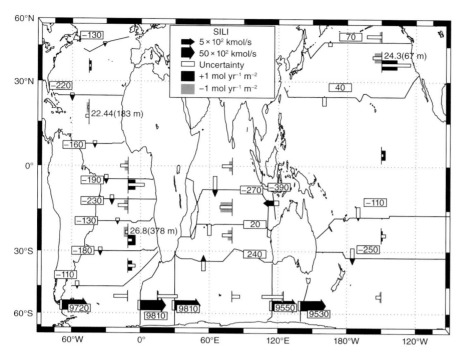

Figure 9.10a: Silicate transports and divergences as estimated in an inverse calculation. Silica bottom-weights the flow field in many regions. A scale from the arrow width of 500 kmoles/s is shown. Solid arrows are the net transport, with the white band representing the estimated uncertainty. Divergences are either top to bottom or to an interface with depth in meters. (From Ganachaud and Wunsch, 2002.)

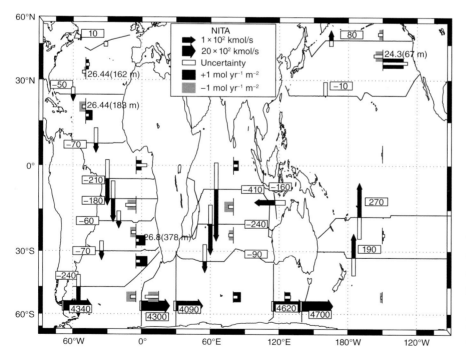

Figure 9.10b: As in (a), but for nitrate transport and divergence, and representing an upper (but not near-surface) weighted circulation. Nitrate tends to be depleted by biological consumption near the surface.

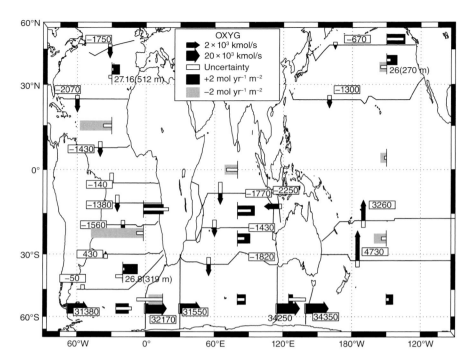

Figure 9.10c: Estimated oxygen flux and divergence. The weighting of the volume transport by the oxygen distribution is a complex one, including mid-depth minima. (From Ganachaud and Wunsch, 2002.)

Because each region requires its own special oceanographic discussion and the number of potentially interesting places is extremely large, we here only list a representative group of regional discussions which include those by Schlitzer (1988), Roemmich and McCallister (1989), and Macdonald et al. (2009), all of whom used the box inverse method. Because of the spatial coverage required, β-spiral estimates are rare. They include particularly, Olbers et al. (1985), who used a time-and-space-averaged climatology. Fukumori (1991) combined the β-spiral and box inverse methods for a North Atlantic circulation estimate. Another variant applied to the North Atlantic is discussed by Bogden et al. (1993). Mercier (1989) and Wunsch (1994) used nonlinear extensions of the basic methodologies in limited regions.

Authors employing the β-spiral and its relations have commonly used one or another of the climatologies discussed previously so as to obtain the required spatial derivatives. Because the climatologies are computed as averages in time and space of temperature, salinity, and/or density, they should be used with equivalent time- and space-averaged equations of motion and in coordinate systems (e.g., geographic or density) consistent with the averaging type. The great difficulty here was already described: averaging the equations introduces terms such as $\langle \mathbf{u} \cdot \nabla \mathbf{u} \rangle$ in the momentum equations and $\langle \mathbf{u} \cdot \nabla \rho \rangle$ in the density conservation equation— "Reynolds terms"—the bracket representing the spatially varying degree of averaging. These terms represent the eddy transports—the subject of intense discussion by numerical modelers. Although probably not important in the open ocean, considerations of such interesting regions as the meandering Gulf Stream or Agulhas Current, or the entire Southern Ocean, readily show that the use of equations not accounting for the Reynolds terms can produce qualitatively incorrect results.

9.4 CONVECTIVE REGIONS

A very large range of dynamically distinct regions exists in the global ocean. Examples are places where deep flows follow the contours of f/h, the tropical current system, sea-ice formation with large discharges of brine into the circulation, hot spots of geothermal heating, special topographic structures, ad infinitum. Many of these areas deserve, and sometimes have, entire books devoted to them. From the perspective of the global circulation and its climate implications, regions of deep convection stand out as of special interest.

"Convective" regions are those places where fluid at and near the sea surface becomes sufficiently dense relative to the underlying ocean, through atmospheric removal of heat and freshwater, so as to become statically unstable (the density increasing upward and producing large Rayleigh numbers). Such unstable configurations cannot persist, and sinking of the dense fluid occurs in a process called *convection*. Where those regions are, why they are located there, how deep the sinking fluid penetrates, how much is entrained, and the rates at which the fluid is exchanged vertically are all the subject of major studies (Marshall and Schott, 1999, is a review). Figure 9.11 is a fortuitous picture of a convective event in the Southern Ocean, most striking because of the very tall, thin nature of the convective plume or chimney.

Regions where surface water becomes dense enough to reach the seafloor have attracted the most attention. These places include the Weddell Sea in Antarctica and the high-latitude North Atlantic in the Greenland and Irminger Seas (see Figure 9.14). That such regions must exist has been known for centuries—ever since it became clear that the abyssal waters in the ocean were very cold (Warren, 1981). On the other hand, regions where convection leads to sinking only to intermediate depths are much more widespread and likely of equal climate significance.

One important physical point emerges from the numerous theoretical and modelling studies that exist for oceanic convection: regions of convection are *not* places with strong downward vertical velocities, w, as is the case with classical convection of a fluid heated from below (Rayleigh-Bénard convection; see Chandrasekhar, 1968). Rather, the vertical property exchange is essentially *diffusive*, with little or no momentum associated with it (Marshall and Schott, 1999). Vertical velocities are displaced laterally by large distances from the convective regions and, in models at least, are confined mainly to the boundary regions (Scott and Marotzke, 2002).

MEDITERRANEAN CONVECTION

Because the strongest convective regions occur at high latitudes under severe winter weather conditions, their direct study at sea is both difficult and disagreeable. For that reason, a considerable effort was directed at the much more accessible convective regions occurring in the Mediterranean Sea, a process manifested in the larger domain by the great salinity tongue emanating from the Strait of Gibraltar (see Figs. 2.3 and 9.18). The eastern Mediterranean in wintertime is the scene of intense evaporation from high winds and cold air, leaving behind an enhanced salinity, rendering the near-surface fluid sufficiently heavy to sink and form the

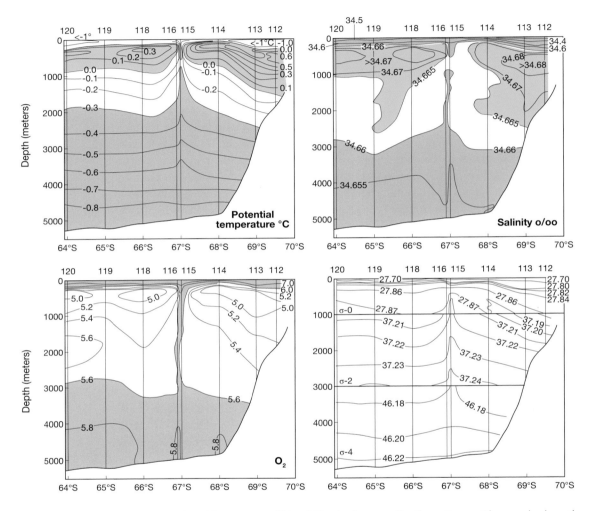

Figure 9.11: An apparent convective *chimney* west of Maud Rise in the open Southern Ocean. The panels show the potential temperature, the salinity, the oxygen content, and the density at fixed pressures in different intervals (denoted $\sigma - 0$ for zero pressure, $\sigma - 2$ for 2000 m pressure, and $\sigma - 4$ for 4000 m pressure). Station numbers are shown. The results are quite remarkable for the extremely narrow width of the convection, and its detection was a matter of luck. This type of structure is very different from classical convection of a fluid heated from below. Note, too, that salinity is important in these high latitudes. (From Gordon, 1978.)

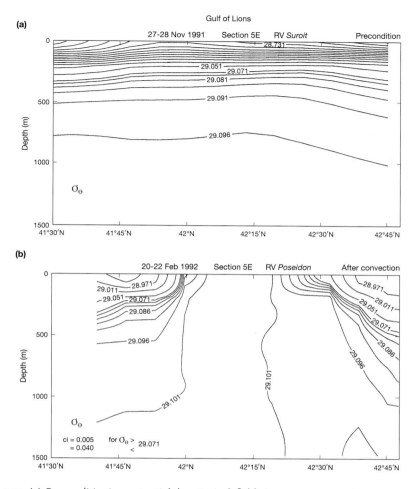

Figure 9.12: (a) Preconditioning potential density (σ_θ) field at zero pressure prior to convection in the Gulf of Lion, with the small upward heave of the isopycnals, and (b) after convection has taken place. (From Marshall and Schott, 1999.)

relatively salty, and slightly warm (relative to its surroundings) Levantine Intermediate Water. This water mass ultimately exits the Strait into the open North Atlantic Ocean and becomes visible as the salt tongue. (See Bryden and Stommel, 1982, for a discussion of the deep-water pathways in the Mediterranean.)

Despite the importance of the eastern basin, most of what is known about convection in the Mediterranean comes from study of the process in the Gulf of Lion south of the French Riviera. There, cold winter winds from the Alps (the mistral) generate a very dense water mass off the coast, one sufficiently heavy to sink. Observations led to the notion of *preconditioning*: the need for an extended period of cold and evaporation to prime the water sufficiently so that when the mistral does set in, the fluid is already close to the point where it will become convectively unstable (see Figure 9.12). That preconditioning in turn occurs only in regions

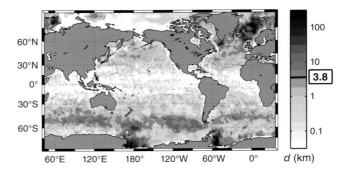

Figure 9.13: The surface sources of global ocean waters. Oceanic volume that has originated in each 2°×2° surface location (11,113 origination sites), scaled by the surface area of each box to make an equivalent thickness, d. The colorscale follows a base-ten logarithm of the field. For reference, the colorbar includes the average depth of the ocean (3.8 km), equal to the area-weighted mean of d. Taken from Gebbie and Huybers, 2011, which shows that a power-law distribution describes this figure, in which the number of areas increases with decreasing contribution.

where the bottom topography (f/h contours) constrains the flow sufficiently to render it dense (Hogg, 1973; Grignon et al., 2010).

MODE WATERS

Historically, the Greenland and Irminger Seas of the North Atlantic, the Weddell Sea, and a few other small regions, were regarded as dominating the production of deep water (water penetrating below about 2000 m) in the world oceans. In more recent years, it has become apparent that convective processes in the seas around on the continental margins of Antarctica (which form denser water than does the North Atlantic) are far more widespread than previously believed (Fukamachi et al., 2010). Gebbie and Huybers (2010; see Figure 9.13) have mapped the apparent origin regions of abyssal waters and have shown the northeast North Atlantic and the Weddell and Ross Seas of the Southern Ocean to be important. Nonetheless, much wider areas distributed around the global ocean are the estimated points of "last contact" between the surface and abyssal oceans. Fluid encountered at depth will generally have had its properties set at multiple regions at the surface and will also have had potentially intense mixing exchanges along its route. All of the processes, including mechanical ones such as Ekman pumping, must be involved in communicating near-surface properties to the deeper interior.

Many of the near-surface processes generate comparatively large volumes of nearly homogeneous (in temperature, salinity, and oxygen concentration) water masses, commonly referred to as *mode waters* (with "mode" referring to the local maximum in a property volumetric histogram). Such named water masses are the North Atlantic 18° Water,[13] Labrador Sea Water, Pacific Intermediate Water, etc. The formation processes of these waters has been studied in considerable detail, and Gladyshev et al. (2003) and Marshall et al. (2009) described two observational programs. Figure 3.25 depicts a global histogram of the volumes in T-S

---

[13] Although it is closer to 17°C in temperature.

space of water in the ocean. Most of the ocean is relatively cold and fresh. Similar diagrams can be constructed for any pairs of properties e.g., temperature and silicate, or salinity and oxygen, and are used to generate further water mass definitions. Diagrams such as Figure 9.13 support the inference that the dominance of the volumetric properties by very cold water are largely set by the near-surface physics of the Southern Oceans and the high latitudes of the North Atlantic, but with nonnegligible contributions from elsewhere.[14]

9.5 OCEAN STATE ESTIMATES

The ocean is now known to be variable on all of its space and time scales—millimeters to 10,000+ km, seconds to thousands of years—involving a great variety of physical processes including wavelike motions, turbulence, directly forced variations, and instabilities. How all of the many physical phenomena outlined in the preceding chapters, some of them seemingly unconnected except as fluid phenomena, fit together into a coherent view of ocean physics is the challenge of ocean-circulation physics.

Synthesizing all of the associated phenomena using any kind of semianalytical theory in the complex geometry of real ocean basins while the system undergoes disturbances from a turbulent atmosphere, freezing and thawing of ice, and freshwater inputs generally has not proven possible, and likely never will be. Numerical solutions of the governing equations for the fluid, associated thermodynamic relationships, and ancillary processes such as the formation of sea ice are necessary. The subject of ocean numerical modelling is a mature and complex field in its own right.[15] A fully realistic model would be as complex as the real ocean, and so its understanding (the goal of science) would be equally difficult. Understanding even quasi-realistic models is *more* difficult than understanding the real ocean because of the possibility of coding errors (not expected in nature) and the use of equations that at best are only approximations to those believed relevant.

With all of their difficulties, numerical models are essential. Many of the properties of most interest (e.g., transports of freshwater or carbon) are not measured but can only be calculated from indirect measurements and various dynamical, chemical, or biological assumptions (models). The best features of models need to be combined with the best understanding of *all* of the observations pertaining to the ocean. Although the task is never complete, the state of the art of carrying out this recipe, circa 2014, is outlined, on the basis of the methods described more fully in Appendix B. A summary of the methodology is, in principle, simple as follows:

Numerical representation of the most complex oceanic general circulation model (GCM) is a set of simultaneous equations coupled in time and space. They commonly have both linear and nonlinear elements and contain errors from truncation, misspecified initial conditions, internal mixing parameters, topography, etc. At any given discrete time, $t = m\Delta t$, the complete oceanic circulation can be described by a *state vector*, $\mathbf{x}\,(m\Delta t)$. It typically includes three components of velocity, the temperature and salinity, and the pressure field, all on a three-

---

[14] Lund et al., 2011, is a schematic discussion of how the surface properties of the Southern Ocean affect the abyssal temperatures and salinities. See Speer and Tziperman, 1992, for a discussion of formation rates in the North Atlantic Ocean.

[15] Relevant textbooks include O'Brien, 1986; Haidvogel and Beckmann, 1999; and Griffies, 2004.

dimensional grid, or equivalent, of varying spacing occupying the full ocean volume.[16] If no external meteorological or other forcing is present, $\mathbf{x}(m\Delta t)$ provides just enough information to predict its value some time increment, Δt, into the future,

$$\mathbf{x}((m+1)\Delta t) = \mathbf{L}(\mathbf{x}(m\Delta t)), \tag{9.24}$$

where the operator $\mathbf{L}$ is a computer code performing the "time-stepping." Because external conditions $\mathbf{q}(t)$, are also present, representing all of the prescribed initial and boundary conditions, and interior sources and sinks, whatever their origin, the full "prediction" statement is

$$\mathbf{x}((m+1)\Delta t) = \mathbf{L}(\mathbf{x}(m\Delta t), \mathbf{q}(m\Delta t)). \tag{9.25}$$

In a truly realistic global model, the dimension of $\mathbf{x}(m\Delta t)$ is immense, and it must be retained over long time intervals. The code for $\mathbf{L}[\cdot]$ can occupy hundreds of thousands of lines.

Any useful scalar ocean observation, $y_j(m\Delta t)$, must be some function, E_j, of the state vector,

$$y_j(m\Delta t) = E_j(\mathbf{x}(m\Delta t), \varepsilon), \tag{9.26}$$

where ε represents the inevitable presence of noise in real observations. A calculation, $\tilde{\mathbf{x}}(m\Delta t)$, from a model will almost always produce a value $y_j^{(m)}(m\Delta t) = E_j(\tilde{\mathbf{x}}(m\Delta t)) \neq y_j(m\Delta t)$ sufficiently different (relative to ε) from the observed value, warranting the conclusion that the model should be modified. If a model reproduced all observations within estimated errors, the subject of physical oceanography would be closed. Because all models also contain errors, Equation (9.25) is further modified to,

$$\mathbf{x}((m+1)\Delta t) = \mathbf{L}[\mathbf{x}(m\Delta t), \tilde{\mathbf{q}}(m\Delta t), \boldsymbol{\varepsilon}_M], \tag{9.27}$$

where $\tilde{\mathbf{q}}$ is used because of discrepancies from the true value, $\mathbf{q}$, producing errors in forcing, initial conditions, and boundary conditions. $\boldsymbol{\varepsilon}_M$ denotes everything in a model that can be partially erroneous, including $\tilde{\mathbf{q}}$, but also internal parameterizations, inadequate resolution, misspecified topography, etc.

Magnitudes of model errors are identified as best can be done, and the goal is to modify the corresponding model elements appropriately such that the data, Equation (9.26), are reproduced within the error estimates; that is, reduction of the differences,

$$\left\| y_j^{(m)}(m\Delta t) - y_j(m\Delta t) \right\|, \tag{9.28}$$

to an acceptable value is sought. "Acceptable" involves understanding of both data and model errors. The vertical brackets most commonly represent the squared difference but could be any plausible difference measure, such as the absolute value. Results discussed immediately below are based upon the use of ordinary least-squares, in which the model (9.25) acts as a constraint and use is made of the *method of Lagrange multipliers* to enforce obedience to the model equations. This problem has huge dimension compared to more familiar least-squares problems, and so it requires considerable technical ingenuity to solve. Here estimates are used that emerged from the ECCO (Estimating the Circulation and Climate of the Ocean) system as summarized by Wunsch and Heimbach (2013b). Many of these results have already been used throughout the book.

---

[16] Use of spectral models without discrete grids does not change the nature of the problem.

GLOBAL DATA SETS

Determinations of the global ocean circulation demand global observations. The technologies that had emerged by about 1990 made plausible the goal of global estimates. Based upon the discussion in Chapter 2 and subsequent material, a summary in Table B.1 is useful.

Meteorological forcing fields obtained from the atmospheric reanalyses are used as preliminary estimates of the air-sea boundary conditions. In data volume terms, altimetry dominates the oceanic constraints; its importance follows from the geostrophic balance equations (9.1), in the Cartesian approximation at the surface,

$$-f\rho v\,(x, y, z = 0, t) = -g\rho\frac{\partial\eta\,(x, y, t)}{\partial x}$$
$$f\rho u\,(x, y, z = 0, t) = -g\rho\frac{\partial\eta\,(x, y, t)}{\partial y},$$

(9.29)

so that up to measurement errors, the surface geostrophic flow is known, and in the classical dynamic method, it has become a level-of-*known*-horizontal-motion. Global ship CTD and underway, velocity data are used in their sporadic space and time coverage. Beginning about 2005, the Argo profiling float network dominates all of the interior measurements above 1-2 km. Many other data types, ranging from local current transport determinations, to moored temperatures and salinities, are employed.

9.6   GLOBAL-SCALE SOLUTIONS

Solutions of this type were first described by Stammer et al. (2002, 2003) and were computed on a $2° \times 2°$ grid with 22 vertical levels, improving with growing computer power. Although some discrepancies continue to exist in the ability to fit some data types in some regions at various seasons, solutions are based on geostrophic, hydrostatic balance over most of the domain and are judged adequate for the calculation of large-scale transport and variability properties. To a very large extent, the state estimate is exploiting the robust character of geostrophy that applies over the majority of the oceanic fluid volume. Limited resolution does mean that systematic misfits are found in special regions such as the western boundary currents, the abyssal overflows, and convective areas. On the other hand, because of the accuracy of the geostrophic interior, poorly resolved boundary currents and related phenomena are nonetheless expected to carry the net transports of volume, heat, salt, potential vorticity, etc., which are necessary to maintain the interiors, even if the processes within them are not dynamically resolved.[17] Often the assumed data-error magnitudes and structures are themselves of doubtful accuracy.

Figures 3.29, 3.30, and 9.14 show two global 20-year average estimates of two elements of the ocean circulation: (1) the estimated surface elevation, and (2) the average vertically integrated flow field. Unlike superficially similar results inferred from individual hydrographic sections, these figures were calculated by averaging the fields from representations known nominally at 1 h intervals (Δt). The major qualitative features of the global-scale flow are

---

[17] The situation is reminscent of the theoretical assumption, fruitfully applied by H. Stommel in estimating the ocean circulation, that the essentially passive boundary currents would take care of mass transport and other imbalances arising from the known geostrophic interior and Ekman flows. The Stommel-Arons theory (Section 10.1.1) is an explicit example of the use of the assumption.

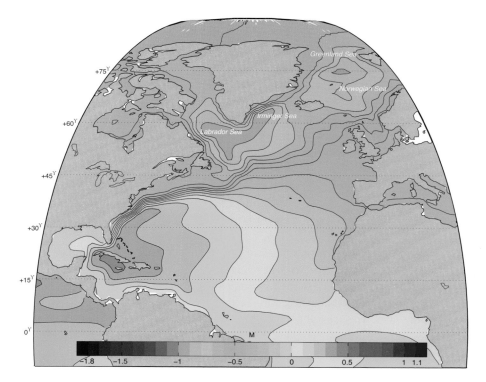

Figure 9.14: Expanded view of the North Atlantic Ocean surface elevation (meters) in Figure 3.29 from two decades of averaging. The subpolar and subtropical gyres emerge cleanly.

all present. These include the subpolar and subtropical gyres and the western and eastern boundary currents. The Gulf Stream, Kuroshio (in the North Pacific), the Agulhas (southwest Indian Ocean), the Antarctic Circumpolar Current, etc. are all visible.

Ganachaud (2003a) inferred that the dominant error in transoceanic transport calculations of properties from individual sections arose from the temporal variability. For example, Figures 9.15 and 9.16 display the global meridional heat and freshwater transport as a function of latitude along with their standard errors computed from the monthly fluctuations. Temporal effects are most conspicuous at low latitudes, but in many ways, the difficulty is greatest at high latitudes: the long time scales governing behavior there mean that the hydrographic structure is very slowly changing, requiring far longer times to produce an accurate time mean. In other words, a 30-year average at 10°N will probably be a more accurate estimate of the longer-term mean than one at 50°N. Even this comment avoids the question of whether stable long-term means exist, or whether the system drifts over hundreds and thousands of years. Providing answers requires knowledge of the frequency spectrum of oceanic variability, which is very poorly known at periods beyond a few years.

For the more than 20 years now available in the global state estimates (as of 2014), most of the large-scale properties, including the time variations, are stable to further data additions and model-code modifications, as a consequence of the fit to the comparatively well-sampled

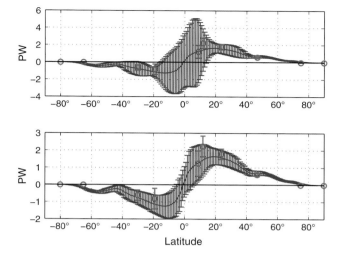

Figure 9.15: Global meridional heat transport in an ECCO state estimate. The upper panel shows standard errors including the annual cycle, and the lower panel, standard errors with the annual cycle removed, as being largely predictable. Possible systematic errors are not included. The red dots with error bars are estimates from Ganachaud and Wunsch (2002). Near-antisymmetry about the equator is apparent. Compare to Figure 9.8. The temporal scattering is roughly consistent with the estimated standard errors of the Ganachaud and Wunsch box inverse estimate. (From Wunsch and Heimbach, 2013.)

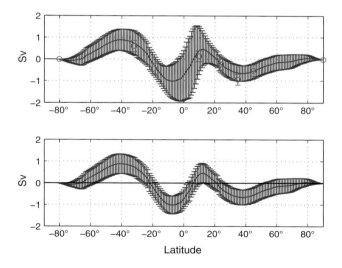

Figure 9.16: Global freshwater transport computed analogously to Figure 9.15 from a state estimate. The upper panel shows standard errors that include the seasonal cycle, and the lower, without the seasonal cycle. The red dots are from Ganachaud and Wunsch (2002).

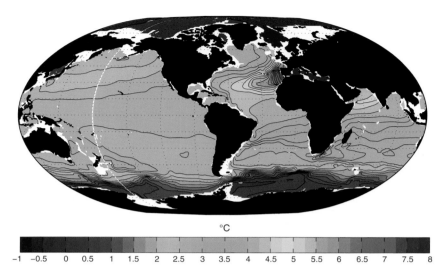

°C

Figure 9.17: Time-average potential temperatures (20 years) at 1517 m, the approximate depth of the Mediterranean salt tongue (Figure 9.18). Temperature and salinity have a strong tendency to produce cancelling contributions to density. As the result of a state estimate, this and other fields are consistent with known model equations of motion and with observations of diverse type.

hydrography and the dense altimetric measurements. The intricacies of the global, time-varying ocean circulation are a serious challenge to the summarizing capabilities of an author. Full discussion of the global state estimates becomes a description of the complete three-dimensional evolving ocean circulation, a subject requiring a large, separate, book encompassing distinctions among time and space scales, geographical position, depth, season, trends, and the forcing functions (controls).

9.6.1 Large-Scale Results

None of the results obtained so far can be regarded as the final state estimate: obtaining fully consistent misfits by the model to the observations has never been achieved. Misfits linger for diverse reasons, including the sometimes premature termination of the descent algorithms before the minimum is reached; misrepresentation of the true model or data errors; or selection of a local rather than a global minimum in the regional nonlinear components of the model. As with all large nonlinear optimization problems, the approach to the "best" solution is asymptotic and finding it is in part an art form. Only a few highlights of the global means are displayed here, as the emphasis throughout has been primarily upon the variability rather than the mean state. The reader is reminded that the kinetic energy of the variability exceeds that of the mean flow everywhere, by up to three orders of magnitude in some places (Figure 3.17).

MEAN-TEMPERATURE AND SALINITY FIELDS

As examples of what is now possible, Figures 9.17 and 9.18 show the 20-year average temperature and salinity at 1517 m depth in the state estimate. The large Mediterranean salt tongue,

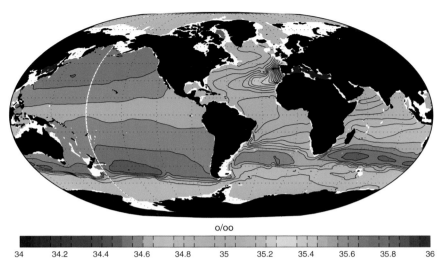

o/oo

| 34 | 34.2 | 34.4 | 34.6 | 34.8 | 35 | 35.2 | 35.4 | 35.6 | 35.8 | 36 |

Figure 9.18: Twenty-year average salinities at 1517 m. Compare the Atlantic Ocean to that in Figure 2.3. The global patterns incidentally reflect the general overall excess salinity in the Atlantic relative to the rest of the world ocean. In general terms, it is the result of an excess of evaporation over precipitation and runoff in the Atlantic Basin (see Warren, 1983, who discusses the contrast in the context of dense water formation). Maintenance of the contrast in a steady state, however, requires an ocean circulation capable of preventing unlimited growth of Atlantic salt by fresh-water dilution from the rest of the oceans.

which is conspicuous in Figure 2.3, is a permanent feature, despite the considerable variability in the Mediterranean convective regions that feed it. The very cold, but comparatively fresh, water at high latitudes is also apparent, as examples of the tendency of temperature and salinity to be partially density compensating.

Figures 9.19–9.24 display the 20-year average temperature and salinity in complete zonal sections at three latitudes, including one wrapping around the Southern Ocean. In a gross sense, these are all conventional, displaying the strong western boundary currents in the northern hemisphere and the weak structures at the eastern boundary. In the Southern Ocean section, the strong thermal and salinity anomalies of the Circumpolar Current, along the west coast of South America, just before it enters the Drake Passage is prominent. That the flow depends upon the spatial derivatives of these fields, not the fields themselves, remains an important context.

VELOCITY FIELDS

The 20-year time-average meridional velocity field, largely geostrophic, is shown in Figures 9.25–9.27 for the same three sections. What is perhaps most striking however, is that even with 20 years of averaging, a great deal of structure persists, particularly in the abyssal regions. How much of this would survive a much longer averaging interval is unknown.

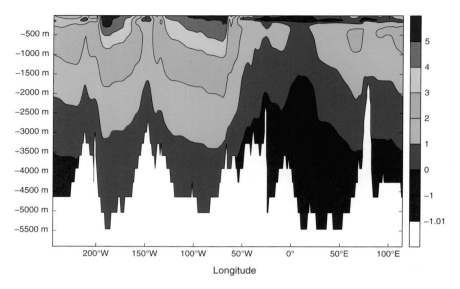

Figure 9.19: Twenty-year average temperatures along a zonal section at 57.5°S running through the Drake Passage (from a global state estimate). High temperatures occur on the upstream and downstream sides of the passage. The Southern Ocean, not properly discussed in this book, is evidently distant from the zonally uniform state often used in theoretical discussions. A higher-resolution result from this region can be found in Mazloff et al., 2010, 2013; see Figure 9.33 below.

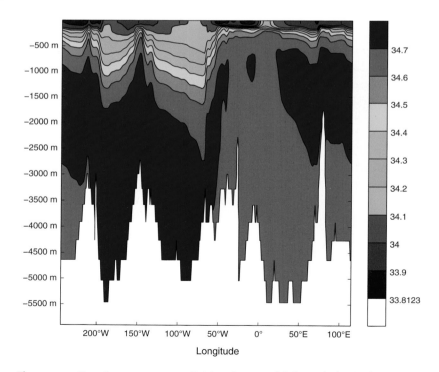

Figure 9.20: Twenty-year average salinities along 57.5°S through the Drake Passage.

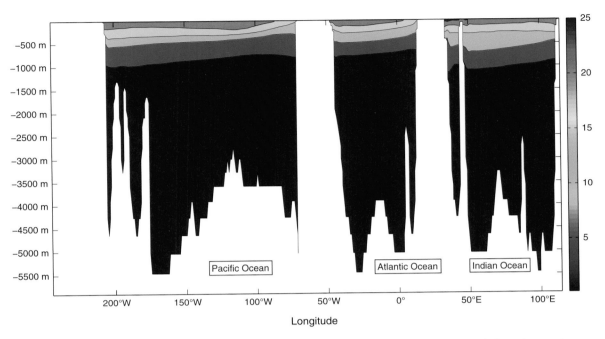

Figure 9.21: Time-average temperatures along the zonal section at 25°S. A characteristic upward tilt to the east is apparent.

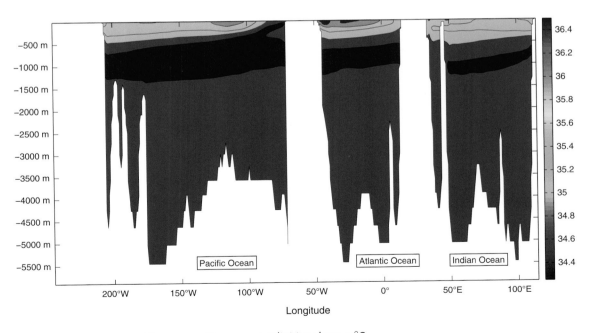

Figure 9.22: Time-mean salinities along 25°S.

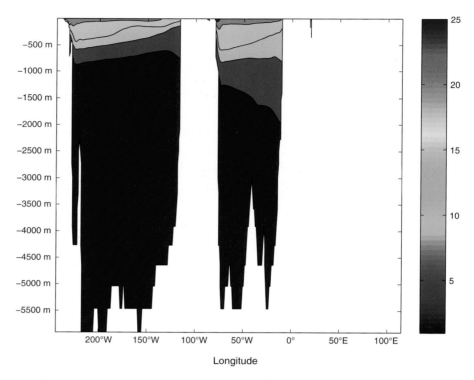

Figure 9.23: Time-mean temperatures along a section at 30°N. The upward tilt toward the east is even more apparent here than in the southern hemisphere.

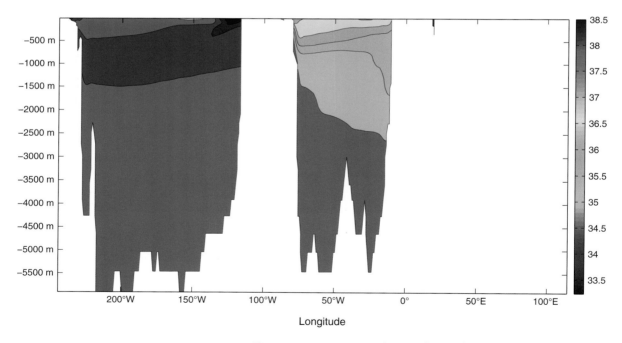

Figure 9.24: Time-mean salinities along 30°N.

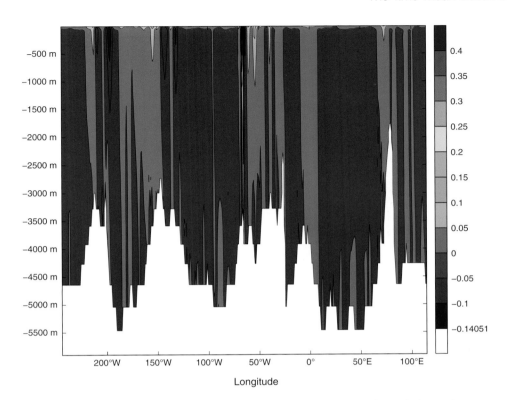

Figure 9.25: Twenty-year average meridional velocities along the section through the Drake Passage. In this region, the tendency toward columnar structures is particularly pronounced.

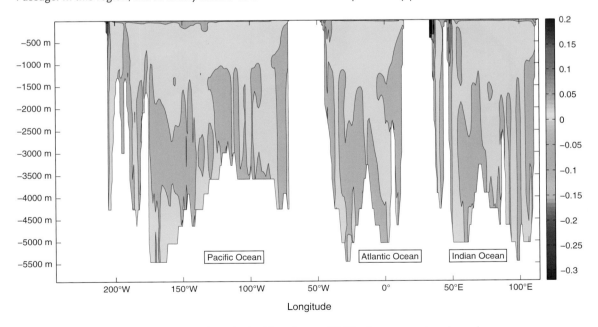

Figure 9.26: Twenty-year average meridional velocities along 25°S. The intense western boundary currents at or near the surface are notable.

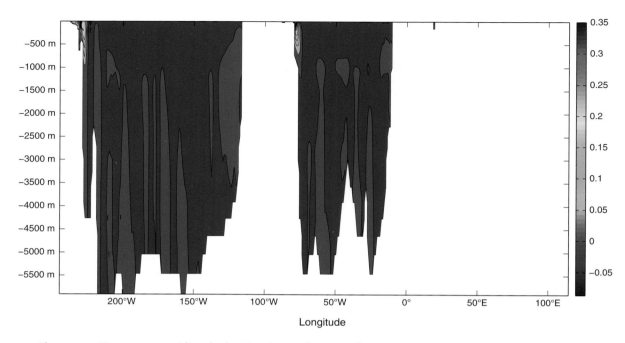

Figure 9.27: Time-mean meridional velocities along 30° N. Here the western boundary currents are the Kuroshio and Gulf Stream.

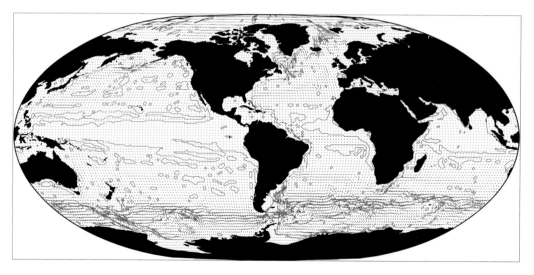

Figure 9.28: Twenty-year mean flow field at 1517 m depth. Time-average velocity vector at every second grid point (2° separations) from a 20-year average at 1500 m. Largest arrow is 25 cm/s. The strong high latitude "overflows" from the Nordic seas are visible, as are fragments of the deep western boundary currents in the Atlantic and South Pacific Oceans. Arrows are shown only where they exceed two times the local temporal standard deviation of the mean, roughly a level of no significance at 95% confidence for Gaussian variables.

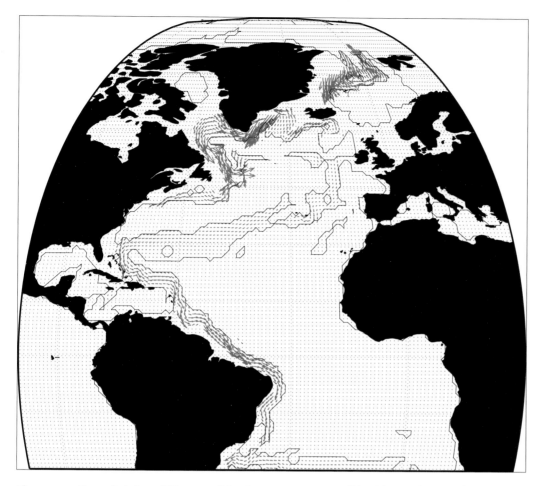

Figure 9.29: Expanded view of Figure 9.28 for the 20-year average of the Atlantic alone at 2° intervals at 1517 m depth. Only arrows whose magnitude exceeds two standard deviations of the temporal variance (outside the black contour) are shown.

As an example of the structure of the time mean currents, Figure 9.28 shows the vector flow at 1500 m as a 20-year average. Arrows are plotted only where their length exceeded the approximate 95% confidence interval for nonzero values. The intervals are based upon the temporal variations at that location. Conspicuously large values occur in the Southern Ocean, and the high latitude North Atlantic shows the strong flows emanating from the marginal seas into the open ocean. The DWBC in the Atlantic is prominent at most, but not all, latitudes and recalls the earlier discussion of the lack of continuity in the DWBC (and see Figure 3.14). Figure 9.29 shows an expanded Atlantic-only view of Figure 9.28, That only restricted regions of significant flow are found after 20 years of averaging demonstrates a fundamental oceanographic difficulty. Even this result is an optimistic one, because it is based upon a state estimate without an explicit eddy field—which dominates the variance—and the assumption that the 240 monthly estimates have uncorrelated errors.

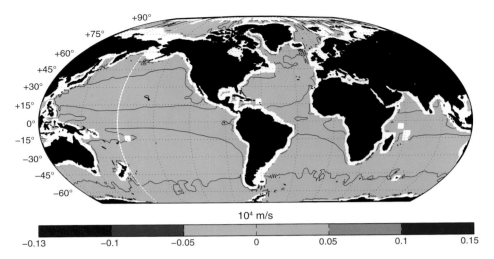

Figure 9.30a: Twenty-year average of $w(z = -195$ m) in units of 10^{-4} m/s, a useful estimate of the Ekman vertical velocity. The subtropical and subpolar gyres roughly correspond to the regions of $w < 0$ and $w > 0$, respectively. Notice the spatially complex pattern relative to analytical examples. About 200 Sv are being pumped downward and an equal amount is being returned, although the calculation is sensitive to how much spatial smoothing is done first. The oceanic near-surface region exchanges massive amounts of fluid with the water below.

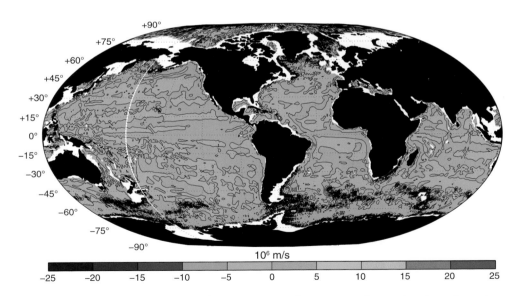

Figure 9.30b: Unsmoothed vertical velocity at 1500 m after 20 years of averaging. Apart from the intense small detail in the Southern Ocean, the result defies simple description.

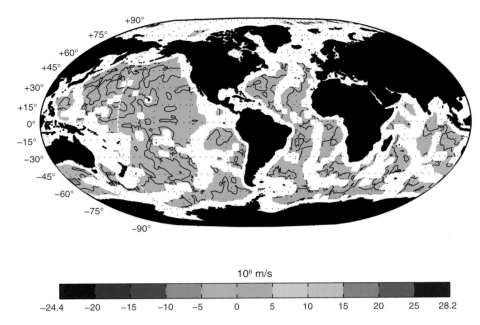

Figure 9.30c: A 20-year spatially smoothed (over 5° of latitude and longitude) estimate of the vertical velocity, w, at 3600 m. The complexity of the field renders moot most assertions about the general applicability of the large-scale steady-ocean circulation theories.

Twenty-year average vertical velocities at three depths are shown in Figures 9.30a–9.30c. The first, (a), at 195 m, can be interpreted as a first-order representation of the Ekman pumping velocity. Subpolar and subtropical gyres, as regions of suction and pumping respectively, are present. The persistence of small-scale structures raises once again the question of whether it would simplify with much longer averaging times. The 1500 m, spatially unsmoothed w in (b) is difficult to summarize. Vertical velocities are computed in existing finite difference ocean models as residuals of the horizontal volume divergence, and no directly constraining observations exist. Consequently w tends to contain a great deal of noise on the grid spacing. Topographic features changing at the grid scale also generate strong apparent vertical divergences which are most evident here in the Southern Ocean. No existing global GCM properly resolves the multiple physical processes involved in the encounter of the large-scale flow with three-dimensional topographic slopes.

VOLUME, ENTHALPY, FRESHWATER TRANSPORTS, AND THEIR VARIABILITY

The most basic elements describing the ocean circulation and its large-scale variability are usually the mass (or volume) transports. An estimate of the time-average transport stream function is shown in Figure 3.30, with the corresponding time-mean surface elevation in Figure 3.29.

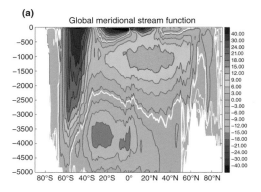

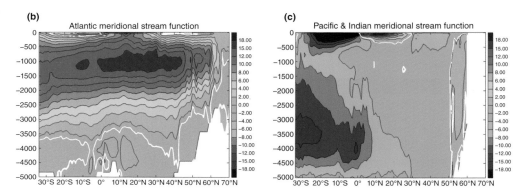

Figure 9.31: (a) Global meridional overturning streamfunction from a 19-year average of a state estimate; see Wunsch and Heimbach, 2013a. (b) The subcomponent of (a) restricted to the Atlantic Ocean; (c) the subcomponent of (a) integrated across the Indo-Pacific oceans. Note that the interpretation of this field differs in the Southern Ocean in the latitude range where no zonal support exists from the continental structures.

Figure 9.31 shows the meridional volume transport as a function of depth integrated zonally across all oceans. The very large degree of temporal variability can be inferred from Figure 9.15 which displays the heat transport values with error bars derived from the temporal variances. An important application of the state estimates has been in estimating the extent of the time variability taking place in the oceans (Wunsch and Heimbach, 2007, 2009). Withheld direct in situ observations in a few isolated regions (Kanzow et al., 2009; Baehr, 2010) are consistent with the inference that even volume transports integrated across entire ocean basins have a large, dominating temporal variability. This variability is apparent directly from the data, including altimetry, floats, moorings, etc. The state estimates synthesize these fluctuations and permit the inference of variations in heat transports and other properties.

9.6.2 Long-Duration Estimates

Although the development of state estimates was a consequence of the need to exploit the relatively dense, global-scale data that first became available with WOCE, the intense interest in decadal-scale climate change has led to attempts to extend the interval to much earlier times. Köhl and Stammer (2008) and Wang et al. (2010) pioneered the application of the least-squares Lagrange multiplier methods to an oceanic state estimate extending back to 1960. Their estimates have the same virtue as the other state estimates, in that they satisfy known model equations of motion and dynamics, with known misfits to all data types. The major problem is the extreme paucity of data in the ocean preceding the WOCE era (see, e.g., Figure 2.26) and the accompanying very limited meteorological forcing observations in the early days. Operational polar-orbiting meteorological satellite data only became available in 1979 (see Figure B.1 and Bromwich and Fogt, 2004) and represent a discontinuity in the meteorological observations, and estimates. Routinely useful altimetry appeared only at the end of 1992.

Some oceanic memory times are very long, and the better-determined later-interval ocean state carries significant information "backward" in time to the earlier, poorly observed time intervals. But which of the myriad components at the earlier times are well determined is not yet known. Published solutions for the interval prior to about 1992 are best regarded as being physically possible, but whose uncertainties, were they known, would necessarily be very much greater than they are in the later, more densely observed times. Even following 1992, the data densities vary significantly, with the arrival of multiple altimetric satellites, the increasing Argo float numbers, and the irregular CTD sampling.

9.6.3 Short-Duration Estimates

Finding a least-squares fit over decades is computationally very demanding, and for some purposes, estimates over shorter time intervals can be useful. In particular, Forget (2010) used the same model and methodology as that of the ECCO system but limited the calculation to three overlapping 18-month periods in the years 2004–2006 at 1° lateral resolution. Model-data misfit is considerably reduced compared to that in the longer-duration solutions, because the number of adjustable parameters (the control vector), particularly in the initial conditions, need track far fewer data.

Solutions of this type are very useful, particularly for upper-ocean and regional oceanographic estimates (see the water mass formation rate application in Maze et al., 2009). They are not climatologies, but to the contrary, they *do* bring us much closer to the ancient oceanographic goal of obtaining an oceanic "snapshot."

9.6.4 Global High-Resolution Solutions

Modelers have been pursuing ever-higher resolution from the very beginning of ocean modelling, and the effort continues. In classical computational fluid dynamics, "numerical convergence" was sought: the demonstration that further improvements in resolution did not qualitatively change the solutions, and preferably that they accurately reproduced known analytical values. Such demonstrations with ocean GCMs are almost nonexistent, global model resolution remaining coarse, and thus a very large literature has emerged attempting to demonstrate the utility of "parameterizations"—constructs intended to mimic the behavior of motions below the resolution capability of any particular model.[18] The most widely used parameterizations significantly improve observational misfits, but major inconsistencies remain. Absent analytical solutions for comparison, parameterizations are mainly tested against each other and regional fragmentary data sets.

---

[18] A recent discussion of the state of the art is that of Marshall et al. (2012).

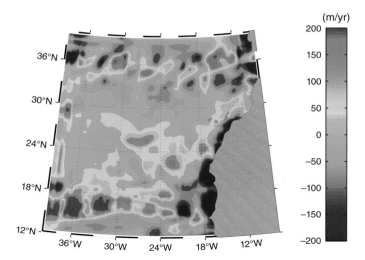

Figure 9.32: Annual average subduction across the mixed-layer interface in a high-resolution state estimate from a region embedded in a larger-scale estimate (From Gebbie, 2007). The ocean's characteristic noisiness reappears once again.

In models of small regions, a few solutions exist that were unchanged under further grid-refinement and which in principle can be used to test the parameterizations. Considerable evidence exists (e.g., Hecht and Smith, 2008; Lévy et al., 2010. See Figure 3.46) that qualitative changes take place in GCM solutions when the first baroclinic deformation radius, at least, is fully resolved. State estimation is a curve-fitting procedure and because of the dominant geostrophic balance, its mass transport properties are comparatively insensitive to unresolved spatial scales—bottom topographic interference being an exception. In data-dense regions, away from boundary currents, results should be robust, even at modest resolution.

Ultimately, however, the boundary current regions in particular must be resolved (no pa-rameterizations exist for unresolved boundary currents) so as to accurately compute transport properties for quantities such as heat or carbon that depend upon the accurate rendering of the second moments in both space and time, $\langle\langle C\mathbf{v}\rangle_o\rangle$, where C is any scalar property, and $\mathbf{v}$ is the velocity. Thus a major effort has been devoted to producing global or near-global state estimates from higher-resolution models (Menemenlis et al., 2005a,b). The same methodologies used at coarser resolution are also appropriate at high resolution—as has been demonstrated in the regional estimates, which will be taken up next. But the computational load rapidly escalates with the state and control vector dimensions. Consequently, available globally constrained models have used reduced data sets, and estimates exist only over comparatively short time intervals.

Because of the short duration, much of the interest in these high-resolution models lies with the behavior of the eddy field rather than in the large-scale circulation. As with ordinary forward modelling, how best to adjust the eddy flux parameterizations when only parts of the eddy field have been resolved is a major unknown.

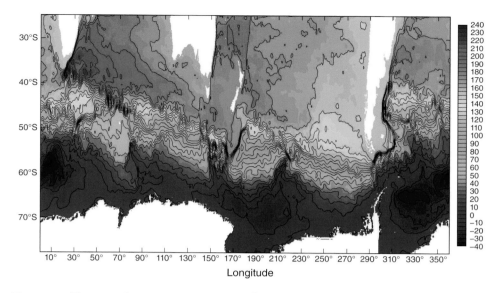

Figure 9.33: The top-to-bottom transport stream function, from a $1/6°$ horizontal resolution state estimate in the Southern Ocean averaged over 3 years. Compare the spatial scales to those perceptible in Figure 3.30. Again, it is the spatial derivatives of these fields which govern the water movement. The narrow boundary currents in the north are particularly important. (From Mazloff et al., 2010.)

9.6.5 Regional Solutions

With the high computational load of improved resolution, efforts have been made to produce regional estimates, typically embedding them in a coarser-resolution global system using appropriate open-boundary conditions. Implementing open conditions is technically challenging, particularly where the velocity field is directly involved—for example, slight barotropic divergences or convergences can produce large volume imbalances.

Gebbie et al. (2006) discussed estimates in a small region of the North Atlantic, and their results were used to calculate (Gebbie, 2006) the eddy contribution to near-surface subduction processes (Figure 9.32). In a much-larger region, the Mazloff et al. (2010) Southern Ocean State Estimate (SOSE) was computed over the restricted time interval 2005–2006 (now being extended) at $1/6°$ horizontal and 42 layer vertical-level resolution. The transport stream function from that result is shown in Figure 9.33 and makes apparent the very small scales in the circulation, which are especially prominent in the Southern Ocean. Comparison may be made to the schematics, such as Figure 9.4.

CHAPTER 10

Large-Scale Circulation Physics

Having summarized some of the observed features of the general circulation, attention is now turned to sketching some of the explanatory theories for the major elements. Determining the extent to which they describe what is observed is not easy. As is evident from the various pictures scattered throughout the book, many different physical regimes make up the ocean circulation, with the flows usually dominated by the temporal variability. The spatial dynamic range is across 10 orders of magnitude. Desirable as it might be to have one, no simple relationship exists between spatial scale and associated temporal scales. The largest spatial scales can shift over hours (surface gravity waves—as tsunamis and tides), but with regions of the intermediate and deep density fields taking as long as 10,000 years to fully change. The 100 km scale can shift temporally as internal waves, fronts, geostrophically balanced eddies and also as topographically controlled features, with all acting simultaneously. In the presence of this variability, testing time-mean properties against existing time-mean theory is difficult. Nonetheless, a start is made.

10.1 THEORIES OF THE WIND-DRIVEN OCEAN

Ekman (1905), Sverdrup (1947), Stommel (1948), and Munk (1950) laid the foundation of modern wind-driven ocean theory, based on the concept of a steady-state ocean. Ekman's solution has been discussed in every textbook written since 1905, and so it is sketched here only for reference purposes.

THE EKMAN LAYER

The geostrophic, hydrostatic balance equations for a resting, constant-density ocean are modified, plausibly, to account for the action of a very large spatial scale, steady wind stress (force/unit area), $\boldsymbol{\tau} = (\tau^x, \tau^y)$ acting at $z = 0$:

$$-fv = -\frac{\partial p'}{\bar{\rho}\partial x} + A_V\frac{\partial^2 u}{\partial z^2}, \tag{10.1a}$$

$$fu = -\frac{\partial p'}{\bar{\rho}\partial y} + A_V\frac{\partial^2 v}{\partial z^2}, \tag{10.1b}$$

describing the perturbations about the resting pressure field. The eddy coefficient, A_V, is assumed to be a constant. From the Reynolds analogy to the molecular case, the boundary conditions are

$$\bar{\rho}A_V\frac{\partial(u,v)}{\partial z}\bigg|_{z=\eta} = (\tau^x, \tau^y) = \boldsymbol{\tau} \tag{10.2}$$

and that at infinite depth, $u, v, p' \to 0$. Ekman sensibly neglected the difference between $z = 0$ and $z = \eta$. Taking advantage of the assumed linear equations, the flow is separated into a geostrophic and an Ekman part: $(u, v) = (u_g + u_E, v_g + v_g)$, so that u_g, v_g balance the pressure gradients. Then,

$$-fv_g = -\frac{\partial p'}{\bar{\rho}\partial x}, \tag{10.3a}$$

$$fu_g = -\frac{\partial p'}{\bar{\rho}\partial y}, \tag{10.3b}$$

$$-fv_E = A_V\frac{\partial^2 u_E}{\partial z^2}, \tag{10.4a}$$

$$fu_E = A_V\frac{\partial^2 v_E}{\partial z^2}. \tag{10.4b}$$

If f is constant as is $\boldsymbol{\tau}$, all x, y dependence vanishes and as Ekman showed, the solution is

$$u_E = \frac{\tau^x}{\bar{\rho}\sqrt{fA_v}}\exp\left(z/\sqrt{2A_V/f}\right)\cos\left(z/\sqrt{2A_V/f} - \pi/4\right) \tag{10.5a}$$

$$v_E = -\frac{\tau^x}{\bar{\rho}\sqrt{fA_v}}\exp\left(z/\sqrt{2A_V/f}\right)\sin\left(z/\sqrt{2A_V/f} - \pi/4\right), \tag{10.5b}$$

where the x-axis is aligned with the wind direction, so that $\tau^y = 0$ (always possible on an f-plane), a diminishing spiral rotating downward from $z = 0$. The surface velocity lies at $45°$ to the right of the wind in the northern hemisphere (to the left in the southern), and the total mass flux,

$$(U_E, V_E) = (0, -\tau^x/f), \tag{10.6}$$

which is at $90°$ to the right of the wind in the northern hemisphere. No pressure gradient is associated with the Ekman layer in this approximation—an important element in the interpretation of altimeter data.[1] Evidently the *direct* effect of the wind stress driving on the ocean, if the representation is adequate, is confined to a distance of order of the *Ekman depth*, $D_E = \sqrt{2A_V/f}$. If $A_V = v \approx 10^{-6}$ m^2/s, had the molecular value, as in laboratory experiments,

---

[1]Ekman's solution is that of a boundary layer. His work preceded that of Prandtl, who is commonly credited with pioneering the concept; see Faller, 2006.

then $D_E \approx 15$ cm at mid-latitudes, which shows that molecular processes have little to do directly with how the wind influences the ocean. If $D_E \approx 100$ m by assumption, then $A_V \approx 0.5$ m$^2/s$ at mid-latitudes.

Although the prediction of a spiral and its apparent explanation of the movement of icebergs in Ekman's original calculation represented a pretty piece of theory, direct observational confirmation of its existence did not take place for many decades. That some flow, acting much like the Ekman spiral, had to exist in the ocean was inferred indirectly—from the existence of the gyre-like circulations, for example, in Munk, 1950. Warm water is piled up into the subtropical gyres like the Sargasso Sea, and as expected from a net transport to the right of the wind stress (compare Figure 3.29). The strong wind-driven upwelling occurring near many eastern boundaries was at least qualitatively supportive of the existence of the lateral transport in Ekman layer encountering a wall.[2] Despite the lack of observational confirmation of a spiral, and serious qualms about the reality of A_V and its possible time and depth dependencies, quantities such as U_E, V_E, the net transports, are independent of A_V itself and of the integration depth as long as it greatly exceeds D_E. Because much interest lies not with the velocities but also with the transports and with the ability of the theory to rationalize the gyres, the Ekman layer became, and remains, one of the building blocks of oceanic physics—despite the many assumptions.

Obtaining direct evidence for Ekman layers and their corresponding spirals had to await technologies capable of producing time series in the upper levels of the ocean. Price et al. (1987) finally showed that a spiral could be seen in open-ocean data, but only after data processing was carried out in a time-dependent wind-turning coordinate system (Figure 10.1), because steady winds are seen only in textbooks. Another recent example of a spiral from the Drake Passage (Lenn and Chereskin, 2009) is shown in Figure 10.2.

Real understanding of the role of the Ekman layer in the general circulation did not come until Stommel (1957) described how it interacted with an otherwise geostrophically balanced ocean. Further discussion is postponed until additional elements of wind-driven ocean current theory are below. described.

SVERDRUP BALANCE

Consider what is called the *Sverdrup balance*. Ekman's equations are modified only by permitting f to vary with y, in the β-plane approximation. Integrating Equations (10.1a) and (10.1b) in the vertical between a depth z_S and the sea surface,

$$-fV(x,y) = -\frac{\partial P(x,y)}{\partial x} + \bar{\rho}A_V \frac{\partial u(x,y)}{\partial z}\bigg|_{z_S}^{\eta} + p'(\eta)\frac{\partial \eta(x,y)}{\partial x} - p'(z_S)\frac{\partial z_S(x,y)}{\partial x}$$

$$fU(x,y) = -\frac{\partial P(x,y)}{\partial y} + \bar{\rho}A_V \frac{\partial u(x,y)}{\partial z}\bigg|_{z_S}^{\eta} + p'(\eta)\frac{\partial \eta}{\partial y}(x,y) - p'(z_S)\frac{\partial z_S(x,y)}{\partial y},$$

[2] A wind blowing parallel to a vertical wall so as to generate an offshore Ekman volume or mass flux could, in a steady state only, acquire the necessary fluid from below. Strong upwelling of cold water from depths is a major feature particularly of many eastern boundary regions, including parts of the African and California coasts, and is very important in the nutrient supply to biological processes; see Gill and Clarke, 1974. An interesting boundary layer structure exists next to a vertical wall, often called *Stewartson layers*; see Greenspan, 1990.

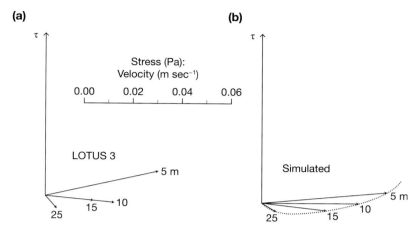

Figure 10.1: The Ekman spiral obtained from near-surface current meter records. (a) Is the result of the data analysis and (b) is the modeled spiral. τ denotes the calculated time-mean wind stress, and it is plotted in a fixed direction, showing flow predominantly at 90° to the right of the time-varying wind. Observations do show a declining spiral with depth albeit different in detail from the simulation. (From Price et al., 1987.)

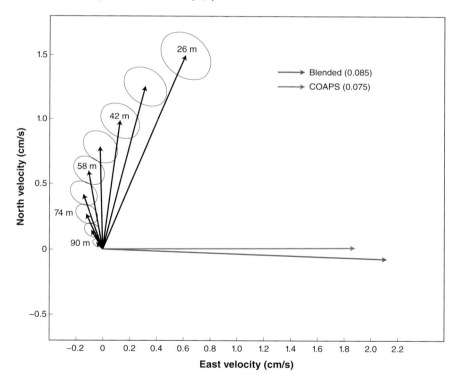

Figure 10.2: Inferred Ekman spiral from direct velocity measurements in Drake Passage. Note the reversed sense in the southern hemisphere, and the flattening relative to the basic theory. "Blended "and "COAPS "refer to two different wind products. (From Lenn and Chereskin, 2009.)

where

$$(U(x,y), V(x,y)) = \int_{z_S(x,y)}^{\eta(x,y)} \bar{\rho}(u,v)\,dz, \quad P(x,y) = \int_{z_S(x,y)}^{\eta(x,y)} p'\,dz, \tag{10.7}$$

and the two "extra" terms on the right side of both equations arise because p' has been placed under the derivative with spatially varying limits of integration.

The Ekman boundary condition (Equation 10.2) permits evaluation of the upper limit of the integrals involving A_V. Then a series of further assumptions is made (see especially Fofonoff, 1962, or Pond and Pickard, 1983), including, particularly, the vanishing of $w(z_S)$ and of all viscous stresses at z_S, and the relatively flat nature of both η and z_S. (Sverdrup assumed that the ocean was at rest below a horizontal z_S, thus meeting all of these requirements.) The equations become

$$-fV = -\frac{\partial P}{\partial x} + \tau^x, \tag{10.8a}$$

$$fU = -\frac{\partial P}{\partial y} + \tau^y. \tag{10.8b}$$

Under those same very strong assumptions, the vertically integrated continuity equation is reduced to

$$\frac{\partial U}{\partial x} + \frac{\partial V}{\partial y} = 0. \tag{10.9}$$

Cross-differentiating Equations (10.8a) and (10.8b), and using Equation (10.9)

$$\beta V(x,y) = \frac{\partial \tau^y}{\partial x} - \frac{\partial \tau^x}{\partial y} = \hat{\mathbf{k}} \cdot \nabla \times \boldsymbol{\tau} \tag{10.10}$$

the *Sverdrup balance*, or the *Sverdrup relation*. This equation states that the meridional mass transport above z_S at any point can be computed from the *purely local* wind-stress components—an extremely powerful statement—if it is true.

From Equation (10.9), the corresponding zonal mass transport is

$$U(x,y) = -\int \frac{\partial V(x,y)}{\partial y}\,dx + G(y). \tag{10.11}$$

Sverdrup famously assumed that $G(y)$ should be chosen to make U vanish on an eastern meridional wall, $x = L_x$, of his model ocean. (Setting the transport U to zero is a great simplification of the real boundary condition, $u = 0$.) Thus from the wind stress alone, a full, steady transport theory for $U(x,y)$, $V(x,y)$ had emerged. The production in this very simple calculation of the zonal equatorial countercurrent, flowing *into* the predominant wind direction, was both striking and convincing (the reversal in the surface gradients visible in Figure 3.29, and labelled "NECC,"North Equatorial Countercurrent in the North Pacific Ocean). Trouble existed when also considering a meridional wall on the west, where $U(x,y)$ would again have to vanish, and the solution failed there. Ignoring that problem, a number of calculations of $U(x,y)$, $V(x,y)$ have been published for the ocean basins under the strong assumptions that Equations (10.10), and (10.11) are correct and that the time-mean wind curl is accurately known. Qualitatively, the gyre-scale flows emerge and show that the physics is at least partially present in the observed circulation.

The Ekman solution has disappeared from view in the Sverdrup relation, and its role is obscure. With a steady wind having no curl (no vertical component of vorticity), Sverdrup balance implies that no *transport* of water exists, $V(x, y) = U(x, y) = 0$. Why that should be, along with other questions, remained to be understood.

EKMAN PUMPING AND SUCTION

As a vertical integral, Sverdrup's elaboration disguises the underlying physics. Much earlier, in seemingly unrelated work, the English mathematicians Hough (1897) and Goldsbrough (1933) had considered the response of a constant-density water-covered spherical ocean (an *aquaplanet*), with a flat bottom in perfect geostrophic balance and subject to idealized patterns of evaporation and precipitation. In a geostrophic flow, meridional mass or volume transports depend only upon the pressure differences between two points, irrespective of the distance between them. Thus (using the β-plane analog), the net meridional mass flux between any two points x_1, x_2 is,

$$V_G = \frac{P(x_2) - P(x_1)}{f(y)}, \tag{10.12}$$

when integrated to an arbitrary depth. With pure geostrophy, flow fields cannot cross lines of constant pressure. In any meridional flow, f would be changing, and thus V_G would also be changing proportionally, necessarily increasing or decreasing depending on the flow direction. Unless a mass source or sink exists, the only possible steady flow would have to be zonal, with $V_G = 0$. In the Hough-Goldsbrough ocean (Figure 10.3), fluid is being added or subtracted by an imbalance of precipitation and evaporation. Then, conversely, Equation (10.12) shows that a perfect, steady-state, geostrophic balance could be maintained *by moving the fluid to a new latitude* where $f(y)$ was correspondingly increased or decreased to accommodate the decreased or increased volume flux. Thus if fluid was being added, V_G would be directed equatorward, where the diminished $f(y)$ would permit it.

Stommel (1957) recognized that the amount of water moving meridionally to the right of the wind stress in the Ekman layer would vary both with f and with any spatial changes in the value of τ. If more fluid was moving meridionally at one latitude in the Ekman layer than could be carried by its transport at a nearby latitude, the fluid could only go downwards—to avoid moving the sea surface in the steady state. Alternatively, if the Ekman layer transport increased with latitude, it could only obtain that fluid from below, sucking it up. Equations (10.4a) and (10.4b) and the continuity equation permit the calculation of the vertical velocity at the base of the Ekman layer induced by a spatially varying Ekman transport:

$$w_E = \frac{\partial}{\partial x}\left(\frac{\tau^y}{\bar{\rho}f}\right) - \frac{\partial}{\partial y}\left(\frac{\tau^x}{\bar{\rho}f}\right), \tag{10.13}$$

which has two parts, one owing to the spatial dependencies in the time-mean wind field; the other from the variations in $f(y)$. (x-dependencies in τ^y can similarly generate flow into or out of the Ekman layer.) By postulating that the oceanic flow below the surface layers is perfectly geostrophic, Stommel showed that it would absorb or provide the Ekman vertical mass transports, analogous to the Hough-Goldsbrough mass sources and sinks. A geostrophic flow below the surface layer has to adapt to its necessarily changing water volume by shifting latitude—moving fluid at a rate V_G. Stommel was thus able to "pick apart" the Sverdrup balance, $V = V_E + V_G$.

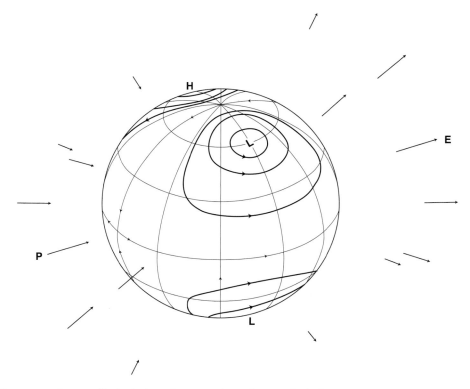

Figure 10.3: Stommel's depiction of an aquaplanet flow in which the eastern hemisphere is one of net evaporation and the western of net precipitation (with zero total imbalance) in geostrophic equilibrium. Adjustment of the flow to the net increase and decrease of mass generates latitude-changing net meridional transports and represents the effects of Ekman layer pumping and suction on the ocean beneath. H and L denote highs and lows in the surface elevation and thus pressure, producing corresponding anticyclonic and cyclonic geostrophic flow fields. (After Stommel, 1957.)

If the wind has no curl, $V = 0$, or $V_G = -V_E$. The total meridional mass transport vanishes, but a flow, possibly intense, still exists. Regions where $w_E < 0$ (Ekman downward pumping) correspond to the subtropical gyres in Sverdrup's calculation where the interior flow is equatorward. Where $w_E > 0$, Ekman *suction* is the subpolar gyres with poleward interior flow. If $w_E = 0$, no interior geostrophic meridional transport exists. The zonal-transport component, $U(x, y)$, of the Sverdrup theory, depending upon $\partial V(x, y) / \partial y$, will still be present. In summary, the picture that emerges is of a wind-driven interior ocean circulation dominated by Ekman pumping with a magnitude on the order of 100+ Sv (see Figure 9.30a). To a good approximation, vertical mass balance requires that the integral

$$\iint \rho w_E dA = 0, \tag{10.14}$$

so that regions of suction nearly balance those of pumping—a result important in understanding air-sea interaction.

THE STOMMEL GULF STREAM

Stommel's (1948) paper was directed at explaining the existence of the Gulf Stream. For present purposes, if Sverdrup's z_S is simply assumed to be the flat bottom of an ocean with constant density, the governing equations become

$$-f\bar{\rho}v = -\frac{\partial p'}{\partial x} + \bar{\rho}A_v \frac{\partial^2 u}{\partial z^2},$$

$$f\bar{\rho}u = -\frac{\partial p'}{\partial y} + \bar{\rho}A_v \frac{\partial^2 v}{\partial z^2},$$

$$\frac{\partial u}{\partial x} + \frac{\partial v}{\partial y} + \frac{\partial w}{\partial z} = 0,$$

the hydrostatic equation being passive with constant $\bar{\rho}$. Integrating vertically, from seafloor to sea surface and assuming both of these surfaces are effectively flat,

$$-f V = -\frac{\partial P}{\partial x} - RU + \tau^x \qquad (10.15a)$$

$$f U = -\frac{\partial P}{\partial y} - RV + \tau^y \qquad (10.15b)$$

$$\frac{\partial U}{\partial x} + \frac{\partial V}{\partial y} = 0 \qquad (10.15c)$$

$$V = \int_{-h}^{0} \bar{\rho}v'\,dz, \quad P = \int_{-h}^{0} p'\,dz, \ldots$$

What would be an Ekman layer at the bottom has been slightly simplified by replacing the corresponding stress terms by a Rayleigh friction, $-R\,(U, V)$.[3] Equation (10.15c) is satisfied by introducing a transport stream function,

$$U = \frac{\partial \psi}{\partial y}, \quad V = -\frac{\partial \psi}{\partial x}.$$

Substituting into Equations (10.15a) and (10.15b) and cross-differentiating to eliminate the pressure field gives rise to

$$\nabla^2 \psi - \frac{\beta}{R}\frac{\partial \psi}{\partial x} = \frac{1}{R}\left(\frac{\partial \tau^x}{\partial y} - \frac{\partial \tau^y}{\partial x}\right),$$

which Stommel solved for a rectangular ocean, $0 \le x \le L_x, 0 \le y \le L_y$, subject to an idealized wind field, $\tau^x = -\tau_0 \cos(\pi y/L_y)$, mimicking the low-latitude trades and higher-latitude westerlies, while having vanishing y derivative on $y = 0, L_y$.

For small R, in approximate form, the solution is

$$\psi(x, y) \approx \frac{\pi \tau_0}{\beta L_y} \sin\left(\frac{\pi y}{L_y}\right)(L_x - x) + \frac{L_x \pi \tau_0}{\beta L_y} \sin\left(\frac{\pi y}{L_y}\right) \exp\left(-\beta x/R\right), \qquad (10.16)$$

the first term of which is, gratifyingly, Sverdrup's solution with its implicit Ekman layer, and the second represents a strong meridional flow trapped at the western boundary. This equation provided the first explanation of western-intensified, Gulf Stream–like flows, showing

---

[3] The bottom Ekman layer differs from that at the surface because the lower boundary condition is one of zero horizontal flow instead of a horizontal stress condition. Stommel did not explicitly consider the bottom flow.

its explicit dependence on the presence of β—the crucial variable. At least three physical explanations of the westward intensification of the circulation exist. Most commonly, vorticity balance is invoked (Stommel, 1965; Pedlosky, 1987, Chapter 5). Veronis (1981) gave an explanation in terms of the behavior of the pressure field. The simplest rationale was provided by Pedlosky (1965) in terms of the way dissipation acts on the energy carried by Rossby waves with eastward and westward group velocities, the former having much shorter wavelengths and thus being more rapidly dissipated.

EXTENSIONS

Munk (1950) and Munk and Carrier (1950) replaced the Rayleigh friction by horizontal viscous eddy terms (Equation 3.29a–e). An important result was the recognition in the second paper that the flow on the western wall was formally a boundary layer—and thus was easily analyzed with the then-new mathematics of singular perturbation theory. Munk (1950) was able to identify qualitatively the major features of the ocean circulation apparent in Figure 3.29, such as the subtropical and subpolar gyres, the equatorial current system, etc.

10.1.1 Stommel and Arons Abyssal Flows

The next major building block for the time-mean flow was the so-called Stommel and Arons (SA) theory (Stommel, 1957) of the abyssal circulation. These ideas were directed at understanding the response of a rotating ocean to the presence of a mass source at high latitudes, in which ocean the injected mass was sufficiently dense to reach the seafloor in a confined region. They sought to analyze the subsequent movement of the dense waters. The deep ocean was idealized to have a constant density given by that of the mass source, one greater than that of the overlying fluid, and to be unresponsive to wind forcing. Again assuming a steady state, the volume of that deep layer has to remain constant despite the continued injection of mass.

The same geostrophic vorticity equation (9.19) is assumed to apply to the ocean everywhere outside the western boundary regions. Let the rate of volume injection of dense fluid be denoted q_{SA}. To balance this volume injection, an upward vertical velocity, w, must exist at the top of the deep layer for a volume transport, wA, where A is its total area. Absent any further information, w might as well be assumed constant, $w = q_{SA}/A$, so that water volume is conserved. (Discussion of what happens to the fluid after it leaves the bottom layer is postponed for the moment—the interface is being assumed not to move despite the finite vertical velocity there.)

Equation (9.19) then demands that

$$V_{SA} = \int_{-h}^{z_I} \bar{\rho} v \, dz = \frac{\bar{\rho} f}{\beta} \left(w(z_I) - w(-h) \right) = \frac{f \bar{\rho}}{\beta} \frac{q_{SA}}{A},$$

assuming a flat bottom where $w(-h) = 0$. In the northern hemisphere where both f, β are positive, the flow and the resulting transport must be *northward*. Thus they obtained, from these very simple ideas, the startling result that with a deep mass source (as is known to exist) in the high-latitude North Atlantic, the interior flow of the deep North Atlantic had to be *toward the source*. That result seemed completely counterintuitive and rapidly led to laboratory tests and analysis (see Warren, 1981; Faller, 1981), vindicating the inference.

To conserve mass across any line crossing an ocean basin, Stommel and Arons assumed that a western boundary current (WBC), of unspecified dynamics would exist, one passively transporting whatever volume flux the interior meridional transport demanded of it. In the North Atlantic, the result predicted a DWBC flow *southward* underneath the Gulf Stream and into the South Atlantic Ocean (see Figure 2.4). Stommel then extended the hypothesis to the global ocean, postulating the presence of mass sources in the northern North Atlantic and in the Weddell Sea of the Southern Ocean. SA theory thus predicted the presence of previously undetected DWBCs in the remainder of the world ocean, along with an estimate of their directions and transports, an inference that has generally, but not in detail, been vindicated (Hogg, 2001).

This simplest form of the SA theory rests on the assumption that w is spatially uniform. In reality, however, a uniform w does not exist; interior motions are clearly more zonal than they are meridional (see Figure 3.15). w is now thought to be dominated by large values in small regions (Figs. 9.30b, 9.30c). Furthermore (e.g., Lozier, 1997), the North Atlantic DWBC appears to be embedded in fields of strong recirculating gyres with which they exchange mass on a continuous basis (see Figure 9.29). Particle pathways involve torturous transfers to and from the interior.

ABYSSAL RECIPES

Munk (1966) gave an explanation of the fate of the water upwelling across z_{SA}. He noted that in the open *abyssal* Pacific Ocean,[4] profiles of temperature, salinity, and carbon-14 all appeared to be both exponential in shape, varying only very slowly horizontally. Consequently he postulated (the abyssal recipe) that below about 1000 m, they all obeyed a local one-dimensional version of the steadyadvection-diffusion equation (Equation 3.29e),

$$w \frac{\partial C}{\partial z} - K_V \frac{\partial^2 C}{\partial z^2} = 0, \qquad (10.17)$$

where C is any of the three scalar variables (for ^{14}C, a decay term, $-\lambda C$, is put on the right). With w, K_V both constant, the solutions to this equation are

$$C = C_0 + C_1 \exp\left(wz/K_V\right)$$

for both temperature and salinity, and Munk successfully fit observed profiles with $C_0 = 0$, $K_V/w \approx 1000$ m. By also solving the radiocarbon (^{14}C) equation with its additional decay term, he was able to separately estimate w, K_V, producing the memorable value, $K_V = 1$ cm^2/s $= 10^{-4}$ m^2/s.[5] With $w = K_V/1000$ m$= 10^{-7}$ m/s, and when multiplied by the oceanic area, $q \approx 30$ Sv, also a plausible value.

[4]The term "abyssal" is another vague one. *The Oxford English Dictionary* notes that it originally referred to the ocean below 100 fathoms (about 200 m), but it now sometimes refers to the very deepest places with no clear upper limit. In common usage for the fluid ocean, it appears to mean generally the volume below about 1000 m, the part above perhaps being labelled the "thermocline region."

[5]Olbers and Wenzel (1989) found that Munk's solution was numerically unstable, arising as it did from an ill-conditioned system. Craig (1969) also showed that the radiocarbon data had been mis-interpreted. Despite these problems, the solutions have stood the test of time. Much later, the problem was revisited by Munk and Wunsch (1998), who more explicitly used the balance as a large area average, rather than as one applying at individual points. That point of view permits at least formal accounting for the mixing coefficients, K_V, being dominated by very large values in regons of topographic roughness. Munk (1966) carefully titled his paper "Abyssal Recipes," but many subsequent authors apparently believed it was meant to apply nearly to the surface. Whether the point balance is accurate anywhere is not clear.

The elegant picture that thus emerged was of a vertical transport inferred from the SA theory being sustained in steady state by a compensating downward diffusion of heat and salt that just balanced the upward advection of cold and/or salty water from q_{SA}. Like many theoretical constructs, the concepts and numerical values are best considered as being sophisticated scale analyses for the orders of magnitude of flows and structural scales, rather than being taken as literal descriptions of the ocean. In particular, if K_V is small, then w is predicted in Equation (10.17) to be correspondingly small, and the geostrophic vorticity equation (9.19) demands that v also be small. Meridional ocean transports are thus directly hostage to vertical mixing rates and the "abyssal recipe" is best regarded as being applicable only as a large-scale areal integral, accounting in that manner for highly localized regions of mixing and of w.

(More recent analyses formulate this balance in isopycnal or neutral-surface coordinates. As long as they are quasi-horizontal, the inferences do not change. "Vertical" mixing is then more properly "diapycnal" mixing and the two can have markedly different values when large isopycnal slopes exist.)

DIRECT BUOYANCY FORCING

The upper-ocean flow field is dominantly the result of wind driving. Apart from efforts to understand the convective regions, and the special behavior of the currents produced locally by the dense fluid running along the bottom (the *overflows*), less attention has been paid to the response of the ocean to buoyancy exchange with the atmosphere. With the rise in recent years of interest in climate change and its implications for shifts in air-sea temperature differences, changes in evaporation and precipitation, etc., the situation has begun to shift. Theory does not produce any relationships as analytically tractable as the Ekman layer or Sverdrup balance. In contrast to the latter, the response to buoyancy exchanges is not interpretable as a local one (although the wind-driven circulation is not purely locally determined either, once meridional boundaries are introduced).

Few complete solutions exist and, in particular, responses are not additive. The results of buoyancy forcing are sensitively dependent upon the nature of the wind-driven circulation that is also present. Some of the flavor of the theory and modelling can be found in Spall's paper (2008). An interesting general framework for discussing the response to buoyancy forcing is in the context of oceanic energetics, and the notion that the ocean is supposedly a form of heat engine, which will be discussed later.

10.2 THERMOCLINE THEORIES

10.2.1 Nonlinear Theories

CONTINUOUS STRATIFICATION

Away from the regions of direct influence of the WBCs, a quite elaborate theoretical structure exists for discussing an oceanic steady state. Areas of direct WBC influence do include the large regions where they have left the coast and are proceeding generally eastward but whose dynamics in those regions are being ignored here. Fundamental to most theories of the interior are the further assumptions that mixing and viscosity acting in the interior of the oceans are vertical. The governing equations, sometimes called the *thermocline equations*, are purely geostrophic, incompressible, and hydrostatic, so that the nonlinear density conservation equation becomes

$$u\frac{\partial\rho}{\partial x} + v\frac{\partial\rho}{\partial y} + w\frac{\partial\rho}{\partial z} = K_V\frac{\partial^2\rho}{\partial z^2}, \tag{10.18}$$

u, v being the geostrophic, hydrostatic flows. The intent of these thermocline theories is *determination* of the density field from the surface boundary conditions: it is not prescribed in the rest state. Solutions are theories of a mean ocean climate.[6]

Finding solutions of these equations has been one of the major foci of theoretical oceanography in the last four or five decades, and the textbooks discuss the equations in detail. Only the sketchiest outline is provided here. Because these theories are directed at the large scale, spherical coordinates are necessary. With $P' = p/\bar{\rho}$, an equation in P' can be derived from Equation (10.18) and geostrophic balance (Veronis, 1969) as,

$$\frac{\partial P'}{\partial\phi}\left(\frac{\partial^3 P'}{\partial z^3}\frac{\partial^2 P'}{\partial\lambda\partial z} - \frac{\partial^2 P'}{\partial z^2}\frac{\partial^3 P'}{\partial\lambda\partial z^2}\right) + \frac{\partial P'}{\partial\lambda}\left(\frac{\partial^2 P'}{\partial z^2}\frac{\partial^3 P'}{\partial\phi\partial z^2} - \frac{\partial^3 P'}{\partial z^3}\frac{\partial^2 P'}{\partial\phi\partial z} + \cot\phi\frac{\partial^2 P'}{\partial z^2}\frac{\partial^2 P'}{\partial z^2}\right)$$

$$= K_V\sin\phi\cos\phi\left(\frac{\partial^2 P'}{\partial z^2}\frac{\partial^4 P'}{\partial z^4} - \left(\frac{\partial^3 P'}{\partial z^3}\right)^2\right), \tag{10.19}$$

a complicated, nonlinear, nonseparable, high-order partial differential equation to be applied below the Ekman layer. Remarkably, a number of solutions to this equation and various equivalents were found beginning in the late 1950s. They generally have a special structure and are known as *similarity solutions*, because they involve specific relationships among λ, ϕ, z in the argument of P'. With of the similarity assumptions, arbitrary surface boundary conditions on density or temperature and the Ekman pumping velocity cannot be satisfied. Despite their restricted nature, some of the known solutions bear a strong visual relationship to the time-mean structure of the main thermocline (e.g., Figure 3.21, 9.19–9.24).

It was recognized almost immediately that solutions with finite values of K_V were almost visually indistinguishable from those with $K_V = 0$, even though the vertical interior physical balances were different. Measurements were then undertaken to directly determine K_V (Gregg, 1987). In general, values in the upper ocean are considerably smaller than Munk's value of 10^{-4} m^2/s and thus generated some consternation.[7] If a sweeping generalization can be made, it is that the upper ocean (the main thermocline and above) is closer to a nondiffusive system (a "perfect fluid") than it is to a diffusive one. The major exception is the Southern Ocean, where many deep density surfaces *outcrop* (reach the surface) and are subject to intense, direct, wind-induced mixing.

In a more general context, and in a problem that plagues the subject to this day, theory produces time-mean, large-scale circulation and temperature and salinity structures that are visually very similar, independent of whether they are dominated by diffusion or are closer to being perfect fluid solutions outside of boundary-mixing regimes (which include the surface). This ambiguity is a consequence of near-geostrophic balance almost everywhere, leaving only small, noisy differences to be interpreted. Disputes over interpretation have tended to move away from those concerning the open-ocean thermocline, which appears to be close to perfect fluid conditions, to those governing the abyssal circulation, where diffusion, topographic mixing, etc., are believed to be dominant.

---

[6] Linearized thermocline theories remain of interest, particularly in discussions of small disturbances to the circulation. See Veronis and Stommel, 1956, and Pedlosky, 1968.

[7] Gregg, 1998.

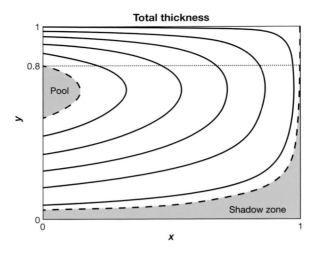

Figure 10.4: Schematic of the flow in a wind-driven, layered ocean plotted as the upper-layer thickness. Apart from the shadow zone and the pool, the circulation is ventilated from the atmosphere. (After Vallis, 2006.)

LAYERED (VENTILATED) THEORIES/HOMOGENIZATION

Equations such as (10.19) are extremely challenging, and the theory hardly progressed after the initial production of the similarity solutions. The problem was reopened when Luyten et al. (1983) reduced the representation from one having a continuous density field, $\rho(x, y, z)$, to one in which density was homogeneous in a small stack of *layers*. Finding the depths, $z_i(x, y)$, proved mathematically much simpler than finding $z(\rho(x, y))$ in the continuous case. At that point, a great expansion of the theory, which is well summarized in textbooks, occurred. The results are known as the *ventilated thermocline theories*, and their structures are intriguingly similar to many (not all) of the observed gross features of the quasi-stable elements of the oceanic thermocline. In the theories, one or more upper-ocean layer is set into motion by the wind, arriving ultimately at Sverdrup balance. They are "ventilated" in the northern parts of the gyre where the layers are at the surface, and thus take on atmospheric conditions, and where downward Ekman pumping sets them into motion. They plunge below lighter layers outcropping further to the south. In the subtropics, the layers thin out and eventually outcrop again at the surface along a line south of which the wind does not act (see Figure 10.4). In the basic theories, the eastern boundary controls the position of that line because of the strong requirement that the geostrophic flow must vanish there, leading to a "shadow zone" as shown in the figure. Another shadowlike region (the "pool") of unventilated water exists along part of the western side where a WBC is expected. Theories attempting to calculate the structures and flows in such regions rely heavily on the notion of conservation of potential vorticity within them and on the deduced actions of the intense eddy fields known to occur there (Rhines and Young, 1982). See Keffer (1985) for the testing of such hypotheses with data.

A number of papers argue that the shadow zone is visible, for example, in the southeastern North Pacific and North Atlantic. As one of the difficulties, note (Figure 9.30a) that the region which in the North Atlantic, and probably in the North Pacific, might be identified as a shadow zone, is a region of Ekman suction. If it is truly unventilated from the north, it will have its own regionally forced circulation. Whether the structure of the long-term, mean

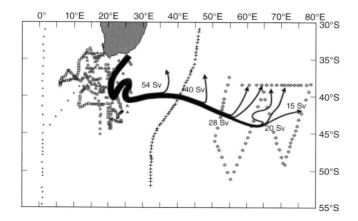

Figure 10.5: Schematic assembled from several decades of data of the mean position of the Agulhas Current and an estimate of the mean transports. The reversal of direction is the retroflection. Data positions are shown with symbols. (From Lutjeharms, 2007.)

wind field producing suction rather than pumping is related to an underlying shadowing is not obvious.

Choice of the layer thickness has a profound effect on the quantitative locations of any outcrops, and no provision is made for the fundamental process by which fluid can change density at the surface or below. As described here, the theories do not encompass the WBCs, nor any ageostrophic flows at the eastern margin, the latter often being associated with wind-forced upwelling and sometimes strong undercurrents (e.g., Barton, 1998). Attempts continue to extend the theories to this and other regions. That gross elements of the observed circulation are similar to major features of the theories is hardly in doubt. But almost no quantitative measure of that resemblance exists, and it is equally true that conspicuous regions of failure, where the theory appears to say nothing much of use, also exist.

10.3 WESTERN BOUNDARY CURRENTS

The basic nature of the major western boundary currents manifested at the sea surface emerged from the Stommel-Munk linear steady-state theories, which then developed into the seemingly more realistic nonlinear, steady-state theories of Charney and Morgan. In a gross sense, the Gulf Stream in the North Atlantic and the Kuroshio in the North Pacific are most closely akin. For reasons that are not entirely understood, the Southern Hemisphere counterparts (the Brazil Current and the East Australia Current) are less well developed, with the latter often characterized as being closer to a series of eddies than to a coherent current. The very powerful Agulhas Current on the east coast of Africa is unique in exhibiting *retroflection*—that is, bending back on itself and running to the east at higher latitudes in the Southern Ocean (see Lutjeharms, 2007 and Figure 10.5). In the process, the Agulhas sheds very powerful eddies (*rings*), which can be tracked across the width of the South Atlantic (Byrne et al., 1995).

Animations (available on the Web) of seasurface height obtained either from altimetry alone or from the state estimate combinations show, in the Gulf Stream and Kuroshio, a complex set of time-dependent meandering pathways over weeks and months (see Figure A.17). These meanders sometimes close on themselves, breaking away, and producing rings initially carrying the full transport of the currents.

One striking phenomenon, which appears to be independent of the complex downstream time-dependent pathway, is the so-called *separation* of the currents from the coast. The Gulf Stream conspicuously leaves the North American coast at Cape Hatteras and heads out to the open sea. Partly because numerical circulation models have struggled to reproduce this separation (they commonly have the current follow the coast much farther north), a lot of attention has been paid to the question of what determines the separation point. In the initial Stommel-Munk theories, it appeared to be the latitude where the windstress curl vanished— corresponding to the position where the incoming zonal Sverdrup interior flow changed sign, to flow outward. With the later nonlinear theories, that hypothesis was no longer adequate, and it was suggested that the topography or the shape of the continent (trending to the northwest from Hatteras) forced the separation. Subsequently, arguments have been made for control by the available volume of warm water (forcing a colder, "lower" layer to surface), by eddy-mean-flow and recirculation interactions, or by interaction at the crossover point with the DWBC, among others. Possibly all of these phenomena act in concert to produce the observed separation point. Ierley (1990) and Dengg et al. (1996) reviewed the subject, which today is primarily the domain of the modeling community.[8] For a description of the Kuroshio, which has two distinct and stable paths, see Taft (1972).

10.4 THE ROLE OF EDDIES

When it became clear that a ubiquitous eddy field existed in the ocean, one whose kinetic energy dominated the motions of the ocean (Figure 3.17), the central question became the extent of its influence on the larger scales. In the years following the Mid-Ocean Dynamics Experiment (MODE Group, 1978; Robinson, 1983), global descriptions in terms of time and space scales and were sought. Determining the extent to which the intense flows would introduce important nonlinear terms such as $\mathbf{v}\cdot\nabla\mathbf{v}$ or $\mathbf{v}\cdot\nabla\rho$ into the governing equations (Equations 3.4), however, became largely problems of theory and modelling instead of direct observation. In particular, the time- and regional space-average values of these terms at depth, $\langle\mathbf{v}\cdot\nabla\mathbf{v}\rangle$, $\langle\mathbf{v}\cdot\nabla\mathbf{v}\rangle_o$, etc. require both data durations and regional measurements well beyond existing capabilities. What field observations have occurred have been focussed primarily on the Southern Ocean, where eddies are particularly intense and where the basic geostrophic balance is suspect at the outset because of the absence of zonal boundaries in the Drake Passage latitude band (LaCasce et al., 2013).

In a weak eddy field, one with wavelike behavior as is seen over parts of the spectrum in Figure 8.10, a Stokes drift analogous to that described for surface and internal waves (Figure 5.7) can be calculated, and the consequences for momentum and heat and related budgets found. With an intense eddy field, however, the nonlinearities are so strong as to preclude

[8] Sachs (2007) interpreted proxy data as showing that under different climate conditions several thousand years ago, the separation point may not have been at Cape Hatteras, but further north. Whether models of the modern ocean that tend inadvertently to put the separation at or near New Jersey, are thus more faithfully describing the early Holocene, 10,000 years ago, is an interesting bit of speculation involving not just the models but also the interpretation of the data.

a perturbation approach. The need to understand the effects has led to a large, technical theoretical and modelling literature aimed at (1) understanding the consequences, and (2) parameterizing the effects in numerical models that do not explicitly resolve the eddies. Very recently, equation sets such as the *transformed Eulerian mean* system have been proposed, in which the primitive equations have been modified, to partially account for eddy effects on the larger scales (see Vallis, 2006). These systems represent the efforts to replace the simple Reynolds-type decomposition with more accurate and explicit representation of eddy effects. Speer et al. (2000) is a clear demonstration of the consequences, but few other direct observational tests of these parameterizations have appeared.

10.5 TESTING MEAN DYNAMICS

Unlike observed data, models in principle calculate the complete oceanic state vector, which can potentially be used to test the various theoretical constructs. Does the thermocline equation (10.19) accurately represent any of the ocean? Many models, however, remain too coarsely resolved or inadequately parameterized to provide useful tests of theory or of the representation of some observations. Thus the Gulf Stream contains structures at and below 10 km horizontal scales, and these are not reproduced by the current generation of large-scale models. Other issues lie directly with numerics. For example, models are often stabilized by using artificially enhanced interior friction. Yet other barriers arise: theories of simple, steady forcing fields (wind curls, for example) are difficult to relate to noisy short-duration observations also displaying major, long-time-persistent spatial structures.

As a consequence of all these problems, direct quantitative testing of quasi-steady circulation models remains rare. Little doubt exists of the broad accuracy of geostrophy, even in obviously time-varying situations. Gyres and western boundary currents and deep western boundary and eastern boundary currents always emerge. Thermocline depths and thicknesses are roughly consistent with those estimated in the theories (although adjustable parameters are available to make them so). Something like a shadow zone of the layered thermocline theory may exist (e.g., Talley, 1988). Here only a bit of the flavor of what can now be done is provided.

IS THE OCEAN IN GEOSTROPHIC EQUILIBRIUM?

As previously described, tests of geostrophic balance have been made with in situ instruments, and no convincing failure is known on scales of about 10 km and greater. What testing of models has been carried out has focussed not directly on such quantities as the thermal wind shear, but instead on indirect tests via the vertical velocities, w, which are diagnosed from the flow field. Model Rossby numbers tend to be small over much of the ocean, which is at least consistent with balanced motion. Simple scale analyses show that geostrophic hydrostatic balance is expected to fail in shallow water as lateral spatial scales contract, and where the continental margin topography (Figure 3.5) spans much of the water depth. The weak abyssal flows near and in strong topographic gradients are, in the models, likely significantly influenced by dissipational and time-dependent effects, but explicit comparisons are uncommon.

The large-scale, slowly varying vertical velocity, w, although unmeasurable directly, plays a crucial role in understanding the ocean circulation. In practice, its values are usually diagnosed from the divergence of the horizontal velocities (Equation 3.29d), in an intrinsically noisy calculation. Magnitudes estimated from plausible balances such as Equation (9.19) are on the order of 10^{-7} m/s or less at mid-depths.

Lu and Stammer (2004) attempted to test the geostrophic vorticity balance from a state estimate, but although some interpretable structures were found, the determination of w was confounded by both the short averaging time available in the presence of variability of decades and longer, and the tendency of all such models to generate a numerical noise in w on the grid spacing.

EKMAN PUMPING

Figure 9.30a shows a 20-year average of w ($z = -195$m) in a state estimate at a depth that is reasonably identified as close to the Ekman suction and pumping (Equation 10.13). Large-scale regions of upward and downward velocities are seen, roughly coincident with the classical gyre boundaries. But a great deal of complex structure is also visible, and it is not easily characterized. How much of this structure is permanent, and how much would vanish with a 100-year average is not known. Whether volume or mass transports are accurately to the right of the wind (northern hemisphere) has not been tested globally, and the many difficulties of determining the effective wind stress must be accounted for in doing so.

THE SVERDRUP BALANCE

The basic notion of Sverdrup balance is a rare example that has been the subject of a number of regional tests beginning with Leetmaa et al. (1977). Some awkward questions immediately appear: (1) The theory depicts a steady state. How long an average is required? How does the answer vary with geographical position? (2) The theory calculates the net meridional transport V between the surface and an integration depth z_S. But z_S is defined (Equation 10.7) as the depth where $w = 0$ (although Sverdrup did not express it that way; he assumed the ocean was at rest below z_S). How is it to be determined—noting that it is not predicted?

Figure 10.6 depicts those regions where diagnostics from a 16-year average state estimate were able to pass a test of consistency with Sverdrup balance. Large regions of apparent consistency do appear, but what is perhaps more striking are the important regions where it fails. Again, questions arise as to whether 16 years is simply too short to produce a static balanced state, or whether the model has shortcomings preventing its proper rendering of the balance, or whether some or all of the physics omitted from the Sverdrup balance dominate the transports. Or is it all of these things? Given the very long adjustment times expected at high latitudes (Figure 4.16), the associated weak stratification producing strong topographic interaction, and the variability of the wind field (Figure 3.35), high-latitude equilibrium is not expected in a two-decade average.[9] Possibly the Sverdrup balance could become an accurate description there over very much longer averaging times, but that has not been demonstrated.

---

[9] The calculation was cut off at the latitude of the Cape of Good Hope owing to the debates concerning the applicability of the Sverdrup balance in the Southern Ocean (Olbers, 1998).

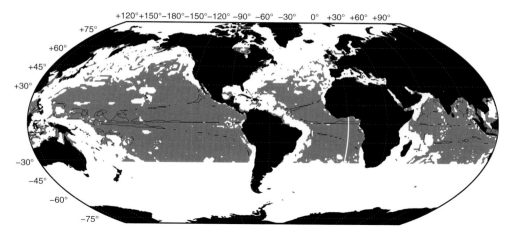

Figure 10.6: The region in gray passes tests for the accuracy of the Sverdrup balance at that location. The black lines are zeros of the residual used to test the physics. Applicability of the theory in the Southern Ocean remains controversial, and some of the flavor of the debate can be obtained from Olbers, 1998. (From Wunsch, 2011, from a 16-year average of an ECCO state estimate.)

10.5.1 The Annual Cycle

Much of the known physics of large-scale variability is a combination of instabilities, forced and free wavelike motions, and turbulent interactions. All of these phenomena contribute to the time dependencies that render charts such as those in Figures 9.28 and 9.29 devoid of regions with statistically significant time-mean flows. Discussion of variability generally, including that beyond the 20-year duration of existing state estimates, is taken up in Chapters 11 and 12. The annual cycle is discussed here because it is nearly a *steady, periodic* oceanic response.

The annual cycle is the response to a very strong underlying forcing function—insolation changes from the moving sun through the year (Equation 6.27) and is by most measures the largest of all climate signals (Figure 12.1). That forcing, however, is mediated through the very complex, annual atmospheric changes, and understanding how and why the ocean shifts seasonally on a global basis is a difficult problem. Vinogradov et al. (2008) mapped the amplitude and relative contributions for salt and heat to the annual cycle in sea level (Fig. 10.7). The importance of the annual variation is visible in Figures 9.15 and 9.16, appearing there as the large contribution to the standard errors.

Despite the relatively very strong seasonal forcing, the annual cycle in the ocean is generally invisible in temperature or velocity below about 300 m—in the rare multi-year records that do exist, with the equatorial region being the major exception. Upper-ocean confinement is a consequence of the long time scales associated with oceanic baroclinic adjustment as reflected in the dispersion curves as shown in Figures 4.16 and 4.17. This description does not describe the barotropic responses, which appear as bottom pressure perturbations at periods of a year (see Ponte and Stammer, 2000). The long time scales generally mean that responses are spatially localized (see Gill and Niiler, 1973), although again, the near-equatorial region is exceptional.

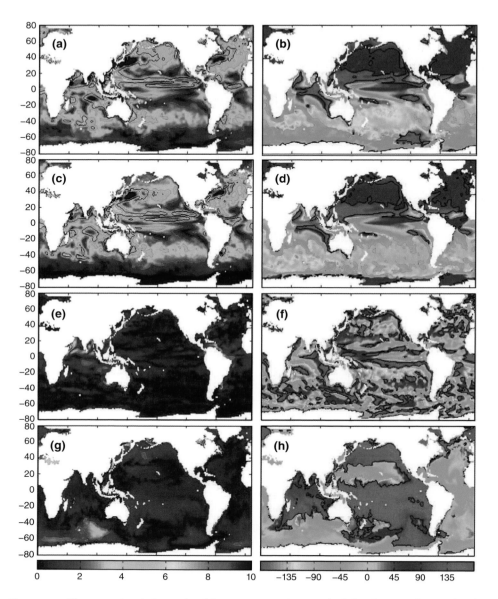

Figure 10.7: The annual cycle in sea level from a state estimate. The left column is the amplitude in centimeters, and the right column is the phase in degrees. From top to bottom, the rows are the surface elevation, the thermal contribution, the salinity contribution, and the bottom pressure. (From Vinogradov et al., 2008.)

10.6   A SUMMARY

The basic physics of the steady circulation that developed beginning with the thermal-wind equations, Ekman's boundary-layer theory, and the works of Sverdrup, Stommel, Munk, and their extensions has developed greatly in recent years. These theories describe in a qualitative, visual sense, much of the observed time-average ocean circulation as far as we can perceive it. Quantitative tests are few. Once again invoking the dictum that if a physical process is possible, it will operate, all of these elements are present, somewhere, in the large-scale circulation. In moving much beyond this bland statement, the science confronts the "noise" of the temporal variability, as well as the ongoing struggle in the modelling community to discern its influence on the poorly depicted time-average flow. That influence extends from the ordinary difficulty in defining accurate means of noisy data to the direct dynamical effects being represented via parameterizations of lumped "eddy" fields. The next few years ought to bring a considerably clearer understanding of the regions and the extent to which the time-mean building blocks describe the observed flow.

CHAPTER 11

Interpreting and Using the Circulation

With useful global, time-varying, state estimates now available, it becomes relatively easy to address a number of topics related to the general circulation, ones that previously could at best be studied only regionally. Many of these topics are the subject of intense ongoing work, and a summary has an implied rapid obsolescence. This and the next chapter overlap somewhat, with this one a bit more focussed on the physics, the second more on the observed long-time-scale temporal change.

11.1 ENERGETICS AND MIXING

All real fluids are dissipative to a degree, however slight. In a statistical steady state, energy sources must resupply them at the same rate as their energy loss. Basic understanding of any physical system normally includes descriptions of the major power suppliers, the mechanisms of dissipation, and the transfers and transformations between the various mechanisms and places of import and export of energy. Reviews can be found in Ferrari and Wunsch, 2009 and Huang, 2010.

Until very recently, surprisingly little attention was paid to understanding the sources maintaining the ocean circulation. If an internally consistent fluid dynamics solution is available, the energetics are diagnostic from the flow and do not provide new information. Also, the intricacies of the thermodynamics of a compressible fluid, with densities dependent upon temperature and salinity over a large pressure range, have only recently come to be appreciated.

On the other hand, in situations like that pertaining to the ocean circulation where much of the workings of the system remain obscure, considerable insight can be obtained by asking questions about the energetics whose answers do not require the complete details of the flow. A discussion of energetics is useful if it proves simple enough to provide physical understanding. No guarantee exists that this simplicity exists, and emerging discussions (e.g., Roquet, 2013;

Tailleux, 2013) suggest that some aspects of the buoyancy and thermodynamic components of power input generally are potentially more complex than insight-producing. In particular, the rates and regions of injection of power from buoyancy forcing apparently depend directly and sensitively on the circulation also in existence from the wind forcing. The latter plays dual roles as both an energy source ("driver") and as a "controller" of the buoyancy responses.

11.1.1 Mean Energy Sources

A list of forces acting on the ocean and conceivably providing power to the large-scale circulation is not very long (Ferrari and Wunsch, 2009):

1. Tides
2. Winds
3. Freshwater fluxes: evaporation, precipitation, river run-off, percolation, ice-melt from both land and sea.
4. Enthalpy (heat) exchange with the atmosphere
5. Geothermal heating (through the seafloor)
6. Atmospheric loading (air pressure changes)
7. Salt-cycling through the mid-ocean ridges and through formation and melting of sea ice
8. Biological processes

TIDES

Tides do not directly impart significant energy to the general circulation in the open ocean: their primary role is as an energy source for the smaller-scale turbulent processes that permit the circulation to stir the fluid. Mixing of a stably stratified fluid raises the center of mass, increasing the system's potential energy. The generation of kinetic energy of fluid flow is a secondary by-product and generally weak. Tides are thus not a driver of the circulation; they are instead an important energy source permitting the circulation to take the form that it does. Without tides and their conversion to baroclinic motions, the circulation would look very different and have different properties. So sources of the gross motions must lie elsewhere.

WINDS

Winds do work on the ocean and vice versa through the Newtonian relationship of a force acting over a finite distance. Power input, P_w, is thus the wind stress (the force per unit area) multiplied by the water velocity. This simple relationship masks a complex coupled-fluid flow problem: because the air-sea boundary is a moving one, on a very large range of time and space scales, computation of the product is complicated. Ripples, gravity waves, swell, Ekman layers, Langmuir circulations, inertial motions, internal waves, fronts, the balanced general circulation, etc. are all generated by the wind itself (Fig. 11.1). Whether a steady state is ever achieved before the wind shifts is unlikely. Determining the quantitative behavior of the moving sea surface, particularly insofar as it generates waves, still remains after hundreds of years of study an active area of investigation.

Reasonably robust theories all strongly support the idea that a very large fraction of the power input by the wind into the ocean goes into gravity wave motions. A small residual enters the Ekman layer and the mixed layer. An even smaller amount penetrates beneath

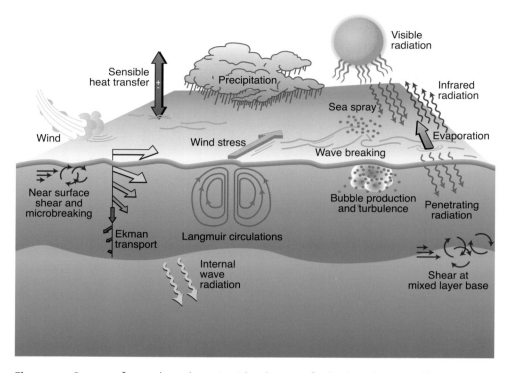

Figure 11.1: Cartoon of upper-layer dynamics. The absence of a depth scale is sensible. Langmuir circulations are long quasi-two-dimensional cells approximately aligned in the downwind direction. They are an important part of the circulation very near the sea surface and have some of the characteristics of turbulence. See for example, Thorpe, 2005, p. 251. (J. Edson et al. [2007]. "The coupled boundary layers and air-sea transfer experiment in low winds." *Bull. Amer. Meteor. Soc.* 88:341–356 © American Meteorological Society. Used with permission.)

these surface layers and accounts for the sustenance there of the great bulk of the oceanic flow owing to the wind; finding this small residual is a considerable problem.

Most estimates of the rate of wind working on the oceans employ the empirical drag law for the stress, Equation (8.4), which was used for bottom drag, except now,

$$\boldsymbol{\tau} = C_D \rho_a |\mathbf{u}_a - \mathbf{u}_o| (\mathbf{u}_a - \mathbf{u}_o), \tag{11.1}$$

with $\rho = \rho_a$ for air, and the velocity is $\mathbf{u}_d = \mathbf{u}_a - \mathbf{u}_o$, the difference between the horizontal air velocity just above the sea surface (typically at a nominal 10 m) and the horizontal velocity of the moving sea surface. If Equation (11.1) is regarded as reliable ,then

$$P_w = \boldsymbol{\tau} \cdot \mathbf{u}_o, \tag{11.2}$$

which is integrated over time and space to produce the power input. Estimates of C_D vary and depend at least upon air-sea temperature differences and the sea state.[1] Plausible arguments

---

[1]See, for example, Behringer et al., 1979: Large and Pond, 1981; Trenberth et al., 1989; Sullivan and McWilliams, 2010.

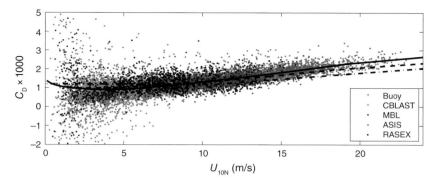

Figure 11.2: Individual measured values of C_d. Notice the large scatter, particularly at low wind speeds. At very high wind speeds, in tropical storms and hurricanes, data are understandably sparse and scattered (e.g., Powell et al., 2003). (From Edson et al., 2013 which should be consulted for the acronym sources.)

give it values approximately equal to 10^{-3}, but how it varies hourly, seasonally, etc., is poorly known. Values are scattered but believed to increase toward both very low and very high wind speeds (e.g., Fig. 11.2).

Much of the time, $|\mathbf{u}_o| \ll |\mathbf{u}_a|$ and $\mathbf{u}_d = \mathbf{u}_a$, reducing Equation (11.1) to,

$$\boldsymbol{\tau}_R = C_D \rho_a \, |\mathbf{u}_a| \, \mathbf{u}_a, \tag{11.3}$$

which one might call the *resting ocean approximation* for the stress. Recently, it was recognized that this approximation leads to a systematic error in $\boldsymbol{\tau}$ that is surprisingly large. The term "resting ocean approximation" refers only to the calculation of the stress—unless the ocean is moving, no work can be done.

Equations (11.1) and (11.3) are nonlinear. That in turn means that use of time or space averages for $\mathbf{u}_a$ or $\mathbf{u}_o$ can introduce large errors because

$$\langle P \rangle = \langle C_D \rho_a \, |\mathbf{u}_d| \, \mathbf{u}_d \cdot \mathbf{u}_o \rangle \neq C_D \rho_a \, |\langle \mathbf{u}_d \rangle| \, \langle \mathbf{u}_d \rangle \cdot \langle \mathbf{u}_o \rangle,$$

even disregarding the time and space variations in C_D. Factor-of-two differences exist between estimates that include synoptic wind systems (for example) and those that do not (Zhai et al., 2012).

THE RESTING OCEAN APPROXIMATION FOR STRESS

Consider Figure 11.3, which shows a large-scale (spatially constant) wind blowing over an oceanic feature. Taking 10 m/s as an estimate of windspeeds, $|\mathbf{u}_a|$, and 10 cm/s for oceanic surface velocities, $|\mathbf{u}_o|$, the latter appears negligible. Consider however, that, as shown in the figure, over part of the feature (an "eddy"), the wind is working against the oceanic flow over half the area, and working with it over the other half. A slight asymmetry exists in the stress: an opposing current increases the drag, a current in the same direction reduces it. Thus large-scale winds act to spin down the eddy and, on average, reduce the work done on the ocean overall. The stress difference with and without $\mathbf{u}_0$ is slight—but it's a big ocean, filled both with eddies and larger-scale structures flowing into the wind, and systematic errors

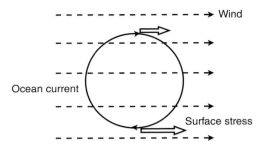

Figure 11.3: Sketch of the effect of oceanic eddy flow on the drag law. The wind is directed from left to right (open arrows), and the oceanic "eddy" surface current is circular as indicated by the small arrows. With the current in the direction of the wind, the stress is reduced. With the current into the wind, the stress is increased, producing a net torque on the eddy, and so systematically reducing its strength. (From Zhai et al., 2012.)

accumulate. Various estimates, of mixed reliability, now exist of this effect (10–30% of the total wind work).[2]

Removing the resting-ocean approximation raises problems for numerical modelers—because nominally *observed* values of wind stress were obtained from oceans that were moving in undocumented ways.[3] How this issue will be overcome remains to be seen; it may ultimately mean that only fully coupled air-sea models can be quantitatively accurate.

WIND-POWER INPUT ESTIMATES

Existing estimates of wind-power input to the ocean are 60–70 TW (see Rascle et al., 2008). Consideration of the wave-generation problem (Komen, 1994) implies that almost all of it goes into generating surface waves and near-surface turbulence, with only about 2 TW remaining to drive the large-scale circulation. Having to deal with the minute residual pertaining to the general circulation is troublesome. Also, indirect contributions of near-surface turbulence to the ocean circulation through internal nonlinearities are likely to be important too. McWilliams and Restrepo (1999) for example, demonstrate that the surface-wave Stokes drift is not a negligible contribution to the general circulation. If one makes the plausible, but important, assumption (Faller, 1968; Stern, 1975) that work on the ageostrophic surface flow is completely dissipated locally within the turbulent near-surface layer, then the time-average power entering the interior geostrophic flow is

$$\langle P_W \rangle = \iint_{ocean} dA \left\langle \boldsymbol{\tau} \cdot \mathbf{u}_g \right\rangle, \tag{11.4}$$

where $\mathbf{u}_g$ is the surface geostrophic velocity. The calculation appears straightforward because the quantities all *seem* well defined and familiar.

[2]Sverdrup's (1947) explanation of the North Equatorial Counter Current (NECC) as an oceanic flow *into* the mean wind stress already implied an energy supply from elsewhere. Published estimates of $\left\langle \boldsymbol{\tau} \cdot \mathbf{u}_g \right\rangle$ all display large areas of negative values (Fig. 11.4).

[3]The observations are labelled "nominal" because many are very indirectly inferred from wind speed and direction, not from the stress itself.

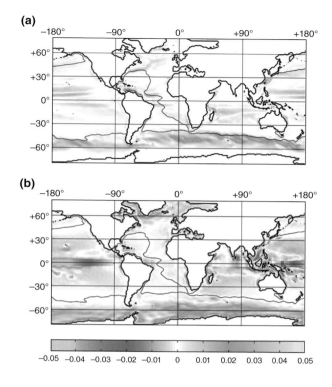

Figure 11.4: (a) Estimate of the rate of wind work on the ocean by the wind computed as $P_W = \langle \tau \cdot \mathbf{u}_g \rangle$, and (b) the pattern input into the geostrophic circulation through upward/downward Ekman pumping. The spatial integrals of the two fields coincide. $\mathbf{u}_g$ is from the short-term state estimate of Forget (2010). The equatorial region is treated specially. (From Roquet et al., 2011.)

One estimate (several exist) is shown in Figure 11.4 and considerable spatial variations appear, including significant regions where the wind is working against the ocean circulation, thus removing energy from it. This power-input pattern depends directly upon the surface circulation that the wind field itself has generated. The Southern Ocean dominates the result, as the established flow tends to be down-wind—right around the globe. Regions of negative work must be supplied by other means with the energy needed to sustain the flow there.

Faller (1968) and Fofonoff (1981) noticed that in a balanced steady state,

$$\langle \tau \cdot \mathbf{u}_g \rangle = \frac{1}{\rho} \langle \mathbf{V}_E \cdot \nabla p \rangle , \qquad (11.5)$$

that is, the rate of work by the wind on the geostrophic flow is identical to the rate of work of the Ekman transport against the surface-pressure force. This result is not obvious but follows from the linkage of Ekman pumping and the interior geostrophic flow (Roquet et al., 2011). Despite the fact that the left-hand side of Equation (11.5) calculates the rate of work on the surface geostrophic velocity, the balanced equations require that the energy transfer so calculated can be interpreted as being *transported laterally by the Ekman layer*. Roquet et al. (2011) showed that the energy enters the general circulation beneath it through the Ekman pumping velocity, w_E, with the rate derived from its product with the local surface pressure,

$\langle g\eta w_E\rangle$. The spatial distribution (Fig. 11.4) of its entry can differ considerably from that where $\langle\boldsymbol{\tau}\cdot\mathbf{u}_g\rangle$ is greatest.

These calculations connect the values of P_w to the ideas of Gill et al. (1974) that involve wind driving of the ocean through the generation of potential energy in the general circulation via Ekman pumping. Interpretation of spatial distributions of energy inputs and transfers does not lead to a uniquely defined picture, and it is worth recalling the principle that *only divergences of energy transports* are observable.

Many uncertainties exist in these calculations, including the appropriate values of C_D; the accuracies of the wind products, which vary strongly as the observing system changes through time; the ocean surface velocity and its temporal variations; and the Faller-Stern hypothesis of complete local turbulence dissipation. Hypothetically, atmospheric models should provide estimates of the regional and global transfers of momentum and energy to and from the fluid ocean beneath. Because of the difficulties in defining exchanges across two coupled, moving, turbulent boundary layers, and known issues with the meteorological estimates, such calculations with useful accuracy are now beyond reach. The entire situation is not very satisfactory, and we leave it with the estimate that direct wind-power input to the ocean circulation is approximately 2 TW—with factor of two uncertainties likely.[4]

ATMOSPHERIC LOADING

Atmospheric pressure fluctuations, p_a, (Fig. 3.36) lead to vertical motions of the sea surface, $\partial\eta/\partial t$. Work can be done (of either sign) at a rate computed from the space-time averages, $\langle p_a\partial\eta/\partial t\rangle$. These prove to be small (Ponte, 2009).

GEOTHERMAL HEATING

Figure 3.43 shows an estimate of the rate of heating through the seafloor. Study of the motions of a fluid heated from below is a mature branch of fluid dynamics, commonly directed at what is sometimes known as the Rayleigh-Bénard (RB) problem (see, e.g., Chandrasekhar, 1968; Turner, 1973) under a wide variety of different circumstances—including strong rotation.[5] The RB problem begins with study of fluid instability: to determine how large a heating rate is required to overcome the stabilizing influences of viscosity and diffusivity; and goes on, in the unstable limit, to determine whether the motion becomes turbulent and what its heat-transfer rates are. Much is known about all of these aspects in many different applications.

The ocean, with its complex geometry, rotation, two-component density control, and large pressure range, is a more complicated system than is dealt with in most theoretical calculations. Thus the most recent estimates of the response of the ocean to heating at the seafloor have been done with general circulation models, often using detailed topography.[6] The major inference is that geothermal forcing can drive motions on the order of 10 Sv, a small but not negligible fraction of the inferred large-scale circulation, albeit not easy to observe directly (but see Reid, 1982). The response is somewhat muted—despite the very high Rayleigh

---

[4] To give an intuitive feeling for such a number, 1 TW is—according to the US Dept. of Energy website—approximately the total world electric power generation capacity. Alternatively, adult humans consume energy at roughly 100 W/individual. So 6 billion adults require 0.6 TW to stay alive.

[5] Much of the literature on spherical geometries deals with stars or the Earth's core and mantle and so is not in the thin-shell limit appropriate to the ocean.

[6] Adcroft et al., 2001; Emile-Geay and Madec, 2009.

numbers—because the heat capacity and lateral-mixing and advection rates of the ocean are sufficiently large that establishing strong thermal gradients is not possible. Extremely hot, salty regions of direct fluid injection also exist, but in volume terms, these are very small, although locally important. Entrainment rates into the hot plumes are considerably larger, but their dynamical consequences do not seem to have been explored.[7] In some regions, the injected hot fluid is marked with excess concentrations of the helium-3 isotope (^{3}He), whose pathways can be tracked over long distances, particularly in the Pacific Ocean (Lupton, 1998).

Global-mean ocean geothermal heat fluxes are on the order of 0.1 W/m^2 or less, compared to surface heating by the sun which is on the order of 1000 W/m^2, and they might have been thought negligible in comparison. However, in the context of the general circulation of the ocean, where transfers of heat to and from the abyssal regions are by fluid advection and diffusion (e.g., as in the Munk "recipe"), 0.1 W/m^2 is not small compared to terms such as $\rho c_p w T$ or $\rho c_p K_V \partial T/\partial z$, which control the heat balance of the deep ocean. A heating rate of 0.1 W/m^2 corresponds to a global integral of about 36 TW, which is far larger than the energy estimated as required to power the general circulations. Conversion of enthalpy fluxes to mechanical energy is very inefficient.

HEATING AND COOLING AT THE SEA SURFACE

Determining the degree to which the ocean circulation is powered by buoyancy forcing requires understanding the behavior of oceanic diffusion. Wind driving can set up a circulation in a completely homogeneous fluid, whereas buoyancy forcing depends upon the establishment of density gradients within the fluid. Rates of input of thermal anomalies at the boundaries grow with diffusive effects, but those in turn dissipate the anomalies as they move away from the boundaries. Whether generation or dissipation "wins" is rarely obvious without detailed calculations.[8] Diagrams such as those in Figure 9.31 are suggestive of the ocean as a heat engine in which high-latitude atmospheric extraction of heat (and concomitant evaporation leading to a salinity increase) makes the fluid dense, which sinks and returns in an overturning cell in which the circulations described by the Stommel-Arons and Munk abyssal recipes theories are the return pathway. The notion has been heavily reinforced by the use of "conveyor-belt" or "ribbon" cartoons of the time mean ocean circulation. Apart from the contradiction between such pathways and the various estimates of the general circulation already described (Figs. 9.4–9.6), basic physical principles preclude it.

Laboratory-scale fluid motions at low Reynolds and Péclet numbers are strongly influenced by the presence of molecular diffusivity, κ. These exist owing to the internal energy of the fluid, which keeps the molecules in continuous active motion. In a geophysical fluid, eddy diffusion tensors $\mathbf{K}_C$, are invoked, having values many orders of magnitude larger. In a stratified fluid in a closed container, diffusion acts to homogenize the stratification, ultimately *raising* its center of mass and hence increasing its potential energy. Diffusive action does produce slight amounts of kinetic energy, and so without being dogmatic, turbulent diffusion per

---

[7] *Entrainment* is the process by which a moving fluid, often turbulent, sets the adjacent resting fluid into motion through viscous and mechanical forces. In oceanography, it is commonly associated with deepening of the mixed layer, and with the interaction of buoyant "plumes" moving through a larger medium at rest. See Turner, 1973, or Phillips, 1977.

[8] Viscous eddy coefficients, A_V, might be thought to analogously control both the rate of input of momentum, through the Ekman layer physics and its destruction in the interior. But as seen, the Ekman layer transport proves independent of A_V, whose value is drastically reduced below the surface boundary layer.

se is not an obvious candidate to provide significant power for general circulation-scale kinetic energy. Instead, the question immediately arises as to what provides the power to sustain the turbulence itself? How much might be generated through instabilities of the buoyancy-driven motion itself as opposed to that provided by external forces (tides, winds, etc.)?

The oceanographic context for this type of discussion is via the general problem of what has come to be known as the study of *horizontal convection* and in which turbulent diffusion is central. In contrast with the Rayleigh-Bénard case, the regions of net cooling (generally the high latitudes) are geographically separated from the regions of heating (the tropics). The problem is usefully divided into three types: (1) The heat source is below the depth of cooling— "unstable horizontal convection"; (2) The heat source lies above the depth of cooling—"stable horizontal convection"; and (3) Heating and cooling are at the same level with an arbitrary horizontal separation—"equal-level horizontal convection"[9] (see Fig. 11.5). Because atmospheric heating and cooling both act at or near the sea surface but in widely separated latitudes, unstable convection does not occur unless some additional mechanical force is available to move the heated fluid below the level of the cold.

Regions of instability, as in the Rayleigh-Bénard problem (Type 1), are relevant in the response to geothermal heating, where an appropriate critical Rayleigh number can be surpassed. In (Type 2), the fluid is normally stable, and although motions must and do occur, they are much weaker than in (Type 1). Most of the oceanic interest is in (Type 3), the equal-level case, although that choice is not entirely obvious: Figures 3.29 and 9.14, strongly suggest that the surface cooling regions are at a lower geopotential than are the heating ones. More generally, convective regimes in the ocean exist in the presence of many other motions, rendering the interpretation and applicability of separate analyses largely irrelevant.

Present theoretical inferences are in something of a muddle, which seems to exist for a number of reasons. The influential work of Sandström (1908), who apparently first called attention to the issues, has been widely misquoted. (See Kuhlbrodt, 2008, for a translation of the original German paper and for a discussion of the fluid and thermodynamical concepts available to Sandström at that time.) The situation was not helped by Bjerknes et al. (1933), who elevated Sandström's approximate inferences to the status of a theorem, which then provided a strawman target for subsequent writers attempting to disprove it.[10] Without obsessing over precisely what Sandström said or meant, his result (as clarified by Jeffreys, 1925) is that in a fluid dominated by turbulent mixing, Type 1, when unstable, will give rise to a stronger circulation than those from Types 2 or 3. Weak motions *are* expected to exist in these latter types. Sandström probably did not believe them to vanish altogether, but the question is moot. His results will be referred to as the "Sandström approximation."

Laboratory studies have been made of equal-level horizontal convection (see Rossby, 1965, and Hazewinkel et al. 2012 for references); measurable motions are observed and are generally understood in both one- and two-dimensional configurations. (The one-dimensional configuration is that of a closed fluid loop, and many of the two-dimensional results have analogs in that even-simpler situation.[11])

---

[9] The labels are intended only to be mnemonics suggestive of the dominant processes. Below a critical Rayleigh number, the Type 1 system is stable, and the label "convection" is not particularly appropriate even though some flow will nonetheless exist. Types 2 and 3 convection could, under some circumstances, be in unstable regimes, but not in the Rayleigh-Bénard sense. The Rayleigh number can usefully be redefined to reflect the horizontal separation.

[10] Bjerknes regarded it as equivalent to his own circulation theorem.

[11] See Wunsch, 2005, for extended references, including those in the engineering heat-exchanger literature.

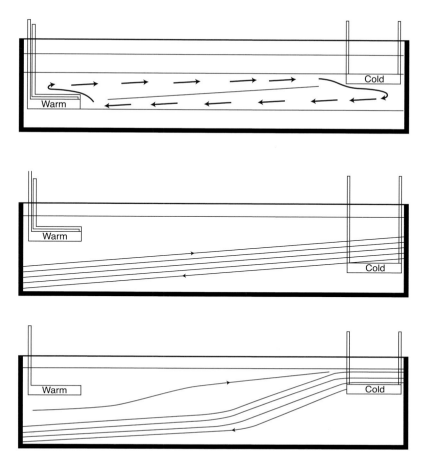

Figure 11.5: Horizontal convection of Types 1, *unstable convection* (upper panel), and 2, *stable convection* (middle panel). In Type 3, *equal-level convection* (lower panel), the heating and cooling sources are at the same geopotential. Temperature contours are purely notional as their form must be computed from the equations of motion and the boundary conditions. (After Defant, 1961.)

In one and two dimensions in the equal-level configuration, three results are worth mentioning: (1) The intensities of motion can both increase and decrease with increasing diffusivity κ or $\mathbf{K}_C$—because of the competing effects of driving strength and dissipation rates. A strong dependence on the Prandtl number (e.g., Phillips, 1970) as well as on the Rayleigh and Reynolds, numbers is expected. (2) The rate at which energy can be communicated from the external buoyancy forcing to the interior flows depends sensitively upon motions generated from the wind stress. (3) The convective response itself appears unable to generate sufficient turbulence to have much effect on the values of κ (Paparella and Young, 2002).

Laboratory and model Reynolds and Rayleigh numbers are many orders of magnitude smaller than in the geophysical case; usually no rotational forces are present at all (Rossby and Ekman numbers are infinite), and they are mostly two-dimensional situations. Enthusiastic extrapolation of laboratory results to the three-dimensional rotating ocean is at present mainly an act of faith.

Barkan et al. (2013) made a start on understanding the three-dimensional, rotating, case, including eddies arising from instabilities. They nonetheless expressed reservations about its application to the ocean—owing to the restricted parameter range they could treat. The β-effect, wind stress, and topographic slopes are all still missing from the discussion and likely to qualitatively change it when introduced.

As Huang (2010) has emphasized, regions of strong convection act to carry dense fluid to depth, thus *reducing* the potential energy of the system. The rate of work by buoyancy forces depends upon the complete circulation arising from all inputs. If a strong buoyancy input exists because of the wind-driven flow, how does one sensibly decide which is "more important"?

Approximate estimates of the wind-power input can be made from knowledge of surface properties alone—the stress and the surface geostrophic velocity. Calculations for the equivalent buoyancy work are less straightforward. A few attempts have been made, starting with Faller (1968), who, using the non-Boussinesq thermodynamic work term, $p\nabla \cdot \mathbf{v}$ (the pressure work against the expanding and contracting water volumes) concluded that the power input was negative. The available data at that time were thin. Huang (2010) calculated the generation of potential energy at the sea surface by buoyancy exchange and found the negative inputs of the convective regions were very important. Although in the global total he obtained a small positive value, the general result was consistent with Faller's estimate. Other estimates (e.g., those of Roquet, 2013; Tailleux, 2013,) involve the technically complex issue of calculation of the generation of available potential energy, which is dependent also on the wind field, and are not easy to interpret. A generalization to three dimensions of the formulas of Huang (2010, Eq. 3.18) or Hazewinkel et al. (2012, Eq. 1.9) would involve both surface buoyancy fluxes and those at the seafloor, including lateral mixing coefficients, and has not been attempted. A more definitive value may eventually be obtained by subtracting the wind-power input from the rate of dissipation. The latter is itself poorly known at present but ultimately may provide a more tractable calculation.

BIOLOGICAL STIRRING AND MIXING

Munk (1966) had suggested in passing that biological processes could power small-scale mixing in the ocean.[12] Although the notion opened an interesting connection between physical oceanography and biology (Dewar et al., 2006), extended calculations show that it is likely a quantitatively unimportant contribution (Kunze, 2011).

11.1.2 Energy Sinks

In discussions of the large-scale circulation, "dissipation" usually means the removal of energy (potential, kinetic and internal) from some conceptual large scale, into a smaller scale assumed to be (1) not part of the general circulation, and (2) in an irreversible process. Thus if a flow is taken, arbitrarily, to have a space scale that is some multiplier of the first deformation radius,

---

[12] The idea was intended as a joke (W. Munk, pers. comm., 2012). As with almost everything else, biological mixing is an indirect use of solar power.

R_{D1}, and it is baroclinically unstable, to balanced eddies arise on a smaller scale, it is assumed that this energy is not returned to the larger scale—that the process is "one-way." Nonetheless, significant mechanisms are known (resonant interactions, critical layers, turbulent cascades, rectification generally) that *can* generate larger-scale time-mean flows from many types of oceanic variability. No convincing estimates yet exist for space-time average rates of these reconversion processes.

On a global basis, the number of known, apparently significant, loss mechanisms for the large-scale circulation (only vaguely defined) is limited:

1. Baroclinic, barotropic, and mixed instabilities
2. Bottom drag
3. Lateral viscous stresses, interior and at the boundaries
4. Topographic interactions generating both internal/inertial and balanced wave-like and eddy-like motions

These possibilities are now briefly considered. As with most estimates of global ocean properties, a large element of almost untested theory and modelling intervenes—the observational base being fragmentary. Whether the theory or modelling has quantitative skill is an omnipresent question.

1. The stability properties of the putative time-mean circulation has a considerable literature (see Fig. 11.6). Existing estimates of the rates of eddy generation are about 0.3–0.8 TW (Ferrari and Wunsch, 2009; Xu et al., 2011) . Some tests of the accuracy of these values becomes possible by comparing them to the rates at which the oceanic eddy field (however defined) is being dissipated—with the argument that in a steady state, energy sources of eddies must balance their rate of loss. An interesting question connected to the arguments over the Sandström approximate solution concern the extent to which motions wholly due to buoyancy forcing can generate smaller-scale turbulence through instabilities at great depth.

2. The bottom boundary layer of the ocean will be turbulent and involve the summation over the dissipative effects of *all* oceanic motions on all time and space scales. Turbulent drag laws are analogous to those used (Eq. 8.4). Thus the bottom stress for flat regions is Equation (8.4), where $\mathbf{u}_B$ is a total near-bottom velocity, which has already been encountered in the discussion of tides, but the velocity is not just the tidal contribution. Then the rate of working on the flow is, from the usual drag law,

$$P_B = \rho C_D \left\langle |\mathbf{u}_B| \, \mathbf{u}_B \cdot \mathbf{u}_B \right\rangle, \tag{11.6}$$

cubic in the bottom velocity. As with the wind-field rate of work, replacing $\mathbf{u}_B$ with its time or space average incurs large errors owing to the suppression of the high frequencies and wavenumbers. Calculations made from cubing empirical, noisy, data are unreliable. Nonetheless, rough estimates, often from fragmentary current meter data (Wright et al., 2012), do also suggest numbers of the order of 0.2–0.8 TW, but ones dominated by the eddy velocities, and the fraction acting directly on the large-scale circulation is not known.

3. Lateral viscous dissipation rates are mainly model-based and thus extremely uncertain. One approach to their estimate is to use viscous theories, such as Munk's (1950), to find eddy-coefficient values from WBC. (see P. 339).

4. One route to dissipation of the circulation is the *loss-of-balance* mechanism already described in which the internal wave terms are included in the equations governing the

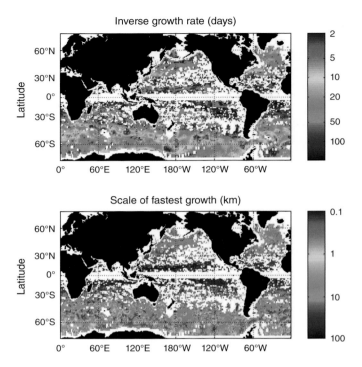

Inverse growth rate (days)

Scale of fastest growth (km)

Figure 11.6: The growth time (upper panel) and space scale (lower panel) for baroclinic instability in the ocean. The Gouretski and Koltermann (2004) climatology was used. Instability was found at all locations, but must be evaluated in the context of other forces acting, such as the wind. (From Smith, 2007.)

otherwise balanced flow. Again, because bottom velocities are dominated by the eddy field, and not by the large-scale circulation, no estimates exist of the corresponding rates, and few direct observations exist. Available measurements are primarily from the Southern Ocean (Sheen et al., 2013, is an example).

11.1.3 Mixing Rates

Oceanic kinetic energy is dissipated at the molecular scale at rates determined by the viscosity controlling the rate of strain components of ε (Eq. 3.33). Mixing (as opposed to stirring) of, for example, temperature takes place at a rate controlled by χ_T (Eq. 3.34), which is directly dependent upon the molecular-scale gradients of the temperature field, in turn dependent upon the u, v, w controlling ε. Thus the rates at which temperature or other scalars can be mixed depends on both the small-scale velocity field and the gradients of the property. Osborn (1980)[13] proposed that for a locally vertically stratified density field, the turbulent eddy diffusion could be calculated as

$$K_\rho = \frac{\Gamma \varepsilon}{N^2},\qquad(11.7)$$

[13] See Davis, 1994b.

where Γ is an empirical constant, called the *mixing efficiency*, that represents the fraction of the rate of turbulent dissipation of energy going into mixing the fluid rather than into Joule heating. Values of $\Gamma \approx 0.2$ are traditional, but it is unlikely that it is a universal constant, with estimated ocean values occupying a wide range.[14] With a nominal constant value of Γ, measurements of ε and of local stratification can be used to calculate values of K_ρ, which are central to the discussions of large-scale ocean circulation, in particular of the overturning circulation crucial in bottom-water formation rates, and the meridional movement of properties such as heat and freshwater. The connections are through the geostrophic vorticity equation, connecting v with w, and through abyssal-recipes type balances in turn connecting w with K_ρ. The meridional flow is a function, $v\left(K_\rho\left(\varepsilon\right)\right)$, thus coupling the very largest and very smallest of oceanic motions.

Instruments capable of direct determination of ε from u, v exist but are rare (see Fig. 2.14), and global average estimates from Equation (11.7) do not exist. Much more common are low-ered acoustic Doppler current profilers (LADCP and CTD or CTD-like instruments) capable of measuring velocity and temperature (and salinity) profiles down to scales approaching, but short of, the relevant molecular scales and requiring extrapolation. Starting with Gregg (1987), but modified in a number of ways subsequently (see Kunze et al., 2006), a semiempirical relationship was proposed for use with measured velocity and thermal (or density) profiles that in one form is

$$K_\rho = K_0 \frac{\langle \xi_z^2 \rangle}{\langle \xi_z^2 \rangle_{GM}} \left\{ \frac{R_\omega\left(R_\omega+1\right)}{6\sqrt{2\left(R_\omega-1\right)}} J\left(f/N\right) \right\},$$ (11.8)

permitting estimates of K_ρ from temperature and salinity microstructure. Here, K_0 is a "background" value of 5×10^{-6} m^2/s, ξ_z is the vertical strain, with its mean-square measured relative to the value obtained from the Garrett-Munk spectrum. R_ω is the ratio of the shear variance to the strain and is thought to lie somewhere between 3 and 7. $J\left(f/N\right)$ is a function of the ratio of the inertial to buoyancy frequencies. Justifying this and related forms, and carrying out these calculations is not easy. (Kunze et al., 2006 concluded that an earlier published calculation had generated values of K_ρ that were an order of magnitude too large, owing to a failure to properly understand instrument noise.) What data are available tend to produce $K_\rho \approx 10^{-5}$ m^2/s, approaching 10^{-4} m^2/s only below about 3000 m. Large positive outliers exist (Fig. 11.7), meaning that undersampling of regions of extreme mixing may be leading to underestimates of the true mean. An upper-ocean map of inferred K_ρ using 5 years of finestructure values of the Argo array is shown in Figure 11.8.

Various estimates have been made with global models in which parameterizations are based on bottom roughness, tidal velocities, and more general abyssal kinetic energy sources (see Decloedt and Luther, 2012). These estimates are crudely consistent with those made from formulas such as Equation (11.8), in being small relative to 10^{-4} m^2/s. Care must be taken to account for what is clearly a very strong depth dependence of average values.

In contrast to these attempts at direct in situ determination of K_ρ are the large-area and volume estimates from the conservation of properties in the inverse calculations using basin-scale geostrophic boxes (Ganachaud, 2003b; Lumpkin and Speer, 2007). In the abyss, these values are usually closer to 10^{-4} m^2/s or greater and thus seemingly in conflict with both the direct calculations and model results (but see Macdonald et al., 2009). The resolution of this conflict is not obvious (Decloedt and Luther, 2012). Systematic errors likely exist both in the

---

[14] See Peltier and Caulfield, 2003; Lozovatsky and Fernando, 2013.

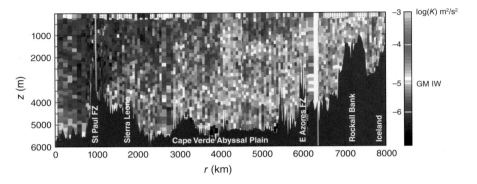

Figure 11.7: The logarithm of the estimated value of K_ρ in a meridional section in the North Atlantic ending at Iceland. Note the extremely large values as well as the very small background estimates. GM denotes the background value inferred from the Garrett-Munk internal wave spectrum. (From Kunze et al., 2006.)

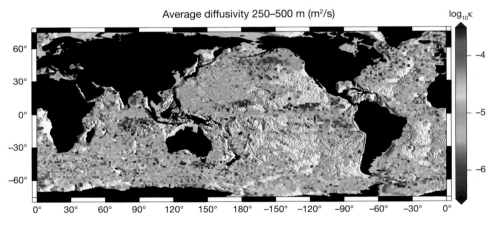

Figure 11.8: K_ρ values estimated using the finestructure measurements from 5 years of Argo profile data between 250 and 500 m. As with many other oceanic fields, the spatial inhomogeneity is marked. Although the Munk (1966) "abyssal recipes" value, 10^{-4} m²/s, does not apply in this depth range, it is close to being an upper bound value, with some outliers being visible. Such estimates involve numerous subtle and important assumptions. The spatial complexity, however, is real. (From Whalen et al., 2012.)

inverse calculations and in the direct measurements. The former make strong assumptions about noise levels—including the temporal aliases described earlier—and their possible effects on the estimated K_ρ. Both formulas, Equations (11.7) and (11.8) are based upon long chains of assumptions, not all of which are well understood, and also encounter the problem of sampling in fields with strongly non-Gaussian probability distributions, particularly ones having long positive tails. At the present time, the nature, rates, and distribution of ocean mixing must be regarded as unsettled, a situation with major implications for understanding of bottom-water formation rates and the physics of climate change. Model parameterizations are physically plausible but still do not account for the full spectrum of abyssal kinetic energy, and so they must be tuned in some way against the sparse observations.

By introducing artificial but easily measured tracers at depth in the ocean and then tracking the shape, depth, and volume of the subsequent time evolution, independent estimates of vertical (and horizontal) diffusion rates have been made. Tracking of a tracer over 13 years produced an estimated value of $K_\rho = 3 \times 10^{-4}$ m^2/s as an average across the very deep Brazil Basin (Rye et al., 2012), similar to the box inversion values. Upper-ocean values (300 m) are much lower, about 0.1–0.2×10^{-4} m^2/s (Ledwell et al., 1998), and are probably compatible with Figure 11.8. These and other experiments are consistent with enhanced values over and near abyssal topography and with very small values in and above the main thermocline. Preliminary results are now becoming available from the Southern Ocean, but in general such direct measurements have been, and likely will remain, spatially and temporally very sparse.

No real summary is possible at the present time of the depth and geographical variations in the overall mixing rates and the regional variations in their generation mechanisms. It does appear that direct and indirect wind influence dominates the upper ocean, particularly in regions of isopycnal outcrops, especially the Southern Ocean. In the abyss, internal/inertial waves, including tide-driven components, and loss-of-balance mechanisms in the geostrophic eddy field are major elements. Regions of rough topography produce extreme values probably governing the spatial averages. Intermediate depths of the ocean, between 1000 and 3000 m, remain an enigma; perhaps all the mechanisms contribute significantly there, but overall mixing rates are likely reduced relative to both the upper and deeper oceans. The consequences of the spatial variations for the general circulation are still being worked out.

11.2 THE MERIDIONAL OVERTURNING CIRCULATION

The ocean circulation is a complex three-dimensional superposition of flows and their properties. These flows vary over all possible coupled time and space scales. In the face of this complexity, summarizing properties are sought, ones depicting the system in shorthand, both as descriptors and as phenomena to be explained. Examples are the gyres, the western and eastern boundary currents, Sverdrup-like interiors, recirculations, etc.

The zonal integral of the meridional volume or, mass, circulation has come to be a standard measure of the influence of the ocean circulation on climate, although almost no direct observational support exists for inferring such a connection. Carrying out the recipe, Figure 9.31 represents what is commonly called the meridional overturning circulation (MOC). In a gross schematic sense, the Atlantic is seen to have an upper ocean moving predominantly northward, with sinking at high latitudes that generates the southward North Atlantic Deep Water (NADW), in turn overlying a patchy layer of northward-moving Antarctic Bottom Water (AABW).

The Pacific Ocean has a visually somewhat similar structure, albeit the northward-moving bottom layer is believed to be somewhat stronger than that in the Atlantic. The structure in the Indian Ocean is similar to that of the Pacific, with the upper levels varying exceptionally with the monsoon winds (see Schott et al., 2009). Visually, the Southern Ocean has a predominantly southward-moving interior overlain by a very strong near-surface northward flow. Underneath is the very dense equatorward-moving bottom water. The geometry of the Southern Ocean is, however, very different from the others, and the zonal integrals do not have the same dynamical relationships as at other latitudes.

In examining plots such as these, it is very important to recognize that no fluid parcel ever follows the indicated transport lines—they move along intricate three-dimensional, time-varying pathways that have been suppressed in the interests of a simplified picture. Recall, too, the difference between the Eulerian representation at fixed points and the Lagrangian particle paths.

The ocean carries heat poleward as part of the coupled atmosphere-ocean system, thus maintaining the gross heat balance of the Earth. By postulating that the zonally integrated circulation is primarily responsible for the ocean transport and that its variations control the climate system, a very appealing story can be developed. Coupled with the idea of the ocean as a heat engine whose circulation intensity varies in time, that story in the minds of many is the central explanatory mechanism for climate variability— past, present, and future.

On the other hand, the MOC is a very limited depiction of the ocean circulation, and indeed one that borders on being uninterpretable. Consider a hypothetical ocean described by the simplest of all circulation patterns: a wind-driven upper region, as described by the Stommel-Munk physics, with a northward-moving western boundary current, and a Sverdrup interior return flow, all overlying a layer at rest. In that theoretical framework, the zonally integrated transport vanishes—the MOC is *zero*. Only if some of the fluid carried by the WBC is not returned by a Sverdrup interior flow, will the MOC have a nonzero value. The "missing" transport must be at depth. As the circulation fluctuates, the total MOC value can increase either because the boundary-current transport grows, without a corresponding increase in the Sverdrup flow, or because the boundary current remains constant, with the Sverdrup flow *decreasing*. Thus *weakening* of the MOC can occur because the upper-level interior transport is *strengthened*. In reality, the regimes of oceanic meridional mass transport occur with a variety of physics and the zonal integration used to define an MOC lumps everything into a single total.

THE ATLANTIC MERIDIONAL OVERTURNING CIRCULATION

The MOC in the Atlantic Ocean has been the focus of most attention (there called the AMOC), and that particular basin is now briefly considered.[15] Figure 11.9 displays a 19-year average

---

[15] See Wunsch and Heimbach, 2013a, for a more extensive account. Wunsch and Heimbach (2009) discuss the global behavior.

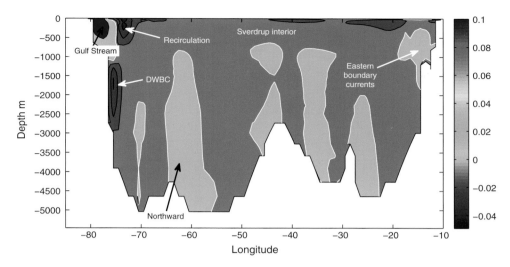

Figure 11.9: Time-average meridional velocity (m/s) at 30°N from a 19-year ECCO estimate. A schematic of dynamical subdomains is overlaid. The Ekman layer is included in the Sverdrup balance region, and other subdomains such as that between the Gulf Stream and the western boundary have been omitted. In this region, the empirical Sverdrup balance integration depth, z_S, ranges between about 1000 and 1500 m. (From Wunsch and Heimbach, 2013a.)

from a state estimate of the transport across the North Atlantic at 30°N, and labels a subset of the flow regimes, including the western and eastern boundary current regions, the Sverdrup interior, etc. Empirically, from altimetric measurements of sea level change, which tightly bound the volume changes, mass is nearly conserved on scales of weeks and longer.[16] Thus any increase in the transport of one component here must be compensated by a change in one or more of the remaining components. Although no consensus exists, a definition of "the" AMOC is, the maximum value obtained by integrating the zonal volume or mass integral downward from the surface. Figure 11.10 shows one such estimate. At 25°N the value is about 15 Sv, a latitude where the Gulf Stream is known to be carrying about 31 Sv poleward (Meinen et al., 2010). The upper ocean is carrying only about one-half of the required return transport—the remainder is found below the defining maximum depth (about 1000 m at this latitude).

A Sverdrup-like interior is returning only about one-half the northward Gulf Stream transport. Some of that "missing" transport is in the deep western boundary current (DWBC), and some of it lies in the deep interior—where competing northward transports are also occurring, as they do in the eastern boundary current regime.

---

[16] That large regional volume changes are not observed is a consequence of the need for a corresponding forcing function, primarily a wind field, capable of supporting the sea-surface pressure gradients that would arise from such changes.

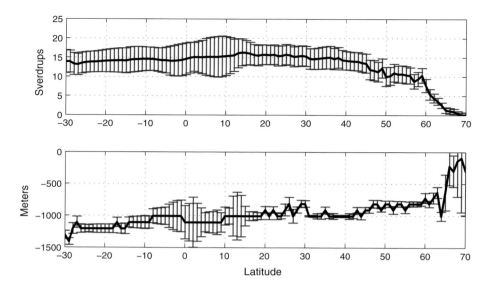

Figure 11.10: Estimated values of the AMOC (upper panel) and of the maximizing depth (lower panel). Error bars are the standard error from the temporal variation in the ECCO estimate.

DIRECT OBSERVATIONS: THE NORTH ATLANTIC OCEAN

A full-depth hydrographic line across the North Atlantic at or near 25°N was first measured in 1957 and then remeasured in 1981, 1992, 1998, and 2004 (see Longworth and Bryden, 2007). From the measured line plus the observed transport in the Florida Current (Fig. 2.24) and the assumption of a nearly closed ocean to the north, box inverse methods can be used to infer the upper-ocean mass, heat, and freshwater transports across the line. Different occupations of the line gave different values, results consistent with the expected variability of the circulation. Temptations to transform those erratically spaced estimates into decadal trend inferences were, unhappily, not resisted and led to a much-hyped claim of gross changes. This claim is best regarded as a classroom example of the pitfalls of temporal aliasing.

In recent years, a mooring array was deployed across the North Atlantic near 25°N, which, in conjunction with the Florida Straits transports, has been producing estimates of the net temporal variability across the line at time intervals as short as one month (Kanzow et al., 2009), as displayed in Figure 11.11. The most striking feature is the very great variability, which is consistent with what had been inferred from the altimetric measurements, and the noise that enters the computation of meaningful time averages. A 20-year mean from the state estimates is also available. For comparison with the direct measurements, Figure 11.11 displays the two together, along with the estimated values at several other latitudes. A robust inference is that with almost 20 years of data, the AMOC is indistinguishable from a stationary Gaussian process. That result in turn is consistent with the inference that *fluctuations in the AMOC are the result of a large number of nearly independent disturbances from diverse sources widely*

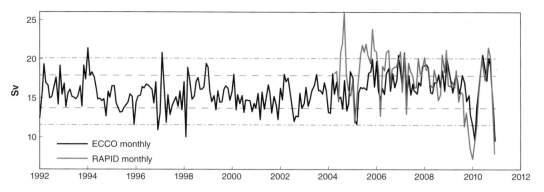

Figure 11.11: The 25°N AMOC as calculated from a state estimate (black curve; from Wunsch and Heimbach, 2013). Monthly values with the annual cycle and harmonics have been removed. The one- and two-standard-deviation levels are shown—these are helpful for inferring the expected occurrence rate of extreme events. The mooring-based RAPID/MOCHA estimate available since 2004 is shown in red. (RAPID, Rapid Climate Change Programme; MOCHA, Meridional Overturning Circulation and Heat Flux Array; see Kanzow et al., 2009). These mooring data were not used in the state estimate. The compressed scale dramatizes the fluctuations. RAPID data include the eddy field and so tend to be noisier than the state estimate, which is generally devoid of eddies.

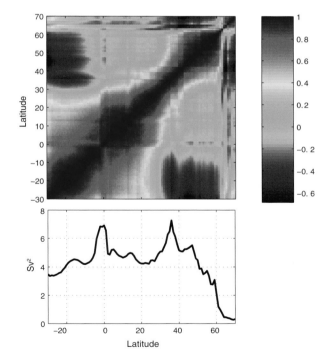

Figure 11.12: Correlations between the annual mean variation in the AMOC at all Atlantic latitudes. The level of no significance is approximately magnitude 0.5. The lower panel depicts the variance at each latitude. (From Wunsch and Heimbach, 2013a.)

scattered in space and time (as expected from wind and buoyancy forcing; fluctuations in sea-ice production; baroclinic and other instabilities; and random-walk superposition of background eddies).[17]

The meridional correlation of the annual mean values for all latitudes is shown in Figure 11.12. The loss of correlation across about 35°N is likely partly associated with the high noise levels in the ocean in the vicinity of the northeastward meandering and moving Gulf Stream. Figures 3.14 and 3.16 suggest a decoupling of the two gyres on the accessible time scales. Many decades of observation will be required to the point where a small correlation could become significant over the entire basin. As it is, the two basins, subtropical and subpolar, are separately *internally* coherent. To the extent that larger-scale trends do exist on longer time scales, they must be extracted from a very noisy background state.

11.3   INTEGRALS AND CHOKE-POINTS

Integrals of the circulation greatly reduce the volume and complexity of numbers needed to represent what is measured. Nature provides some natural integrals in the ocean—places such as the Drake Passage, the Indonesian Throughflow, and the Straits of Florida, where the large-scale flows are reduced to narrow regions and whose volume and other transports are comparatively easy to measure. Some notable time series have been obtained from these places (see, e.g., Figs. 2.24 and 11.13) and that stir the imagination. But like the integrals making up the AMOC, the resulting fluctuations are summations from vast regions (in principle, the entire global ocean) and times (back to the extended memory of the circulation) and diverse dynamical regimes. Wind-field and other forcing over the global oceans have major stochastic elements. Instabilities and other interior fluctuations of the ocean have a similar character. These random increments are summed in the choke-point regions. Ordinary random walks (see Chapter 3) produce large, impressive excursions. Major fluctuations are expected on the time scales related to the storage capacity of the basins and of the ocean as a whole. As with any random walk, attributing some particular event to a deterministic cause can be hopelessly misleading.[18] Thus interpretation of the time series from these "choke points" and integrals, like the MOC, becomes a chimera—changes are seen, but almost no information is available about the why of it or for determining the regions and times that are most important.

The "choke-point fallacy" is the claim that the comparatively simple and cheap measurement systems required in such regions as the Indonesian Throughflow or other monitored sections are a substitute for description of the entire global fluid. Would that it were true. Change can be documented in the choke points, but understanding recedes. Global processes demand global observations.

---

[17] See Wunsch and Heimbach, 2013a.

[18] Some isolated, powerful event lying many standard deviations beyond the usual stochastic mean can, of course, be perceived when the result reaches the choke point. No such events have been observed.

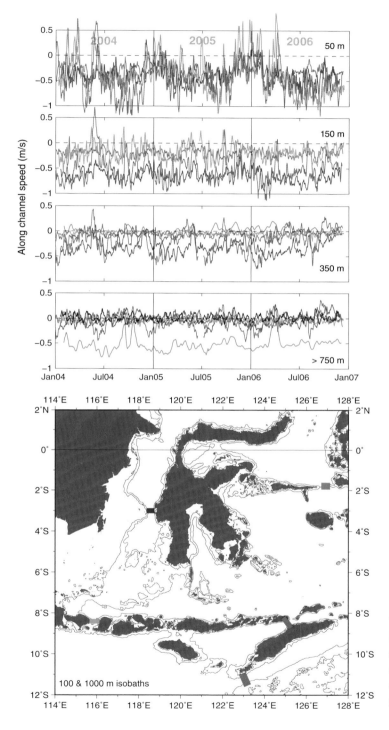

Figure 11.13: The along-channel flows within several of the major pathways within the Indonesian Archipelago. Values are color coded to the particular narrow conduits. (From Gordon et al., 2010; cf. Wunsch, 2010.)

11.4 TIME SCALES

Many of the details of the ocean circulation and its changes remain obscure, and even simplified model results can be difficult to interpret. To some extent, the details can be avoided by examining the basic time, space, and magnitude scales governing changes.[19] Order-of-magnitude estimates appropriate to the global scale are sought here—for the context they provide. Basin time scales are some fraction of the global ones. Emphasis is placed here on the time scales rather than on the space scales, because the major challenge of observational oceanography today is that of achieving adequate record duration. In contrast, observational techniques exist, at least in principle, covering all oceanic spatial structures from the global to the molecular.

Before proceeding, consider the analog of ordinary surface gravity waves, whose time scale is approximately 1 sec. Scientific study of these waves dates back to Newton's time, and yet many important aspects of their physics remain poorly understood (e.g., Craik, 1985; Munk, 2009). The ratio of the duration of observations (about 300 years using instruments) to their basic time scale of 1 sec, is about 10^{10}. A central problem in the study of many oceanic phenomena are that observations span only small *fractions* of their basic time scales—not large multiples of them—and even when a multiple of the time scale—durations are too short to use simple statistics.

Intense interest exists in the ways and rates at which the ocean responds to external and internal disturbances. Some of these issues are dynamic: if the wind curl undergoes a step change over some large area of ocean, how long before a new (quasi-) steady state is achieved? This sort of question is, generically, sometimes called the *spin-up* problem. Other problems relate to chemical issues: if a step change occurs in atmospheric carbon concentration, how long before a new equilibrium is reached with the ocean?

The time-scale issue is made a bit more concrete by Figure 11.14, which although depicting a meteorological variable (temperature of central England), shows the strong tendency for natural red-noise processes to exhibit long-duration *apparent* trends. No instrumental oceanic time series comes close to this duration, not even the longest unbroken tide-gauge records.

A number of time scales appearing in the general circulation are sketched below. These are based only partly on observations: also included are theoretical constructs often derived from observations (e.g., eddy mixing and viscosity coefficients). The listing is incomplete because, for example, it does not include the lifetime of bubbles injected by wave breaking at the sea surface, which is important in carbon uptake and other mechanisms.

11.4.1 Time Scales Based on Volume

The oceanic volume is about 1.3×10^{18} m^3. One set of time scales arises from calculating how long it would take to replace all of the fluid. In reality, parts undergo extended recycling so that they are renewed multiple times, whereas other parts are not renewed for far longer periods. Nonetheless, the orders of magnitude are useful.

---

[19] So-called *Fermi problems* are complex ones where helpful order-of- magnitude insights can be gained through comparatively crude numerical estimates (see Mertens, 2008). A number of oceanic change problems fall into that category.

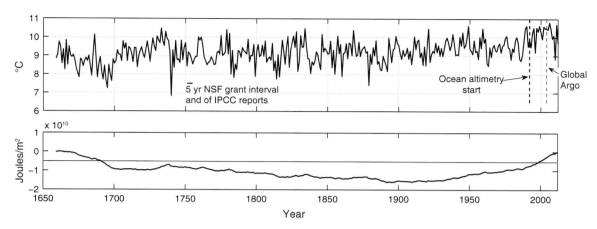

Figure 11.14: (Upper panel) The temperature time-series of central England, perhaps the longest continuous instrumental climate record, pieced together by Manley (1974) and subsequent investigators and now maintained by the UK Meteorological Office. Extended intervals of apparent trend appear, but they ultimately reverse. Whether the visible overall warming trend is part of global warming and will persist indefinitely, or whether it too will ultimately reverse, is a central question in climate change. Also shown for reference are the extent of a five-year research grant, roughly the interval, too, of the Intergovernmental Panel on Climate Change (IPCC) reports. The recent start of altimetry and global Argo-float coverage appear near the end. A statistical discussion of this record is by Tung and Zhou (2013). (Lower panel) The accumulating sum of the central England temperature record in the upper panel converted into an equivalent heat flux. If the ocean has a physical memory of prior heating rates, it will accumulate heat from long-past surface forcing variability (recall Hasselmann, 1976, and Fig. 3.45). It is assumed that heat is stored in a 100 m thick mixed layer and that the heat transfer depends only upon the anomaly of air temperature about the record mean (9.2°C). That the net change over the entire record is zero is an artifact of using the record mean to find the sign of the heat uptake. A more sophisticated handling of such physics is described by Frankignoul and Hasselmann (1977), but the long time scales emerging from accumulation are apparent. (The horizontal line is arbitrarily placed to emphasize that use of the mean temperature is an arbitrary reference. Only the "reddening" of the behavior by summation is significant.)

Precipitation and river runoff, (which are almost balanced by net evaporation), are about 12 ± 6 Sv (Xie and Arkin, 1997; and see Table 3.1).[20] Dividing the oceanic volume by the nominal net precipitation rates produces a cycle time through the ocean of about 3500 years. Subsurface percolation from the continents into the ocean has been estimated as 0.07 Sv (Moore, 2010), producing a very long renewal time. Cycling of water by entrainment through the ridge-crest hydrothermal plumes is estimated as requiring about 3000 years (Kadko et al., 1995). This value corresponds to about 14 Sv, which is comparable to surface precipitation rates. Net (downward or upward) Ekman pumping or suction is on the order of 100 Sv.[21] A cycling time for the Ekman flux through the entire ocean is about 300 years. These and other time scales are shown in Figure 11.15.

[20] Schanze et al. (2010) suggest a much smaller uncertainty, but their value rests upon strong assumptions about spatial error covariances.

[21] From integrating the smoothed downward-directed portions of Figure 9.30a. This value is sensitive to the existence of a strong model grid-scale noise near the continental margins and how it is smoothed; thus only the order of magnitude is quoted.

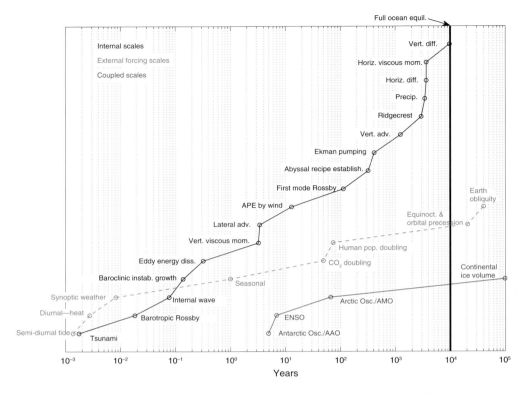

Figure 11.15: Rough estimates of the time scales of processes potentially important in the ocean circulation. Some are internal to the ocean, such as diffusion times (solid black line), some lie with the forcing functions (dashed red line), and some are the result of coupling with the atmosphere (dash-dot blue). The list is incomplete, and values are nominal because almost all occupy a range, sometimes an extremely broad one. Even the astronomical frequencies occupy small but finite intervals. Internal time scales can be regarded as defining a "memory" over which the ocean remembers previous states, whenever that time scale comes into play. External and coupled time scales define processes that may look to the eye as though representing true trends but which are nonetheless oscillatory in some way over a much longer time. No extended interval is free of an important time scale. ENSO, El Niño Southern Oscillation; AMO, Atlantic Multidecadal Oscillation; APE, available potential energy. "First mode Rossby" refers to the basin-crossing time at the group speed.

Biogeochemical properties are fluxed through the system at varying rates and are not discussed here (see Kadko, 1993, or Peacock, 2010).

11.4.2 Time Scales Based on Propagation and Advection

The area of the ocean is about 3.6×10^{14} m^2 $= \pi L^2$ m^2 giving a gross length, $L \approx 10^7$ m. With a mean depth of about 3.7×10^3 m, the long-wave propagation time, defined from the velocity, $\sqrt{gh}$, is about 27 h—the time for a tsunami signal to cross the ocean once. An equivalent depth (see Fig. 4.13) of 2 m produces a crossing time for a long first-mode internal wave of about 30 d (although no such wave could survive that long in the presence of interactions, topographic scattering, and instabilities). Baroclinic Kelvin waves traversing the circumference of a circular basin with this area (radius $\approx 10^7$ m) would take about 150 d.[22] The fast equatorial Kelvin waves travel as internal waves and do sometimes survive trans-Pacific crossings on a similar time scale.

A barotropic Rossby wave at midlatitudes would cross in about 7 d. The horizontal phase and group velocities of a Baroclinic Rossby wave are strongly latitude dependent. At $30°$ latitude a first-mode transit time (based upon the fastest group velocity) would be about 100 years, but much longer at higher latitudes and much shorter at lower ones (see Fig. 4.16).

Time-average oceanic vertical velocities are of order $w \approx 10^{-7}$ m/s (3 m/year). Thus the vertical transit time, should any parcel ever be carried fully upward or downward by advection, would take about 1000 years. Lateral advection speeds vary enormously, but taking 10 cm/s as an upper bound, the cross-basin advection time is on the order of 3 years.[23]

11.4.3 Time Scales from Diffusion and Viscosity

Diffusion terms in equations governing the evolution of tracers or momentum (viscosities) have the highest-order derivatives and usually control the time to final equilibrium while they remove the residual spatial gradients. Scalar turbulent vertical diffusivities, K_z ($K_z = K_{33}$ in the full tensor) in the ocean are, on the spatial average, thought to range from about 10^{-5} m^2 to about 3×10^{-4} m^2. The corresponding diffusion time scale, h^2/K_z, ranges from about 1500 to 45000 years at the two extremes but is plotted as one intermediate time in Figure 11.15. (The molecular diffusion time exceeds 4 million years.) Vertical viscosity, A_V, estimates are few and far between. Some have been based upon the estimated observed thickness of Ekman or Ekman-like layers (Lenn and Chereskin, 2009) and are on the order of 10^{-2}–10^{-1} m^2/s. But these values would apply only to the turbulent flows directly in contact with the wind field and must represent a very large upper bound on interior values. If they are, nonetheless, applied to the entire water column, time scales are on the order of 45 years. All these times are for values to decay to e^{-1} of their initial values and are not times to equilibrium, which are much longer.

Estimates of lateral viscosity, A_H, are poorly constrained and no single satisfactory value appears appropriate. Values have been calculated on the basis, for example, of the widths of western boundary currents (Munk, 1950) and range over 10^2–10^4 m^2/s. Hence the horizontal momentum diffusion time $L^2/A_H \approx 1200$ years at the minimum and much longer at the maximum. Scalar lateral eddy diffusivities, K_H, were calculated from a state estimate by Ferreira et al. (2005) with values in the range of 500–1000m^2/s, producing similar e-folding diffusion times, L^2/K_H, on the order of 1000 years. These values are particularly prone to systematic errors from coefficient estimates based upon coarse-resolution ocean models. The use of "hyperviscosities" (P. 99) using ad hoc biharmonic terms in the momentum equations introduces yet more time scales, as do various assumptions about the influence of eddies, which are functions of depth, lateral variations, and "mean" flows.

---

[22] Marshall and Johnson (2013) discuss the theoretical speeds of boundary-trapped baroclinic waves, for which dissipation proves important.

[23] A useful rule of thumb is that 1 cm/s corresponds to a bit less than 1 km/day.

TRACER RATIO CONCENTRATION

Consider two tracers, C_1, C_2 satisfying equations like Equation (11.12),

$$\frac{\partial C_{1,2}}{\partial t} + \mathbf{v} \cdot \nabla C_{1,2} - \nabla \cdot (\mathbf{K} \nabla C_{1,2}) = -\lambda_{1,2} C_{1,2} + q_{1,2} \qquad (11.9)$$

assuming they advect and diffuse at the same rates and with their appropriate boundary conditions, often in a steady state. Not uncommonly, geochemists find that measurements of the concentration ratio, $r_{12} = C_1/C_2$ are much more accurate than those of the separate values, and it is r_{12} that is mapped and discussed. The most common examples are for the isotopic ratios, for example, neodymium, Nd, $r_{12} = \left[^{143}\text{Nd}\right] / \left[^{144}\text{Nd}\right]$ (e.g., Tachikawa et al., 2003) or the oxygen isotope ratio $\left[^{18}\text{O}\right] / \left[^{16}\text{O}\right]$, the bracket denoting concentration in suitable units. Ratios are also used for parent and daughter tracers, in which the former decays into the latter. Tritium (^3H), introduced into the ocean by atmospheric nuclear-weapons testing beginning in 1945, and helium-3 (^3He) are examples, with tritium having a half-life of about 12 years, the one acting as a source for the other. Thus,

$$\frac{\partial \left[^3\text{H}\right]}{\partial t} + \mathbf{v} \cdot \nabla \left[^3\text{H}\right] - \nabla \cdot \left(\mathbf{K} \nabla \left[^3\text{H}\right]\right) \;=\; -\lambda_{^3\text{H}} \left[^3\text{H}\right], \qquad (11.10)$$

$$\frac{\partial \left[^3\text{He}\right]}{\partial t} + \mathbf{v} \cdot \nabla \left[^3\text{He}\right] - \nabla \cdot \left(\mathbf{K} \nabla \left[^3\text{He}\right]\right) \;=\; +\lambda_{^3\text{H}} \left[^3\text{H}\right]. \qquad (11.11)$$

The ratio $r_{12} = \left[^3\text{He}\right] / \left[^3\text{H}\right]$ is sometimes used to produce a measure of elapsed time by computing $\tau_{\text{He}} = -1/\lambda_{^3\text{H}} \ln(r_{12})$, and with the same interpretation problems as with ^{14}C (see p. 343). Chlorofluorocarbons (Freons), which are artificial, stable, chemical compounds can be analyzed in analogous ways by measuring the temporally varying ratios by which the different forms entered the oceans (e.g., Fine et al., 2008).

Unfortunately, unless the denominator, C_2, is effectively constant in space and time, the equation governing the space-time evolution of $r_{1,2}$ is a complicated nonlinear one, ultimately requiring specification of C_1, C_2 separately in any case. A good exercise for the reader is to find such an equation (see Jenkins, 1980, or Wunsch, 2002).

Some ^3He exists in the ocean from mantle sources and not as the decay product of nuclear-bomb-induced ^3H. Most of the latter has now vanished from the oceans because several half-lives have passed since it was injected into the ocean by nuclear weapons tests.

11.4.4 Time Scales Based on Energetics

The available potential energy (APE) in the ocean is not easy to estimate (see e.g., Winters and Young, 2009; Roquet, 2013; Tailleux, 2013; or the extensive discussion in Huang, 2010, Chap. 3). Most values are roughly $(200–600) \times 10^{18}$ J (200–600 EJ), although Huang (2010) lists one estimate of 1900 EJ. The amount of energy input to the ocean required to sustain the circulation is on the order of $2–3 \times 10^{12}$ W. Thus the time scale defined by the APE reservoir and the rate of energy input is 3–60 years—a remarkably short period. Much of the APE is found in the Southern Ocean, where the great bulk of wind-energy input occurs.

If, instead of the total energy input, that fraction arising from buoyancy forcing in all its forms is used, a net *loss* of about 10^{12} W is found (principally from convective responses; Huang, 2010), and the time scale is about 70 years, the time required to run the APE reservoir down by net convective processes. "Shutting down" convection, which is much discussed in the context of deglaciation and climate change would, paradoxically, appear to suppress this removal of APE, thus theoretically permitting a larger response to wind forcing in the upper ocean.

The eddy energy reservoir was estimated as containing about 10^{19} J. If it is assumed that much of the windpower input to the large scale is dissipated via instabilities generating geostrophic eddies at a rate of about 10^{12} W, the renewal timescale for the eddy field is a few months.

11.4.5 Time Scales Based on Tracers

Assume that a tracer, $C(\mathbf{r}, t)$, obeys a conventional advection-diffusion equation of form

$$\frac{\partial C}{\partial t} + \mathbf{v} \cdot \nabla C - \nabla (\mathbf{K} \nabla C) = -\lambda C + q, \tag{11.12}$$

that is, including possible source-sink terms, q, and terms, λC, as are appropriate for ^{14}C, etc. Most generally, $\mathbf{K}$ is a 3×3 tensor with significant nondiagonal terms. Boundary conditions at the sea surface are normally some combination of concentration and flux, depending upon the tracer and the region. The term q includes important in-falling particulate contributions for tracers like radiocarbon, remineralization chemistry, nonlocal mixing processes, and biological terms for oxygen and nutrients.[24]

The simplest time-scale discussions involve so-called passive tracers—ones not directly modifying the density field or, indirectly, $\mathbf{v}$, $\mathbf{K}$—such as tritium, helium, or carbon, and in particular those for which interior values of $q \approx 0$, such as tritium, helium, or the chlorofluorocarbons (see the box "Tracer Ratio Concentration"). All of the dynamical time scales listed above potentially enter the solution to Equation (11.12) which governs the time evolution of the concentration. Additional time scales arise from the value of λ, rates derived from q—which could reflect photosynthesis or remineralization, depending upon the choice of C. Even more emerge from the structure of the equation itself. The number of time scales arising from this equation, even in one dimension, is quite impressive. For example, K_z/w^2 is the time scale required to establish a boundary layer of thickness K_z/w, but only where $w > 0$ (labelled "abyssal recipes establishment" in Fig. 11.15). Time scales can also be based upon *signal* velocities, that is, determining when the presence of a tracer disturbance is first detectable—doing so depends upon instrument sensitivity and background-noise levels.

Diffusion terms, which usually control the final equilibrium when residual spatial gradients are removed, produce time scales proportional to $L^2/norm(\mathbf{K})$. Figure 11.16 displays an estimate of the time in years required for a tracer to reach 90% of its equilibrium value at 2000 m as calculated from a state estimate. Boundary conditions were $C(\lambda, \phi, z = 0, t) = 1, t > 0$, so that the final equilibrium is $C = 1$ everywhere. Values of $\mathbf{v}(\lambda, \phi, z, t)$, $\mathbf{K}(\lambda, \phi, z)$, $\lambda, q(\lambda, \phi, z)$, as well as the particular boundary conditions, all enter. But the very long time scales, particularly in the mid-depth Pacific Ocean, are a robust feature and are consistent with the orders of magnitude calculated from the diffusion rates, although some very long advection times exist as well.

[24]See, for example, Sarmiento and Gruber, 2004; Schlitzer, 2007; Williams and Follows, 2011.

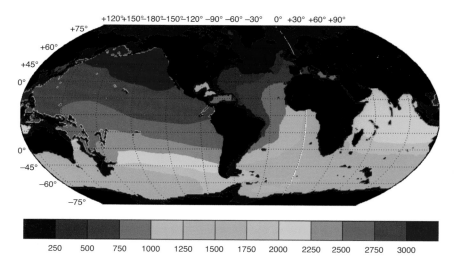

Figure 11.16: The estimated time in years to 90% of equilibrium near 2000 m of a passive (i.e., not influencing the density) dye tracer fixed globally at $z = 0$ at $t = 0$ as derived from a state estimate. The mid-depth North Pacific Ocean appears to be the "end-of-the-line" taking considerably longer than either the abyssal or near-surface layers to reach equilibrium. (From Wunsch and Heimbach, 2008.)

Active tracers, which influence the density field, and hence **v** and **K**, greatly complicate the discussion, because modifications of the flow and mixing fields will introduce even more time scales. Some of the most important and interesting tracers, such as temperature and freshwater, are active. A plausible assumption is that small disturbances to these fields can be analyzed by treating them as passive, but little guidance as to the robustness of that assumption is available. Some problems—such as the paleoceanographic discussion of a massive ice melt that changes the oxygen isotope concentration ratio, $\left[^{18}O\right]/\left[^{16}O\right]$, itself a passive tracer—are associated with the injection of sometimes massive amounts of freshwater, and so must be treated as active.

AGE TIME SCALES

The search for simple descriptors of the ocean circulation has led to a focus on the "age" of a fluid parcel, usually meaning the time since it left the sea surface, where it perhaps equilibrated with some property of interest (oxgyen, or carbon, for example). If that number corresponds to the time taken by the parcel to reach the observed position, it can be used to compute biogeochemical reaction rates. An example would be the difference between the oxygen concentration observed at depth and the value with which it began at the sea surface, divided by the water age, thus producing an *apparent oxygen utilization rate*.[25]

Much of the discussion of ages begins with radiocarbon (^{14}C) and similar unstable isotopes. Living organic matter takes up ^{14}C from the atmosphere with the value C_0. When the organism

---

[25] AOUR; see Sarmiento and Gruber, 2004.

dies, the ^{14}C begins to decay to a reduced concentration $C(t)$ with a half-life of about 5730 years. The equation governing tracer evolution in the body is

$$\frac{dC(t)}{dt} = -\lambda C, \lambda = 1/8267 \text{ year}, \tag{11.13}$$

with solution $C = C_0 \exp(-\lambda t)$, where $t = 0$ is defined as the time of death of the organic body. Thus the elapsed time is

$$t_e = -\frac{1}{\lambda} \log_e \left(\frac{C(t)}{C_0} \right). \tag{11.14}$$

Radiocarbon dates from t_e are important in archeology and many other fields. Even for such purely terrestrial applications, many problems intrude, including the important nonconstancy of C_0 through time, and lead to multivalued determinations (see Buck and Millard, 2004).

The oceanographic determination of an age has been through direct application of Equation (11.14): measure the ^{14}C content: assume C_0, the originating surface concentration, is known, and calculate the water mass "age" from the concentration observed at any place. Unhappily, the calculation is based upon a false analogy: although Equation (11.14) does produce a time parameter (from the multiplication by $1/\lambda$), the governing equation is *not* Equation (11.13), but some form of Equation (11.12); see the Box on "Tracer Ratio Concentration" . The putative age is just the logarithm of the concentration of ^{14}C, with its dependence on all of the velocities, mixing coefficients, and boundary and initial conditions. If Equation (11.12) produces three-dimensional time-dependent concentration fields that can be interpreted through a single parameter—a time labelled as the last contact with the surface—it would constitute a most remarkable mathematical inference about the solutions to a partial differential equation. Note too, that because $\log(C_1 + C_2) \neq \log(C_1) + \log(C_2)$, spatial average values of t_e have no particular meaning. The use of C directly, instead of its logarithm, is much simpler.

For example, the Green function solution to Equation (11.12) can be obtained at any position, $\mathbf{r} = (\lambda, \phi, z)$, within the ocean from a passive tracer disturbance, one occurring at any surface position $\mathbf{r}_0 = (\phi_0, \lambda_0, z = 0)$. Resident concentrations at depth at any time, t, have originated from all surface positions (Fig. 9.13), but with very different relative contributions, having arrived by advection and diffusion over many different times and pathways. For any finite ocean volume, $a^2 \Delta\phi \Delta\lambda \Delta z$, the fraction at time t of tracer originating from surface values can be regarded as a space-time probability density, sometimes known as a *transit time distribution*.[26] By integrating over all possible surface origination areas and times, a *mean transit time* at $\mathbf{r}$ can be calculated as a rough determinant of the time since the fluid particles were at the surface, but the spread of times and of originating regions in the distribution can be very large.

---

[26] Holzer and Hall, 2000. These solutions are the boundary Green functions. Using those functions, the concentrations at any moment in time at any fixed position is found as an integral over the entire time and space history of the surface concentrations.

An ultimately more satisfactory use of observed $C(\mathbf{r}, t)$ is to infer the physical parameters of the ocean circulation—including $\mathbf{K}$, u, v, w, and the boundary and initial conditions—using inverse or state estimation methods, perhaps applied directly to the Green function solutions. With best estimates of those properties, equations analogous to Equation (11.12) can be solved for oxygen, silicate, or carbon distributions without gross oversimplification. In any case, age-tracer values have not been included in Figure 11.15.

11.4.6 Dynamical Time Scales

Oceanic kinetic energies are strongly dominated by the geostrophic eddy field. From the inference that the larger-scale ocean circulation is unstable via several mechanisms, various time scales emerge. As idealized, these mechanisms are commonly labelled as "baroclinic," "barotropic," and "mixed" and give rise to characteristic exponential growth times depicted, for example as in Figure 11.6 where they were calculated from a climatology (Smith, 2007). Characteristic times are days—uncomfortably close to the wind time scales visible in Figure 3.35. Figure 11.6 depicts many different dynamical regimes, including, for example, the western boundary currents, where apparent instabilities are visible in altimetric and sea-surface temperature satellite data. More generally, observations are restricted to the apparent statistical steady state of the resident eddy field.

Many fluid flows are nominally stable to infinitesimal disturbances—the most usual form of instability study—but some can produce *non-normal* responses to forcing, in which otherwise stable mode-like responses can grow to produce finite amplitude instability (Farrell, 1989; Trefethen, 1997). Quantification of this process in the ocean is unavailable.

11.5 CONSEQUENCES OF MEMORY

Systems with memory longer than the time scales of variability in stochastic forcing can have counterintuitive behavior because the response can depend primarily upon the time integral of the forcing rather than on the forcing itself. A schematic demonstration of this phenomenon is shown in Figure 11.14, with the assumption that the ocean responds to the time integral of the temperature anomaly in the central England record—simply storing the accumulated thermal disturbances over the full record (350 years). Three hundred fifty years is a short interval compared to many of the time scales in Figure 11.15. The "red" character of the result is easily understood from the Hasselmann (1976) discussion of a random walk (and see p. 106).

Another demonstration of the consequences of memory, in this case dynamical, can be seen in Figure 11.17, where using surface winds found in an atmospheric renanalysis and assuming a fixed oceanic surface flow, the power input to the ocean is computed and integrated through time. Multidecadal and longer time scales again emerge if the ocean, as expected, tends to integrate the input power over its long-memory components.

In a general sense, extended memory means that initial conditions used, for example, when calculating the impacts of ongoing climate change, will be a consequence of integrals over the complete past history of the ocean circulation as far back as previous disturbances are "remembered" by the system. "Modern" calculations likely involve initial conditions far from equilibrium.

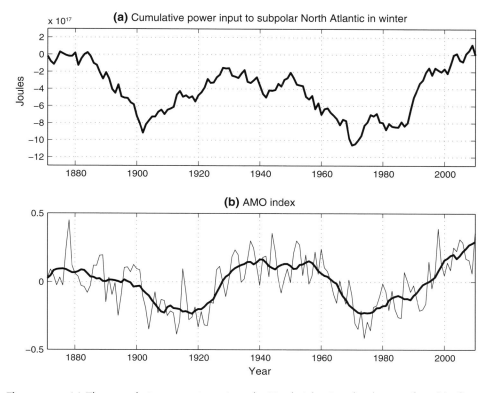

Figure 11.17: (a) The cumulative power input into the North Atlantic subpolar gyre found by fixing the ocean circulation and assuming the system remembers the prior input for an interval long compared to the record length. The memory "reddens" the response to wind forcing. (b) The so-called Atlantic Multidecdal Oscillation (AMO; see Enfield et al., 2001) in sea-surface temperature. The smoothed line is a 10-year running average. Notice that the record is long enough only to barely define a single dominant variation, one with a visual similarity to the cumulative wind-power input. Inadequate data exist to reject the hypothesis of coincidence in the visual similarity of the variations in power input and AMO. (From Zhai and Wunsch, 2013.)

11.6   OTHER PHYSICAL ASPECTS AND REGIMES

BOTTOM BOUNDARY LAYERS

Munk and Wimbush (1971) and Armi and Millard (1976) described what are usually interpreted to be typical of the bottom boundary layer on flat, horizontal regions (abyssal planes). The latter might be summarized as being essentially turbulent boundary layers over a rigid bottom influenced by rotation and stratification. Monin and Yaglom (1975) provided the theoretical background for the nonrotating case; and Thorpe (2005) included additional context.

Sloping ocean boundaries are difficult places to observe low-frequency physics because the effects of internal waves encountering slopes greatly amplify the noise background. A considerable theoretical literature exists on the applicable physics (e.g., Garrett et al.,

1993). More generally, time-dependent balanced motions, commonly involving combinations of gravity (Kelvin-wave-like features) and bottom topography control, act to produce intensified motions.

The full influence of arbitrary topographic features on the large-scale circulation is poorly understood. That the tendency of geostrophic flows to follow f/h contours leads to a strong coupling between the flow and *bottom torques* induced by the interaction of the pressure field and h is clear from models (e.g., Hughes and de Cuevas, 2001). A full physical discussion, however, involves three-dimensional time-dependent and time-mean boundary layers driven by encroaching balanced eddies, internal and inertial waves, tides, and any mean flow. Flow separation must take place under some circumstances, and lee wave regimes will exist. All of these and other processes will coexist and compete. Although an extensive modeling literature exists, little direct observational evidence supports it, apart from the specialized literature on continental margins.

SEA ICE

The importance of sea ice to both the ocean circulation and climate more generally has become much more conspicuous in recent years. Sea-ice models have been developed within the state estimation framework as fully coupled subsystems influenced by and influencing the ocean circulation (Menemenlis et al., 2005a; Losch et al., 2010). A step toward full coupled ocean–sea ice estimation, in which both ocean and sea-ice observation were synthesized, was made by Fenty and Heimbach (2013) for the Labrador Sea and Baffin Bay.

ICE SHEET–OCEAN INTERACTIONS

The intense interest in sea level change and the observed acceleration of outlet glaciers spilling into narrow deep fjords in Greenland and ice streams feeding vast ice shelves in Antarctica (e.g., Rignot et al., 2011) has led to inferences that much of the ice response may be due to regional oceanic warming at the glacial grounding lines. Recent, and as yet incomplete, model developments have been directed at determining the interactions of changing ocean temperatures and ice-sheet response and for the purpose of their inclusion in the coupled state-estimation system (Losch, 2008; Straneo and Heimbach, 2013). Continental ice sheets have time scales (memory) far longer even than the ocean—some of the Antarctic ice has ages approaching 1 million years before present (Jouzel et al., 2007).

AIR-SEA TRANSFERS AND PROPERTY BUDGETS

By definition, state estimates permit calculations up to numerical accuracies of global budgets of energy, enthalpy, etc. As an example, Figure 11.18 is an estimate by Stammer et al. (2004) of the net air-sea transfers of fresh water. That paper compares this estimate to other more ad hoc calculations, and evaluates its relative accuracy. Large and Yeager (2009) comprehensively discuss the methods and results used with the diverse data sets required to estimate air-sea fluxes. See also Figure 9.9.

As examples of more specific studies using the state estimates, we note only Piecuch and Ponte (2012), who examined the role of transport fluctuations on the oceanic heat-content distribution, and Roquet et al. (2011), who used them to depict the regions in which mechanical forcing by the atmosphere enters into the interior geostrophic circulation.

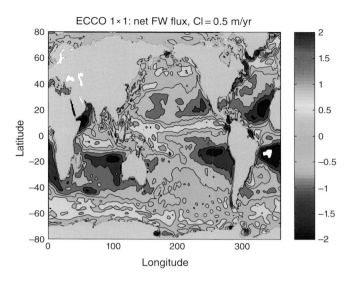

Figure 11.18: An estimate of the multiyear average freshwater transfers between ocean and atmosphere. The intertropical convergence zone (ITCZ) is particularly conspicuous as the narrow band of intense precipitation across the low-latitude Pacific Ocean, but it extends to all oceans. Large subtropical regions of intense evaporation are conventionally known. Otherwise, an intricate pattern of varying net water flux exists even with a multiyear average. (From Stammer et al., 2004.)

THE MIXED LAYER

Over much of the ocean, measured temperature profiles display a region at and near the surface of much-reduced vertical gradient—thus commonly termed the *mixed layer* (MLD)—the subject of a large specialized literature in its own right. Figure 11.1 is a cartoon displaying some of the many processes active at and near the sea surface. The mean and maximum seasonal depth of this layer have been mapped globally (see Figs. 11.19 and 11.20), and efforts have been made to understanding its physics. That it is not truly fully mixed (a residual nonzero vertical gradient is almost always present) sometimes plays an important role in its evolution (Fig. 8.12). Lateral gradients within the mixed layer can be large. Discussion of the physics of this specialized region is left to the references, including Price et al., 1986; Gaspar et al., 1990; and Large et al., 1994.

SUBDUCTION AND OBDUCTION

Oceanic mixed layers undergo very strong seasonal cycling—deepening in the autumn and winter, thinning again in the spring and summer. As depicted in Figure 11.19, these vertical movements can extend over 1000 m or more. Water forced vertically out of the mixed layer by Ekman pumping can be overtaken and "captured" again by the seasonally deepening mixed layer. A picturesque discussion of this process, in terms of an "Ekman demon," was provided by Stommel (1979). The spatial structure over the course of a year in one small region can be inferred from Figure 9.32. "Obduction" is the term used by Huang (2010) to label rates at which fluid is entrained into the mixed layer from below.

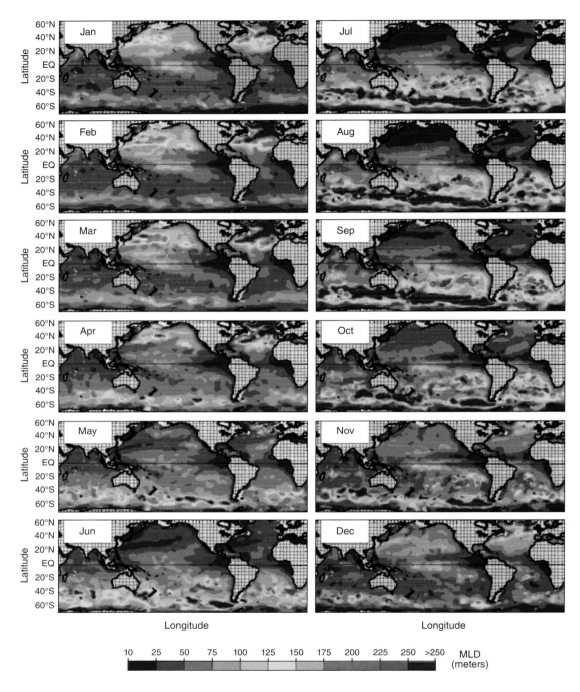

Figure 11.19: Estimated monthly climatological average mixed-layer depth (MLD). The result is from a combination of observations and a simple mixed-layer model. At high latitudes, the winter time depths can greatly exceed the cutoff value of 250 m in the plots. Lateral temperature gradients within the layer can be very strong. (From Kara et al., 2003.)

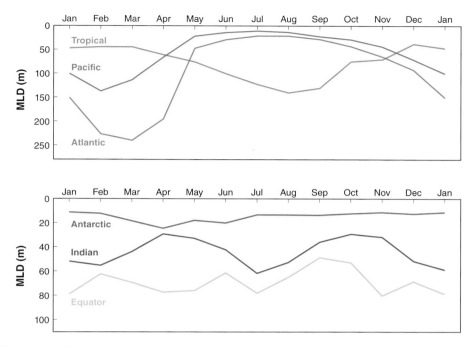

Figure 11.20: Example of regional variations in mixed-layer depth over the year. Mixed layer depths can, regionally, become much deeper than shown when convective processes are active. Many oceanic processes will be modified when such layers change thickness by hundreds of meters, from on the order of 10 to several hundred meters. (From Kara et al., 2003.)

11.7 SEA LEVEL, THE GEOID, AND RELATED PROBLEMS

Determining and explaining the form of Earth gravitational equipotentials is traditionally a part of geodesy and was of little concern to physical oceanographers.[27] With the rise of altimetry as central to determining the ocean circulation, the two fields have become entwined and a brief look at their connections is useful.

As in Equation (2.1), altimetric satellites measure the shape of the sea surface, $S_a (\lambda, \phi, t)$, relative to the center of the Earth. Apart from measurement errors,

$$S_a (\lambda, \phi, t) = \eta (\lambda, \phi, t) + N_g (\lambda, \phi),$$

where N_g is the height of the gravitational equipotential called the geoid. N_g defines "mean sea level." As will be seen in Chapter 12, mean sea level is poorly known, ambiguous, and changes with time. A more oceanographically useful definition is a hypothetical one—as the surface to which the ocean would have conformed were it at rest with a mean density chosen for some particular date, t_N. It is thus a *level surface*, and η is the deflection of the sea surface relative to the local equilibrium value owing to water movement in the ocean circulation.

[27] But see Sturges, 1974.

With the nonlinear equation-of-state no consensus exists on how to define the mean density. In practice then, the oceanic geoid height is a somewhat arbitrary approximation to a true gravitational equipotential on an ill-defined date.[28]

Spatial variations in N_g (Fig. 3.2) exceed 100 m. Although N_g is being written as though it is time independent, it does change, owing to glacial isostatic adjustment (GIA), which is significant over decades. That trend is visible in the GRACE measurements, as are regional changes resulting from major earthquakes. Tides can be regarded as part of the time-dependent geoid, but in practice, they are best dealt with separately. (A zero-frequency tide also exists; it is most simply absorbed into the time-mean geoid.)

Ideally, N_g would be known extremely accurately from measurements independent of altimetric values, S_a. Simple subtraction then determines the absolute sea surface,

$$\eta\left(\lambda, \phi, t\right) = S_a\left(\lambda, \phi, t\right) - N_g\left(\lambda, \phi\right), \tag{11.15}$$

which if available, is a powerful constraint on estimates of the ocean circulation. Unfortunately, the situation is not so favorable. Suppose at the opposite extreme, η were very well known: then the best estimate of the geoid height is

$$N_g\left(\lambda, \phi\right) = S_a\left(\lambda, \phi, t\right) - \eta\left(\lambda, \phi, t\right), \tag{11.16}$$

and the marine geodetic problem would be solved.

Equations (11.15) and (11.16) represent two unreasonable extremes—one with no knowledge of the ocean circulation, the other with no knowledge of the marine geoid. A great deal is already known about both of these fields, albeit it varies with geographic position and space and time scales. In particular, the form of η has important consequences for oceanic behavior and must be in accord with ocean circulations that are quantitatively sensible. Similarly, the geoid is well determined at long wavelengths from the corresponding gravity field and its influence on satellite orbits. Great improvements came with the gravity satellites, GRACE and GOCE. A major interest has lain with detecting the *time dependence* in N_g—for inferences about changing land-ice and liquid-water storage, as well as ocean bottom pressures (see Tapley et al., 2004, and Chapter 12). The very shortest wavelengths, below order a few hundred kilometers, are known only regionally from error-prone shipboard and land measurements of the gravity field, although GOCE has led to considerable improvements in the time-mean values. (See textbooks for discussions of the connection between local gravity g, and the behavior of N_g.)

Spatial variations in S_a are on the order of 200 m peak-to-trough (Fig. 3.2), and those in η are on the order of 3 m. Setting $\eta = 0$ would thus provide a geoid height estimate, $\tilde{N}_g$, from S_a (Equation 11.16) accurate to about 2%. A great deal *is* known about the ocean circulation and hence any information about η in Equation (11.16) can be used to improve $\tilde{N}_g$ further. Difficulty comes from the uncertainties in estimates of $\tilde{\eta}$ that in practice are a strong function of both spatial scale and geographical position. Use of Equation (11.15) to estimate η encounters a similar range of problems, including the requirement that the implied circulation is physically sensible.

---

[28]Additional problems arise in extending N_g over land, where it can lie well below the surface topography.

Ultimately, all knowledge of both fields is combined, along with the dynamical equations for the oceanic movement, to make best estimates of N_g simultaneously with estimates of η that lead to physically sensible circulations. The state estimation machinery provides the means to do so; Figure 3.29 displays $\langle \tilde{\eta} \rangle$ averaged over 16 years from such a combined estimate. Because of the great difficulty in assigning useful error estimates to the results, no corresponding uncertainty estimates have been published for $\tilde{N}_g$. In contrast, the geodetic community has a long and laudable history of always requiring them.

CHAPTER 12

Low-Frequency, Time-Varying, Global-Scale Flow

12.1 BACKGROUND

With the rise of interest in climate change, documenting and understanding how the ocean changes over time scales spanning decades and beyond has become important. The challenge to the oceanographer, and to climate scientists more generally, is the need to describe a complicated system with few direct observations. Theory and modelling results untestable by observations ultimately become competing stories. The purpose of this chapter is to depict what appears to be known of oceanic variability in the period roughly corresponding to that of instrumental observations, whose earliest limit might be taken as roughly 1650—when recognizably modern thermometers and barometers were developed.

Prior to about 1992 and the World Ocean Circulation Experiment, only fragmentary elements scattered in time and space were measured. A handful of observations can always be averaged and then proclaimed to be a regional or even the global mean value—but with little understanding of its significance.

Before discussion of the general problem, recall the depiction of temporal and spatial variability as seen in the spectra displayed in, for example, figures 3.12 or 8.10. These spectra are best characterized as representing a frequency and wavenumber continuum: apart from the line-frequency components of the tides, the energy has a stochastic character and exists at all space and time scales. If spectral gaps existed, they could have supported a hypothesis of decoupling between variability at high frequency or high wavenumbers and that at lower frequencies or wavenumbers. Necessarily, a low-wavenumber cutoff exists at the diameter of the Earth, and a low-frequency one occurs at the age of the oceans. Beyond these two unhelpful inferences, little can be said.

A conceptual difficulty lies with the descriptive theories based on Reynolds-type decompositions in which velocities and other properties are separated into two (or more) parts,

$\bar{u} + u'$. $\bar{u} = \langle u \rangle_L$ is an average over a long time and/or space interval, L, where "long" is measured relative to the space-time scale of variations in u'. By definition, $\bar{u}' = 0$ on that particular averaging interval. This representation is substituted into the equations of motion, and then averages are formed of those equations, again over L, arguing that $\langle \bar{u}u' \rangle_L = 0$, etc., thus dropping these "cross terms." If motions on time or space scales short compared to L are decoupled from those on scales of L and longer, then the cross terms do vanish. Ideally, L would correspond to an identifiable gap in the spectrum—implying no energy transfer across it. However, if the choice of L is made arbitrarily, corresponding to a scale in the midst of the spectral continuum, cross terms will *not* vanish, and the covariance they represent could dominate the physics. Spectral gaps are not necessary for a separation of distinct physics—an example would be coherent vortices embedded in a stationary random wave field with overlapping scales—but the Reynolds decomposition with an arbitrary scale based on record length or domain size is the one usually made without comment (see Davis, 1994b).

No evidence exists for gaps in the wavenumber spectrum. The minimum visible, for example, in the current meter frequency spectra between the inertial peak and the general rise into the balanced motion band, is the only observed indication of a potential decoupling of high-frequency motions ($f \leq \sigma$) from the extended continuum below. A growing body of literature however, implies that significant coupling across the minimum does exist—through a variety of physical mechanisms, that include the loss-of-balance- and lee-wave-generation mechanisms already mentioned.

12.1.1 Forcing Fields

As with many elements of the climate system, some strong inferences can be made with limited data (see Hartmann, 1994): (1) Basic wind patterns of latitudinally varying belts of westerlies and easterlies are very robust, stable, and well-understood components of the atmospheric circulation. (2) Dominant regions of desert and of heavy precipitation are largely consequences of the land and ocean distribution and so are stable for thousands and sometimes millions of years. (3) Seasonal cycles of temperature and salinity existed into the indefinite past.

To go beyond these basic notions and to answer such questions as, "What was the over-ocean meteorology before 1990, and how accurately was it known?" is much more involved. The first part of the question is usually answered by referring to the so-called meteorological reanalyses. These products suffer from the shortcomings outlined in Appendix B. Figure B.2 is an example of the discrepancies between different reanalyses over the recent past in the southern hemisphere—a region that is comparatively data poor. Some fields, such as Southern Ocean precipitation, show differences of over 50% in these products.

Luterbacher et al. (2002) have made estimates of regional surface pressure over western Europe dating back to 1500, preceding even the invention of the barometer and thermometer. Quantification of the accuracy of these remains incomplete, rendering inferences about possible forced ocean changes almost impossible to make. Using the global surface-pressure observations (Compo et al, 2011), have produced an estimate of the global atmospheric state extending from 1871. Some major trends are found in their estimates, but the number and distribution of measurement sites and methods of observation changed greatly over that interval, leaving open the question, what is a real trend and what is an artifact of the changing observation distribution?

PATTERNS OF ATMOSPHERIC VARIABILITY

The search for a simplifying description of the ways in which the atmosphere varies through time has led to the identification of a number of "modes" of variability, sometimes labelled as "oscillations." Evidence for their existence varies greatly. The annual cycle dominates almost everywhere (one measure of relative amplitude can be seen in the spectrum of temperature in Fig. 12.1). Beyond that, El Niño–related (called the El Niño-Southern Oscillation, or ENSO) variability patterns in the atmosphere can be identified (e.g., Alexander et al., 2002) for global-scale precipitation and other pattern changes. Most of these modes are identified by computing correlations in atmospheric variables. Some of the more conspicuous ones have been given names and acronyms: Southern Annular Mode (SAM); Arctic Oscillation (AO), see Figure 12.2 and North Atlantic Oscillation (NAO). Controversy exists (see Thompson and Wallace, 2000) about whether the latter two are separate phenomena. The so-called Pacific Decadal Oscillation (PDO; Fig. 12.3) has been described as not decadal, not an oscillation, and not confined to the Pacific! It may be a secondary part of ENSO.

Some of these structures are so weak as to be possibly figments of spurious correlations from records that are too short, or they comprise such a small fraction of the total variance as to convey little information. In understanding how the ocean might respond to them requires determining the fraction of the variance they contain, because the ocean responds to all meteorological forcing, not just to identifiable organized patterns. Signal-to-noise ratios become the essence of the discussion, as does the chronic problem of records that are short compared to the dominant time scale of variability.

These patterns occur on time scales where the ocean itself is responding, thus likely changing the sea-surface temperature and generating significant, even controlling, feedbacks onto the forcing itself.[1] Most conspicuous and famous of these coupled modes is the ENSO (Philander, 1990; Sarachik and Cane, 2010) on a time scale of 4–9 years (Fig. 11.15). The other labelled phenomena such as the AO, NAO, and PDO are less well documented, the available records being generally far too brief to provide definitive descriptions. Rationalizations often invoke mechanisms that cannot be documented (supposed MOC variations; movement of convective regions; etc.). Many records are indistinguishable from large-scale, red-noise processes, and the integrating properties of the long-ocean memory further redden the response spectrum (see Fig. 11.17).

12.1.2 Atmospheric Spectra

Although most attention has focussed on apparent spatial structures (modal patterns) in atmospheric variables, another approach is to regard atmospheric forcing as simply another set of time-stationary stochastic fields disturbing the ocean circulation.

Frankignoul and Müller (1979) produced a few examples of atmospheric wavenumber spectra. Frequency-wavenumber spectra of the atmosphere have not been commonly computed. Vinnichenko and Dutton (1969) showed that at periods longer than a day, the wind field has a nearly white frequency spectrum (Fig. 12.4), and Gage and Nastrom (1985) created a composite wavenumber spectrum of winds (Fig. 12.5). Care is required in interpreting atmospheric spectra because they do depend upon the time of day, proximity to the ground, season, etc. (cf. Xu et al., 2011b). The wind-variation time scales shown in Figure 3.35 are a crude summary description of the higher-frequency variability.

---

[1]See Appendix C.

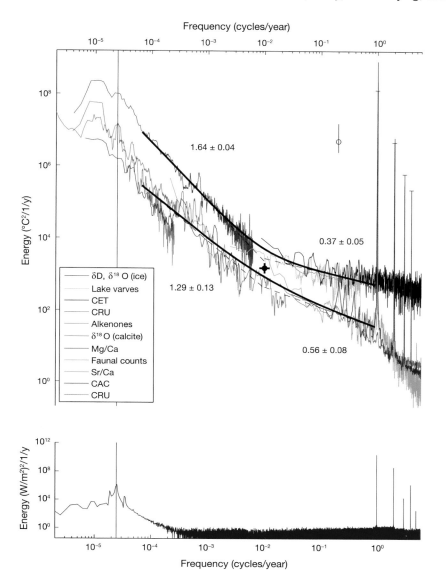

Figure 12.1: An estimated northern- hemisphere temperature spectrum out to 100,000+ years. The upper curve represents high latitudes from ice-core data and related proxies. The lower curve is the estimate for low-latitude sea-surface temperature (SST). The strongest element is the annual cycle and its harmonics. A weak, broad peak near 100,000 years, capping the red-noise power law, represents the glacial-interglacial cycles of the last 800,000 years or so. The curve at the bottom of the figure is the power in the Milankovitch forcing at 65°N, sometimes chosen as the defining representation of Equation (6.27). A transfer function representation of the upper curves in terms of the forcing function is obviously problematic, given their very different frequency structure. The cross lies midway between the annual cycle and the Milankovitch forcing. An approximate 95% confidence interval is shown. Records used include the central England temperature (CET) and those of the Climate Research Unit (CRU) and the Climate Analysis Centre (CAC). δ D and δ^{18}O are deuterium and oxygen isotope anomalies. (Huybers and Curry, 2006.)

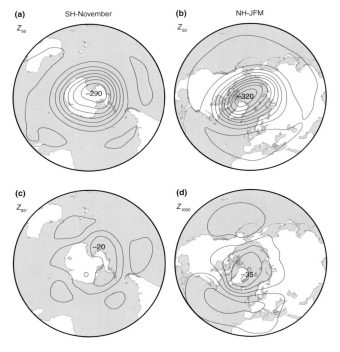

Figure 12.2: Correlation patterns at 50 mb height (a) in the southern hemisphere (SH) in November, and (b) for January–March in the northern hemisphere (NH). These patterns are the leading singular vectors (EOF). (From Thompson and Wallace, 2000.)

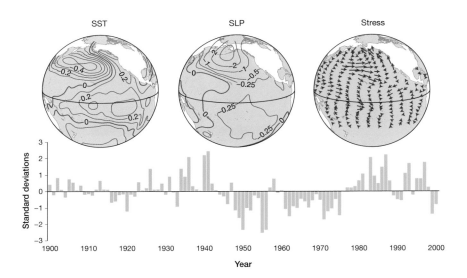

Figure 12.3: The nominal Pacific decadal oscillation (PDO). Displayed are the anomaly patterns of sea-surface temperature (SST), atmospheric sea level pressure (SLP), and the estimated wind stress. The bottom bar graph shows the amplitude of the anomaly through time, which has the character of an ordinary red-noise process. The record is too short relative to the dominant visual time scale to properly characterize this variability. Arguments have been made that the PDO is primarily a remnant part of the ENSO, but the issue remains open. (From Mantua and Hare 2002.)

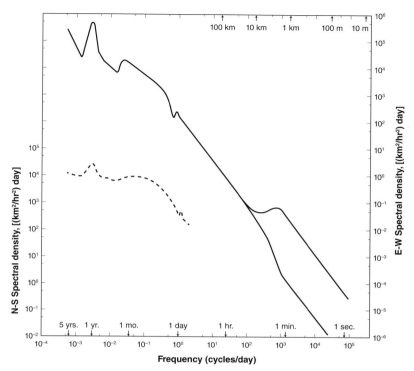

Figure 12.4: Estimates of the free atmospheric wind frequency spectra. Both east-west and north-south (dashed line) component spectra are shown (on different scales, right and left, respectively). The split at high frequencies in the east-west spectrum represents strong and light wind conditions (After Vinnichenko and Dutton, 1969.)

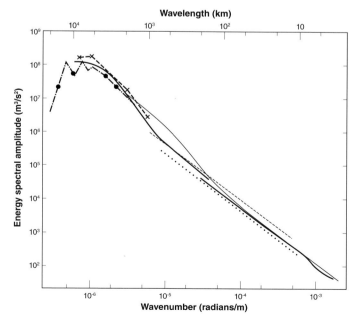

Figure 12.5: A composite wavenumber power spectrum of atmospheric winds; At scales longer than about 10,000 km, a general power reduction is seen, and which must vanish entirely at the circumference of the Earth. Otherwise, the spectrum has the red-noise character common in geophysical processes, with an approximate −2 power law. (From Gage and Nastrom, 1985.)

Surface air-temperature spectra have been estimated by Huybers and Curry (2006: Fig. 12.1) extending back past 100,000 years by relying upon preinstrumental "proxy" data for the low frequencies. A recognizable spectral continuum exists with no indication of a spectral gap. The powerful annual cycle and its overtones dominate the record variance—even when ice ages are included. A wavenumber breakdown of the variability is not available.

12.2 DECADAL-SCALE OCEAN VARIABILITY: THE RECENT PAST

Approximately 20 years of data have accumulated since global-scale ocean observations became available, and that data can be used for calculating interannual to decadal variability. Figure 2.24 is a concrete example on the decadal scale but is in a very limited regional form—the Gulf Stream transport changes. The main difficulty was described in the Chapter 11 as the choke-point problem—western boundary current transports are summations over all time and space of all flows. In spectral terms, the process appears to have a weakly red-noise character.

12.2.1 Hydrographic Results

Attempts at estimating changes in the total amount of heat and freshwater stored in the ocean have become a cottage industry in the past decade or so for two major reasons: (1) the continued difficulties in accounting for the total heat budget of the Earth under global warming, and (2) the need to understand their relative contributions to estimated past and future sea level changes.[2] Figure 2.27 (from Worthington, 1981) was constructed using strong data quality requirements. About 1978, he concluded that only about 46% of the ocean had been sampled in temperature and salinity *even once* in all of oceanographic history. A different depiction of the same problem can be seen in Figure 2.26 which shows the historical temperature data (over 100 years ending circa 2005) that reached to at least 300 m. Much of the world ocean, including most of the volume below about 300 m, was unsampled. For salinity, the situation is far worse.

Given the ongoing global warming and an inferred net injection of freshwater from melting land ice, it is plausible that the ocean has both warmed and become fresher. As with many such inferences, however, changes in measurement technology and in the distribution of the data make some of the claimed accuracies less than convincing. An oceanic heat-content change generated by a heating rate of 1 W/m^2 corresponds to an approximate volume mean temperature change of $0.002°C/y$. (Existing estimates of net global warming are roughly 0.5 W/m^2 over the past few decades.)

A few top-to-bottom trans-oceanic hydrographic lines were obtained in the Atlantic in the 1920s by the German ship *Meteor* (Wüst, 1978), a few more were done there during the International Geophysical Year (1958–1959), and then data were collected only regionally until the WOCE era. A single top-to-bottom North Pacific line was pieced together by Kenyon (1983); the first such measurements in the South Pacific were obtained in the 1970s (Stommel et

---

[2]The intense interest in these questions can be gauged from the continuing series of reports from the Intergovernmental Panel on Climate Change (IPCC). These should be consulted for references to the many attempts at estimating heat and freshwater content changes.

al., 1973) and not repeated until WOCE. A number of attempts have been made to determine temperature and salinity changes in the oceans by comparing such lines even when the latter are separated by decades in time. The difficulties are many: a major change occurred in technology from reversing thermometers for the determination of temperature and depth to thermistors and strain gauges; spatial patterns of sampling were different; prior to the GPS era navigation errors could be large; and early sections failed to spatially resolve the unrecognized eddy scales, leading to wavenumber aliasing into the regional means.

Some of the first recognizable hydrographic measurements in the modern sense were obtained from the RRS *Challenger* in 1873–1875. One section ran from New York to Puerto Rico and is displayed in Figure 12.6 along with a late 1950s near-repetition of the measurements. In a related test, Rossby et al. (2010) could find no significant difference between the Gulf Stream today and that observed by *Challenger*. The differences result from some combination of changed technology, spatial offsets, real long-term changes, and temporal shifts in the noise field. Quantifying those contributions is the central problem of low-frequency oceanography.

That large-scale changes have taken place in temperature (and in salinity) data is no surprise. Whether they represent true long-term trends, or simply large-scale red-noise processes, or aliased eddy noise is the subject of much debate. Bindoff and McDougall (1994), Purkey and Johnson (2010), and several others discussed these apparent changes and what they might mean. From the spectral content described and shifts in the observing systems, changes *must* occur in the ocean on all space and time scales; determining whether they are true secular trends is problematic.

Using the global RRS *Challenger* data, Roemmich et al. (2012) have produced the most convincing demonstration thus far of a long-term mean change in oceanic temperatures, as shown in Figures 12.7a and 12.7b. By comparing the Argo float profiles to the temperature measurements made on the *Challenger* Expedition in the nineteenth century, they found a significant warming since the early 1870s in the upper ocean, corresponding to a net heating of $0.2 \pm 0.1 \, \text{W/m}^2$. Particular care was required to render the nineteenth-century data comparable to the modern and to derive a plausible error estimate. A similar calculation is not possible for the salinity field.

The temperature and inferred density changes found are sufficiently small that any changes in the dominant thermal-wind balance over decades are small perturbations. Missing entirely from that discussion is any description of changes in the nearly three-quarters of the volume of ocean below 1000 m, and whether it warmed or cooled on average is unknown, and may well remain so. But no evidence exists anywhere of a *dynamically* significant change.

An estimate, restricted to 19 years during the post-WOCE data-dense era is shown in Figures 12.8a and 12.8b. Shifts in regional heat content, most likely primarily the result of changing wind patterns, render the determination of accurate global mean changes very difficult. The region below about 2000 m has very few in situ measurements for this interval yet has slightly less than one-half the ocean volume. Figures 12.8a and 12.8b show that any inferred net warming of the ocean must be computed as a small residual of the competing regions of heating and cooling and represents much of the challenge to observational systems. In the particular estimate shown, the net heat uptake was $0.2 \pm 0.1 \, \text{W/m}^2$ (Wunsch and Heimbach, 2014), the same as Roemmich et al.'s (2012) estimate for their much longer interval. Some recent papers claim to have determined global oceanic-heat-content changes over varying time intervals with precisions better than $0.01 \, \text{W/m}^2$. Both the spatial structure of decadal change and the intense temporal noise suggest that one should retain considerable

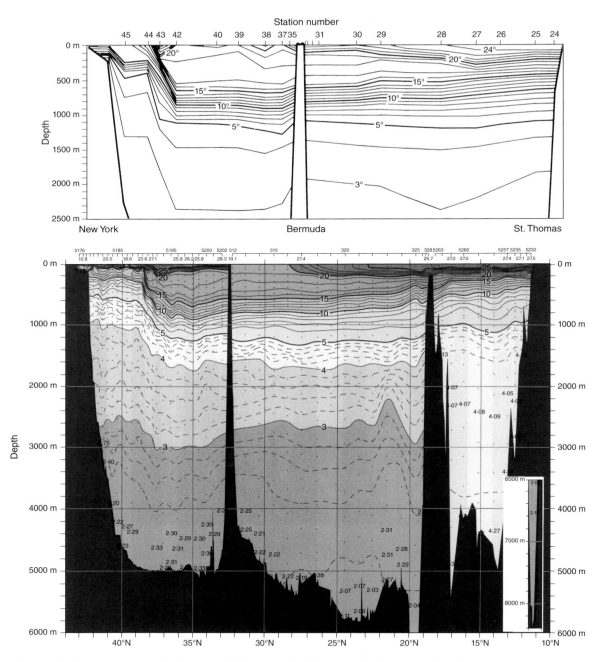

Figure 12.6: The section made by the RRS *Challenger* in 1873 from New York to Bermuda and Puerto Rico (upper panel) and an analogous section (not exactly the same position) made from Woods Hole to Bermuda and to Puerto Rico as a composite from 1954 and 1958. Station numbers are shown at the top of each section. Qualitatively, the two sections are indistinguishable, while quantitatively they differ in many subtle ways. Thermometer technology with implications for both temperatures and depth determination changed in the elapsed interval. In the lower plot, Bermuda is at about 33°N. A near-surface horizontal discontinuity occurred because of the seasonal change while the ship stayed in port. See also Rossby et al., 2010, for a discussion of the lack of obvious changes in the Gulf Stream component and compare to Roemmich et al., 2012, Figure 1. (From Wunsch, 1981.)

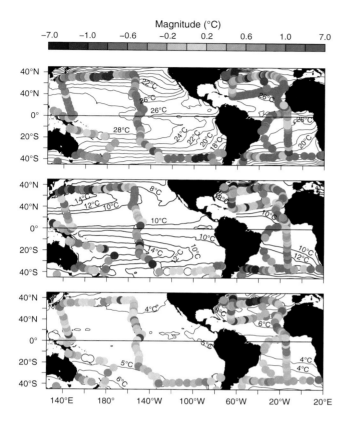

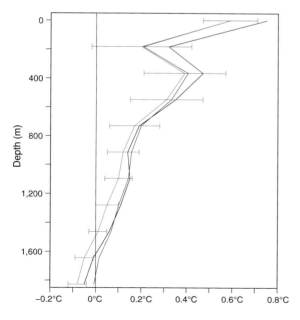

Figure 12.7a: Background contours indicate the mean temperature (2004–2010) estimated from the Argo data at (top) the sea surface, (middle) 366 m and (bottom) 914 m (c). Dots indicate the Argo minus *Challenger* temperature differences. A preponderance of warmer values is visible. (From Roemmich et al., 2012.)

Figure 12.7b: Spatially averaged Argo minus *Challenger* temperature differences as a function of depth. The black line is a simple mean over all stations with data at 183 m (100 fathoms) intervals. The red line uses values for the Atlantic and Pacific Oceans in a weighted mean, with weights proportional to the area of the two oceans. The blue line applies a correction to the weighted mean (see Roemmich et al., 2012, for further explanation). These results become a direct test of estimates of how much heat the upper ocean has gained during the latter half of the Industrial Revolution. Error bars are approximately one standard deviation, showing the marginality of the change at any particular depth, albeit with systematic sign. Unfortunately, no data exist permitting a similar estimate for the ocean below these depths or of salinity change at any depth.

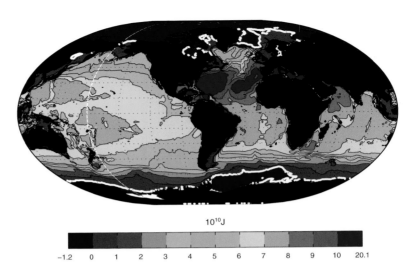

10^{10}J

| −1.2 | 0 | 1 | 2 | 3 | 4 | 5 | 6 | 7 | 8 | 9 | 10 | 20.1 |

Figure 12.8a: Estimated average top-to-bottom heat content of the oceans in the global ocean from a state estimate of 1993 to 2011. Any shifts in the ocean circulation, whether forced or free, will redistribute this heat, and unless the observing system is spatially and temporally uniform, apparent changes in the total will result. This average integrates over the time change interval seen in Figure 12.8b. Based on a degree Celsius scale. (From Wunsch and Heimbach, 2014.)

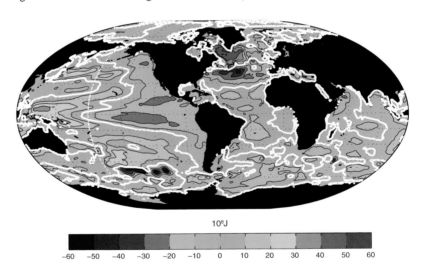

10^{8}J

| −60 | −50 | −40 | −30 | −20 | −10 | 0 | 10 | 20 | 30 | 40 | 50 | 60 |

Figure 12.8b: The estimated change in heat content over the full water column, 2011 minus that in 1993. The complex pattern of heating and cooling represents a major challenge to observation. Compare to (a), noting that the scale of change is of the order of 1% of the range in the time-mean values. A net warming of about 0.2 ± 0.1 W/m^2 is found. The value is the small difference from the sum of large positive and negative numbers integrated over the global ocean and despite the formal error bar from the temporal variability alone, it is unlikely to be statistically significant. Most of the observed structure arises from the internal redistribution of heat, not from air-sea exchange. Spatial scales of change vary over several orders of magnitude even after nearly 20 years of integration.

skepticism about such claims when they are converted into equivalent mean-temperature shifts (and see Figs. 3.25 and 3.26).

To reinforce the sampling issue, Figure 12.9 shows the temperature anomaly on one day during the year 2000 at both 100 m and 2000 m. As expected, and made obvious in animated temporal versions of these figures, multiple dynamical and kinematical regimes exist in the ocean, including an intense small-scale structure that persists even to the seafloor. Direct determination of the average temperature of the oceans to useful accuracies is an extremely challenging observational problem.

12.2.2 In Situ Measures of Circulation

Multiyear current meter records are extremely rare. Müller and Siedler (1992) reported a 9-year long current meter record from the deep eastern Atlantic with no statistically significant mean flow. Records there now nominally extend past 20 years (Siedler et al., 2005) but are broken in various ways, rendering them difficult to interpret.

In the same area, Fraile-Nuez et al. (2010) described a 9-year-long set of current meter records from passages in the Canary Islands (Figs. 12.10a and 12.10b). Even at one standard deviation, *none* of them shows any indication of a trend.

At the present time, beyond the inferences that can be made from more than 20 years of altimeter data, the lateral differences among a few isolated tide-gauge records and from the Argo-*Challenger* results described above, no evidence exists for significant large-scale transport changes in the ocean. Changes are seen to occur everywhere on all time scales, but no bulk multi-decade or multi-century changes in mean dynamic topography, the thermal wind balance, or property transports have taken place beyond the ordinary fluctuations of space-time red-noise processes.

12.2.3 Global Sea Level Change and Heat Uptake

Determining and understanding "sea level" change brings together many of the most interesting and important issues in physical oceanography as they relate to climate change, exploits many of the analysis tools, and is simultaneously a subject of intense public interest. Quotation marks are used in the previous sentence because in common usage the term is vague, at least insofar as the modifier "level" is concerned. For this book, it means specifically the dynamical quantity η appearing in the equations of motion. This definition makes explicit that sea level is simultaneously a dynamical variable, a boundary condition, and a quantity that runs the physical gamut from capillary waves through tides and tsunamis to low-frequency baroclinic motions to geological time-scale shifts on a global basis. In the scientific context, distinction must be made between values of η measured relative to the center of the Earth and tide-gauge measurements relative to the moving seafloor—the latter being what society perceives.

Because of the very long time scales appearing in the sea level problem, a bit of context, more properly the subject of paleoceanography, is useful. Figure 7.21 displays an estimate of *global* sea level change reconstructed from geological sources over the last 450,000 years. The very largest changes are well over 100 m and can only have arisen from the formation and melting of major continental glaciers. One interpretation, germane to what follows, is that global sea level rose rapidly beginning about 15,000 years ago, slowed down about 10,000 years ago, and that we are living through a period representing the tail end of the deglacial process.

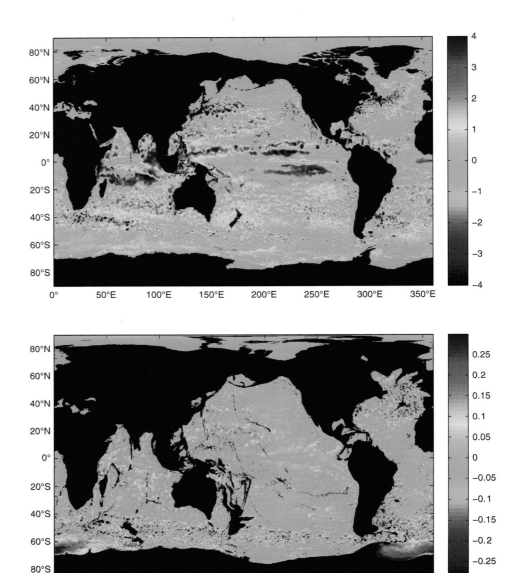

Figure 12.9: Estimated anomaly of temperature, °C, relative to a 20-year average at 100 m (upper panel) and at 2000 m (lower panel). The state estimate is for 27 June 2000. Note particularly the intense small-scale structures at depth. Animations, available online, show that all these structures have an intense temporal variability. Sampling such a fluid to obtain extremely accurate global averages is challenging.

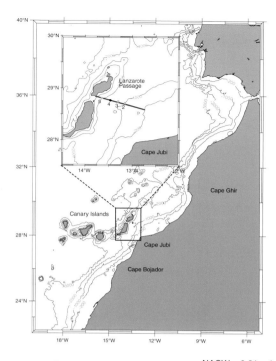

Figure 12.10a: The position of a 9-year records of velocity in the Canary Islands chain used to compute the transport values in Figure 12.10b. (From Fraile-Neuz et al., 2010.)

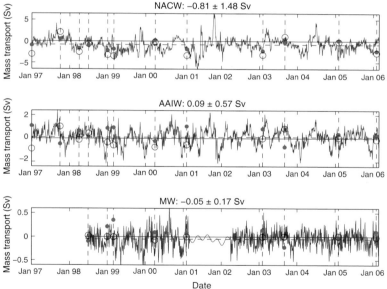

Figure 12.10b: Nine-year estimates of transports of various North Atlantic water masses. From moorings in the passage shown in Figure 12.10(a). None of them shows any statistically significant change. Red dots show a comparison with a geostrophic, level-of-no-horizontal-motion estimate. NACW, North Atlantic Central Water; AAIW, Antarctic Intermediate Water; MW, Mediterannean Water.(From Fraile-Nuez et al., 2010.)

The addition of approximately 130 m of freshwater over the past 15,000 years ($\approx$ 1 cm/year $\approx$ 0.1 Sv) is a major cumulative event, one taking place on a time scale strongly overlapping many of the processes in Figure 11.15 and one with numerous dynamical and kinematical consequences. The geological literature commonly refers to *eustatic* sea level change, meaning a hypothetical globally uniform shift. As will be seen, however, that concept is at best an approximation whose accuracy is dependent upon both the magnitude and time interval of the change (see Figs. 12.8a and 12.8b for the thermal contribution).

MODERN CHANGE

Estimates of recent global mean sea level rise (the altimeter period) hover around 3 mm/year and are frequently quoted with values implying precisions of a few hundredths of a millimeter/year.[3] These values come primarily from altimetric records, usually starting in 1992 and continuing to the present—an elapsed time interval usefully compared to Figure 11.14 or Figure 7.21. The actual values, the precision with which they are known, the physics governing the changes, and possible future values, are all of societal interest. Scientific interest involves maintaining an extremely complex quasi-global observation system and coping with some of the most difficult of all modelling problems.

A bit of the flavor of the challenges can be obtained by simply noting that the ocean has a mean depth of about 3800 m and a change in its volume owing to sea-level shifts of 1 mm/year implies ocean volumetric accuracies of about 1 part in 10^7 in any given year. Determining whether such accuracies, or the precision in a change estimate, for the global mean circulation, are possible with realistic existing, observation and modelling systems is challenging. (See the listing on p. 28 of the corrections necessary for using altimeter data, and Fu and Haines, 2013.) The numerical modelling problem has been described by Losch et al. (2004) and Griffies and Greatbatch (2012).

THE GLOBAL MEAN

A limited number of physical processes can produce time-dependent shifts in $\langle \eta \rangle_o$, the global spatial mean. These include *changes* in

1. Net heating and cooling
2. Total input or removal of freshwater
3. The physical volume of the oceans

Heating and cooling are dominated by enthalpy exchanges with the atmosphere but also include direct radiational inputs. Issues of global mean heating are today central to the debates over atmospheric global warming through greenhouse-gas increases. Effects conventionally thought to be small, such as the temperature difference between the ocean and melting ice (originating as either floating sea ice and land-supported glaciers), need to be evaluated if accuracies below 1 mm/year are sought. A 1 mm/year change in $\langle \eta \rangle_o$ requires a mean ocean temperature change of about 0.002°C/year. (See p. 369.)

The addition and subtraction of freshwater involves melting glacial ice, runoff from rivers and elsewhere, subsurface percolation from continents, and fluctuations in atmospheric storage. Salt input and output is, on modern time scales, probably confined to exchanges

---

[3] For example, see Alverson, 2012, Figure 1; or King et al., 2012.

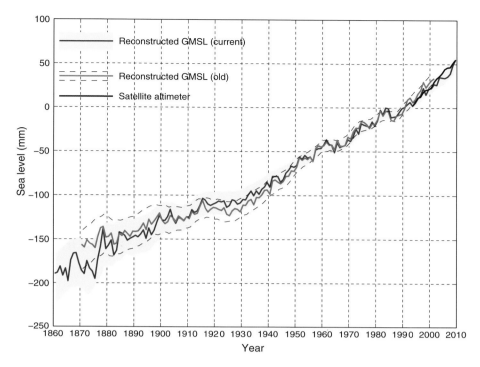

Figure 12.11: Two estimates of global mean sea level (GMSL) from satellite altimetry (from 1992 onward) and tide gauges (see Fig. 12.15). That a visible change in slope occurs with the advent of the altimeter has raised caution flags discussed elsewhere. (From Church and White, 2011.)

with sea ice. If the sea ice is not completely fresh, its melting puts salt back into the fluid in solution, where it is no longer supported by Archimidean forces, and hence changes the sea level. The melting of completely fresh sea ice changes the ocean salinity, but not the sea level.

Both (1) and (2) vary on many different time scales. Although net precipitation and evaporation over the global ocean are believed to (nearly) balance when integrated over a year (Minster et al., 1999), they tend to be out of balance over the course of the seasonal cycle. Table 3.1 provides some basic numbers. Of particular importance is that the numerical values of estimated *increases* in Greenland and Antarctic ice melt are dwarfed by the background runoff and estimated over-ocean precipitation. Should these background values shift by 10% (an arbitrary but not unreasonable reference value), the change would greatly exceed the increased ice-melt numbers. Caution is advised in the attribution of very small observed changes, and the treacherous nature of random-walk processes from cumulative change is worth remembering. An estimate of the global mean is shown in Figure 12.11, with the data coming from tide gauges prior to 1992.

REGIONAL CHANGES

Sea level responds to the three global effects and also to their regional variations. Localized changes are important even for the global mean, because they set the noise level of the spatial averaging. Omitting the tides, several additional factors, not changing the global mean to a first approximation, become very important:

4. The large-scale wind field generates and supports a range of about 3 *meters* of variation in the spatial structure of the surface ocean elevation (Fig. 3.29), which is dominated by quasi-permanent gyres and large-scale currents. Thus changes in the wind field, which take place on all known time and space scales, shift these features and will strongly influence regional sea level at the millimeter-per-year rate. Subtropical gyres tend to be warmer in the upper ocean than are subpolar ones, and so some fraction of the total variation is thermal in origin, but the thermal properties of the abyss must also be accounted for. Much (not all) of the heat content changes visible in Figures 12.8a and 12.8b correspond to regional sea level changes and are wind induced.

5. Regional variations in atmospheric pressure change the load on the sea surface and over large scales and long times are expected to follow the inverted barometer response of 1 cm of water/100 Pa (1 millibar = 1 hectopascal) of air pressure. Figure 3.36 suggests that long-term shifts of tens of centimeters will occur when atmospheric circulation patterns change.

6. Gravity-field changes. Glacial isostatic adjustment (GIA), or post-glacial rebound, generates gravity-field shifts from the moving mantle below the continental plates. Land ice has a nonneglible gravitational attraction that disturbs the geoid, and thus melting or formation of glacial ice changes the geoid essentially instantaneously (Conrad and Hager, 1997; Mitrovica et al., 2001). GIA takes place on time scales of thousands of years, but its signature is visible with modern instruments, for example, the GRACE satellite, and through vertical land motions in some places on the order of 1 m in 100 years. The breakup of *floating* ice sheets can be very fast (days to weeks) but has no first-order gravity-field influence.

THE COMBINED PROBLEM

A globally uniform, or eustatic, component is a minor, perhaps even nonexistent, element in the modern decadal records. As an equilibrium response it involves a global adjustment of the ocean elevation under a complex interplay of dynamical processes taking, at a minimum, decades to achieve (see, e.g., Stammer, 2008; Cessi et al., 2004). In the interim, the system is subject to further disturbances, and given the time scales in Figure 11.15, an equilibrium change is probably never obtained.

Exceptions to that long time scale do exist: abrupt changes in volume, such as in an earthquake, produce global equilibrium within days through tsunami propagation. *Bottom pressure* changes from loading owing to removal and addition of freshwater will also occur on barotropic time scales of hours to days—and will immediately influence measurements of the Earth's polar motion and rotation measurements.

The global average addition and removal offreshwater or expansion and contraction by heating and cooling are convenient summary numbers. They gloss over the far larger regional contributions that form the global means.

MEASUREMENTS AND ATTRIBUTION

The relatively high accuracy TOPEX-POSEIDON-Jason series of spacecraft reached nominal latitudes of about 63°, omitting the entire Arctic Sea and a band around the coast of Antarctica. The somewhat less accurate ERS-1, ENVISAT, and other altimeters, which were generally calibrated by comparison to TOPEX-POSEIDON-Jason, did reach higher latitudes, although with a much less frequent temporal sampling interval (typically every 35 days). No altimeter works properly as a water-level gauge in the presence of sea ice; thus even the best of them has a *seasonal bias*, omitting measurements whenever ice is present in the altimeter "footprint."

Many of the corrections made to the altimetric data (p. 28) are themselves susceptible to significant drifts. The attribution problem—determining the contributions to sea level change—encounters several issues. The Argo array, which has revolutionized upper-ocean global-scale measurements, did not become quasi-global until about 2004. It measures only the upper ocean, now above 2000 m. Before Argo, fragmentary hydrographic and XBT (expendable bathythermograph) measurements were heavily biassed toward the northern hemisphere, shipping lanes, and fair weather; the few available predecessor (ALACE) floats somewhat less so. Shifts in the ice sheet volume have been known only in very recent years and remain very uncertain (Shepherd et al., 2011; King et al., 2012).

Much attention has been directed to budgeting the global mean values. The total ocean mass is

$$M_O = \int_{-\bar{h}(x,y)}^{\bar{\eta}(x,y)} \rho(x, y, z)\, dz dA = (\eta + h)\, \bar{\rho} A,$$

Where A is the oceanic area (ignoring its dependence upon depth; see Fig. D.1). If the density changes by heating, the total mass is unchanged (to a good approximation) and hence,

$$\Delta M_0 = 0 = \left[\Delta\bar{\rho}\left(\bar{\eta} + \bar{h}\right) + \bar{\rho}\Delta\eta\right] A \approx \left[\Delta\bar{\rho}h + \bar{\rho}\Delta\eta\right] A,$$

and

$$\frac{\Delta\eta}{\bar{h}} = -\frac{\Delta\bar{\rho}}{\bar{\rho}}.$$

For the observed small perturbations (Eq. 3.7),

$$\Delta\bar{\rho} = \bar{\rho}\left(-\alpha\Delta\bar{T} + \beta_S\Delta\bar{S}\right),$$

or

$$\frac{\Delta\eta}{\bar{h}} = \alpha\Delta\bar{T}.$$

Using $\alpha = 10^{-4}/°C$ in the middle range of values listed in Appendix D, a 1 mm change in η requires that

$$\Delta\bar{T} = \frac{10^{-3}\mathrm{m}}{10^{-4} \times 3800\mathrm{m}} = 0.0026°, \quad \text{or} \frac{\Delta\bar{T}}{\bar{T}} = 7 \times 10^{-4}. \tag{12.1}$$

In 20 years, if the change is 3 mm/year for a total of 60 mm of water, the implied mean temperature change is still only $0.16°C$.[4] Determination to that precision in the face of the wide temperature distribution in Figure 3.25, the structures in Figure 12.9 and the historically sparse measurements, particularly of the huge mass of abyssal water near the mean, is problematic.[5]

A parallel development in salinity, with $\beta_S = 0.8 \times 10^{-4}$, produces

$$\Delta\bar{S} = -\frac{10^{-3}\mathrm{m}}{\beta_S \times h} = 0.003, \quad \frac{\Delta\bar{S}}{\bar{S}} = 8.6 \times 10^{-5}, \tag{12.2}$$

as the mean salinity change. Given the even sparser salinity measurements, with their own changing instrument biases, attempts to calculate decadal variations with existing data at

---

[4]This number is crude, because the temperature dependence of the coefficient of thermal expansion implies a strong dependence upon where the heating takes place.

[5]Even in the much better sampled upper ocean, measurements continue to be plagued with systematic errors (e.g., DiNezio and Goni, 2011; Abraham et al., 2013) in addition to the high noise levels.

accuracies consistent with putative sea level change seem quixotic. Salinity change attribution can also be attempted from examination of the freshwater input to the oceans (Table 3.1). Apparent observed increases in freshwater input remain minute compared to the inferred interannual and interdecadal variability of the natural background, including over-ocean precipitation changes. Even though floating freshwater ice does not change sea level when melted, it does change the oceanic salinity (Munk, 2002).

Post-glacial rebound (GIA) as well as associated tectonic adjustments have been estimated (e.g., by Peltier, 1994) to be increasing the ocean volume by about 0.3 mm/year. The accuracy of this estimate is not easy to determine.[6]

Another complication at the level of accuracies being sought involves compressibility effects in the equation-of-state of sea water. Based purely upon theory (no direct observations exist), estimates are that as much as 6 mm/year of equivalent sea level change today is suppressed by compression of the water column as fluid is advected and mixed; (McDougall and Garrett, 1992; Schanze et al., 2013; also see Gille, 2004). Piecuch and Ponte (2013) discussed the potential for changes in that number as the circulation shifts through time.

Hamlington et al. (2011) and many other group wrestled with the problem of the signal-to-noise ratio in records that, in normal science, would be regarded as of too short a duration to address the problem being considered.

ALTIMETRIC-ERA TRENDS

Figures 12.12 and 12.13 display estimates of the *regional* trends obtained from the altimeter era, one using altimetry alone, the other from a state estimate combining the altimetry with many other data types (Table B.1). The strong regional dependence is clear, including regions (over this time interval) of falling sea level.

ALTIMETRIC VARIABILITY

Signals, typically on the order of centimeters per year, can be used to infer changes in the geostrophic flow. The regional patterns of change, both positive and negative, defy simple description beyond the fact of their spatial complexity. Altimetric measurements of η directly define the surface flow, but the dynamics show that η is a reflection of structures deep within the water column, sometimes all the way to the seafloor.

Figure 9.14 is a 19-year average of the surface elevation η from a global state estimate but it is confined for pictorial clarity to the Atlantic: Figure 12.14 is the difference in the annual mean of η over a 16-year interval from that estimate. The shifts seen are complex, particularly at high latitudes, but are generally small relative to the background time average. Expansion of other regions in the global trend chart (Fig. 12.13) shows an equivalent level of spatial structure—reflecting the intricate relationship between η and the oceanic circulation. Regional trends cannot persist indefinitely unless the wind stress or another force exists to maintain the ever-increasing pressure gradients. Prediction of regional sea level change inevitably becomes in large part that of prediction of the wind system.

---

[6] See Mitrovica et al., 2005, for discussion of the problem of coupling measurements of glacial isostatic adjustment, sea level, and Earth rotation.

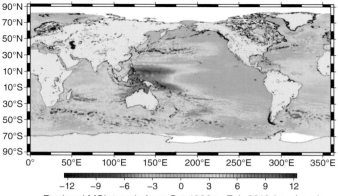

Regional MSL trends from Oct 1992 to Feb 2012 (mm/year)

Figure 12.12: Trends, in millimeters per year, of a determination from temporal mean sea level (MSL) at a (nominal) 1/3° resolution from altimetric data alone—but with strong assumptions about the space-time covariance model. While some large-scale patterns are apparent, including significant areas of *fall*, the result is notable for its noisiness. Obtaining an accurate global average from such a result is an involved process. Note such troubling regions as that surrounding the Gulf Stream, which would suggest development of a strong cyclonic recirculation pattern for which there is no independent evidence (is mass or volume conserved overall?). Compare to Figure 12.13 which uses dynamics and many additional data types (From AVISO website.)

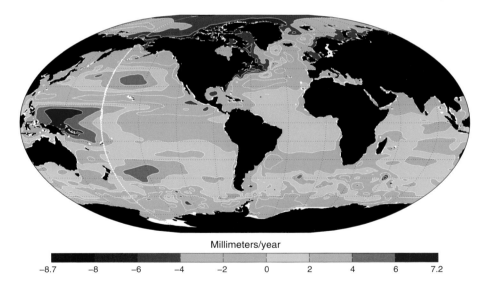

Millimeters/year

Figure 12.13: The trend in sea level from an ECCO state estimate in millimeters per year. This result combines the altimetric data with hydrography and many other data types, including atmospheric forcing data, along with the physics of the model. The global average value is a small residual of this complex spatial pattern. Volume is conserved by construction but accounts for fresh water exchange. Compare to Figures 12.12, 12.8a, and 12.8b. Use of data in addition to the altimetry reduces the influence of the expected measurement errors in the latter.

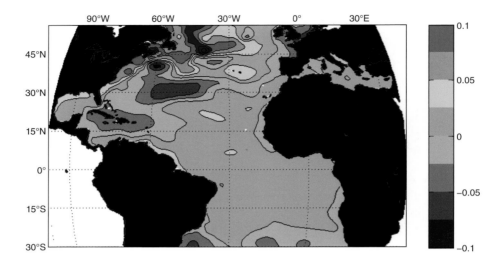

Figure 12.14: The difference in meters of the annual mean surface elevation in years 17 (2009) and 3 (1994) corresponding to Figure 9.14. Changes are of the order of 5% of the time-average value and can thus be regarded as perturbations.

Altimetric and Argo profile measurements are the only existing near-global data representing in any direct way the interior-dynamical-flow regime. On time scales of days to a year or two, analysis of current meter moorings shows a very strong response in the nominal barotropic and first baroclinic modes, although higher modes are strongly suppressed (Eq. 4.89). At very short periods (a few days), motions are strongly barotropic, particularly at high latitudes, where Rossby wave theory means that little or no baroclinic adjustment is possible.

Unfortunately, almost nothing is known directly and generally about the vertical structure of oceanic adjustments on time scales beyond about one year. Theory (e.g., Anderson et al., 1979) suggests a spatially complex summation of vertical structures asymptoting slowly toward a new equilibrium.

12.2.4 The Tide-Gauge Era and Multidecadal Sea Level Changes

The longest oceanic time-series records available have been obtained from tide gauges, and despite their extreme geographical confinement, it is worth exploring some of the implications of these data. Figure 12.15 shows the locations with nominally continuous records exceeding 100 years. The Honolulu, Hawaii, record is displayed in Figure 12.16 and is typical of open-ocean records once the tides have been removed and with the residual averaged by month. Difficulties of calibration were touched on in Chapter 2, and always raise certain questions: (1) Was the setting, including tectonics, harbor works, position, etc., stable? (2) Did the changing technology introduce spurious shifts and trends? (3) Was the calibration of uniformly high quality over the record length?

With spatially sparse data, the question of the regional representativeness of the records becomes paramount. In some extreme cases in paleoclimate, it is simply assumed that one

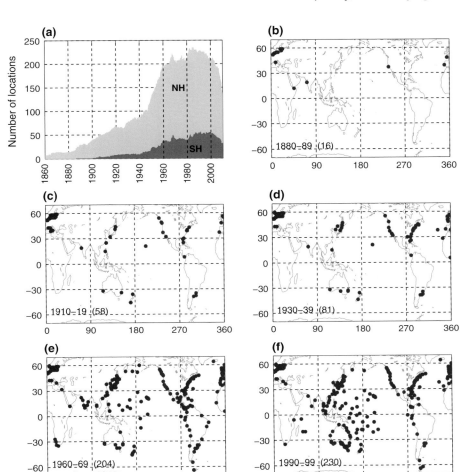

Figure 12.15: Tide-gauge distribution used by Church and White (2006) to discuss the multide-cadal sea level rise and its possible acceleration (see Fig. 12.11). The difficulties of finding physical state trends in the presence of the greatly varying spatial sampling properties will be apparent. Comparison of these charts to Figures 3.31, 12.8a or 12.8b illuminates the sampling issue.

location is representative of an entire hemisphere or even the entire globe. A usually unstated postulate is that the lower the frequency of any variability, the larger the spatial scale on which it must be occurring. The assumption is reasonable, but counterexamples are not difficult to find, ones often tied to proximity to permanent geographical features. Thus a record of sea level made adjacent to the Antarctic ice sheet might plausibly show variability very different from a record in the equatorial Indian Ocean, even on time scales of thousands of years and longer. At what time scale (or, even if) two such records become coherent is an open question.

Notice the strong decadal fluctuations in Figure 12.16. If an overall linear trend is removed, the spectrum appears to become white noise in character on time scales of about 3 years and longer and is almost certainly dominated by variations in wind curl.[7]

---

[7] Sturges et al., 1998; Wunsch et al., 2007.

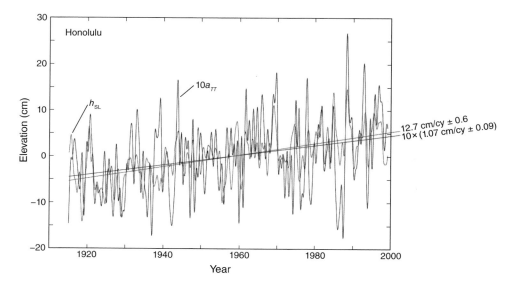

Figure 12.16: The monthly mean sea level (h_SL) record at Honolulu, Hawaii, after removal of the tidal contributions and correction for tectonic effects. Decadal-scale variations are visible. As regional changes, they are much larger than any plausible global average value. The curve labelled $10a_{TT}$ is the amplitude of the M_2 tide (multiplied by 10) and shows the apparent and surprising correlation between the "mean" sea level and the tidal range. The rise over 100 years (cy) is 12.7 cm, and 1.07 is the corresponding increase in M_2 amplitude. (From Colosi and Munk, 2006.)

Figure 12.15 shows the distribution of tide-gauge records used by Church and White (2006) to calculate the global mean sea level rise extending back to 1870 and should be compared with the spatial complexity of sea level rise during the altimetric era shown in Figures 12.12 and 12.13. A common approach to extrapolating undersampled fields for obtaining global estimates is to employ the singular vectors (EOFs) as computed from comparatively dense and widespread modern coverage, and to assume that they are appropriate in the thinly observed past. Thus the EOFs of the recent altimetric coverage are then fit to the sparse coverage by tide gauges as shown, for example, in Figure 12.15. Amplitudes of a few dominant singular vectors can be determined from a relative handful of the tide-gauge records extending into the past. The procedure is a good one if the EOFs are stable; but they are sensitive to the empirical covariance matrices derived from short records used to calculate them. Also the physics, and hence the singular vectors, are assumed to remain fixed as durations are extended—with no new physics entering the global covariances at longer periods—a surprising inference given Figure 11.15.

SURFACE GRADIENTS

Regional features in η and their changes are represented in the dynamical equations of motion. For that reason, the sea level change problem is inextricably one of ocean dynamics and has led to the determination of η using the state estimation machinery. Before global data sets became available, temporal changes in sea level slopes computed from long tide-gauge records

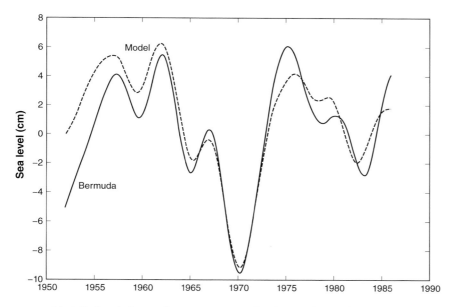

Figure 12.17: Modelled and observed variations in sea level (note the small range) at Bermuda. The model is a first-baroclinic-mode Rossby wave excited by wind forcing along the entire latitude line. Higher modes would not be visible in the tide gauge record. (From Sturges and Hong, 1995.)

were used in a number of studies, albeit all are highly regional with little global generalization possible. An early study by Groves and Zetler (1964) of the difference between Honolulu and San Francisco showed a mildly red continuum difference spectrum over a duration of about 50 years, with a small, but apparently significant, coherence between the two records. Such differences and relationships must reflect what is now known of the presence of a variety of physical phenomena, including geostrophic eddies, wind-field variability (the Pacific Decadal Oscillation?), boundary currents, ENSO cycles, and the like. The elevation difference does provide an upper bound on the extent to which the surface geostrophic flow might have changed over that period.

Sturges and Hong (1995) showed that they could rationalize much of the multidecadal variability as seen in the Bermuda tide gauge by invoking wind forcing of a first-mode baroclinic Rossby wave (see Fig. 12.17). The dynamics (Eq. 4.89), which suppress the signals from higher modes at the free surface, simplifies interpretation of the observations. Higher modes are likely very important, as would be anticipated from theoretical study of forced motions at these periods, but their detection with tide gauges requires much longer records than are now available.

Nowhere over the entire historical written record has any change in sea level slope been detected depicting a qualitative shift in oceanic circulation. Apart from the general stochastic red noise, the slight warming of the last century or so, and regional episodes in the meteorology such as the Little Ice Age (circa 1600–1850), the most plausible working hypothesis is that the ocean circulation has been in a statistically steady state throughout the instrumental period beginning in the seventeenth century.

12.3 SEA-SURFACE TEMPERATURE

Measurements of sea-surface temperatures (SST) began almost as soon as the thermometer was invented. Because of its use as a controlling lower boundary condition on atmospheric weather-forecast models, it has for many decades been the subject of intense scrutiny (see Chapter 2) and of attempts to determine its changing character over many different time scales. The term itself is somewhat ambiguous, as the temperature at the sea surface itself can, under some circumstances, differ significantly from that even a few centimeters below (the skin-temperature effect).

Because the very upper layers of the ocean are so strongly influenced by the atmospheric state (and in turn influence that state), the very indirect coupling between SST and the deeper ocean circulation, and the apparent strong influence of anomalies on day-to-day weather (particularly tropical SST; see, e.g., Nobre and Shukla, 1996), it is perhaps best regarded as primarily a meteorological variable rather than an oceanographic one. Most inferences about the longer-term behavior of SST necessarily come from models (e.g., Deser et al., 2003). Models, again, have been used to attempt inferences about the relationship between SST anomaly patterns and fluctuations in horizontal oceanic heat transports. Records are, however, so short that the inferences remain generally untestable against observations.

A Brief Afterword

Any reader who has made it this far will have recognized what was stated in the Introduction—many major topics have been omitted, including equatorial physics; the special oceanography of the Southern Ocean; details of the near-surface and near-bottom boundary layers; the circulation of the shallow, adjacent seas; the physics of the overflow plumes; bottom water formation; the influence of the waxing and waning of sea ice; and the transfers to and storage within the ocean of carbon. Even this long list is incomplete. At best, this book represents an exposition of some of the most interesting aspects of the modern ocean circulation and its variability.

My intention, originally, had been to include a chapter on physical paleoceanography, a large and fascinating subject, full of controversy. The ballooning size of the book to cover only the modern ocean, and the ticking of the clock, have precluded anything more than passing mention of paleoceanographic observations and inference. A subsequent volume may one day emerge. For anyone seeking entree into the field, the review by Huybers and Wunsch (2010) should be of help.

A Primer of Analysis Methods

Many methods are used for analyzing and making sense out of the noisy, spatially and temporally complex data obtained by oceanographers. The reader of this book is assumed to know what a spectral estimate is, how to interpret it, what is involved in objective mapping, and the meaning of "degrees-of-freedom," among other fundamental concepts. Some of the more commonly used methods are sketched here—to at least define the vocabulary. Many good books on these subjects written by true experts are available at all levels of sophistication. In practice, understanding the analysis of real data—as with learning to drive a car or learning to swim—comes only with the experience, not from words in a book.

A.1 EXPECTATION AND PROBABILITY

The proper definitions for probability densities and distributions and their interpretation is a large and interesting subject discussed in depth in many books. Here, only some notations and definitions are written. Let $p_q(q')$ be a probability density for a physical variable q taking on possible numerical values q'. Heuristically, it can be thought of as constituting a normalized histogram built up asymptotically from a large number of perfect measurements of the variable, and with numerical values distinguishable at an infinitesimal level. Then in the limit, from the histogram assumption, $p \geq 0$, and

$$\int_{-\infty}^{\infty} p_q(q')\, dq' = 1,$$

defining the normalization. The theoretical average, or mean, of q is defined,

$$\langle q \rangle = \int_{-\infty}^{\infty} q'\, p_q(q')\, dq', \tag{A.1}$$

and the average of any function, $f(q)$, is

$$\langle f(q) \rangle = \int_{-\infty}^{\infty} f(q')\, p_q(q')\, dq'. \tag{A.2}$$

The bracket operation is linear and distributive, $\langle ax + y \rangle = a \langle x \rangle + \langle y \rangle$, where x, y are random variables and a is a constant.

For a variety of reasons, the most-used probability density is the Gaussian, or normal,[1] which is written

$$p_q(q') = \frac{1}{\sqrt{2\pi}\sigma} e^{-(q'-m)^2/2\sigma^2}, \tag{A.3}$$

whose mean is

$$\langle q \rangle = \int_{-\infty}^{\infty} q' \frac{1}{\sqrt{2\pi}\sigma} e^{-(q'-m)^2/2\sigma^2} dq' = m,$$

and its *variance*

$$\left\langle (q-m)^2 \right\rangle = \int_{-\infty}^{\infty} (q'-m)^2 p_q(q') dq' = \sigma^2.$$

Textbooks, or the reader, can confirm that the ordinary *sample mean*,

$$\tilde{q} = \frac{1}{M} \sum_{i=1}^{M} q_i, \tag{A.4}$$

will asymptote to $\langle q \rangle$ as M becomes large enough, and where q_i are *perfect* measured values. (Formally, the probability that $|\tilde{q} - m| > \varepsilon$ goes to zero with M, for any finite ε.)[2] Discussions here of concepts such as *ergodicity*, which would assure that a space or time average is asymptotically identical to a theoretical value obtained by integrating over the probability density, are being omitted. The property is commonly taken for granted. Demonstrating convergence of sample means to true population means can become much more complicated if the samples q_i are contaminated by noise—which is usually the case. The average of the difference, $\langle \tilde{q} - m \rangle \neq 0$, is the *bias*.

Textbooks all explain how to use change of variable rules for finding probability densities for functions of random variables. A particularly useful one derived from the Gaussian is that for the behavior of the square of a Gaussian, $X = q^2$, where the mean of q is $m = 0$, and its variance $\sigma^2 = 1$, producing

$$p_\xi(X) = \begin{cases} \frac{1}{\sqrt{2\pi X}} e^{-X/2}, & X \geq 0 \\ 0, & X < 0 \end{cases}, \tag{A.5}$$

called *chi-squared*, and written χ_1^2 with the subscript denoting the square of a single Gaussian variable (it has *one-degree-of-freedom*).

If two or more random variables, $x, y, z \ldots$ are being discussed, their multivariate probability densities can be defined,

$$p_{xy\ldots}(x', y', z', \ldots) \geq 0, \tag{A.6}$$

$$\int_{-\infty}^{\infty} \cdots \int p_{xy\ldots}(x', y', z', \ldots) dx' dy' dz' \ldots = 1.$$

As used here, heuristically, the reader can simply regard joint probability densities as the continuous limiting cases of empirical histograms of jointly occurring values as the number of samples becomes infinitely large.

---

[1] Apart from its comparative analytic tractability, the Central Limit theorems show the tendency in many situations (various assumptions) for sums of many independent non-Gaussian variables to become asymptotically Gaussian as the number of independent elements becomes large. It isn't always true, but often is.

[2] ε should not be confused with its definition elsewhere as the fluid rate of viscous dissipation.

An important theoretical concept is that of a *stationary* process:[3] one in which the probability densities do not depend upon time. In the above, if $x, y, z, \ldots$ are measurements of a variable, perhaps temperature in time or space, and if the variable is stationary, then its mean and variance and so-called cross-moments, such as $\langle xy \rangle$, should all be time independent. The bracket implies integration over the corresponding multivariable probability density. Processes are often assumed to be stationary, as it represents a great simplification of the statistical description.

Two or more variables $x, y, \ldots$ for which $\langle (x - m_x)(y - m_y) \rangle = 0$ are said to be *uncorrelated*, where $m_{x,y}$ are the corresponding means. If zero correlation implies that $p_{xy}(x', y') = p_x(x') p_y(y')$, then the variables are also *independent*. Independence implies zero correlation, but the opposite is not necessarily true (although it is true for the Gaussian). A common, but important, unstated assumption in claiming that $\bar{q}$ is close to the true mean in Equation (A.4), is that the q_i are independent, or that "enough" of them are.

Let $x, y, z, \ldots$ be ν uncorrelated (and thus independent) Gaussian random variables with zero-means and unit variances. Then the probability density of $\xi = x^2 + y^2 + z^2 + \ldots$ is a chi-square (χ_ν^2) variable with ν *degrees-of-freedom*,

$$p_\xi(X) = \frac{1}{2^{\nu/2} \Gamma(\nu/2)} X^{\nu/2-1} \exp(-X/2), \quad X > 0 \qquad (A.7)$$
$$= 0, \quad X \leq 0,$$

whose mean is ν and variance is 2ν. The square root of two such elements, $\sqrt{x^2 + y^2}$, has a *Rayleigh* probability density.

A.2   TIME SERIES ANALYSIS: FOURIER METHODS

Nature separates many natural processes by their time scales. Thus in the ocean, what are called internal waves exist in a finite band of frequencies that is almost everywhere disjoint from that of the phenomenon of Rossby waves. In contrast, these waves have overlapping wavelengths, and no simple means exists to separate them on the basis of their spatial structure alone.[4] Even more generally, the ability to describe motions with a dominant time scale is a powerful depiction of the ocean. Processes occurring on time scales of 1 s (surface gravity waves) have a very different set of physical or chemical properties from those occurring over decades. Fourier methods are the most widely used approach for time-scale separation, and it is worth a digression to describe the major elements of their practical use—which differ in some respects from their description in mathematics or physics courses.[5]

---

[3] The term should not be used for a *steady* (time-independent) system.

[4] Sweeping generalizations like this one are treacherous. The frequency-dependent relationship between two horizontal components of velocity, u, v, would, if known, provide powerful insight into the underlying physics.

[5] Among the many possible references, Körner, 1988, is a stimulating and particularly readable account of conventional Fourier analysis. For practical application to time series, Bracewell, 2000, has a useful treatment of Fourier analysis; Percival and Walden, 1993, is useful for spectral estimation; and Priestley, 1981, is a general broad reference incorporating both mathematical and practical issues. (Percival and Walden do not treat coherence, whereas Priestley does.) Among the older books (pre–fast Fourier transform, FFT), Jenkins and Watts, 1968, is outstanding and still useful for the basic concepts. For time-domain methods, Box et al., 2008, is a standard reference. A more complete discussion can be found at dash.harvard.edu (C. Wunsch, *Time Series Analysis: A Heuristic Primer*).

Although the point will not be constantly reiterated, "time," t, is an index or coordinate that can be replaced by a coordinate, r, when spatial-scale discrimination is sought. Most methods touched on here have generalizations into two or more dimensions—an important feature in dealing with combined space-time variability.

Anyone using supposedly more "advanced" methods, such as maximum entropy or wavelets, is urged to first master these more basic ideas. It is not easy to identify any example of importance in oceanography that emerged from the use of more complex approaches—one that was not more readily determined by the classical methods.

A.2.1 Process Types

Many of the descriptions of various signal types originate from radio engineering, where the need for distinguishing them first arose. *Deterministic* processes are those whose value can be notionally predicted perfectly at any time, t, in the past or future. Examples are a straight line, $y = at + b$; exponentials, $\exp(-\alpha t)$; and periodic ones such as the sines and cosines or their complex counterparts, $A \cos(2\pi t/T - \phi_1)$, $B \sin(2\pi t/T - \phi_2)$, $\exp(i(2\pi t - \phi_3))$, etc. Psychologists and evolutionary biologists have written about the human need for conceiving of the world as understandable and predictable, and judging from much of the literature, many scientists want desperately to believe that the world is periodic or otherwise determined in advance. Apart from purely astronomical phenomena, including tidal and solar insolation forcing, almost nothing in nature is truly deterministic, much less periodic. Commonly the deviations from determinism are the only element of real interest. On geological time scales, even the astronomical phenomena are not periodic, but do remain deterministic for tens of millions of years., and of course, the concept of time extending to infinity is only a convenient mathematical fiction.

Stochastic, or *random*, processes are those for which some or all of the values can be predicted only in a probabilistic sense. *White noise*, described below, is by definition completely unpredictable, except for its time-mean moments, and becomes a building block for more interesting time series that are partially predictable. Special cases of random processes include those that are *narrowband*, meaning that their energy is largely confined to a small region in frequency space; *broadband*, meaning that their energy occupies a wide portion of frequency space; *red noise*, meaning that most of the energy is at low frequencies; and others that will be defined later as needed. Terms such as "narrow," "broad," and "low-frequency" are relative and context dependent.

A.3 BASIC FOURIER REPRESENTATIONS

Suppose $x(t)$ is periodic with period T:

$$x(t) = x(t + T).$$ (A.8)

The complex Fourier *coefficients* are

$$\alpha_n = \frac{1}{T} \int_0^T x(t) \exp\left(\frac{-2\pi i n t}{T}\right) dt.$$ (A.9)

Under very general conditions, $x(t)$ can be written as a Fourier series:

$$x(t) = \sum_{n=-\infty}^{\infty} \alpha_n \exp\left(\frac{2\pi i n t}{T}\right).$$ (A.10)

The function is being represented with a discrete set of circular frequencies, n/T, each a harmonic of the period, T. (Usually, $x(t)$ will be real-valued, but it need not be.)

$x(t)$ can also be written as a Fourier cosine and sine series without the use of complex notation:

$$x(t) = \frac{a_0}{2} + \sum_{n=1}^{\infty} a_n \cos\left(\frac{2\pi nt}{T}\right) + \sum_{n=1}^{\infty} b_n \sin\left(\frac{2\pi nt}{T}\right), \tag{A.11}$$

with

$$a_n = \frac{2}{T} \int_0^T x(t) \cos\left(\frac{2\pi nt}{T}\right) dt, \quad b_n = \frac{2}{T} \int_0^T x(t) \sin\left(\frac{2\pi nt}{T}\right) dt, \tag{A.12}$$

so that $\alpha_n = (a_n + ib_n)/2$. Except at special times when, for example, $x(t)$ takes a jump, the representations in Equations (A.10) and (A.11) are exact.[6]

We now make a seemingly trivial, but extremely important observation: The Fourier series representation, Eq. (A.10) or (A.11) can be used to represent any *nonpathological function,* exactly, over any interval of length T: the function need not be periodic outside that interval and, indeed, need not be specified there at all. A Fourier series in a finite interval can often be extremely useful in understanding the physics within that interval without specification of what happens otherwise. Use of a Fourier series does *not imply periodic data.*

The Parseval theorem for Fourier series is

$$\frac{1}{T} \int_0^T x(t)^2 \, dt = \sum_{n=0}^{\infty} \left(|a_n|^2 + |b_n|^2\right) = \sum_{n=-\infty}^{\infty} |\alpha_n|^2. \tag{A.13}$$

By convention, the left-hand side is the "power" in $x(t)$, and the right-hand side shows that it is made up of all of the Fourier coefficients.

Some authors use a Fourier transform (FT), which taken over the finite interval $0 \le t \le T$, would produce a pair:

$$\hat{x}(s) = \int_0^T x(t) \exp(-2\pi i st) \, dt, \tag{A.14}$$

$$x(t) = \int_{-\infty}^{\infty} \hat{x}(s) \exp(2\pi i st) \, ds. \tag{A.15}$$

As with the Fourier series, excluding jumps, the Fourier transform representation is exact, whatever the behavior outside the interval being used. The operation, $\mathcal{F}(\cdot)$ is defined as the action of taking a Fourier transform,

$$\mathcal{F}[x(q)] = \hat{x}(p), \tag{A.16}$$

and its inverse,

$$\mathcal{F}^{-1}[\hat{x}(p)] = x(q), \tag{A.17}$$

where p, q are arbitrary independent variables. Unlike the Fourier series frequencies, $s_k = k/T$, the Fourier transform exists at *all* frequencies, including those for the Fourier series coefficients, but also all of them in between. Notice that the Fourier series coefficients at the Fourier series frequencies, $s_k = k/T$, are

$$\alpha(s_k) = \frac{1}{T}\hat{x}(s_k). \tag{A.18}$$

---

[6] In mathematics, the type of functions for which Equation (A.10) is valid would be carefully defined. Here the reader is assumed to know enough to exclude functions, such as $\sin(1/t)$, that are not encountered in nature.

Thus, the Fourier transform values,

$$\hat{x}(s_k) = \frac{1}{1/T}\alpha(s_k), \tag{A.19}$$

can be thought of as an amplitude *density*, as $1/T$ is the frequency separation of the Fourier series values. The Parseval theorem is readily modified for the Fourier transform.

It might appear that many more values $\hat{x}(s)$ are required—at s_k plus all of the values in between—to make a Fourier transform than does a Fourier series. The Fourier coefficients are known from the Fourier transform at $s = s_k$, but in any finite interval, the *information* represented by the Fourier series coefficients contains everything in the original series. It should come as no particular surprise that $\hat{x}(s)$ can be reconstructed *exactly* at any frequency from the sparser $\alpha(s_k)$. *No more information content exists in the Fourier transform* than there is in the Fourier series coefficients. The representations can be interchanged for convenience without any loss of information.

THE FAST FOURIER TRANSFORM

A major event in numerical analysis during the twentieth century was the development of the *fast Fourier transform*. Calculation of the Fourier coefficients involves a number of additions and multiplications that increases as M^2—where M is the number of data points. In the mid-1960s, it was recognized that during numerical Fourier calculations, many of these operations were being performed multiple times. Cooley and Tukey (1965) showed that the number of computations could be reduced to $M\log_2 M$, if $M = 2^p$, where p is an integer, a much smaller value than M^2. Almost immediately their algorithm was generalized to cope with values M that were products of powers of small prime numbers, p_i, so that $M = 2^{p_1}3^{p_2}5^{p_3}\dots$ Fourier transform calculations were revolutionized. The original 2^p algorithm is not difficult to understand, but the reader is referred to numerical analysis textbooks for its discussion.

DELTA FUNCTIONS

A number of special functions have useful known Fourier transforms (e.g., the Heaviside, or unit, step function). Special attention is given here only to the Dirac-delta function, $\delta(t - t_0)$, which Bracewell (2000) exploited heavily and usefully. By definition,

$$f(t) = \int_{-\infty}^{\infty} f(t')\,\delta(t - t')\,dt', \tag{A.20}$$

as long as $f(t)$ is "nicely" behaved at $t = t'$. $\delta(t - t_0)$ is envisioned as being infinitely high and infinitely thin, centered at t_0, so that it has area unity.[7] To maintain dimensions, it must have the units of $1/t$. Delta-function properties are most easily developed by defining a sequence of functions approaching, with some parameter, that infinitely high and infinitesimally narrow limit. Any doubts about its meaning in a specific case can usually be resolved by defining such a sequence and taking the limit. It follows that the Fourier transform,

$$\mathcal{F}(\delta(t - t_0)) = \exp(-2\pi i s t_0). \tag{A.21}$$

[7] Lighthill (1958) provides a brief readable account of the mathematics that caused trouble when delta functions were first introduced into physics and describes how the difficulties were resolved.

$\delta(t)$ has an FT that is a constant in frequency, and the FT of $\delta(t - t_0)$ has a constant magnitude and linear phase with frequency. Interchanging t, s, we have usefully that the FT of a function constant in time is $\delta(s)$. The FT of a cosine in t is

$$\hat{f}(s) = \frac{1}{2}[\delta(s - s_0) + \delta(s + s_0)] \tag{A.22}$$

and with many similar results. Among other uses, delta functions permit the calculation of Fourier transforms of periodic functions.

A.3.1 Splitting

Consider the product of two cosines (or of a sine and cosine, etc.) in time:

$$\cos(2\pi s_a t)\cos(2\pi s_b t) = \frac{1}{2}\cos(2\pi(s_a - s_b)t) + \frac{1}{2}\cos(2\pi(s_a + s_b)t), \tag{A.23}$$

that is, the product is the same as two cosines in which s_1 is "split" into two new frequencies, $s_a \pm s_b$. In the tidal potential described in Chapter 6, a product of a cosine of a frequency of $s_a = 1$ cycle/day, and of a sinusoid of frequency $s_b = 2$ cycles/year occurs. Thus this product can be rewritten as the sum of two terms, at 1 cycle/day plus and minus 2 cycles/year. Further splitting occurs from products of periodic terms at 8.9, 18.6, 21,000 years, and longer. In general, any nonlinear function of two or more sinusoids will produce variations at all of the sum and difference frequencies that arise, including those that arise from self-interactions, $s_a \pm s_b = 2s_a, 0$.

A.3.2 Discrete Forms

Today, almost all data, or model outputs, are discrete rather than continuous in time. Thus the variable $x(t)$ is known only at times, $x(t_n = n\Delta t)$, where n is a finite set of integers. The time origin can always be shifted so that time starts at $t_0 = 0\Delta t$, introducing only a phase change in the Fourier coefficients. Then the discrete version of Equation (A.11) is (Hamming, 1973, p. 512)

$$x(m\Delta t) = \frac{a_0}{2} + \sum_{k=1}^{M/2-1} a_k \cos\left(\frac{2\pi k m\Delta t}{T}\right) + \sum_{k=1}^{M/2-1} b_k \sin\left(\frac{2\pi k m\Delta t}{T}\right) + \frac{a_{M/2}}{2}\cos\left(\frac{2\pi M m\Delta t}{2T}\right), \tag{A.24}$$

where

$$a_k = \frac{2}{M}\sum_{p=0}^{M-1} x_p \cos\left(\frac{2\pi k p\Delta t}{T}\right), \quad k = 0, \ldots, M/2 \tag{A.25}$$

$$b_k = \frac{2}{M}\sum_{p=0}^{M-1} x_p \sin\left(\frac{2\pi k p\Delta t}{T}\right), \quad k = 1, \ldots, M/2 - 1, \tag{A.26}$$

and $x_p = x(p\Delta t)$, with an equivalent complex form. For simplicity, M is assumed to be an even number, and $T = M\Delta t$ (the data duration is $(M - 1)\Delta t$).[8] It is sometimes convenient to

---

[8]The Fourier series sines and cosines or corresponding complex exponentials have the unusual and useful property of being exactly orthogonal when integrated over the interval T, and remain so when summed on discrete points over that interval.

use the amplitude and phase: $\sqrt{a_k^2 + b_k^2}$, $\phi_k = \tan^{-1}(b_k/a_k)$. (*Notation note:* often, as above, $x(m\Delta T)$ is written x_m, etc., because it saves a bit of space.)

This representation is *again exact*. a_n, b_n have a character depending upon the nature of $x(t)$ as they represent it exactly over interval T, be it deterministic or stochastic or both, again with no implication of periodicity. *Interpreting* their numerical values depends directly upon the nature and structure of $x(t)$. As with the continuous-time case, a numerical Fourier transform could be computed at all continuous frequencies, s, but the value can be exactly determined from the Fourier series coefficients at the frequencies, s_k, alone.

Note the important features: (1) that the number of coefficients a_k, b_k is exactly equal to the number of x_m; (2) that the highest frequency appearing is $s_{k=M/2} = 1/2\Delta t$; and (3) the Parseval relationship becomes

$$\frac{1}{M}\sum_{m=0}^{M-1} x(m\Delta t)^2 = \frac{a_0^2}{4} + \frac{1}{2}\sum_{p=1}^{M/2-1}(a_p^2 + b_p^2) + \frac{a_{M/2}^2}{4}, \tag{A.27}$$

Property (1) implies a conservation of information: exactly the same number of numbers is required in the time and frequency domains to perfectly reconstruct the series. The power in the record is the sum over all of the squared Fourier frequencies and thus presents a convenient way to depict the relative importance of any single frequency or group of frequencies, $s_k = k/T$, for example,

$$\frac{1}{2}\sum_{r=p-n}^{p+n}(a_r^2 + b_r^2) / \frac{1}{M}\sum_{m=0}^{M-1} x(m\Delta t)^2 = \frac{1}{2}\sum_{r=p-n}^{p+n}(a_r^2 + b_r^2) / \frac{1}{2}\sum_{p=1}^{M/2-1}(a_p^2 + b_p^2) \tag{A.28}$$

is the fraction of the total record power lying in the band of frequencies from $(p-n)/T$ to $(p+n)/T$. (The last equality drops the zero and Nyquist frequency terms, which are ordinarily made negligible before analysis.)

Property (2) has important consequences: if the original underlying record contains frequencies higher than $s_{M/2} = 1/2\Delta t$ (the *Nyquist* frequency), what happens to them in the exact representation Equation (A.24)? The answer is not difficult to find (see Figure A.1): if the original frequency is $s_0 > 1/2\Delta t$, it will appear instead at the lower frequency,

$$s = s_a = s_0 \pm m/\Delta t, \tag{A.29}$$

where the integer m is determined by the requirement $|s_a| \le 1/2\Delta t$. (Values can be negative, but for real processes, only the absolute magnitude is of concern.) Thus, a high frequency will appear to be a lower one—an effect known as *aliasing*, because the original frequency masquerades as a lower one, it has already been encountered in Chapter 2 and Figure A.2. As examples, Figure A.3 shows the alias period in hours of the oceanic inertial frequency, f, when sampled at 24 h intervals at different latitudes. ($f = 2\Omega \sin(\phi)$, where Ω is the Earth rotation rate, and ϕ is the latitude.) Because many Earth-orbiting satellites are in so-called sun-synchronous orbits—returning to each latitude and longitude at a fixed time every day, $\Delta t = 24$ h—aliasing inevitably occurs, including that of the sometimes powerful daily cycle itself. Jayne and Tokmakian (1997) discussed published model studies of oceanic variability, which when subsampled at 3 d intervals, produced qualitatively erroneous results. Defant (1950) clearly recognized the problem. Figure 7.1 is a list of tidal alias periods from a satellite in a 10 d repeat orbit (cf. the box on "The Sampling Theorem").

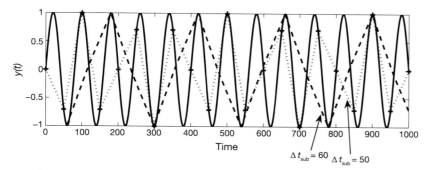

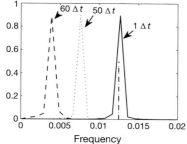

Figure A.1: (Upper panel) A sine wave of period $80\Delta t$, and subsampled at $\Delta t_{sub} = 60\Delta t$ and $50\Delta t$. (Here, $\Delta t = 1$ unit.) (Lower panel) Periodogram $(a_k^2 + b_k^2)$ of the subsampled time series showing that the inadequately sampled pure frequency peak appears at an alias quite distinct from the correct value. $60\Delta t$, $50\Delta t$, and $1\Delta t$ denote the sampling intervals giving rise to the corresponding peaks at the aliased frequencies. The dashed vertical segment is at frequency $1/80\Delta t$.

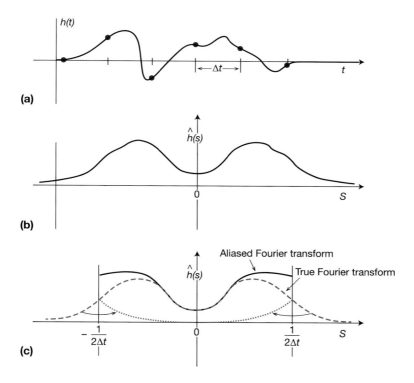

Figure A.2: A function (a) with Fourier transform as in (b) is sampled as shown at intervals Δt, producing a corrupted (aliased) Fourier transform as shown in (c). The phenomenon of aliasing is not limited to periodic signals. (After Press et al., 2007.)

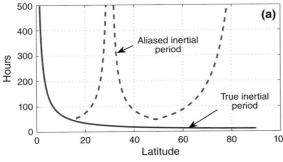

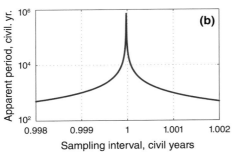

Figure A.3: (a) The period of an inertial oscillation $(2\pi/f)$, $f = 2\Omega\sin(\text{latitude})$, where Ω is the Earth's rotation frequency, in hours, compared to the apparent (aliased) period if any such motion is sampled at 24 h intervals. The rapid change with latitude is notable. Because inertial oscillations commonly dominate the velocity field, measurements, for example, of surface thermal gradients by sun-synchronous satellites, will be subject to aliasing into much lower frequencies. (b) Aliasing characteristics of a periodicity at the tropical year of 365.242190419 d, when sampled at intervals at and near the civil year are very sensitive. Alias periods approach 1 million years but drop sharply away to periods of hundreds of years depending upon the precise sampling interval. This extreme sensitivity to the actual sampling interval of the apparent alias period is an issue, for example, in the interpretation of spectra from the Greenland ice-core records (see Wunsch, 2000).

THE SAMPLING THEOREM

The *sampling theorem* (sometimes *Shannon's sampling theorem*) states the aliasing problem in a different way. Assuming a continuous time series, $y(t)$, is band limited so that $|\hat{y}(s)| = 0$, $s \leq 1/(2\Delta t) = s_c$, then $y(t)$ can be perfectly reconstructed from its samples, $y(m\Delta t)$, alone via the formula

$$y(t) = \sum_{m=-\infty}^{\infty} y(m\Delta t) \frac{\sin((\pi/\Delta t)(t - m\Delta t))}{(\pi/\Delta t)(t - m\Delta t)}. \qquad (A.30)$$

The sampling theorem is thus a statement about perfect interpolation. Equation (A.30) is never used for interpolation in practice, as the assumption of an infinite number of perfect samples cannot be met. "Smoothing" (or "mapping") of finite, noisy sampled data is better handled in other ways. The rule of thumb that one must sample at intervals $\Delta t \leq 1/(2\Delta s_c)$ is, however, a practical one. Interchanging s and t, and $1/T$ and Δt, shows how to calculate the Fourier transform values at any frequency from the discrete Fourier coefficients.

At another extreme, Figure A.3 also shows the apparent alias period of the tropical year of 365.242190419 d (the dominant element in the seasonal cycle) when subsampled at intervals surrounding that period, for example, when Δt is the civil year of 365.25 d. The aliased annual frequency is extremely sensitive to minute changes in Δt and can be a major issue in paleoclimate studies.

Aliasing is a serious, ubiquitous, unforgiving problem. A widely quoted paper "demonstrating" a major trend in the North Atlantic circulation over decades is a textbook example of a poorly sampled record, one dominated by an aliased artifact. Most real records are subject to aliasing—the burden lies with the analyst to show that it is negligible or accounted for.

A corollary of the Fourier transform representation shows that any truly discrete time series occurring at intervals Δt will have a periodic Fourier transform, and that it would be indistinguishable from a *sampled, but otherwise time continuous, function*, whose Fourier transform is undefined outside the *baseband* $|s| \le 1/2\Delta t$. The heights of audience members in theater seats do not exist in between the seats, and so are intrinsically discrete.

The relationships above, between Fourier series coefficients and Fourier transform values, all have analogs in the discrete cases. Infinities from δ-functions in the continuous case become large finite values in the discrete one.

That a Fourier series representation does not imply a periodic function should be examined by the reader, for example, by producing random numbers, $x\,(m\Delta t) = \xi_m$, on a computer.[9] The reader should confirm, using a_k, b_k as defined above, and the summation in Equation (A.24), that the original random numbers are reproduced to machine precision. The a_k, b_k being defined then as (weighted) sums of random numbers will themselves be random numbers (if $x\,(m\Delta t)$ is Gaussian, a_k, b_k will be also). Can anything useful be inferred from such values?

A recapitulation in slightly different terms is useful. The Fourier representation in Equations (A.24)–(A.26) demonstrates that the coefficients a_k, b_k can have very different meanings, depending upon the nature of $x\,(m\Delta t)$. They can be (1) deterministic as derived from any true periodic function; (2) deterministic, representing *any* nonperiodic function in a fixed interval; (3) stochastic variables if $x\,(m\Delta t)$ are random; and (4) a mixture, depending upon the frequency dependence, of properties 1–3. Conventional *spectral estimation* theory, elaborated below, is directed at situations where a significant stochastic element exists.

A.3.3 Convolution

Next to the various Fourier inversion theorems, the *convolution* operation is the most powerful and useful of the time-series machinery. Consider any two functions in continuous time, $f\,(t), g\,(t)$. A third function, $h\,(t)$, is said to be the *convolution of f with g*, if

$$h\,(t) = \int_{-\infty}^{\infty} f\,(t')\,g\,(t - t')\,dt' = \int_{-\infty}^{\infty} g\,(t')\,f\,(t - t')\,dt', \qquad (A.31)$$

a form already used to define the δ-function. Textbooks show the *convolution theorem* that,

$$\hat{h}\,(s) = \hat{f}\,(s)\,\hat{g}\,(s). \qquad (A.32)$$

If two functions are convoluted, it is equivalent to multiplying their Fourier transforms (coefficients). Conversely, *convolution in frequency space implies multiplication in the time domain*. In discrete time, for two sequences, f_n, g_n existing at the same time intervals, Δt, their convolution gives rise to a third sequence, h_n,

$$h_n = \sum_{m=-\infty}^{\infty} f_m g_{n-m} = \sum_{m=-\infty}^{\infty} g_m f_{n-m}, \qquad (A.33)$$

[9] Random numbers produced on computers are generated from formulas and are technically only pseudorandom. Demonstrating failure of true randomness is a subtle business (see, e.g., Morokoff and Caflisch, 1994, or Press et al., 2007). It can become an important issue in some Monte Carlo simulations.

and Equation (A.32) remains correct, using Fourier series or transform values from the discrete forms. Equation (A.33) describes the coefficients from ordinary polynomial multiplication,

$$\sum h_p z^p = \left(\sum_q f_q z^q\right)\left(\sum_{q'} g_{q'} z^{q'}\right),$$

both a useful mnemonic and a demonstration that discrete Fourier analysis corresponds to polynomial manipulations. Determining f_n from knowledge of h_n and g_n leads to the very important problem of *deconvolution*.

Convolution and δ-functions are very important in the use of a powerful tool called a *Green function*. Books (e.g., Duffy, 2001) are devoted to this subject, and for present purposes, note only that they are commonly the solution, G, to differential or partial differential systems from a disturbance given by a δ-function $\delta(t - t_0)$ in an ordinary differential equation, or $\delta(\mathbf{r} - \mathbf{r}_0)\delta t - t_0)$ in a partial differential system, where t is commonly time, and $\mathbf{r}$ a vector of position. In discrete models represented by difference, rather than differential, equations, $\delta(t - t_0)$ is replaced by the Kronecker delta, δ_{tt_0}. Sometimes a distinction is made between a disturbance in the interior of the system and one occurring in the boundary conditions, leading to the definition of *boundary Green functions*. Typical Green functions can be written as $G(\mathbf{r}, \mathbf{r}_0, t - t_0)$ under the assumption, sometimes a good one, that the solution only depends upon the time, $t - t_0$, since the disturbance occurred.[10] In an infinite spatial domain, the simplified case $G(\mathbf{r} - \mathbf{r}_0, t - t_0)$ would appear. If G is known, then a general disturbance, $q(t)$, in time at $\mathbf{r} = \mathbf{r}_0$, produces the solution by convolution,

$$\psi(\mathbf{r}, \mathbf{r}_0, t) = \int q(t')\, G(\mathbf{r}, \mathbf{r}_0, t - t')\, dt', \tag{A.34}$$

or in the equivalent discrete form. Multidimensional convolution becomes available for the infinite spatial domain case. A number of studies have been directed at inferring G from ψ and q and are direct applications of deconvolution methods. *Transit time distributions* occurring in tracer studies are forms of Green functions.

A.3.4 Miscellaneous Notes on Fourier Methods

LEAKAGE

Numerical Fourier analysis involves a number of slightly technical elements. An important one concerns the accuracy of the coefficient determination if one or more of the underlying coefficients is much larger than the others. "Energy" can *leak* from the strong frequencies to the weak ones, corrupting the calculated values. At least two approaches can be used and often both are: (1) Remove any known large amplitude components. Those components usually include the one for zero-frequency—the sample average—which can be many orders of magnitude larger than anything else in the record. Tides are often removed when they dominate the data. (2) Multiply the record by a *window*, a function, $w(m\Delta t)$, before doing the Fourier analysis. Invoking Equation (A.32), the Fourier coefficients or transform values will be a convolution of $\hat{w}(s)$ with that of $\hat{x}(s)$. A suitable choice of w_m, for example, a function that tapers down the two edges of the finite record, can substantially reduce the leakage, although a loss of resolving power results.

---

[10] The assumption fails if the system dynamics depends upon time; for example, if time appears in the coefficients of a differential system.

VECTOR PROCESSES, ROTARY COMPONENTS, AND RANDOM FIELDS

All of these representations can be generalized to *vector time series*, $\mathbf{x}(t) = [x_1(t), x_2(t), \ldots]^T$, as in the two- or three-dimensional components of a velocity field. For two-dimensional vectors, an important special set of tools is obtained by defining the complex time series, $\mathbf{x}(t) = x_1(t) + ix_2(t)$. With algebraic tools borrowed from optics, components rotating in clockwise or counterclockwise directions can be separated, ellipses of motion deduced, etc.[11]

Other processes might represent spatial *fields*, $u(t, \mathbf{r})$, where $\mathbf{r}$ is a continuous or discrete position vector of two or more dimensions. Such random fields may be stationary, or not, in space and/or time, and have covariances among various positions. A propagating wave is a special case. A great deal is known about random fields, and the reader is referred to the texts by Vanmarcke (1983) and Adler (2010).

RESOLUTION, SUPER-RESOLUTION, AND LEAST-SQUARES

The so-called Rayleigh criterion states that two frequencies differing by less than $1/T$—a difference precisely the frequency separation of the ordinary Fourier series elements—cannot be separated or "resolved." This statement is, however, only a rule of thumb: for example, one can write

$$
\begin{aligned}
x(m\Delta t) \;=\; & c\cos(2\pi s_1 m\Delta t) + d\sin(2\pi s_1 m\Delta t) + \\
& e\cos(2\pi s_2 m\Delta t) + f\sin(2\pi s_2 m\Delta t) + \varepsilon(m\Delta t),
\end{aligned}
\tag{A.35}
$$

where $|s_1 - s_2| \ll 1/T$ and $\varepsilon(m\Delta t)$ is a residual. Then minimizing the sum,

$$
J = \sum_m \varepsilon(m\Delta t)^2,
$$

with respect to the unknown coefficients c, d, e, f presents an ordinary least-squares problem (see Appendix B). Such an approach leads to the idea of *super-resolution* (Munk and Hasselmann, 1964). The ability to carry out such a recipe with useful results is sensitively dependent upon the signal-to-noise ratio in the observations. The magnitude and phase of a 21,000-year tidal period might be determined by least-squares from a handful of perfect data points separated by a few minutes in time. Such essentially perfect data do not, however, exist.

The coefficients in Equation (A.24) can also be determined from least-squares, and so with uniformly spaced data, the least-squares solution coincides with Equations (A.25) and (A.26). Their determination from Equation (A.12), either directly or through the FFT, can be regarded as mathematical tricks for obtaining the solution to the least-squares equations by exploiting the special nature of the matrix that needs to be inverted. The least-squares relationship is especially helpful if data are *non*uniformly spaced, and methods exist for solving that problem efficiently (one is named for Lomb and Scargle).[12]

A.3.5 *Randomness and Spectral Estimation*

Of what use are the Fourier series coefficients of random numbers? If the random values, ξ_m, above are Gaussian independent (uncorrelated) variables with zero mean, the Fourier

---

[11] See, for example, Gonella, 1972.
[12] Press et al., 2007.

coefficients are sums of Gaussian variables and are themselves (1) Gaussian of zero-mean, (2) uncorrelated with (and thus independent of) each other; and consequently, (3) their phases $\phi_k = \tan^{-1}(b_k/a_k)$ are completely random, being unpredictable and uniformly distributed as

$$p_{\phi_k}(\phi) = 1/2\pi, \quad -\pi \le \phi \le \pi. \tag{A.36}$$

The values of

$$P_k = a_k^2 + b_k^2 \tag{A.37}$$

are the sum of the squares of two independent Gaussian variables, and hence are distributed in the chi-square distribution (χ_2^2—with $v = 2$ degrees-of-freedom, as in Eq. A.7).

P_k is the *periodogram* and it is a random variable in k. The average, or expected, value of P_k is constant, independent of k if it is computed from uncorrelated ξ_m. Then the *expected* (true mean) value of the *squared* Fourier series coefficients is constant—leading to the label for ξ_m as a zero-mean *white-noise* process having on average equal "power" at all frequencies. Discrete white noise is a rather boring time series, but it is the basic building block of more interesting objects.

Suppose, as an example, a damped mass-spring oscillator is described by a differential equation,[13]

$$\frac{d^2 y(t)}{dt^2} + r\frac{dy(t)}{dt} + \gamma y(t) = q(t), \tag{A.38}$$

m, r, γ being constants. It can be rendered discrete with a one-sided difference, so that

$$\frac{y((m+1)\,\Delta t) - 2y(m\Delta t) + y((m-1)\,\Delta t)}{\Delta t^2} + r\frac{y(m\Delta t) - y((m-1)\,\Delta t)}{\Delta t} +$$
$$\gamma y(m\Delta t) = q(m\Delta t). \tag{A.39}$$

If terms are rearranged and combined, and $\Delta t^2 q(m\Delta t) = \xi(m\Delta t)$,

$$y((m+1)\,\Delta t) + (r\Delta t - 2 + \gamma\Delta t^2)\,y(m\Delta t) + (1 - r\Delta t)y((m-1)\Delta t) = \tag{A.40}$$
$$\Delta t^2 q(m\Delta t) = \xi(m\Delta t),$$

then Δt is chosen to be sufficiently small to produce an accurate rendering of the known analytical solution. An example of such a calculation—from starting values $y(0) = y(1\Delta t) = 0$, and with $\xi(m\Delta t)$ being zero-mean white noise—is shown in Figure A.4. Given a set of measurements of $y(m\Delta t)$, its Fourier series coefficients can be calculated. Equation (A.40) is linear in $y(m\Delta t)$ and so is a linear combination of Gaussian white-noise variables. Its Fourier series coefficients are sums of Gaussian variables, and therefore will themselves be Gaussian variables.

Figure A.4 depicts an example (a *realization*) of $y(m\Delta t)$, the corresponding random P_k, and the phases ϕ_k, calculated from one set of $\xi(m\Delta t)$. The average value $\langle P_k^2 \rangle$ is unknown but is sought because the best-fitting m, r, γ might be determined from its shape and magnitude.

---

[13] Specification of a random $q(t)$ and the handling of the solution $y(t)$ in continous time is the domain of stochastic differential equations and their partial differential extensions. The rigorous mathematics becomes quite elaborate, requiring detailed attention to the meaning of the differentials, dq, or dy (see Gardiner, 1985). Here almost all of these considerations are avoided by working with the finite, discrete representations used in digital computers. Whether the limit of the time-step, $\Delta t \to 0$, has *observable* consequences for the real ocean has to be determined. In continuous time, even the definition of white noise requires using considerable mathematical machinery associated with Brownian motion, Wiener processes, etc., which is not necessary here.

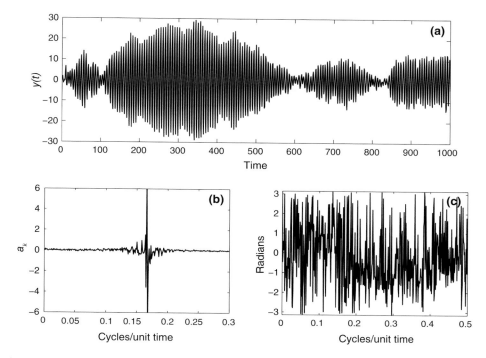

Figure A.4: (a) The time-series of displacement of a simple mass-spring oscillator, driven by white noise, and computed numerically such that $r = 0.1$, $\gamma = 1$ with $\Delta t = 1$. An amplitude modulation is visible as the beating of neighboring frequencies. In (b) are some of the displacement's Fourier coefficients, a_k, whose average is plausibly zero, and (c) shows $\phi_k = \tan^{-1}(b_k/a_k)$, which is evidently uniformly distributed in $-\pi \le \phi \le \pi$.

How can the average be obtained? Make a working hypothesis that over small ranges of k, meaning small frequency ranges, the average value of P_k changes little with k. In that case, a reasonable estimate can be obtained from averaging P_k in the neighborhood of k (Fig. A.5). The result has significant structure; indeed, it begins to look like a "peak." The maximum of the peak is not far from the frequency $2\pi s_0 = \sqrt{\gamma}$—the resonant frequency of the underlying oscillator, something that can be proven for the true average. The width of that peak, suitably defined, is dependent upon r, the value of the dissipation rate (and see the box "The Quality Factor," defining its measure, called Q).

The periodogram, P_k, is the *unaveraged* squared Fourier coefficients. The *power spectrum* is defined as the *theoretical expected value*, $\langle P_k \rangle = \Phi(s_k)$, in the definition of Equation (A.2).

Obtaining *estimated* values $\tilde{\Phi}(s_k)$ requires some form of averaging. In all cases, such sample spectral estimates, for example, as shown in Figure A.5, are *finite sample* averages of the random variables, P_k, and themselves remain random variables. If the estimate is made over L-pairs of squared Fourier coefficients, the distribution of $\tilde{\Phi}(s_k)$ is approximately χ^2_{2L} (with 1 degree-of-freedom from each of the a_k, b_k). Frequency-band averaging is only one approach, and finding and manipulating such estimates is the concern of the subject of spectral estimation. Percival and Walden (1993) describe a particularly useful method, usually called *multitaper estimation*. Periodograms, $P_k^{(j)}$, are formed from "tapered" values of

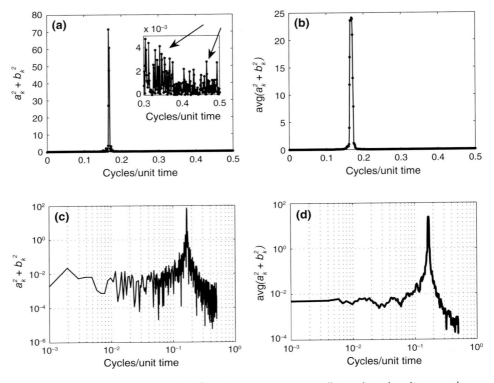

Figure A.5: (a) The periodogram $(a_k^2 + b_k^2)$ of the mass-spring oscillator plotted on linear scales, and (c), on logarithmic scales. Note that the apparent very sharp peak is actually made up of several nearby coefficient sets whose phases are random with respect to each other. Arrows in the inset plot point to "peaks" that are expected statistical accidents and have no significance. Panels (b) and (d) show the periodogram averaged over nine adjacent values on both sets of scales to produce power spectral estimates with approximately 18 degrees-of-freedom. Because neighboring averages differ by only one value, they are not uncorrelated, and conventionally one retains only the ones that are disjoint. (Some analysts prefer to use tapered, overlapping weights so that some degree of correlation is retained.) Linear and logarithmic scales give different impressions of the structure of the underlying time series.

the data, $w_p^{(j)} x_p$, and then averaged as $\tilde{\Phi}(s_k) = 1/L \sum_{j=1}^{L} P_k^{(j)}$. Tapering is familiar from the discussion of spectral leakage above. The orthogonal tapers, or windows, $w_p^{(j)}$, are specially chosen, and this method is now probably the most common default estimate. Because of the orthogonality, $\sum_p \left(w_p^{(j)} x_p \right) \left(w_p^{(j')} x_p \right) \approx 0$, and the tapered data are almost uncorrelated.

One of the major points of historical confusion in the literature has been the misconception that the periodogram $P_k = a_k^2 + b_k^2$ is nearly equal to its average value, $\langle a_k^2 + b_k^2 \rangle$, and the related assumption that the more data one has, the closer they correspond. The inference is false: no matter how much data exist, $a_k^2 + b_k^2$ remains a χ_2^2 random variable, and it has no more relationship to its mean value than does any other instantaneous measurement of a variable correspond to its true average. Several ways exist to understand why this must be the case. Consider that if the record length of a discretely measured variable is doubled, that the number of Fourier coefficients that must be estimated is also doubled—see the Parseval relation (Eq. A.27). Every pair of them differs by exactly 1 cycle/record length from each of its neighbors at higher and lower frequency—and all of the data have to be used to separate them. Much of spectral estimation concerns the determination of whether local maxima ("peaks") in $\tilde{\Phi}(s_k)$ are physically significant, or simply expected fluctuations that must be be present in a random variable (see the arrows in Fig. A.5) leading to construction of confidence limits and other pieces of analysis machinery.

THE QUALITY FACTOR: Q

A useful measure of dissipation in an oscillatory system is the *quality factor*, or Q, which is borrowed from electrical engineering. Consider a system—such as a pendulum or mass-spring oscillator or organ pipe or oscillating solid Earth—set into motion from an initial disturbance. If it oscillates for a long time it is said to have a "high Q." A "low Q" system could come to rest within a single, incomplete cycle.

Define Q (e.g., Jackson, 1999) as

$$Q = \sigma_0 \frac{\text{stored energy}}{\text{energy loss rate}} = 2\pi \frac{\text{stored energy}}{\text{energy loss/cycle}} \qquad (A.41)$$

where σ_0 is the natural frequency in the absence of dissipation. If U_0 is the energy initially found in the system, then Equation (A.41) implies

$$U(t) = U_o e^{-2\pi s_0 t/Q},$$

for the energy as a function of time. If ξ is a measure—for example, of the angle of a pendulum, or the position of an oscillating mass, or the pressure disturbance inside an organ pipe—and ξ_0 is its initial value, then, equivalently,

$$\hat{\xi}(s) = \int_0^\infty \xi_0 e^{-2\pi s_0 t/2Q} e^{i2\pi(s-s_0)t} dt$$

is its Fourier transform, and

$$\left|\hat{\xi}(s)\right|^2 \propto \frac{1}{(2\pi s - 2\pi s_0)^2 + (2\pi s_0/2Q)^2},$$

shown in Figure A.6 for various values of Q.

The periodogram and derived spectral estimates historically may have been the focus of more misinterpretation than almost any other statistical quantity. Thus the large literature that asks questions: such as, is there an x-period oscillation in my variable of choice? which is almost inevitably answered by yes, because it can be proven that no physically realizable time series can have vanishing power even in an infinitesimal frequency interval.[14] Appropriate statistical tests and/or physics have to be brought to bear to demonstrate the importance of the energy that is found, relative to the surroundings. Proclamations of "peaks" are rampant; many of them prove to be statistical accidents of small samples.

---

[14] A theorem of Paley and Wiener. "Physically realizable" requires that the starting time should be finite.

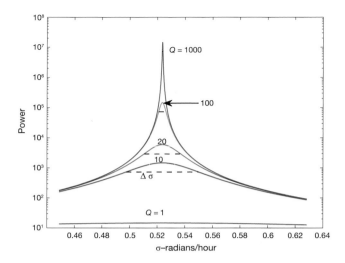

Figure A.6: The power density spectrum as a function of Q, showing how the "quality" of the peak increases. Q is defined in Equation A.41.

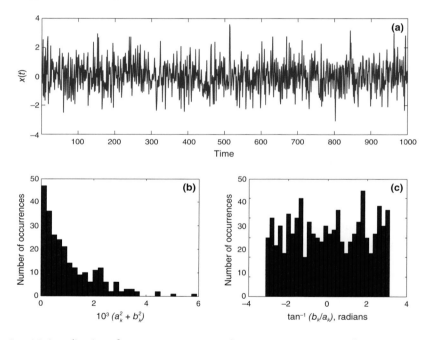

Figure A.7: (a) A *realization* of stationary Gaussian white-noise time series with zero mean and unit variance. (b) Histogram of the squared-amplitude of the Fourier transform of a 1024-point- sample of the time series. The histogram should follow a χ_2^2 distribution and it does, qualitatively. Note the long tail representing the occurrence of a number of comparatively extreme values. These have no particular physical significance. (c) The histogram of phases from the same Fourier series co-efficients. Asymptotically, as the number of coefficients goes to infinity, the histogram should ap-proach a constant value between $\pm\pi$.

Discussion of the periodogram and derived power or power density estimates ignores the phase information, $\phi_k = \tan^{-1}(b_k/a_k)$ in the Fourier coefficients, because for a linear, stationary, random process, they are completely random with no structure as a function of k (Eq. A.36). The assumption of their complete randomness can be tested by examining the actual distribution. Figure A.7 shows a realization of a discrete Gaussian white-noise process whose squared Fourier coefficients, $a_k^2 + b_k^2$, are tabulated there in a histogram. That they are derived from a χ_2^2 distribution with a long positive tail is plausible. Similarly, the same figure shows the histogram of phases of the Fourier coefficients that are derived from a uniform distribution. That the coefficient amplitudes and phases follow neither probability density exactly is expected—and required—of finite sample numbers.

Much of the time, the estimated power spectrum is normalized by dividing the values by the width of the frequency band over which the average has been taken. Such a *power density* spectral estimate has the property, for random processes like the one analyzed, of being effectively independent of the record length. (The power *density* of a pure sine wave will, in contrast, grow with the record length.) The terms "spectrum" and "spectral estimate" are used indiscriminately for both power and power density versions, and they differ by a constant multiplier implicit in the context.

Figure A.8 shows the periodogram of the time series in Figure A.7. Some obvious large values appear, but because it is known to have a white spectrum, none of them is significant. The figure also shows a power spectral estimate, $\tilde{\Phi}(s_k)$, obtained from the periodogram by averaging the values in groups of 11 (5 from each side of the central frequency plus the central value itself), thus producing an estimate with approximately 22 degrees-of-freedom. A 95% confidence interval is shown as computed by assuming that the estimates have a χ_{22}^2 distribution, in which, by convention, the intervals above and below the estimated value have equal probability. The periodogram produces a frequency breakdown with maximum resolution, albeit with maximal statistical fluctuations (see Fig. A.5). As the frequency-band averaging width increases, the averages become more stable, but frequency resolution is lost. Part of the art of time series analysis is deciding on the *trade-off* between a wish or need for high resolution and for statistical stability. Commonly, during the course of an analysis, several such estimates are made, and indeed, the trade-off for some parts of the frequency band may not be the best choice for another part—the physics being different. Because of the changing physics with frequency, and with the underlying true spectrum often having different shapes and trends with s, the usual interpretation of the confidence interval is as a test of local spectral extrema relative to a background spectrum assumed to be *locally* white.[15] In frequency regions of rapid spectral change, or the presence of deterministic energy, the confidence interval as computed here will be unreliable.

PLOTTING POWER DENSITY SPECTRA

The choice of display of spectra (and periodograms) is worth a brief examination. For most geophysical purposes, the default form of plot is commonly a log-log one where both the estimated power density and the frequency scale are logarithmic. Several reasons make it useful: (1) Many geophysical processes produce spectral densities that are near power laws, s^{-q}, $q > 0$, over one or more ranges of frequencies. These appear as straight lines on a log-log plot

[15] On p. 211 is a discussion of a spurious spectral peak that led to generation of a theory of a nonexistent phenomenon.

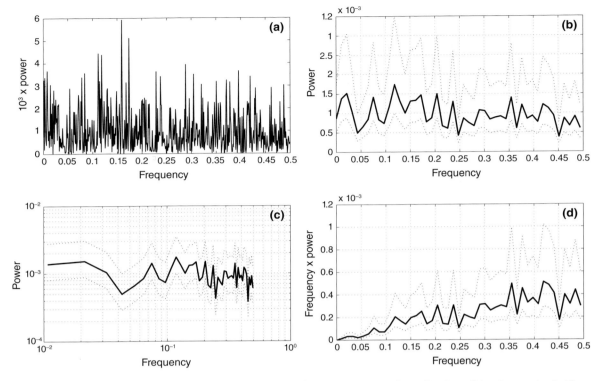

Figure A.8: (a) A periodogram (positive frequencies only) from 1024 points of a realization of the time series in Figure A.7. None of the peaks (or troughs) has any statistical significance. (b) The power spectral estimate, $\tilde{\Phi}(s)$, as the average of 11 neighboring frequencies in (a) along with (dotted lines) an approximate 95% confidence interval computed from a χ^2_{22} distribution. Only every eleventh point is plotted, so that the estimated values are approximately independent. Notice that larger peaks have a wider confidence interval about them (Jenkins and Watts, 1968, p. 254). Because power is bounded below by zero, the interval below the estimated value is smaller than that above. A horizontal line can be drawn within the confidence interval consistent with a hypothesis that the spectrum is white. With a 95% confidence interval, about 5% of the values *should* lie outside the confidence band. (c) Same as (b) except plotted on a log-log scale. The width of the 95% confidence interval is now constant with frequency. (d) Same as (b) except in the area-preserving form, $s\tilde{\Phi}(s)$, compensating for the frequency compression in (c).

and so are easily recognized. A significant amount of theory describes them: the most famous being Kolmogoroff's wavenumber $\kappa^{-5/3}$ rule for turbulent kinetic energy, but other theories producing various power laws exist as well. (2) Long records are often the most precious, with particular interest lying in the behavior of the spectrum at the lowest frequencies. The logarithmic frequency scale compresses the high frequencies, rendering the low-frequency portion of the spectrum more conspicuous. (3) The "red" character (often called *red noise*) of many natural spectra, which decrease with frequency, produces such a large dynamic range in the spectrum that much of the result is invisible without a logarithmic power scale. (4) Confidence intervals (or standard errors; discussed below) are proportional to $\tilde{\Phi}(s_k)$ and become nearly constant intervals when plotted on a log-power scale.

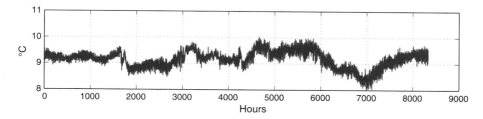

Figure A.9: A temperature time series, from 844 m at 27°N, 41°W, sampled every 32 min for almost a year and whose spectral density is displayed in various forms in Figure A.10

But logarithmic plots can be very misleading. The Parseval relationship shows that all frequencies are on an equal footing, and the human eye is not effective at compensating for the crowding of the high frequencies. To give a better pictorial representation of the energy distribution, some investigators prefer to plot $s_k \tilde{\Phi}_k (s_k)$ on a linear scale against the logarithm of s_k. This form is sometimes known as an *area-preserving plot* because it compensates for the squeezing of the frequency scale by the multiplication by s (simultaneously reducing the dynamic range in red spectra by suppression of the low frequencies) rendering the fractional area under the curve proportional to the fraction of the total power. Consider how this behaves for white noise (Fig. A.8). The log-log plot is flat within the confidence limit, but it may not be immediately obvious that most of the energy is at the highest frequencies. The area-preserving plot looks "blue"—although this spectrum is that of white noise.

Beginners often use linear-linear plots, as it seems more natural. This representation is acceptable over limited frequency ranges, but it becomes problematic when used over the complete frequency range of many real records.

Consider Figure A.9, whose spectral density estimate (Fig. A.10) is plotted on a linear-linear scale. The linear plot suggests that the semidiurnal band tides (labelled M_2) are the only significant energy at high frequencies. But they contain only a fraction of the record variance, which is largely invisible solely because of the way the linear plot suppresses the smaller values of the larger number of continuum estimates. Again too, the confidence interval is different for every point on the plot. Which frequencies dominate the record variance is very difficult to distinguish by eye, and it is not always true that the largest amplitudes have the greatest physical importance. (Another such plotting comparison can be seen in Fig. 7.9.)

STATIONARITY, AUTOREGRESSIVE, AND MOVING-AVERAGE PROCESSES

As noted above, a time series is said to be *stationary* if its statistics (in the form of all of the underlying probability densities) are independent of time. Thus if the mass-spring oscillator equation defines the entirety of a process, with the statistics of the forcing $q(t)$ being unchanging with time, t being discrete, then any joint probability density $p(x_t, x_{t'}, x_{t''}, \ldots)$ will not depend upon time, but only upon the temporal separations $t - t'$, $t - t''$, etc. (see Eq. A.6). For many purposes, only the mean and covariances are assumed to be independent of time—with the behavior of the higher-order moments being unspecified. Such a process is known to mathematicians as *weakly stationary*, and to engineers as *wide-sense stationary*. If

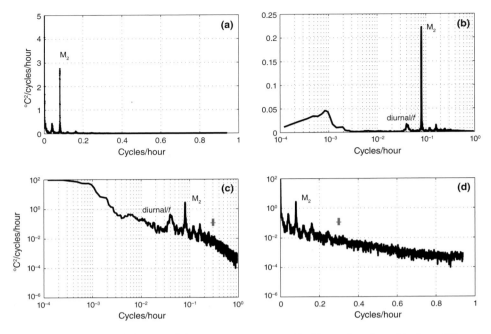

Figure A.10: The spectral density of the temperature record in Figure A.9 plotted in four different ways. (a) A linear-linear plot. Note that the vertical axis should extend to 100 at the first nonzero frequency but has been truncated so some structure is visible. The variance of these estimates (not shown) is proportional to the square of the depicted values. M_2 denotes the semidiurnal tide band. (b) An area-preserving plot log-linear in which the area under the curve is proportional to the fractional variance contributed to the whole and that shows that the low frequencies dominate the total (apart from the tidal line). "Diurnal/f" labels the diurnal tides which on this plot are indistinguishable from the inertial frequency, f. (c) A log-log plot that makes clear the resemblance to a "power law" at high frequencies, and a different one in parts of the low frequencies. The vertical bar is an estimated approximate 95% confidence interval. (d) A semilog plot which emphasizes the high frequencies, again with an estimated confidence interval (identical). The spectral continuum, away from the peaks, is of great physical importance.

the process x_t is both wide-sense stationary and Gaussian, it is completely specified by its mean, and second moments, $R_\tau = \langle x_t x_{t+\tau} \rangle$, and is thus also strictly stationary.

By the Wiener-Khinchin theorem (described below), R_τ is the Fourier transform of $\Phi(s)$, and thus either the spectrum or the second moment (*covariance*) matrix, along with the mean value, totally specify this class of time series—one of the several reasons spectral densities play such a central role. If the process is neither Gaussian nor stationary, the spectral density is still highly useful, but as the first term in a sequence of spectra involving cubic, quadratic, etc. expectations of the Fourier transform (called *bispectra, trispectra, ...*). McComas and Briscoe (1980) and Elgar et al. (1995) discussed some of the issues in computing these and related estimates.

The theoretical behavior of stationary Gaussian processes is highly developed, and in particular, detailed calculations can be made of the occurrence of extreme values, prediction possibilities, rates of the crossing of the mean value, etc. (Vanmarcke, 1983). With real data, distinguishing between the presence of nonstationarity and non-Gaussian behavior can be very difficult. See Priestley (1988) for some help.

Fourier representations are only one class of analysis method for understanding a time series. The other major approach avoids the use of frequency altogether and constructs representations purely in the time domain. One class of representations is the *autoregressive* (AR) form,

$$y(m\Delta t) = \alpha_1 y((m-1)\Delta t) + \alpha_2 y((m-2)\Delta t) + \ldots + \alpha_{L_A} y((m-L_A)\Delta t) + \theta(m\Delta t),$$
(A.42)

where m, L_A are integers written AR(L_A), where $\theta(m\Delta t)$ is zero-mean white noise. (These α_j should not be confused with the Fourier coefficients.) An AR(1) is the simplest version, with $\Delta t = 1$,

$$
\begin{aligned}
y(m\Delta t) &= \alpha_1 y((m-1)\Delta t) + \theta(m\Delta t), \\
&= \alpha_1 y_{m-1} + \theta_m, \quad \text{integer} m
\end{aligned}
$$

$y(m\Delta t)$ will be stationary as long as $|\alpha_1| < 1$. Determining L_A, the values of the α_i, and of the $\theta(n)$ is part of the estimation problem. More generally, the α_i must satisfy some simple rules to assure stationarity. Nature only rarely produces records accurately described by an AR(1) and having only two parameters.

Another standard form of random process built out of white noise is the moving average (MA(L_M)),

$$y(m\Delta t) = \beta_0 \theta(m\Delta t) + \beta_1 \theta((m-1)\Delta t) + \ldots \beta_M \theta((m-L_M)\Delta t),$$
(A.43)

$$y_m = \beta_0 \theta_m + \beta_1 \theta_{m-1} + \ldots + \beta_M \theta_{m-L_M}, \quad \beta_0 = 1.$$
(A.44)

An AR of a stationary process can always be converted into an MA, and vice versa, and sometimes combined forms are used for efficiency of representation (the ARMA). For prediction problems, MA representations are particularly convenient.

Time domain methods are very powerful, and switching back and forth between time and frequency domain representations is often convenient. This chapter has emphasized the frequency representations, both because of their immediate time-scale separation and because their behavior is somewhat less intuitive than ARMA and related processes.

AR processes can also give rise to peaks (recall the difference equation A.40, which is an AR(2)). Consider the stationary time series, $y(m\Delta t)$, given by the rule

$$y_m = \alpha_1 y_{m-1} + \alpha_2 y_{m-2} + \alpha_3 y_{m-3} + \alpha_4 y_{m-4} + \theta_m,$$
(A.45)

$$\boldsymbol{\alpha} = [0.4455, -0.7774, 0.0320, -0.1054]^T,$$

where θ_m is again zero-mean white noise. The α_i would be obtained from a physical model of a process. A realization of the process, started by time-stepping it from $t = 0$ with all zero values to start, is displayed in Figure A.11 along with its estimated power spectrum. The latter shows a broad excess of energy around 0.2 cycles/unit time without any periodicities.

ALIASING IN SPECTRA

Although aliasing of pure frequencies is the most intuitively understandable case, as Figure A.2 shows, the phenomenon applies just as much to stochastic processes. Distortion of power

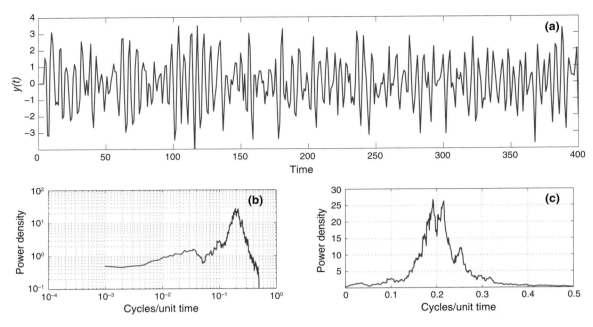

Figure A.11: Segment of the time series of an AR(4) process (Eq. A.45) and its estimated power density spectrum on both logarithmic and linear scales. The "peak" has a finite bandwidth and does not connote a periodic process. Although the 95% confidence interval is not shown, its magnitude (on the log power scale) is readily inferred visually from the range of the small-scale variations there.

law spectra s^{-q} by aliasing can be calculated analytically.[16] A qualitative inference is that if $q \geq 2$, aliasing can be ignored except very close to the Nyquist frequency, but that if $q < 2$, the entire spectral shape may be distorted beyond recognition.

TRANSFER FUNCTIONS

The linear mass-spring oscillator provides a prototype of the notion of a *transfer function*. If the forcing, $q(t)$, has a Fourier transform $\hat{q}(s)$ and if the response $y(t)$ has transform $\hat{y}(s)$ (either discrete or continuous forms), then the transfer function is defined as

$$\hat{T}_r(s) = \frac{\hat{y}(s)}{\hat{q}(s)}, \tag{A.46}$$

as long as $\hat{q}(s)$ does not vanish anywhere. A little algebra shows that $\left|\hat{T}_r(s)\right|^2$ can be written in terms of the model parameters and, in particular, can be represented using Q. Knowledge of the transfer function in very general systems, even at only a few frequencies, sometimes permits inferences of the magnitude of Q and hence of the dissipation rate.

DETERMINISTIC PEAKS

Peaks sometimes arise because a true sinusoid exists in the data. If $\sin(2\pi s_0 m \Delta t)$ is added to a random $y(m \Delta t)$, an isolated sharp maximum will appear in the spectrum (if the amplitude

[16] Wunsch, 1972a.

is large compared to the background random values, and most cleanly if the record length is manipulated so that $s_0 = p_0/T$ for some integer p_0). Many finite-duration records show such "lines," for example, for the annual cycle. True periodicities are, however, extremely rare in nature. Even the principal lunar tide is split into multiple peaks. Conventional discussions of spectral probabilities and statistics, including biases, must be modified at and near such deterministic structures in a Fourier estimate (Middleton, 1960).

QUANTIZATION NOISE

Consider a sinusoid of unit amplitude,

$$y(t) = \sin(2\pi st),$$

that is sampled at time intervals $\Delta t = 1$ with a measurement accuracy of Δy. If the variations in $y(t)$ are too small relative to Δy to be detected at some measurement threshold, its presence will remain unknown. So a sinusoid of amplitude of 1 mm will be imperceptible in perfect measurements rounded to the nearest 1 cm. However, ironically, if the signal is noisy enough, sufficiently common passages through the threshold of Δy will permit it to be detected and its amplitude, frequency, and phase to be determined. (This subject is that of *quantization error*, or *least-count noise*; see, e.g., Gold and Rader, 1983.)

These considerations become important in determining the smallest signals detectable in a particular measurement. If the quantization level is q (e.g., 1 cm, 1 mm, etc.) and the measurements of a time series $y(n\Delta t)$ are correctly rounded up or down with no bias, then the textbooks show that the energy of rounding error is $\langle q^2 \rangle/12$. This energy has, under these assumptions, a white-noise character and so is spread uniformly over the entire power spectrum as $\langle q^2 \rangle/12M$, where M is the total number of points in the spectrum. If the power in $y(n\Delta t)$ falls below $\langle q^2 \rangle/12M$ in any frequency band, the quantization error appears as a white noise "floor" (Fig. A.12). By making M large enough (obtaining a long enough record), this error can be reduced to arbitrarily small values. Results such as this one depend upon the assumption that the rounding is not directly dependent upon the signal itself—an assumption that will fail in general nonlinear systems. Real systems can be a bit more complex, as in Figure A.13, a comparison of the spectra of two space series derived from independent altimetric missions at the same time and location.

A.4   AUTOCORRELATIONS AND AUTOCOVARIANCES

For practical purposes, the power spectrum was defined as the average of the periodogram

$$\langle P(s_k) \rangle = \langle a_k^2 + b_k^2 \rangle, \tag{A.47}$$

and the power density, $\Phi(s_k)$, as the limit

$$\Phi(s_k) = \lim_{T \to \infty} \langle P(s_k) \rangle / (1/T), \tag{A.48}$$

so that its values, for a random discrete time series, are independent of the data length, T. As $T \to \infty$, the s_k become arbitrarily close together, are then continuous variables, and the subscript is omitted. (Here $\Delta t = 1$ for notational simplicity).

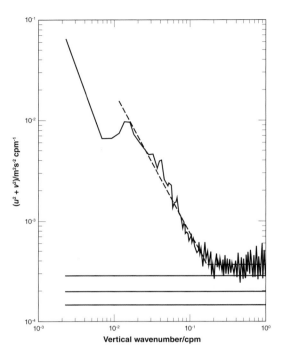

Figure A.12: A wavenumber spectral density estimate of 2×(kinetic energy) in the Arctic Sea. At high wavenumbers, a white-noise floor is seen, which is consistent with the estimated instrument quantization limit. Top and bottom horizontal lines are an estimated 95% confidence interval. (From D'Asaro and Morehead, 1991.)

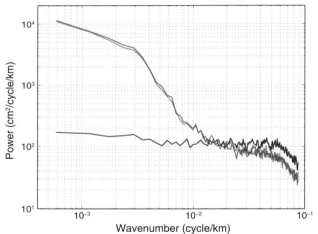

Figure A.13: The spectra of measurements from two coincident altimetric satellite missions computed along their tracks. The red and green spectra densities are from the TOPEX and Jason altimeters flying a few seconds apart in the same orbit (see Chap. 2). The spectrum of the difference (blue curve) is essentially white noise but becomes a factor of two above the individual spectra at wavelengths of about 100 km and shorter. At short scales, the measurements are dominated by noise in the two instrument systems, which is uncorrelated between them. That lack of correlation leads to the power in the difference time series at high frequencies becoming the sum of the powers in the two instruments. (From Fu and Ubelmann, 2014.)

Some mathematical difficulties accompany these definitions (but they are still useful in practice), largely having to do with the interpretation of a_p, b_p calculated as weighted sums of quantities, x_t, over intervals $0 \le t \le T$ as $T \to \infty$. These values are random, and no limiting values are reached as the data length becomes arbitrarily large. To make everything mathematically satisfactory, again define the *autocovariance* as the average value

$$R_\tau = R(\tau) = \langle x_t x_{t+\tau} \rangle, \tag{A.49}$$

and which for a wide-sense stationary process is independent of t, as assumed. Note that $R(-\tau) = R(\tau)$. (Here time series are assumed to have zero mean values.) Then the *Wiener-Khinchin Theorem* shows that

$$\Phi(s) = \sum_{\tau=-\infty}^{\infty} R(\tau) \exp(2\pi i s \tau) = R(0) + 2 \sum_{\tau=1}^{\infty} R(\tau) \cos(2\pi s \tau). \tag{A.50}$$

Nothing in Equation (A.50) is random (because the averaging operator makes everything deterministic), so no limit issues arise. The autocovariance is often normalized to produce the *autocorrelation* $\phi(\tau) = R(\tau)/R(0)$, $\phi(0) = 1$. It is not difficult to show that the limit in Equation (A.48) is consistent with the definition in Equation (A.50).

The autocovariance can be *estimated* as

$$\tilde{R}(\tau) = \frac{1}{M} \sum_{n=0}^{M-|\tau|-1} x_t x_{t+\tau}, \quad |\tau| \le M - 1 \tag{A.51}$$

Note that the number of terms in the averaging sum is $M - |\tau|$, yet the division is by M. The purpose of this *biassed estimate* is to avoid giving large weight to values, $\tilde{R}(\tau)$, calculated from a very small number of terms (only one, when $\tau = M - 1$). The Fourier transform of $\tilde{R}(\tau)$ is

$$P(s_k) = \tilde{R}(0) + 2 \sum_{\tau=1}^{M-1} \tilde{R}(\tau) \cos(2\pi s \tau) \tag{A.52}$$

and equal to the periodogram, P, a statistically unstable estimator of the power density spectrum. ($\tilde{R}(\tau)$ is statistically unstable for large τ and corresponds to the instability of P.) The so-called Blackman-Tukey method of spectral estimation produces an average on the left-hand side of Equation (A.52) by introducing weights on the right-hand side to give

$$\tilde{\Phi}(s_k) = w(0) \tilde{R}(0) + 2 \sum_{\tau=1}^{M-1} w(\tau) \tilde{R}(\tau) \cos(2\pi s \tau), \tag{A.53}$$

The effect of the weights is found from the discrete convolution theorem above (see Jenkins and Watts, 1968). An important virtue of Equation (A.53) was that by using weights (*windows*),[17] $w(\tau)$ of lengths, $M' \ll M$, the calculation of the transform is reduced to M' points—a very important consideration before the FFT algorithm became available. The smaller is M'; the wider is the averaging interval in frequency—leading to elaborate discussions of the trade-offs of variance reduction, resolution needs, bias, numerical efficiency, and other requirements, such as assuring the nonnegativity of $\tilde{\Phi}(s_k)$. As a numerical algorithm, this method is obsolete

---

[17] This use of windows differs from the application described above for reducing leakage in the Fourier transform.

and should not be used except under special circumstances. The covariance, if required, is readily computed instead as the Fourier transform of the power density spectral estimate.

As a transform pair, the autocovariance and power density spectrum are equivalent. For understanding, however, the power density is usually the more useful form, because the physics commonly separates by frequency. The autocovariance is a weighted sum over all frequencies, and the human eye is poorly adapted to interpreting it. For computational purposes though, for example, making objective maps (discussed below) or doing linear predictions, $R(\tau)$ is generally more convenient—because the combined effects of all frequencies contributing are sought.

For simple processes, such as the AR or MA or combined ARMA processes, theoretical power densities or autocovariances are easily found. Consider the AR(2),

$$y_n = \alpha_1 y_{n-1} + \alpha_2 y_{n-2} + \theta_n, \quad y_n = y(n\Delta t), \text{ etc.} \tag{A.54}$$

with $\langle \theta_n \rangle = 0$, $\langle \theta_n^2 \rangle = \sigma_\theta^2$, Then its complex-form Fourier transform is, with $\Delta t = 1$,

$$\hat{y}(s) = [\alpha_1 \exp(2\pi i s) + \alpha_2 \exp(4\pi i s)] \hat{y}(s) + \hat{\theta}(s), \tag{A.55}$$

and hence,

$$|\hat{y}(s)|^2 = \frac{|\hat{\theta}(s)|^2}{|1 - \alpha_1 \exp(2\pi i s) - \alpha_2 \exp(4\pi i s)|^2}. \tag{A.56}$$

Taking the average, using the white noise property of θ_n, and noting that everything except $|\hat{\theta}(s)|^2$ is nonrandom,

$$\Phi(s) = \frac{\sigma_\theta^2}{|1 - \alpha_1 \exp(2\pi i s) - \alpha_2 \exp(4\pi i s)|^2} \tag{A.57}$$

is the power density spectrum (*not* an estimate of it). If Δt is reinserted, $s \rightarrow s\Delta t$, all information is confined to the *baseband*, $|s| \leq 1/2\Delta t$. (This form is that of the spectrum of the mass-spring oscillator already considered.)

The Fourier transform of $\Phi(s)$ is $R(\tau)$, but $R(\tau)$ can also be computed directly:

$$\langle y_n^2 \rangle = \alpha_1 \langle y_n y_{n-1} \rangle + \alpha_2 \langle y_n y_{n-2} \rangle + \sigma_\theta^2 \tag{A.58}$$

$$\langle y_n y_{n-1} \rangle = \alpha_1 \langle y_{n-1} y_{n-1} \rangle + \alpha_2 \langle y_{n-1} y_{n-2} \rangle. \tag{A.59}$$

That is,

$$R(0) - \alpha_1 R(1\Delta t) - \alpha_2 R(2\Delta t) = \sigma_\theta^2 \tag{A.60a}$$

$$(1 - \alpha_2) R(1\Delta t) - \alpha_1 R(0) = 0 \tag{A.60b}$$

$$-\alpha_2 R(0) - \alpha_1 R(1\Delta t) + R(2\Delta t) = 0, \tag{A.60c}$$

which are three equations in three unknowns for $R(j\Delta t)$ and are easily solved (and also for all larger lag values); see, for example, Box et al. (2008). Notice that if estimates $\tilde{R}(j)$ are known, Equations (A.60a–c) can be used to calculate estimates of the α_i and σ_θ^2. The equations are named for Yule and Walker.

A.5 COHERENCE AND MULTIPLE TIME SERIES

Suppose two zero-mean wide-sense stationary random time series, $x\,(m\Delta t)\,=\,x_m$ and $y\,(m\Delta t)\,=\,y_m$, are measured and thought to be related in some, but not necessarily all, frequency bands. How are are these to be distinguished? The *cross-correlation coefficient*,

$$c_{xy} = \frac{1/M \sum_{i=1}^{M} x_i y_i}{\sqrt{\left(1/M \sum_{i=1}^{M} x_i^2\right)\left(1/M \sum_{i=1}^{M} y_i^2\right)}}, \tag{A.61}$$

is familiar, producing an estimate such that $-1 \le c_{xy} \le 1$. If the magnitude of c_{xy} is close to one, the two time series are said to be correlated and thus related. The fraction of the variance in one of the records that is linearly related to the other is c_{xy}^2. A limitation of this approach is that it lumps together the physics in *all* of the frequency bands making up x_t, y_t. But many real time series contain disparate physics, with contributions to the variance as a function of frequency differing by many orders of magnitude. In processes where some band of frequencies has most of the energy, c_{xy} may carry little or no information about the physics of the weaker bands. Testing the hypothesis that the relationship between x_t, y_t differs by frequency band becomes essential.

Hypothesis tests can be produced through the concept of *coherence*. Although it is not the only approach, and another is used below, a convolution relationship can be postulated:

$$y_m = \sum_{-\infty}^{\infty} \alpha_k x_{m-k} + \varepsilon_m, \tag{A.62}$$

with α_k constant, where the residual, or noise, ε_m, is *uncorrelated with* x_m, $\langle \varepsilon_m x_p \rangle = 0$ (assuming that y_n, x_m are sampled at the same interval, Δt, over the same time span.) The α_k are not random. Equation (A.62) encompasses all possible linear relationships between two time series, including simple delay, $y_m = x_{m-q} + \varepsilon_m$, or an averaging operation. By the convolution theorem,

$$\hat{y}\,(s) = \hat{\alpha}\,(s)\,\hat{x}\,(s) + \hat{\varepsilon}\,(s). \tag{A.63}$$

Because x_t, y_t are random, the amplitudes $|\hat{y}\,(s)|, |\hat{x}\,(s)|$ and the corresponding phases, $\phi_x\,(s), \phi_y\,(s)$, are also random.

Suppose for the moment that $\hat{\varepsilon}\,(s) = 0$ for some band of s. Then the ratio of amplitudes $|\hat{y}\,(s)|/|\hat{x}\,(s)|$ should *not* be random—because it must then equal $|\hat{\alpha}\,(s)|$—which is being assumed to be deterministic and a smooth function of s. Similarly, the difference of the phases $\phi_x\,(s) - \phi_y\,(s)$ is the phase of $\hat{\alpha}\,(s)$—also a smooth, nonrandom, function of s). In the opposite extreme, if $\hat{\alpha}\,(s)$ should vanish in some band of s, and unless the magnitude of $\hat{y}\,(s)$ is truly zero (almost never true), then the ratio of the amplitudes of $\hat{x}\,(s), \hat{y}\,(s)$ and their phase differences must remain completely random. In between the two extremes, some nonrandom structure can be present, depending upon the relative magnitudes of $\hat{\alpha}\,(s)\,\hat{x}\,(s)$ and $\hat{\varepsilon}\,(s)$.

Using these assumptions about smoothness in frequency, textbooks discuss tests of the hypothesis of connections such as Equation (A.63). These permit the explicit estimate of $\hat{\alpha}\,(s), \hat{\varepsilon}\,(s)$ and thus of a_k, ε_k. Useful definitions, employing smoothness implicitly, are the *cross-power* (or *cross-power density* depending upon the normalization), $\Phi_{yx}(s_n) = \langle \hat{y}(s_n) \hat{x}^*(s_n) \rangle$, and the *coherence*,

$$C_{yx}(s_n) = \frac{\Phi_{yx}(s_n)}{\sqrt{\Phi_{yy}(s_n)\Phi_{xx}(s_n)}}. \tag{A.64}$$

The notation is that $\Phi_{yx}(s) = \langle \hat{y}(s)\hat{x}(s)^* \rangle$, etc., where $|C_{yx}(s_n)| \le 1$. Then,

$$\hat{\alpha}(s_n) = \frac{\Phi_{yx}(s_n)}{\Phi_{xx}(s_n)} = C_{yx}(s_n)\sqrt{\frac{\Phi_{yy}(s_n)}{\Phi_{xx}(s_n)}}, \tag{A.65}$$

determining the α_k. Coherence is a complex quantity, whose theoretical phase is that of $\hat{\alpha}(s_n)$. If the coherence vanishes in some frequency band, then so does $\hat{\alpha}(s_n)$, and in that band of frequencies, no linear relationship exists between y_t, x_t, with the computed phase being meaningless. Should $C_{yx}(s_n) = 1$, then $y_t = ax_t$, a constant at those frequencies. (Some authors use the nonstandard term "coherence" for $|C_{yx}|^2$.) Figure A.14 shows an example applied to the measured sea level curves at Eniwetok and Kwajalein Islands.

It follows that

$$\Phi_{yy}(s_n)\left(1 - |C_{yx}(s_n)|^2\right) = \Phi_{\varepsilon\varepsilon}(s_n). \tag{A.66}$$

That is, the fraction of the power in y_t at frequency s_n, that part *not related to* x_m, is just $\left(1 - |C_{yx}(s_n)|^2\right)$ and is called the *incoherent power*. It vanishes if $|C_{yx}| = 1$, meaning that in that band of frequencies, y_t would be perfectly calculable (predictable) from x_m. Alternatively,

$$\Phi_{yy}(s_n)|C_{yx}(s_n)|^2 = |\hat{\alpha}(s_n)|^2 \Phi_{xx}(s_n), \tag{A.67}$$

which is the fraction of the power in y_t that is related to x_t. This product is called the *coherent* power, the part related to or predictable from x_n, so the total power in y_n is the sum of the coherent and incoherent components. (These values are frequency domain generalizations of the conventional correlated and uncorrelated variances.) The frequency structure of the power densities and the coherence often differ considerably; an example is visible in Figure 8.14. In an analog of the Blackman-Tukey method, coherence can be estimated from the Fourier transform of the cross-power, although that is now almost never an advisable approach.

Significance tests for the coherence amplitude are well known. The most important point is that the estimated coherence amplitude *never* vanishes, even if the two time series are known to be unrelated: the coherence amplitude thus has a *positive bias*. To determine it, a *level of no significance*, $\gamma_C \le 1$ can be used. γ_C is a probability: if the magnitude of the estimated coherence lies below γ_C, no evidence exists for true coherence, and the estimated phase would be meaningless.[18] If the magnitude lies above γ_C, it is taken as evidence of true coherence, and the phase interpreted as the phase of $\hat{\alpha}(s)$. Because the estimated amplitude is a random variable, the possibility of a false positive always exists (the value lying above γ_C might be a statistical accident). Similarly, false negatives are expected. In most geophysical contexts, a level of $\gamma_C = 0.95$ is commonly adopted—meaning that about 5% of the estimated values *should* lie falsely above the level of no significance. (Some desperate authors have used values as low as $\gamma_C = 0.8$, meaning that about 20% of the estimated values are expected to give false significance. The probability that an artifact is being addressed is then disconcertingly large.) If

[18]This statement does *not* imply that no coherence exists; merely that it has not been detected.

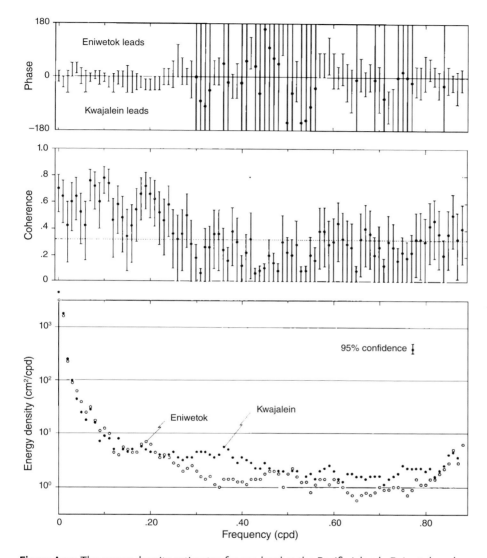

Figure A.14: The power density estimates for sea level at the Pacific islands Eniwetok and Kwajalein, and the amplitude and phase of their coherence. Where the amplitude of the coherence is above the 95% level of no significance at low frequencies, there is no detectable phase shift between them. Groves and Hannan (1968) describe how the confidence intervals were placed on both the amplitude and phase of the coherence. Note the use of logarithmic power, but linear frequency, scales. (From Groves and Hannan, 1968.)

such outliers are *not* present, it is likely that something about the assumed statistics is mistaken. When one record is possibly completely random, and the other partially deterministic, as in tidal and Milankovitch theory studies, a special statistical test has to be constructed (see Munk and Cartwright, 1966, appendix). This subject is that of *confidence intervals*, briefly described below. All of these methods can be readily generalized to vector processes involving, simultaneously, relations among three or more time series, leading to the notions of *multiple and partial coherence*.

A.6   POWER AND COHERENCE IN A WAVE FIELD

Alternatives exist to the use of the convolution relation to discuss coherence, although all can be proven equivalent. Spatially separated time series in a wave field are a common measurement type in oceanography. Consider a pure travelling sinusoid

$$q = \alpha \exp\left(2\pi i \kappa_0 r - 2\pi i s_0 t - i\phi\right). \tag{A.68}$$

The variance, $\langle qq^* \rangle = \alpha^2$, is independent of space and time. Note the use of complex conjugation in products when complex time functions are discussed. The temporal covariance between two locations $r, r + \Delta r$ is

$$\langle q(r,t)q(r+\Delta r, t+\tau) \rangle = \left\langle |\alpha|^2 \right\rangle \cos\left(2\pi s_o \tau\right) e^{2\pi i \kappa_0 \Delta r} \tag{A.69}$$

independent of r, t. By the Wiener-Khinchin theorem, its frequency Fourier transform is a δ-function at $s = s_0$, and the coherence (Eq. A.64) between these two locations has magnitude 1 and radian phase $2\pi \kappa_0 \Delta r$ at frequency s_0.

Suppose that two such waves are present with slightly different wavenumbers, $\kappa_0, \kappa_0 + \Delta \kappa$:

$$q = \alpha \exp\left(2\pi i \kappa_0 r - 2\pi i s_0 t\right) + \beta \exp\left(2\pi i \left(\kappa_0 + \Delta\kappa\right) r - 2\pi i s_0 t\right), \tag{A.70}$$

and that are not correlated with each other, requiring $\langle \alpha\beta \rangle = 0$ (a random wave field). Now the cross-covariance is

$$R\left(r, r+\Delta r, t, t+\tau\right) =$$
$$\left\langle |\alpha|^2 \right\rangle \exp\left(-2\pi i \kappa_0 \Delta r - i s_0 \tau\right) + \left\langle |\beta|^2 \right\rangle \exp\left(-2\pi i \kappa_0 \Delta r - 2\pi i \Delta\kappa \Delta r - 2\pi i s_0 \tau\right). \tag{A.71}$$

(If $\langle \alpha\beta \rangle \neq 0$, the power density would be a function of position, r, from cross-terms that would appear in Equation (A.71), and it could not be stationary in space.)

Dividing by the powers, the squared coherence magnitude at frequency s_0 is

$$|coh\left(\Delta r, s_0\right)|^2 = \left(1 + \frac{\left\langle |\beta|^2 \right\rangle}{\left\langle |\alpha|^2 \right\rangle}\right)^{-2} \left[1 + \frac{\left\langle |\beta|^2 \right\rangle}{\left\langle |\alpha|^2 \right\rangle} + 2\frac{\left\langle |\beta|^2 \right\rangle}{\left\langle |\alpha|^2 \right\rangle} \cos\left(2\pi \Delta\kappa \Delta r\right)\right]$$

$$= \left(1 + \frac{\left\langle |\beta|^2 \right\rangle}{\left\langle |\alpha|^2 \right\rangle}\right)^{-2} \left[1 + \frac{\left\langle |\beta|^2 \right\rangle}{\left\langle |\alpha|^2 \right\rangle} + 2\frac{\left\langle |\beta|^2 \right\rangle}{\left\langle |\alpha|^2 \right\rangle} \left(1 - \frac{\left(2\pi \Delta\kappa \Delta r\right)^2}{2} + \ldots\right)\right] \tag{A.72}$$

This latter expression has magnitude unity when $\Delta r = 0$ and decays quadratically as $|\Delta r|$ grows. $\Delta \kappa$ is the wavenumber *bandwidth*, and it leads to the ability to interpret the loss of coherence with spatial separation between two observing points as implying a summation of waves having different wavenumbers; see Munk and Phillips (1968).

THE RADON TRANSFORM

The Radon transform is defined as the straight-line integrals of some property, C, through a closed area, as depicted in Fig. A.15 at all possible angles, α,

$$\tilde{C}_R(\alpha) = \int_{-\infty}^{\infty} C\left[x(t), y(t)\right] dt, \quad [x(t), y(t)] = [t(\sin\alpha, -\cos\alpha), s(\sin\alpha, \cos\alpha]$$
(A.73)

(Here t, s are geometric functions.) Then if C is a plane wave travelling at a fixed angle across the domain with a dominant phase velocity, $\tilde{C}_R(\alpha)$ will be a maximum when the angle corresponds to the dominant phase velocity, $\alpha_{max} = \tan^{-1}(x/t)$. This calculation has been used by Chelton et al. (1996) and others to discuss phase velocities in altimetric data and has found widespread use in tomography, particularly in the medical version (references are given by Munk et al., 1995, p. 285–287). The inversion formula is derived by writing the function in terms of its two-dimensional Fourier transform and then manipulating it (Eq. A.73). The information content is therefore identical to that of the Fourier transform, and as the latter is generally more informative—because it permits the frequency separation distinguishing differing physical regimes—little reason exists to use the Radon transform except when the interior values are not directly accessible, as in the tomographic or seismic applications.

Useful generalizations such as multidimensional Fourier transforms exist. Thus in both frequency and one-dimensional wavenumber, a new transform pair is

$$\hat{f}(\kappa, s) = \iint_{-\infty}^{\infty} f(r, t)\, e^{-i2\pi\kappa r - 2\pi i s t}\, dr dt$$
(A.74)

and

$$f(r, t) = \iint_{-\infty}^{\infty} \hat{f}(\kappa, s)\, e^{i2\pi\kappa r + 2\pi i s t}\, d\kappa ds.$$
(A.75)

If $f(r, t)$ is real, various symmetries among κ, s exist. This and higher-dimensional versions all have discrete finite equivalents, and coherences, space-time autocovariances, etc. can all be defined. A special case of the two-dimensional transform, where both coordinates are spatial, is the Radon transform, sometimes applied to tomographic and altimetric measurements (see the box " The Radon Transform").

Higher-dimensional Fourier transforms permit generalization of the wavefield coherence to wavefields having arbitrary, but spatially stationary, Fourier transforms:

$$coh(\Delta r, s) = \frac{\int_{-\infty}^{\infty} \left\langle |\hat{f}(\kappa, s)|^2 \right\rangle \exp(2\pi i \kappa \Delta r)\, d\kappa}{\int_{-\infty}^{\infty} \left\langle |\hat{f}(\kappa, s)|^2 \right\rangle d\kappa}.$$
(A.76)

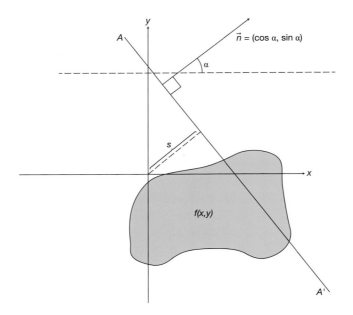

Figure A.15: Geometry of the Radon transform, in which integrals are measured along all straight lines passing at all angles, α, through an object.

A.7 A NOTE ON CONFIDENCE LIMITS

The *standard error* of an estimated variable is determined from the fluctuations of the measurements contributing to the value, as computed from the sample standard deviation. Thus a sample average of M values, $x_i = m + \xi_i$, where m is the true average sought, and ξ_i are the "noise," having assumed zero-mean value and variance σ_ξ^2 and to be uncorrelated, is

$$\tilde{m} = \frac{1}{M} \sum_{i=1}^{M} x_i,$$

(Eq. A.4) and with a *sample variance*, about the sample average,

$$\tilde{s}_d^2 = \frac{1}{M-1} \sum_{i=1}^{M} (x_i - \tilde{m})^2, \tag{A.77}$$

If, in addition, the ξ_i are Gaussian variables, then so is $\tilde{m}$, and in particular $\tilde{m}$ has a probability density with a mean value equal to the true mean m (which isn't known). How reliable is $\tilde{m}$ as an estimate of m?

Assuming $\tilde{s}_d \approx \sigma_x$, then for the Gaussian case, the true value, m, lies, with about 68% probability, in the interval $\tilde{m} \pm \tilde{s}_d / \sqrt{M}$. It lies with about 95% probability in the two-standard-deviation interval $\tilde{m} \pm 2\tilde{s}_d / \sqrt{M}$. The quantity $\pm \tilde{s}_d / \sqrt{M}$ can be interpreted as an approximate *68% confidence interval*, and $\tilde{m} \pm 2\tilde{s}_d / \sqrt{M}$, as a *95% confidence interval*. One interpretation is that in the first case, the true mean lies outside the interval with about a 32% probability, and in the latter, there is only a 5% chance that it does so. (Slightly more sophisticated methods account for the use of $\tilde{s}_d$ without assuming it equal to σ_ξ—in Student's t-test.)

The "standard error" of the mean, $\tilde{s}_d/\sqrt{M}$, can be made arbitrarily small by letting M become arbitrarily large. But consider an extreme, *reducio ad absurdum*, case: the average temperature at one point in an ordinary room is sought, and a large number of measurements is obtained, perhaps 10^8, in a span of 1 s. Intuition rebels at an uncertainty calculated as $\sigma_x/\sqrt{10^8}$, because the measurement errors would be essentially identical, with the true number of independent ones being perhaps only one. Data manipulation often drastically reduces the number of degrees-of-freedom. In particular narrow-band filtering of a time series in the limiting case reduces the time series to one having but 2 degrees-of-freedom—the amplitude and phase of the resulting pure sinusoid. Estimation of the equivalent number of independent noise measurements becomes the essential step.

Approximate confidence intervals are commonly computed for spectral estimates, using the χ_ν^2 distribution, and for coherence amplitudes and phases involving more complicated probability densities described in the textbooks. These are used to determine if some values are outside the limits and are thus of *significance* (e.g., a spectral peak relative to a background spectrum). If a too-narrow confidence interval is being used, the probability of a false positive can be uselessly large. Confidence intervals are only guidelines. For example, claiming an apparently statistically significant deviation from a patently false nominal value is wrong. Numerous papers proclaim "peaks" in spectra measured as a deviation from the spectrum of an AR(1) process—often when the latter is not conceivably an acceptable physical model.

Many, if not most, oceanographic measurements are *not* independent or uncorrelated, and typically the number of degrees-of-freedom is less, and sometimes much less than the number of data points. Priestley (1981, Sec. 5.3) has a useful discussion of how to estimate true degrees-of-freedom from sample autocovariances. As one example, his Equation (5.3.25) asserts that for sufficiently large numbers, M, of sample data points, the variance of the autocovariance estimate $\tilde{R}$ itself depends upon the true autocovariance R,

$$var\left[\tilde{R}\left(\tau\right)\right] = \frac{1}{M} \sum_{n=-\infty}^{\infty} \left[R^2\left(n\right) + R\left(n+r\right)R\left(n-r\right)\right], \quad \Delta t = 1, \qquad (A.78)$$

which might be roughly estimated from $\tilde{R}\left(j\right)$ itself. Strongly autocovarying processes have much reduced equivalent degrees-of-freedom (consider the limit of a pure sinusoid).

In some fields, notably drug tests, 99%, or even higher, confidence intervals are required. Oceanographers rarely have the luxury, or need, for demanding such degrees of confidence, but always being on the alert for the possibility of an artifact is good advice. Seife (2000) has provided an interesting discussion of the need for having very high confidence before proclaiming new discoveries, and of apparently robust examples that turned out not to exist.[19] Comparisons of two estimates of some quantity are often carried out by asking whether their standard errors or confidence intervals overlap, with an overlap being considered evidence of consistency. As Lanzante (2005) and others emphasize, this apparent consistency can be spurious. In recent years, the availability of cheap, fast computers has made it possible to determine confidence limits and related estimates of uncertainty using brute-force resampling schemes without the need to work through the analytical expressions. These methods are known as the *bootstrap* and *jackknife* (Efron and Tibshirani, 1993).

---

[19] The search for interesting, positive results also strongly biases the literature. Lehrer's article (2010) is a non-technical introduction to a widespread problem. Diaconis and Mosteller (1989) discussed the inevitability, and misinterpretation, of coincidences.

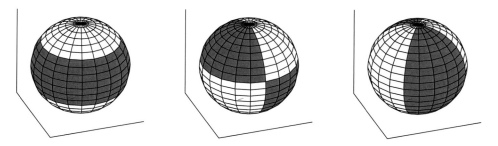

Figure A.16: Real parts of three of the lower-degree and lower-order spherical harmonics, $Y_2(\theta)$, $Y_2^1(\theta, \lambda)$, $Y_2^2(\theta, \lambda)$, respectively. Colatitude, θ, is measured from the north pole.

A.8 SPHERICAL HARMONICS

Spherical harmonics are the natural generalization to spherical coordinates of the Fourier sines and cosines of Cartesian coordinates. In this book, their primary application is in the representation of the tidal and insolation forcing functions. They are of the form

$$Y_n^m (\theta, \lambda) = A_n^m P_n^m (\cos \theta) \, e^{\pm im\lambda}, \tag{A.79}$$

which uses colatitude, $\theta = \pi/2 - \phi$, and longitude, λ, as in Figure 6.4 (ϕ is latitude). The $P_n^m (\cos \theta)$ are the so-called associated Legendre polynomials, and almost always, n, m are integers called the degree and order, respectively. $|m| \le n$. (A_n^m are normalizing factors. See the box "Spherical Harmonics" and Figure A.16.) Equation (A.79) shows that the functions are essentially ordinary Fourier sines and cosines or equivalent complex exponentials in the zonal direction but are more structured in the meridional one owing to meridian convergence.

The A_n^m are chosen from an array of different normalizations in different fields, such that

$$\iint |Y_n^m (\theta, \lambda)|^2 \cos \theta d\theta d\lambda = M,$$

where M is a constant. When integrated over the sphere, any two Y_n^m with at least one differing index are orthogonal. The normalizations in geodesy (of importance in satellite altimetry, sea level change, paleoclimate,...) differ from those in geomagnetism (which is important in navigation and in some vector current-measuring devices) and from those in classical physics. Physics chooses $M = 1$, which renders them *orthonormal*.

Assuming orthonormality,

$$F (\theta, \lambda) = \sum_{n=0}^{\infty} \sum_{m=-n}^{n} \alpha_n^m Y_n^m (\theta, \lambda), \tag{A.80}$$

for any physically reasonable function F where

$$\alpha_p^q = \int_0^{2\pi} \int_0^{\pi} F (\theta, \lambda) \, Y_p^q (\theta, \lambda) \cos \theta d\theta d\lambda. \tag{A.81}$$

The real and imaginary parts of the Y_n^m produce cosine and sine components, respectively, so that $\mathrm{Re} \left\{ Y_n^m (\theta, \lambda) \, e^{-i\sigma t} \right\} = A_n^m P_n^m (\theta, \lambda) \cos (m\lambda - \sigma t)$ is a travelling wave, and similarly for the imaginary part.

SPHERICAL HARMONICS

Spherical harmonics are the generalization to a sphere of the Cartesian space cosine and sine functions used in Fourier analysis. Following the treatment given by Jackson (1999), define them here as

$$Y_l^m(\theta, \lambda) = \sqrt{\frac{(2l+1)(l-m)!}{4\pi(l+m)!}} P_l^m(\cos\theta) e^{im\lambda},$$

where $l = 0, 1, 2, \ldots, |m| \le l$, $P_l^0 = P_l$ are the Legendre polynomials and when $m = \pm 1, 2, \ldots$ the $P_l^m(\cos\theta)$ are the associated Legendre polynomials as defined in textbooks as

$$P_l^m(x) = (-1)^m (1-x^2)^{m/2} \frac{d^m}{dx^m} P_l(x), \quad m \le l.$$

The first few $P_l(\cos\theta)$ are

$$P_1(\cos\theta) = 1, \quad P_1(\cos\theta) = \cos\theta, \quad P_2(\cos\theta) = \left(3/2\cos^2\theta - 1/2\right),$$
$$P_3(\cos\theta) = \left(5/2\cos^3\theta - 3/2\cos\theta\right). \quad (A.82)$$

The physicist's normalization produces

$$\int_0^{2\pi} \int_0^{\pi} |Y_l^m(\theta, \lambda)|^2 \sin\theta d\theta d\lambda = 1.$$

The different $Y_l^m(\theta, \lambda)$ can be shown to be orthogonal, hence *orthonormal*, over the sphere.

The first few are:

$$Y_0^0 = 1/\sqrt{4\pi} \quad (A.83)$$
$$Y_1^0 = \sqrt{3/4\pi}\cos\theta, \quad Y_1^{\pm 1} = \mp\sqrt{3/8\pi}\sin\theta e^{\pm i\lambda} \quad (A.84)$$
$$Y_2^0 = \sqrt{5/4\pi}\left(\frac{3}{2}\cos^2\theta - \frac{1}{2}\right), \quad Y_2^{\pm 1} = \mp\sqrt{15/8\pi}\sin\theta\cos\theta e^{\pm i\lambda} \quad (A.85)$$
$$Y_2^{\pm 2} = \sqrt{15/32\pi}\sin^2\theta e^{\pm 2i\lambda}$$

On a water-covered sphere (*aquaplanet*), the Y_n^m are a complete set, meaning that an arbitrary function, F, can be represented exactly as in Equation (A.80). (As with ordinary Fourier methods, mathematicians can construct pathological functions for which these expansions fail, but I am unaware of any physical situation in which they would exist.) Also, as with the transition from continuous representations to discretized ones in ordinary Fourier series, fast transform methods have been developed (e.g., see Spherepack on the website of the National Center for Atmospheric Research), accompanied by some important numerical details that do not change the physical interpretation.

On a sphere containing land regions, the situation is more complex. That n is an integer is dictated by the requirement that the representations should be finite at one or both poles, $\theta = 0, \pi$, and m is an integer only if the domain extends over 360° of longitude. In realistic oceanic geometries, spherical harmonics are neither orthogonal nor scale selective, and a number of methods have been proposed for alternative representations, none of which are particularly compelling nor has found widespread use. Singular vectors (empirical orthogonal functions, EOFs; see Appendix B) are an orthogonal basis but provide no control over spatial scale and have a sometimes confusing mix of time and space. Some interesting attempts have been made using wavelets generalized to spherical cap geometries,[20] but the subject remains incomplete and no widely used method has emerged.

A.9 MAKING MAPS

A central problem in employing spatially distributed data concerns the problem of mapping it. Maps have at least two distinct uses: (1) visualizing scattered noisy measurements and (2) providing numbers to numerical models written on regular spatial grids. A map drawn for purpose (1) need not be on a regular grid as in (2), but it often is the choice. The problem is taken up in summary here, because it is so central to the interpretation of much oceanographic data.

Traditionally, measurements were placed as point values on a geographical map, and the values were hand contoured according to some explicit, or more commonly, implicit, aesthetic process. Defant (1950) and Fuglister (1955) called attention to the very different outcomes that were possible (see Fig. A.17).

Recognition of this ambiguity problem, as well as the swelling volume of observations that was becoming too large for human contouring, led to methods for so-called objective mapping. A distinction is made between "objective" methods and "optimal" ones and between "mapping" and "interpolation."

Consider a one-dimensional problem: temperature, ξ, at a fixed depth is measured by a ship along a line at positions r_i. The hypothetical true values are $\xi(r_i)$, but because perfect measurements do not exist, what is obtained are the samples $y_i = \xi(r_i) + \varepsilon_i$, where ε_i is the noise in the data. For simplicity, ignore errors in r_i, which will also always be present, but with good GPS navigation, they might be negligible. Suppose too, that temporal changes during the measurement process are negligible. (All of these assumptions can be removed.) $\xi(n\Delta r)$ is sought, where Δr is a uniform spacing, and $1 \le m \le M$ produces a regular grid, $\tilde{r}_p = p\Delta r$. One or more $m\Delta r$ might coincide with an observation position

A conventional approach is to linearly interpolate the data, y_j, between neighboring samples as depicted. At any place where $m\Delta r = r_j$, the value will equal the observed one, and this defines true "interpolation." More complicated ordinary interpolation schemes are widely available and used (splines, quadratic curves, Aitken-Lagrange; see, e.g., Press et al., 2007; Wunsch, 2006, p. 156). Such rules are "objective" in the sense that, given the same data, two different investigators will produce the same gridded ("mapped") field.

Somewhat more elaborate techniques that have come to be widely used in the last thirty to forty years (Bretherton et al., 1976) are based upon the need for additional information not used in ordinary interpolation, and to gain more information about the result. Consider one possibility: a sample, y_k, happens to lie on a grid point, $m\Delta r$, but it is clustered with

---

[20] Percival and Walden, 2000; Slobbe et al., 2012.

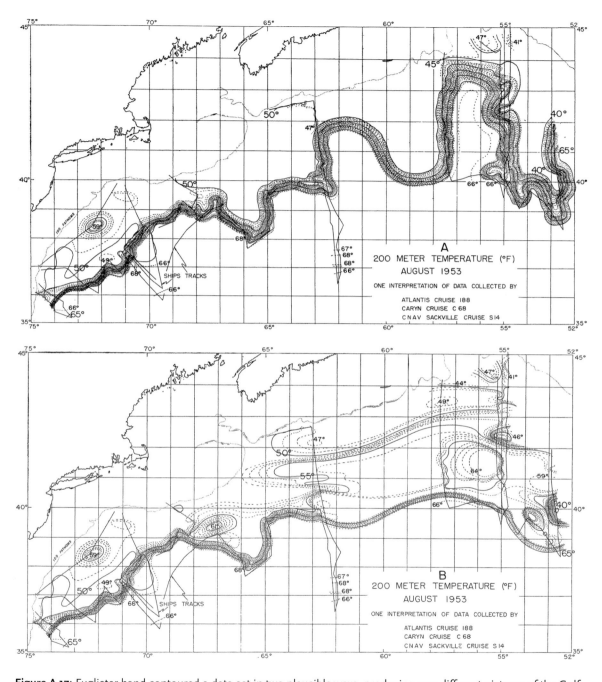

Figure A.17: Fuglister hand contoured a data set in two plausible ways, producing very different pictures of the Gulf Stream from the same data. Such pictures tend to be dependent upon the particular investigator doing them and so are not regarded as "objective," a term that can be interpreted to mean "reproducible" but that is not the same as "correct" or" accurate." The ambiguities can be regarded as resulting from space-time aliasing of the underlying fields. (Used with permission from F. C. Fuglister [1955]. "Alternative analyses of current surveys." *Deep-Sea Res.* 2: 213–229.)

many other samples. Using interpolation, the value assigned at the grid location will be the measured one, including its noise component. Averaging the local samples might produce a better estimate of $\xi\left(m\Delta r\right)$ than does y_k by itself. The result will no longer be interpolation in the conventional definition. But what constitutes "local"? And should the data points be given different weights in the average? That information is provided by the spatial covariances of *both* the noise and temperature fields.

Most of the methods in use are some variation or simplification of what is often called the *Gauss-Markov theorem*. Commonly some information is available concerning the magnitudes and spatial covariances (correlations) in the true field, ξ, and in the noise of measurement ε_i. That information can be exploited in making estimates both of the field of interest, as well as the underlying noise, which in many cases is really the signal of interest.

For simplicity, assume that all fields have zero spatial averages, which can usually be arranged with acceptable accuracy simply by removing the sample mean. More sophisticated methods are available, if needed, often under the label *kriging*. Again employing a bracket, $\langle g \rangle$, to denote the true average of a quantity, the information about spatial extents and variances is all contained in the covariance functions $\mathbf{R} = \langle \varepsilon_i \varepsilon_j \rangle$, and $\mathbf{S} = \langle \xi\left(r_p\right)\xi\left(r_m\right)\rangle$, which are both ordinary covariance matrices. All quantities here are being treated as though they were random—often a good assumption—with the true average being definable from an underlying, plausible probability density.) Then the variances, with the information about their magnitudes, are specified with $i = j$, and $n = m$, that is, the diagonals of the matrices. In many cases, that is the only information an investigator may feel comfortable in providing. If even these variances are unknown, it raises the question of whether any estimate should be attempted—as there would be no way to evaluate its quality and would imply that arbitrarily large or small numerical values had to be accepted.

Assuming that the estimate of $\xi\left(\tilde{r}_p\right)$ will be a linear combination of *all* of the data, not just the nearest neighbors,

$$\tilde{\xi}\left(\tilde{r}_p\right) = \sum_{j=1}^{M} b\left(\tilde{r}_p, r_j\right) y\left(r_j\right) = \sum_{j=1}^{M} b\left(\tilde{r}_p, r_j\right)\left(\xi\left(r_j\right) + \varepsilon_j\right).$$

A tilde, ˜, is used to distinguish the estimated value from the true one and also to denote the positions, $\tilde{r}_p$, where the estimate is to be made. Using $\mathbf{R}, \mathbf{S}$, the Gauss-Markov theorem dictates the choice of the weights $b\left(\tilde{r}_p, r_j\right)$, depending upon both the position where the mapped point is sought and where the observations are made, in such a way as to make the theoretical mean square difference $\left\langle \left(\tilde{\xi}\left(\tilde{r}_p\right) - \xi\left(\tilde{r}_p\right)\right)^2 \right\rangle$ as small as possible (resulting in what is called a *minimum error variance estimate*). The same machinery provides information about just how big this difference will be (the error variance) and how the errors in the mapped values are correlated spatially. Spatial correlations of errors are of major importance for mapped fields that will undergo differentiation, for example, in models.

Often the covariances $\mathbf{R}, \mathbf{S}$ are poorly known, beyond perhaps some knowledge of the diagonals (variances) and a plausible guess about the scale of decorrelation. What is commonly done is to make a reasonable guess of their diagonal values and decorrelation distance and proceed. The result is then objective, but it is *not* optimal. Much recommends this procedure (it is repeatable), and it is only objectionable when it is forgotten that the result does depend directly upon the information contained in $\mathbf{R}, \mathbf{S}$. Mapped values, in regions of dense data, tend to be insensitive to $\mathbf{R}, \mathbf{S}$, but the estimated uncertainties can be sensitive to them. Checking that the resulting estimated fields are not qualitatively inconsistent with $\mathbf{R}, \mathbf{S}$ is an important safeguard against blunders. Although the two different maps in Figure A.17 were made subjectively, with artistic license, they do represent, qualitatively, two different assumptions about the structures of the spatial covariances.

The methodology generalizes readily to two, three, and four dimensions and is now the common default procedure. For observations involving highly non-Gaussian, for example, bimodal, fields where near-average values are consequently rare, it may be distinctly inappropriate.

Inverse and State Estimation Methods

B.1 INVERSE METHODS AND INVERSE MODELS

The basic ideas of inverse methods, including the subset we call *state estimation*, as applied in physical oceanography are sketched here. Greatly extended discussions are available in the references, but the underlying ideas are simple. Some of the confusion exists because of terminology, and so to begin it is helpful to distinguish an inverse *method* from an inverse *model*.[1]

An inverse model is one having a negative definition: it is one, be it analytical or numerical, that differs from what is conventionally defined as a *forward* model. Forward models are encountered in elementary calculus. One example: given a differential equation, find $y(t)$, where

$$\frac{d^2 y(t)}{dt^2} + \alpha^2 y(t) = q(t), \ t \geq 0, \tag{B.1}$$

(in continuous time) subject to $y(t=0) = y_0$, $y'(t=0) = dy(0)/dt$, and the constant α^2 and the function $q(t)$ are known. The independent variable t is suggestive of a time dependence, but nothing precludes defining it as a spatial coordinate, r. Elementary methods show that this problem is *well-posed*, meaning that just enough information is available to calculate a unique solution, one that has several desirable properties such as continuity or differentiability. Standard methods exist for finding the solution for these and related forms (e.g., replacing known $y'(t=0)$ by known $y(t_1)$, $t_1 > t_0$). Because it belongs to the canon of mathematical physics, it is a *forward problem*, or *model*.

Once observations are involved, the tidy framework of forward problems becomes inadequate; in the real world, this problem is modified in many ways. For example, suppose $y(t)$ were known and $q(t)$ was sought. A simple solution procedure exists: substitute $y(t)$ into Equation (B.1) and the problem is solved. Although solvable by an almost trivial operation,

---

[1] A more extended account is Wunsch, 2006.

determining $q(t)$ is labeled as an *inverse* problem because it differs from the conventional forward one, and the equation has become an inverse model.

Or, suppose the initial conditions have an error, stipulated as $y(t = 0) = y_0 \pm \Delta y_0$, and the derivative $y'(t = 0) = y_0' \pm \Delta y_0'$, where the meaning of the Δ is often a standard deviation. Possibly $q(t)$ and/or a^2 are also only approximately known. Problems are encountered in which the starting time, $t = 0$, is uncertain or the times used to describe $q(t)$ are only approximate. Some knowledge of $y(t)$ might be available over a limited domain, $t_s \leq t \leq t_f$, with or without errors. Any combination of these things can be encountered simultaneously. A unique solution no longer exists, and some of the information is likely to be contradictory so that classically, by convention, no solution exists. This collection of *ill-posed problems*, which sensibly includes all of the ancillary information, such as the error sizes, requires methods different from those used in the usual mathematical physics texts.

Inverse *methods*, including ones based on least-squares, are able to cope with ill-posed problems, ones in which there may be too much or too little, and/or contradictory or noisy information. Problems arising from the use of observations are almost inevitably ill posed.

B.2 LEAST-SQUARES

Least-squares is so familiar that its great power and some of the nuances are often overlooked. Much interesting mathematics starts with continuous measures of squared differences between two quantities,

$$J = \int_a^b (f(q, p_1) - g(q, p_2))^2 \, dq,$$

where f, g can be models, data, or some combination, and the p_i are functions. Commonly, J is to be made as small as possible. Handling such quantities—for example, to find p_i so as to make the integral a minimum, perhaps subject to auxiliary constraints—is a problem in infinite dimensional spaces and functional analysis. The calculus of variations is one of the tools available for such problems. The ubiquitous use of computers for handling models and data, however, means that then nothing is ever infinite dimensional, that all quantities so represented are necessarily discontinuous, and that the problems all reduce computationally to forms of least-squares where minimizing,

$$J = \sum_{j=1}^{M} (f(q_j, \mathbf{p_1}) - g(q_j, \mathbf{p_2}))^2 \, \Delta q_j,$$

is an example, the index j representing a discrete variable, the $\mathbf{p}_i$ are vectors and M is always finite.

The choice of the square as a measure of difference (the *2-norm*) is not the only possibility. A *1-norm* problem of minimizing,

$$J_1 = \sum_{j=1}^{M} |f(q_j, \mathbf{p_1}) - g(q_j, \mathbf{p_2})| \, \Delta q_j,$$

or the *infinity norm*,

$$J_2 = \max |f(q_j, \mathbf{p_1}) - g(q_j, \mathbf{p_2})| \, \Delta q_j,$$

have definite advantages in some circumstances.

The 2-norm is, however, intimately connected with Gaussian probability distributions and, for the very large problems of oceanography, has mature associated software designed for existing computer architectures. One-norms, in the guise of linear programming, have an equally large and mature software, but oceanographic applications of that method remain rare.

BASIC IDEAS

Most applications of least-squares in oceanography, meteorology, and climate are directed at the solution of sets of simultaneous equations, ones that arise from models, both time dependent and static, and from data. Consider first the fundamental ideas, using the standard matrix-vector notation in the box "Vector Notation and Basic Machinery." Let

$$\mathbf{A}\mathbf{x} = \mathbf{b} \tag{B.2}$$

be a set of M conventional simultaneous linear equations in L unknowns, $\mathbf{x}$, with knowns, $\mathbf{b}$. The most familiar situations are $M = L$, or $M > L$, but in practice, often the relevant situation is $M < L$, or even $M \ll L$. Whatever the ratio M/L, the system of equations can be inconsistent, and a more realistic representation of Equation (B.2) is

$$\mathbf{A}\mathbf{x} + \boldsymbol{\varepsilon} = \mathbf{b}, \tag{B.3}$$

where $\boldsymbol{\varepsilon}$ is an error vector. If $M > L$, $\boldsymbol{\varepsilon} = 0$ almost never. In problems involving models and/or data, zero-noise solutions are mainly irrelevant except as theoretical limits.

Calculus and physics courses instruct students to seek the minimum squared error solution,

$$\min : J = \boldsymbol{\varepsilon}^T \boldsymbol{\varepsilon} = (\mathbf{b} - \mathbf{A}\mathbf{x})^T (\mathbf{b} - \mathbf{A}\mathbf{x}). \tag{B.4}$$

Taking the partial derivatives of J with respect to $\mathbf{x}$ and setting them to zero results in

$$\tilde{\mathbf{x}} = \left(\mathbf{A}^T\mathbf{A}\right)^{-1}\mathbf{A}^T\mathbf{b}, \quad \tilde{\boldsymbol{\varepsilon}} = \mathbf{b} - \mathbf{A}\tilde{\mathbf{x}}, \tag{B.5}$$

where tildes are placed on $\mathbf{x}, \boldsymbol{\varepsilon}$ to distinguish them from other possible solutions. The conventional solution, Equation (B.5), raises several immediate questions: (1) Why should the minimum error solution be the appropriate one to choose? Suppose the magnitude of $\tilde{\boldsymbol{\varepsilon}}$ proves to be much smaller than anticipated? (2) When does $\left(\mathbf{A}^T\mathbf{A}\right)^{-1}$ exist? And what should be done if it doesn't? (3) What if some of the elements, x_i, are much larger or smaller than anticipated? (4) Suppose some linear combinations of x_i or ε_j are known exactly; can solutions be found that are consistent with those constraints? (5) Just how sensitive are $\tilde{\mathbf{x}}, \tilde{\boldsymbol{\varepsilon}}$ to disturbances in $\mathbf{b}$? (6) Perhaps variances and covariances $\langle x_i x_j \rangle$, or $\langle \varepsilon_i \varepsilon_j \rangle$ are known; can that information be used? (7) If $M > L$, the system is usually labelled *overdetermined*. Notice, however, that Equation (B.5) has provided values for $M + L$ variables, that is M of the ε_i, and L of the x_i. Why are such problems labelled overdetermined when more unknowns than knowns exist?

All of these questions have answers, and they are explored in the references (e.g., Lawson and Hanson, 1995; Wunsch, 2006a). Methods that address these and other questions are usefully defined as *inverse methods*, because they extract information from formally underdetermined, contradictory information; exploit ancillary information such as solution statistics; and provide a full understanding of what has been determined and how well. Least-squares, if properly used, is a very powerful inverse method. Without attempting to reproduce those extended accounts, we sketch *one* approach to providing general machinery, because of its great utility even beyond the least-squares context.

SINGULAR VALUE DECOMPOSITION

Consider any rectangular matrix $\mathbf{A}$ of dimension $M \times L$. A theorem exists (named for Eckart, Young, and Mirsky) that says

$$\mathbf{A} = \lambda_1 \mathbf{u}_1 \mathbf{v}_1^T + \lambda_2 \mathbf{u}_2 \mathbf{v}_2^T + \ldots + \lambda_K \mathbf{u}_K \mathbf{v}_K^T = \sum_{j=1}^{K} \lambda_j \mathbf{u}_j \mathbf{v}_j^T \tag{B.6}$$

exactly. Here K is no larger than the minimum of M and L, written $K \le \min(M, L)$. λ_i are all positive (or zero, but then they do not contribute to $\mathbf{A}$) and are the *singular values*, and the $\mathbf{u}_i, \mathbf{v}_i$ are *singular vectors* of dimension M, L, respectively. Singular vectors are mutually orthonormal, $\mathbf{u}_i^T \mathbf{u}_j = \delta_{ij}$, $\mathbf{v}_i^T \mathbf{v}_j = \delta_{ij}$, and are found from solutions of

$$\mathbf{A}\mathbf{A}^T \mathbf{u}_i = \lambda_i^2 \mathbf{u}_i, \quad \mathbf{A}^T \mathbf{A} \mathbf{v}_i = \lambda_i^2 \mathbf{v}_i, \tag{B.7}$$

and/or

$$\mathbf{A}\mathbf{v}_i = \lambda_i \mathbf{u}_i, \quad \mathbf{A}^T \mathbf{u}_i = \lambda_i \mathbf{v}_i, \quad i = 1, \ldots, K. \tag{B.8}$$

The equations in (B.7) are a pair of eigenvalue problems involving square symmetric matrices $\mathbf{A}\mathbf{A}^T$ and $\mathbf{A}^T\mathbf{A}$ of dimension $M \times M$ and $L \times L$, respectively, and whose solutions are discussed in numerical analysis textbooks.

The solution minimizing J in Equation (B.4) is

$$\tilde{\mathbf{x}} = \sum_{i=1}^{K} \frac{\mathbf{u}_i^T \mathbf{y}}{\lambda_i} \mathbf{v}_i + \sum_{i=K+1}^{L} \alpha_i \mathbf{v}_i, \tag{B.9}$$

and the minimum noise in this norm is

$$\tilde{\boldsymbol{\varepsilon}} = \mathbf{y} - \mathbf{A}\tilde{\mathbf{x}} = \sum_{i=K+1}^{M} \mathbf{u}_i \left(\mathbf{u}_i^T \mathbf{y} \right). \tag{B.10}$$

The α_i are unknown and represent components of the solution that cannot be determined from the available data. Tildes have been written on $\mathbf{x}, \boldsymbol{\varepsilon}$ because the solution depends upon the choice of K and the unknown α_i. If the latter are set to zero, the special solution, $\tilde{\mathbf{x}}$, which is also of minimum size, is found as well.

Solutions in this form can be manipulated in a great many useful ways to provide uncertainty estimates, sensitivity of the solutions to particular data points, etc. Crucially, no matter what the relative magnitudes of M, L, the determinable elements of the solution can always be extracted, and the indeterminate structures (second term on the right of Equation (B.9) specifically described. The singular value decomposition (SVD) solution(s) provide least-squares solutions to simultaneous equations and are readily shown to reduce to familiar, standard solutions when matrix inverses such as $(\mathbf{A}^T\mathbf{A})^{-1}$, $(\mathbf{A}\mathbf{A}^T)^{-1}$ do exist. By using the singular vectors, a great deal of information about the system of equations can be found that is not obvious in more conventional solutions.

EMPIRICAL ORTHOGONAL FUNCTIONS

The Eckart-Young-Mirsky theorem applies to *any* matrix. In some oceanographic applications, those not involving equations, the matrix rows correspond to fixed times and the columns to fixed geographical positions. So, for example, **A** might be constructed from altimetric measurements on a set of latitude-longitude points in the ocean with each point assigned to a column containing the time-dependent measurements. In that configuration, the $\mathbf{v}_i$ are often called the *empirical orthogonal functions* (EOFs; a set of spatial patterns) with coefficients, or *loadings*, $\mathbf{u}_i$, that depend upon time. *Principal components* and similar terms are closely related to the EOFs (which involve differing normalizations).

A corollary of the Eckart-Young-Mirsky theorem is that the truncated form,

$$\mathbf{A} \approx \lambda_1 \mathbf{u}_1 \mathbf{v}_1^T + \lambda_2 \mathbf{u}_2 \mathbf{v}_2^T + \ldots + \lambda_{K'} \mathbf{u}_{K'} \mathbf{v}_{K'}^T = \sum_{j=1}^{K'} \lambda_j \mathbf{u}_j \mathbf{v}_j^T, \quad K' < K, \tag{B.11}$$

is the best possible representation in K' pairs of orthogonal vectors. Because the squared norm (magnitude) of **A** is $\sum_{j=1}^{K} \lambda_j^2$, and because sometimes the magnitudes of λ_j^2 fall sharply with j, an expansion as in Equation (B.11) can be a very accurate representation of **A**, sometimes with as few as two or three vector pairs.

The physical interpretation of the singular vectors (or EOFs) arises from the equations in (B.7), showing that they are the eigenvectors of the measurement covariances that are the columns and rows of $\mathbf{A}^T \mathbf{A}$ or $\mathbf{A} \mathbf{A}^T$ and are thus pattern correlations. Von Storch and Zwiers (2001) provide an extended discussion of EOFs, their manipulation, and extensions.

MODEL STRUCTURES

Although the details are omitted here (see any book on numerical methods), static linear partial differential equations such as

$$\nabla^2 \phi = \rho,$$

in discrete form become coupled linear difference equations such as

$$\frac{\phi_{i+1,j} - 2\phi_{ij} + \phi_{i-1,j}}{\Delta x^2} + \frac{\phi_{i,j+1} - 2\phi_{ij} + \phi_{i,j-1}}{\Delta y^2} = \rho_{ij}, \tag{B.12}$$

where i, j are grid indices. Mapping i, j into a one-dimensional index, α, these are evidently a set of simultaneous equations in ϕ_α and can be written as

$$\mathbf{C}\phi = \rho, \tag{B.13}$$

where ϕ, ρ are vectors formed from ϕ_{ij}, ρ_{ij}. Similar versions can be written in three or higher dimensions. **C** is usually very sparse, and sometimes

$$\phi = \mathbf{C}^{-1}\rho. \tag{B.14}$$

$\mathbf{C}^{-1}$ conventionally exists only if it is square and nonsingular. In Equation (B.12), fewer equations than unknowns exist, and **C** will be square only if the correct number of boundary conditions are provided. Methods such as the SVD can always be used. If a model is nonlinear, a set of simultaneous equations can still be constructed, but the solution is usually found by iteration instead of immediate matrix inversion.

VECTOR NOTATION AND BASIC MACHINERY

Vectors are boldface lowercase, $\mathbf{q}$. Unless specified otherwise, they are all column vectors. Matrices are written as boldface, uppercase letters, $\mathbf{A}, \mathbf{B}$. Their dimensions are written as $M \times L$, where M is the number of rows, and L, the number of columns. Transposition, $\mathbf{A}^T$, interchanges rows and columns rendering the matrix as $L \times M$. When applied to a column vector, $\mathbf{q}^T$, it becomes a row vector, and vice versa. Inverses, $\mathbf{A}^{-1}$ are defined only for square matrices, $M = L$, and do not necessarily exist. Products $\mathbf{AB} = \sum_{j=1}^{P} A_{ij} B_{jk}$, $\mathbf{Aq}$, imply correct dimensions to make sense. Thus if $\mathbf{A}$ is $M \times P$, $\mathbf{B}$ must be $P \times L$.

Vector derivatives are defined such that

$$\frac{\partial \mathbf{q}^T \mathbf{x}}{\partial \mathbf{x}} = \mathbf{q}, \quad \frac{\partial \mathbf{q}^T \mathbf{x}}{\partial \mathbf{q}} = \mathbf{x}, \tag{B.15}$$

$$\frac{\partial \mathbf{Ax}}{\partial \mathbf{x}} = \mathbf{A}^T. \tag{B.16}$$

The third equality is a consequence of the first two (noting that $\mathbf{Ax}$ is a vector of the dot products of the rows of $\mathbf{A}$ with $\mathbf{x}$) and is the connection between transposes and adjoint operators.

The eigenvalue-eigenvector problem is

$$\mathbf{Ax}_i = \lambda_i \mathbf{x}_i, \quad i = 1, \ldots, M$$

for $M \times M$ square matrices.

The norm, or size, of a matrix can be defined in a number of ways. A useful one is the $2 - norm$, $\|\mathbf{A}\|_2 = \sqrt{\text{maximum eigenvalue}(\mathbf{A}^T \mathbf{A})}$.

Obtaining ϕ from a matrix inversion as in Equation (B.14) is unlikely to be carried out for large problems, simply because the $\mathbf{C}$ matrix is both extremely large and sparse, and various efficient numerical methods exist for solving sparse linear equations. But their use is equivalent to having found $\mathbf{C}^{-1}$, and recognition that the solution is identical to that in (B.14) is important.

Consider now a linear, time-dependent model, such as the mass-spring oscillator equation with two initial conditions, a position and a velocity,

$$\frac{d^2 y}{dt^2} + r \frac{dy}{dt} + ky = q(t), \; y(0) = y_0, \quad y'(0) = y'_0 \tag{B.17}$$

that can be discretized in many ways. One is (as used in Appendix A),

$$y(t + \Delta t) - 2y(t) + y(t - \Delta t) + r\Delta t (y(t) - y(t - \Delta t)) + \Delta t^2 ky(t) \tag{B.18}$$
$$= \Delta t^2 q(t), \quad t = n\Delta t,$$

or collecting terms,

$$y(t + \Delta t) + \left(\Delta t^2 k - 2 + r\right) y(t) + (1 - r\Delta t) y(t - \Delta t) = \Delta t^2 q(t).$$

Δt is as small as is necessary for requisite accuracy. Writing it out for a few time-steps, Δt,

$$
\begin{aligned}
y(0) &= y_0 & \text{(B.19)} \\
y(1\Delta t) - y(0\Delta t) &= \Delta t\, y_1' \\
y(2\Delta t) + \left(\Delta t^2 k - 2 + r\right) y(1\Delta t) + (1 - r\Delta t)\, y(0) &= \Delta t^2 q(1\Delta t) \\
y(3\Delta t) + \left(\Delta t^2 k - 2 + r\right) y(2\Delta t) + (1 - r\Delta t)\, y(1\Delta t) &= \Delta t^2 q(2\Delta t) \\
y(4\Delta t) + \left(\Delta t^2 k - 2 + r\right) y(3\Delta t) + (1 - r\Delta t)\, y(2\Delta t) &= \Delta t^2 q(3\Delta t) \\
&\ \ \vdots \\
y(M\Delta t) + \left(\Delta t^2 k - 2 + r\right) y((M-1)\Delta t) + (1 - r\Delta t)\, y((M-2)\Delta t) &= \Delta t^2 q((M-1)\Delta t)
\end{aligned}
$$

The first two equations are the initial conditions. This system is again a set of simultaneous equations that can be solved by fully constructing it and then inverting the coefficient matrix. In practice, $y(j\Delta t)$ is usually more sensibly found by time-stepping the system, starting with the first three equations and then sequentially calculating from those that follow. The result is equivalent to inverting the matrix, but one taking advantage of its special form. Should one of the initial conditions, perhaps the second equation in (B.2), be replaced by, for example, a known terminal condition, $y(M\Delta t)$, ordinary time-stepping is no longer possible, and it is a *shooting problem*. Direct matrix inversion could then become the most effective way to solve it.

With $\mathbf{x}(t) = [y(t), y(t - \Delta t)]^T$, the simultaneous equations can be rewritten in the *state vector form*,

$$
\mathbf{x}(t + \Delta t) = \mathbf{A}\mathbf{x}(t) + \mathbf{q}(t), \quad t = \Delta t, \ldots, (M-1)\Delta t, \tag{B.20}
$$

where

$$
\mathbf{A} = \left\{ \begin{array}{cc} -\left(\Delta t^2 k - 2 + r\right) & -(1 - r\Delta t) \\ 1 & 0 \end{array} \right\}, \quad
\mathbf{q}(t) = \begin{bmatrix} \Delta t^2 q(1\Delta t) \\ \Delta t^2 q(2\Delta t) \\ . \\ \Delta t^2 q((M-1)\Delta t) \end{bmatrix}
$$

plus the first two special equations. $\mathbf{x}(t)$ is the *state vector*, and it includes just the right amount of information about the state variable, $\mathbf{x}(t)$, to make it possible to calculate the state-one Δt into the future in the absence of outside disturbances (here called $\mathbf{q}(t)$). The assertion is made, without proof, that *any* linear system, including partial differential ones, can be written as

$$
\mathbf{x}(t + \Delta t) = \mathbf{A}(t)\mathbf{x}(t) + \mathbf{B}(t)\mathbf{q}(t). \tag{B.21}
$$

The square *state transition matrix*, $\mathbf{A}$, can be time dependent and a slightly more general externally prescribed term $\mathbf{B}(t)\mathbf{q}(t)$ has been used. A *nonlinear* time-dependent system can be written as

$$
\mathbf{x}(t + \Delta t) = \mathbf{L}(\mathbf{x}(t), \mathbf{B}(t)\mathbf{q}(t), t), \tag{B.22}
$$

where $\mathbf{L}$ is any operator. A time-dependent model is a time-stepping rule.

Suppose the state vector at different times is concatenated into one, conceivably gigantic, vector, $\boldsymbol{\xi} = \left[\mathbf{x}(0)^T, \mathbf{x}(1)^T, \ldots \mathbf{x}(t_f)^T\right]^T$, as is the forcing vector,

$$
\boldsymbol{\eta} = \left[\Delta t^2 \mathbf{q}(1\Delta t), \Delta t^2 \mathbf{q}(2\Delta t), \ldots, \Delta t^2 \mathbf{q}((M-1)\Delta t)\right]^T.
$$

Then once again, the system (B.22), along with any initial and boundary values, is the set of simultaneous equations,

$$\mathbf{C}(\boldsymbol{\xi}, \boldsymbol{\eta}) = 0, \tag{B.23}$$

where $\mathbf{C}$ is matrix operator, $\boldsymbol{\xi}$ includes *all of the* $\mathbf{x}(t)$, and $\boldsymbol{\eta}$, *all* of the prescribed parameters and boundary values, and Equation (B.23) is thus a potentially giant set of linear or nonlinear equations. If the entire system is linear, Equation (B.23) is equivalent to

$$\mathbf{C}_1 \boldsymbol{\xi} = \mathbf{D} \boldsymbol{\eta}, \tag{B.24}$$

where $\mathbf{C}_1, \mathbf{D}$ are ordinary matrices and matrix inversion is again a method of solution in principle.

Static problems are a specialized subset of time-dependent ones. In practice, dimension growth in time-dependent models leads to methods taking advantage of the specialized forms of Equations (B.21) or (B.22).

MODEL ERRORS

Consider a model, whether static or time dependent, having acknowledged errors. Models always have errors—otherwise they would be reality, not models. Errors occur in the physics of the supposed underlying partial differential equations; in the numerical operators used to form the discrete system; in the initial and boundary conditions; and in the internal parameterizations. Usually all are present. They can be represented generically as $\Delta \mathbf{q}(t)$ in

$$\mathbf{x}(t + \Delta t) = \mathbf{L}(\mathbf{x}(t), \mathbf{B}(t)\mathbf{q}(t), \mathbf{B}(t)\Delta\mathbf{q}(t), t), \tag{B.25}$$

or in a linear system all at once as

$$\mathbf{Cx} = \mathbf{b} + \Delta\mathbf{b}. \tag{B.26}$$

Equation (B.26) is a linear system of the form (B.3) and could be solved by SVD or another method; otherwise it has become a nonlinear problem,

$$(\mathbf{C} + \Delta\mathbf{C})\mathbf{x} = \mathbf{b} + \Delta\mathbf{b}, \tag{B.27}$$

for which various numerical algorithms exist for solution.

DATA ERRORS

Observations, $y_j(t_i)$, of almost any kind are a linear combination of the elements of a model state vector, so that $y_j(t_i) = \sum_q E_{jq}(t_i) x_q(t_i)$. The elements $E_{jq}(t_i)$ can select observations including—for example, a temperature measured at a single point, so that $E_{jq}(t_i) = 0, j \neq j_0$— or complex weighted spatial averages over depth ranges, regions, or the entire volume of combinations of data types. In practice, no observation is without error, and so an arbitrary observation should be written

$$y_j(t_i) = \sum_q E_{jq}(t_i) x_q(t_i) + \varepsilon(t_i), \tag{B.28}$$

where $\varepsilon\,(t_i)$ is a noise element.[2] The complete collection of observations is, in matrix-vector form,

$$\mathbf{Ex} + \boldsymbol{\varepsilon} = \mathbf{y}. \tag{B.29}$$

Collecting Equations (B.25) and (B.29) into one large system,

$$\mathbf{Cx} + \boldsymbol{\delta} = \mathbf{d}, \tag{B.30}$$

where $\mathbf{C}$ includes both $\mathbf{B}, \mathbf{E}$; $\boldsymbol{\delta}$ contains both model and noise error; and $\mathbf{d}$ is the collective knowns. Any useful method for solving such a system, including the SVD, can be used if it is computationally practical.

B.3 STATE ESTIMATION

State estimation is the process of combining observations with models, be they steady or time evolving. With an origin in orbit determination problems in the hands of Legendre and Gauss (Sorenson, 1970), such combinations are now used in many fields. In its specific applications to geophysical fluids, the subject in recent years has been dominated by the effort devoted worldwide to numerical weather prediction—an application that sometimes has led to misunderstanding and misuse in other fields. Weather forecasting is a special case of the prediction problem, but forecasting is distinct from the general problem of making the best possible estimate of the state of some system over a finite interval.

What is now widely known as *data assimilation* (DA) developed in numerical weather forecasting, beginning the in 1950s, and then into specialized methods of model-data combination (e.g., Talagrand, 1997; Kalnay, 2003; Evensen, 2009). A brief review of what these techniques do and the more general problems of oceanography and climate is useful.

Consider any time-evolving model of a physical system satisfying known equations, written generically in discrete time, rewritten from Equation (B.25),

$$\mathbf{x}\,(t + \Delta t) = \mathbf{L}\,[\mathbf{x}\,(t)\,, \mathbf{Bq}\,(t)\,, \boldsymbol{\Gamma}\mathbf{u}\,(t)]\,, \quad 0 \le t \le t_f = (M - 1)\,\Delta t, \tag{B.31}$$

where $\mathbf{x}\,(t)$ is the "state" at time t. $\mathbf{B}\Delta\mathbf{q}\,(t)$ has been replaced here by $\boldsymbol{\Gamma}\mathbf{u}\,(t)$ to denote all elements subject to adjustment and termed independent, or *control, variables* (or simply "controls"). $\mathbf{L}$ is again a matrix operator and is usually defined only implicitly in a computer code working on arrays of numbers. (Notation is approximately that of Wunsch, 2006a.) For steady models, $\mathbf{x}\,(t + \Delta t)$ is replaced by $\mathbf{x}\,(t)$ in (B.31). Thus the box inverse methods and their relatives, such as the β-spiral, are special cases of the estimation problem. For computational efficiency, however, the equations are then normally rewritten so that time does not appear. Useful observations are in the form of the subcomponents of Equation (B.29) and again, in a steady-state formulation, parameter t would be suppressed.[3]

Given noisy observations scattered over the time interval $1 \le t \le t_f$ (perhaps highly irregularly), a time-evolution model such as Equation (B.31) and an estimate of the noise in

---

[2]On rare occasions, data are a nonlinear combination of elements of the state vector: an example would be a speed measurement in terms of two components of the velocity, or if a frequency spectrum is specified. Methods exist (not discussed here) for dealing with such observations.

[3]Observations relating to $\mathbf{u}\,(t)$ may exist, and one approach to using them is to redefine elements of $\mathbf{u}\,(t)$ as being part of the state vector.

data and model, the estimation and control literature distinguishes several problems. Among them are (1) Determining the best estimated values of $\mathbf{x}(t)$, $0 \le t \le t_f$, all t. This problem is called "smoothing" and is here referred to as *state estimation*. (2) Finding the best estimate of $\mathbf{x}(t + \tau)$, $\tau > 0$, prediction, as distinct from (1). In both cases, determining the accuracy of the solutions is important. Problem (1) is directed at understanding and is the present focus. Problem (2) is commonly directed at practical applications, such as weather forecasting, but the results should not be confused with those for Problem (1).

Important note: In Problem (1), $\tilde{\mathbf{x}}(t)$ and $\tilde{\mathbf{u}}(t)$, are intended to exactly satisfy Equation (B.31), but because $\tilde{\mathbf{u}}(t)$ can have large magnitudes (and it usually includes internal parameters and boundary and initial conditions), the final version of the model may be considerably different from the starting estimate (with $\mathbf{u}(t) = \mathbf{0}$). *But the adjusted model is known, fully specified, exactly satisfied, and run freely forward over the time interval being considered to obtain the estimated solution.* Prediction methods do not usually produce solutions satisfying these requirements.

Some knowledge of the statistics of the controls, $\mathbf{u}(t)$, and observation noise, $\boldsymbol{\varepsilon}(t)$, is essential and is commonly represented by the first- and second-order moments,

$$\langle \mathbf{u}(t) \rangle = \mathbf{0}, \quad \left\langle \mathbf{u}(t) \mathbf{u}(t')^T \right\rangle = \mathbf{Q}(t)\, \delta_{tt'} \quad 0 \le t \le t_f - \Delta t = (M - 2)\,\Delta t, \qquad \text{(B.32a)}$$

$$\langle \boldsymbol{\varepsilon}(t) \rangle = \mathbf{0}, \quad \left\langle \boldsymbol{\varepsilon}(t) \boldsymbol{\varepsilon}(t')^T \right\rangle = \mathbf{R}(t)\, \delta_{tt'} \quad 1 \le t \le t_f = (M - 1)\,\Delta t \qquad \text{(B.32b)}$$

The brackets denote expected values, and superscript T is the vector or matrix transpose. These matrices specify roughly how large the state and control vectors might be—the minimum necessary information. The assumption of zero time correlation is not fundamental.

In generic terms, the state estimation problem is one of *constrained estimation*, in which the minimum of both the expected normalized quadratic model-data differences,

$$\left\langle (\mathbf{y}(t) - \mathbf{E}(t)\tilde{\mathbf{x}}(t))\, \mathbf{R}^{-1}(t)\, (\mathbf{y}(t) - \mathbf{E}(t)\tilde{\mathbf{x}}(t))^T \right\rangle \qquad \text{(B.33)}$$

and the independent variables (the controls),

$$\left\langle \mathbf{u}(t)\, \mathbf{Q}^{-1}(t)\, \mathbf{u}(t)^T \right\rangle, \qquad \text{(B.34)}$$

are sought, usually by summing them, requiring the minimum of

$$J_1 = \left\langle (\mathbf{y}(t) - \mathbf{E}(t)\tilde{\mathbf{x}}(t))\, \mathbf{R}^{-1}(t)\, (\mathbf{y}(t) - \mathbf{E}(t)\tilde{\mathbf{x}}(t))^T \right\rangle + \left\langle \mathbf{u}(t)\, \mathbf{Q}^{-1}(t)\, \mathbf{u}(t)^T \right\rangle. \qquad \text{(B.35)}$$

For data sets and controls that have unimodal probability distributions (of which a Gaussian is a special case), the problem as stated is equivalent to weighted least-squares minimization of the scalar,

$$J = \sum_{m=1}^{M} (\mathbf{y}(t) - \mathbf{E}(t)\tilde{\mathbf{x}}(t))^T\, \mathbf{R}^{-1}(t)\, (\mathbf{y}(t) - \mathbf{E}(t)\tilde{\mathbf{x}}(t)) + \qquad \text{(B.36)}$$

$$(\tilde{\mathbf{x}}(0) - \mathbf{x}_o)^T\, \mathbf{P}(0)^{-1}\, (\tilde{\mathbf{x}}(0) - \mathbf{x}_o) + \sum_{m=0}^{M-1} \mathbf{u}(t)^T\, \mathbf{Q}^{-1}(t)\, \mathbf{u}(t), \quad t = m\Delta t,$$

subject to Equation (B.21), replacing the theoretical (bracket) average with time sums. The second term explicitly separates the wish to fit to an uncertain initial-condition value, $\mathbf{x}_o$.

Minimizing J with a solution satisfying the time-stepping model is *a constrained least-squares problem* and is nonlinear if the model is nonlinear. Terms expressing knowledge of some elements of $\mathbf{x}(t)$ for any or all t can be easily included by adding terms analogous to that used for initial conditions.

One approach to solving such problems is the method of Lagrange multipliers (MLM). In summary, the model equations are "adjoined" to J using vectors of unknown Lagrange multipliers, $\boldsymbol{\mu}(t)$, to produce a new objective function,

$$J' = \sum_{m=1}^{M} \left(\mathbf{y}(t) - \mathbf{E}(t)\mathbf{x}(t)\right)^T \mathbf{R}(t)^{-1} \left(\mathbf{y}(t) - \mathbf{E}(t)\mathbf{x}(t)\right) \tag{B.37}$$

$$+ \left(\tilde{\mathbf{x}}(0) - \mathbf{x}_o\right)^T \mathbf{P}(0)^{-1} \left(\tilde{\mathbf{x}}(0) - \mathbf{x}_o\right) + \sum_{m=0}^{M-1} \mathbf{u}(t)^T \mathbf{Q}(t)^{-1} \mathbf{u}(t)$$

$$- 2 \sum_{m=1}^{M} \boldsymbol{\mu}(t)^T \left\{\mathbf{x}(t) - \mathbf{L}\left[\mathbf{x}(t - \Delta t), \mathbf{Bq}(t - \Delta t), \boldsymbol{\Gamma}\mathbf{u}(t - \Delta t)\right]\right\},$$

$$t = m\Delta t.$$

Textbooks show how the problem can now be treated as a conventional, unconstrained least-squares problem in which the $\boldsymbol{\mu}(t)$ are found as part of the solution. In principle, vector differentiation is done with respect to all of $\mathbf{x}(t)$, $\mathbf{u}(t)$, and $\boldsymbol{\mu}(t)$, setting the results to zero and then solving the resulting *normal equations*.[4] J and J' are very general, and internal model parameters, such as mixing coefficients, water depths, etc., are readily added as further parameters to be calculated. Because the solution amounts to solving sets of simultaneous equations, any relevant method could be used. The special nature of the model time-stepping rule (B.21) permits efficiencies of computation and storage and special iterative approximation methods that are the domain of numerical analysis and engineering.

The entire problem of state estimation reduces to finding the stationary values of J'. The large literature on what is commonly called the *adjoint method* ("4DVAR" in weather forecasting, where it is used over short time spans) is thus a form of least-squares and simultaneous equations.

B.4   THE OBSERVATIONS

As is true of any least-squares solution, no matter how it is obtained, the results depend upon the weights or error variances, $\mathbf{R}(t), \mathbf{Q}(t), \mathbf{P}(0)$, assigned to the data sets. An overestimate of the error corresponds to the suppression of useful information; an underestimate, to imposing erroneous values and structures. Although an unglamorous and not well-rewarded activity, a quantitative description of the errors is essential and is often where oceanographic expertise is most central. Partial discussions are provided by Forget and Wunsch (2007), Ponte et al. (2007), Ablain et al. (2009), and Fu and Haines (2013). A list of the diverse data types and sources used in a recent state estimate can be seen in Table B.1. Little is known about the space-time covariances of these errors, information, which if it were known, could improve the solutions. Model errors, which dictate how well estimates should fit to hypothetical perfect data, are poorly known and forms of $\mathbf{Q}(t)$ are often little more than informed guesses.

---

[4] Written out in Wunsch, 2006a and Wunsch and Heimbach, 2007.

Table B.1: Condensed table of data types used in the ECCO state estimates as of about 2012 (from Wunsch and Heimbach, 2013b, which should be consulted for more information).

Observation	Instrument	Product/Service	Area	Period	dT
Mean dynamic topography (MDT)	GRACE SM004-GRACE3	CLS/GFZ (A M. Rio)	Global	Time-mean	Mean
	EGM2008/DNSC07	N. Pavlis/ Andersen & Knudsen	Global		
Sea-level anomaly (SLA)	TOPEX/POSEIDON	NOAA/RADS & PO.DAAC	65°N/S	1993–2005	Daily
	Jason	NOAA/RADS & PO.DAAC	82°N/S	2001–2011	Daily
	ERS, ENVISAT	NOAA/RADS & PO.DAAC	65°N/S	1992–2011	Daily
	GFO	NOAA/RADS & PO.DAAC	65°N/S	2001–2008	Daily
Sea-surface temperature (SST)	Blended, AVHRR (O/I)	Reynolds & Smith	Global	1992–2011	Monthly
	TRMM/TMI	GHRSST	40°N/S	1998–2004	Daily
	AMSR-E (MODIS/Aqua)	GHRSST	Global	2001–2011	Daily
Sea-surface salinity (SSS)	Various in situ	WOA09 surface	Global	Climatology	Monthly
In situ Temperature and Salinity	Argo, P-ALACE	IFREMER	"Global"	1992–2011	Daily
	XBT	D. Behringer (NCEP)	"Global"	1992–2011	Daily
	CTD	Various	Sections	1992–2011	Daily
	SEaOS	SMRU & BAS (UK)	SO	2004–2010	Daily
	TOGA/TAO, PIRATA	PMEL/NOAA	Tropics	1992–2011	Daily
Mooring velocities	TOGA/TAO, PIRATA	PMEL/NOAA	Tropics	1992–2006	Daily
	Florida Straits	NOAA/AOML	North Atlantic	1992–2011	Daily
Average Temperature and Salinity	WOA09	WOA09	"Global"	1950–2000	Mean
	OCCA	Forget (2010)	"Global"	2004–2006	Mean
Sea-ice cover	Satellite passive microwave radiometry	NSIDC (bootstrap)	Arctic, SO	1992–2011	Daily
Wind stress	QuickScat	NASA (Bourassa)	Global	1999–2009	Daily
		SCOW (Risien & Chelton)		Climatology	Monthly
Tide gauge Sea-surface height (SSH)	Tide gauges	NBDC/NOAA	Sparse	1992–2006	Monthly
Flux constraints	From ERA-Interim, JRA-25, NCEP, CORE-2 variances	Various	Global	1992–2011	2 d–14 d
Balance contraints			Global	1992–2011	Mean
Bathymetry		Smith & Sandwell, ETOPO5	Global	–	–

B.5 DATA ASSIMILATION AND REANALYSES

In prediction, minimization of J_1 is replaced by minimization of

$$\mathrm{diag}\left\langle \left(\tilde{\mathbf{x}}\left(t_{now}+\tau\right)-\mathbf{x}\left(t_{now}+\tau\right)\right)\left(\tilde{\mathbf{x}}\left(t_{now}+\tau\right)-\mathbf{x}\left(t_{now}+\tau\right)\right)^{T}\right\rangle, \qquad (\text{B.38})$$

that is, the variance of the state about the true value at some time *future* to t_{now}. Consider an analysis time, $t_{now} = t_{ana} + \tau$, $\tau > 0$, when data have become available and where t_{ana} is the previous analysis time, typically 6 h earlier. The (weather) forecaster's model has been run forward to make a prediction, $\tilde{\mathbf{x}}(t_{now}, -)$, with the minus sign denoting that newer observations have *not* been used. The new observations are $\mathbf{E}(t_{now})\mathbf{x}(t_{now}) + \boldsymbol{\varepsilon}(t_{now}) = \mathbf{y}(t_{now})$. With some understanding of the quality of the forecast, which is expressed in the form of an uncertainty matrix (second moments of the estimated value about the truth) called $\mathbf{P}(t_{now}, -)$, and of the covariance matrix of the observational noise, $\mathbf{R}(t_{now})$, the best combination of the information of the model and the data is the minimum of

$$J_2 = \left(\mathbf{x}(t_{now}) - \tilde{\mathbf{x}}(t_{now}, -)\right)^T \mathbf{P}(t_{now}, -)^{-1}\left(\mathbf{x}(t_{now}) - \tilde{\mathbf{x}}(t_{now}, -)\right) + \qquad (B.39)$$
$$\left(\mathbf{y}(t_{now}) - \mathbf{E}(t_{now})\mathbf{x}(t_{now})\right)^T \mathbf{R}(t_{now})^{-1}\left(\mathbf{y}(t_{now}) - \mathbf{E}(t_{now})\mathbf{x}(t_{now})\right).$$

The least-squares minimum of J_2 for a linear model is given rigorously by application of the *Kalman filter* numerical algorithm, and the updated model is then run forward in time for a new forecast. In DA practice, only some very rough approximation to that minimum is sought and obtained. True Kalman filters are never used for prediction in real geophysical fluid flow problems because they are computationally overwhelming. Notice that J_2 assumes that a summation of errors is appropriate, even in the presence of strong nonlinearities.[5]

DA was developed for the purposes of forecasting over time scales of hours to a few days. Little or no attention was paid to considerations such as conservation of heat or fresh water or mass, because they have little or no impact on weather forecasts made a few hours or days in advance. With the exception of groups attempting the prediction of storm surges, or mesoscale eddies—typically for military purposes—oceanographers more commonly seek state estimates capable of being used for global-scale energy-, heat-, and water-cycle budgets over finite time intervals. The ability to close global budgets is crucial to the understanding of climate change. Construction of closed budgets is also rendered physically impossible by the forecasting goal: solutions jump toward the data at every analysis time, usually every 6 h, introducing spurious sources and sinks of basic properties, as well as probably introducing artificial variability.[6]

Meteorological "reanalyses" are of interest here for a couple of reasons: (1) they are often used as an atmospheric "truth" to drive ocean, ice, chemical, and biological models; and (2) a number of ocean state estimates have followed their numerical engineering methodology. In meteorological jargon, the "analysis" consists of an operational weather model run in prediction mode, which is analogous to the simple form just described, adjusted to come close to the data, thus displaying physical discontinuities at the analysis times.

In a reanalysis, computation is done with the same prediction methodology as previously used in the analysis, but with the differences that (1) the model code and combination methodology are held fixed over the complete time duration of the calculation (e.g., over 50 years), thus eliminating artificial changes in the state from model or method improvements over the decades; (2) many data are included that had arrived too late to be incorporated into the real-time analysis (see Kalnay, 2003; Evensen, 2009); and (3) New data types appear. Notable among these are the meteorological satellites, which, when operational, induced very large shifts in the calculated atmospheric states (Bromwich et al., 2007; see Figure B.1). No account is taken, however, of the changing observational networks.

---

[5] The Kalman filter is a *prediction* method, as is apparent in Kalman's (1960) paper, which, reading between the lines, was directed at inferring ballistic-missile impact locations. Contrary to widespread misimpression, on its own it is inappropriate for estimating past states.

[6] A more extensive discussion of these and related problems is in Wunsch and Heimbach, 2013b.

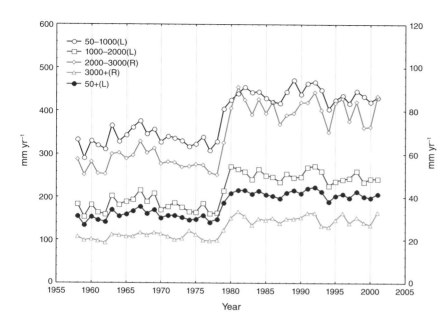

Figure B.1: The mean annual precipitation minus evaporation over the Antarctic as a function of time in the ECMWF reanalysis ERA-40 showing the impact of new observations, in this case the arrival of the polar orbiting satellites. Different curves are for different elevations. The only simple inference is that the uncertainties must exceed the size of the rapid transition seen in the late 1970s. L and R identify whether the left or right axis is to be used for that curve. The inclusion of new or more extensive data types commonly introduces step like changes into data-assimilation products. (From Bromwich et al., 2007.)

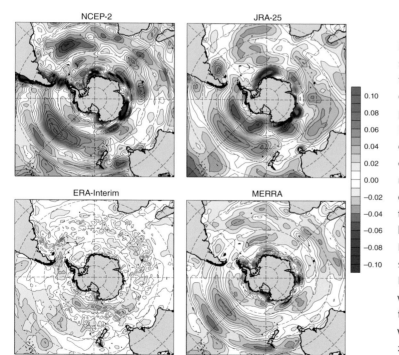

Figure B.2: Calculated trends (meters/second/year) in the 10 m zonal wind fields at high southern latitudes from four different atmospheric reanalyses . Note particularly the different patterns in the Indian Ocean sector and the generally discrepant amplitudes. Because of the commonality of data sets, forecast models, and methodologies, the differences here must be lower bounds on the true uncertainties of trends. The panel labelled NCEP was provided by the US NOAA; JRA is from the Japanese weather services; ERA is from the ECMWF; and MERRA is a NASA product. See Bromwich et al., 2011, for a description of the four different estimates. (From D. Bromwich and J. P. Nicolas, pers. comm., 2010)

Finally, the information content of the observations of the formal *future* evolution of the state is usually ignored so that the analysis time jumps remain. Although as already noted above, clear warnings have appeared in the literature—that spurious trends and values are artifacts of changing observation systems (see Elliott and Gaffen, 1991; Marshall, et al., 2002; Thompson et al., 2008)—the reanalyses are often used inappropriately. Figure B.2 and other similar ones show that reanalyses using much of the same data and models display significant qualitative disagreements on climate time scales. General agreement among them where data are dense (e.g., the continental northern hemisphere) and serious divergences where data are sparse (the southern hemisphere), shows that as currently used, model skills can be of secondary importance relative to the observational coverage.

Problematic Terms and Concepts

Science evolves by building upon previous understanding, experiments, and concepts. The outcome of this process is not predictable, and sometimes certain concepts evolve in directions that were unexpected—with the consequence that the earlier, often simpler, terminology is so far removed from current thinking, that the result is widespread confusion outside the inner club of experts. Fields like climate and physical oceanography are by nature interdisciplinary and require that both scientists and nonscientists, specialists and nonspecialists alike, should understand what is being discussed. Science is difficult enough without the added burden of semantic confusion.

Popular usage can seriously corrupt a scientific meaning. A "quantum jump" was originally unimaginably small; in the vernacular, it has come to mean something extremely large. The word "theory" typically means something completely different to the public than it does to scientists. Somerville and Hassol (2011) provide a list of examples where a scientific term has come to mean something very different to the public at large—a different issue from the way words have different meanings within the scientific community (Table C.1, with some additions).

Simple Definition Issues

Sometimes the problem is simply the definition of a word. An oceanographic example is the expression "barotropic flow." At least six different usages of this term exist:

1. autobarotropic (giving rise to the original definition), meaning a fluid in which the pressure p is a function of the density ρ, alone ($p = p(\rho)$); not common in the ocean
2. the oscillatory linear motion of an unstratified, flat-bottom ocean, or the corresponding lowest vertical normal mode of a vertically stratified ocean (mode 0)
3. the vertical average of the horizontal velocity over the full water depth
4. the geostrophic velocity at the seafloor (Fofonoff, 1962)
5. the geostrophic velocity at the sea surface
6. the reference level velocity in the dynamic method

Table C.1: Terms that have different meanings for scientists and the public. After Somerville and Hassol, 2011, with some additions.

Scientific Term	Public Meaning	Better Choice
Aerosol	spray can	tiny atmospheric particle
Anomaly	abnormal occurrence	change from a long-term average
Bias	distortion, political motive	offset from observations
Dihydrogen monoxide incursion	chemical pollution	flood
Driver	controller (as in automobile)	power source
Enhance	improve	intensify, increase
Error	mistake, wrong, incorrect	difference from exact, true, number
Fast ice	rapidly moving ice	stuck or unmoving ice
Manipulation	illicit tampering	scientific data processing
Model	fashion display person	analytical/numerical representation; statistical assumptions
Positive feedback	good response, praise	self-reinforcing cycle
Positive trend	good trend	upward trend
Quantum jump (very small)	very large change	discontinuous change
Scheme	devious plot	systematic plan
Sign	indication, astrological sign	plus or minus sign
Stationary	not moving	steady movement or statistical equilibrium
Streamline	remove obstacles	fluid flow pathway
Theory	hunch, speculation	scientific understanding
Trick	misleading analysis	novel mathematical method
Uncertainty	ignorance	range
Values	ethics, monetary value	numbers, quantity

Other definitions probably exist. But the above list suffices: these usages are inconsistent and incompatible.

Physical oceanographers often refer to the "interior" flow when discussing the fluid outside a boundary layer (such as the surface Ekman layer). Fluid dynamicists will commonly instead refer to the "exterior" flow as that not residing within the boundary layer, so that "interior" and "exterior" both refer to the same region! Although such contrary usage is usually readily discovered, being as explicit as possible is recommended.

"Drivers" of the Ocean Circulation

In everyday usage, the term "driver" has several distinct and conflicting meanings. A hammer is a driver of a nail in the sense of transmitting power, but in a car, the engine is the provider of power, with the automobile driver being instead the "controller." In an oceanographic or climate context, the "driver" of the circulation sometimes means a controller, and sometimes the source of power keeping the system moving—two very different roles.

Thermohaline Circulation of the Oceans[1]

Movements of temperature and salt can be and are often radically different (their surface boundary conditions are distinct). The large-scale ocean circulation is observationally almost indistinguishable from a geostrophically balanced one. Most of the mass or volume fluxes are almost exactly proportional to gradients in the density field. One cannot assert, as some writers do, that the density gradients are driving the mass transports—all one can say is that they balance each other (with equal logic, the mass transports could be asserted to be driving the density gradients).

The term "meridional overturning circulation" (MOC) is now more commonly used to denote the zonally integrated volume or mass flux in the ocean, (e.g., Fig. 9.31) as it is neutral relative to energetics and driving forces. Because circulation features such as the Gulf Stream are major components of the meridional flow, the wind-driven circulation is part of the MOC.

Salinity

Historically, salinity was represented as the fraction of a kilogram of water contributed by the dissolved salts, multiplied by 1000. Thus a salinity value of 35.41 represents a mass of about 3.5% or 35 o/oo or 3.541 g/kg from salt. Sufficiently accurate measurement of salinity is a surprisingly difficult matter, and many definitional and technology changes occurred over the last 100+ years. Until very recently, the working definition of salinity was as a nondimensional quantity (see Millero et al., 2008) that when multiplied by density produces the same mass fraction. Numerically, the shift from the original definition is sufficiently small that for theoretical or intuitive purposes, it usually suffices to continue to interpret the numerical value as the mass fraction.

Unfortunately, many oceanographers found nondimensional salinities distasteful and so made up a "practical salinity unit," usually written PSU, as though this were a dimensional quantity. This misunderstanding has sometimes exasperated the experts (see Millero, 1993) who worked hard to develop a self-consistent definition and methodology and who have pointed out that there is no such thing as a PSU. It is sensible to say salinity has been measured on the practical salinity scale, to distinguish the slightly different value from earlier ones based upon other definitions. As these words are being written, the international community has reinstated salinity as g/kg (parts/thousand) as the basic definition, to be labelled "absolute salinity"; see the Intergovernmental Oceanographic Commission, 2010. This development is a welcome one, but history suggests that the story has not ended.

Mesoscale Eddies

When oceanic eddies were first clearly depicted about 1973, they were called "mesoscale," as an analog of the atmospheric weather systems. In the atmosphere, however, "mesoscale" usually refers to such ageostrophic phenomena as thunderstorms and fronts, while geostrophically

---

[1]A published essay on this topic is given by Wunsch (2002).

balanced motions are commonly referred to as "synoptic" scale. Apart from the misleading parallel with atmospheric dynamics, no great harm is done, but this book generally refers to "geostrophic" or "balanced" eddies.

Deconvolution

"Convolution" is a specific mathematical operation that defines a third function, $h(t)$, or sequence as in Equations (A.31) or (A.33) and whose great importance, among other reasons, lies with the convolution theorem, Equations (A.32) and (A.33); that the Fourier transform of h is the product of the Fourier transforms of f, g.

"Deconvolution" is the process of determining f from a knowledge of h, g; that is, given $h(t)$ or h_m and $g(t)$ or g_m, to determine in the best possible way what the original function f was. The process is extremely important in diverse fields where elaborate methodologies have been developed in signal processing for dealing with incomplete and noisy data. Use of the terminology "deconvolution" to mean "explaining" or "rationalizing" a record is unhelpful unless the specific mathematical operation of undoing the operations in Equations (A.32) or (A.33) is intended.

Validation and Verification

The word "validate" is loosely used apparently to mean confirming that a model or hypothesis is "correct," usually in an unspecified sense. It is sometimes applied to the testing of some novel observation set against more conventional data. The term raises all kinds of questions. As a number of authors have noted (e.g., Oreskes et al., 1994), models can be shown to be consistent with observations, or inconsistent, but are never demonstrably "valid" in the sense of having been shown to be correct. A single new observation can *invalidate* a model, but validation is a hopeless goal. One can usefully speak of *calibrating* or evaluating a model in the same sense that any other piece of machinery is calibrated—the model will have some range in which it has useful skill, but normally also a range in which the errors are unacceptably large. By definition, no model is ever completely error free.

The term *verification* is also loosely used, sometimes interchangeably with "validation." It can usually be understood as a code for "my model is at least qualitatively consistent with the data" or "my new data type is consistent with the older, more familiar, data." In some circumstances, model or data calibration can take place if a real standard is known.

Data

The observational community has lost control of the word "data," which has come to be used, confusingly, for the output of models, rather than having any direct relationship to instrumental values. In the context of reanalyses and state estimates involving both measurements and computer codes, the word generally no longer conveys any information. In this book, "data" are always instrumental values of some sort, and anything coming out of a GCM is a "model value," "model datum," "state estimate value," or similar label. Models are involved in all real observations, even in such familiar values as those coming from, for example, an aneroid barometer, in which a physical displacement is converted into a pressure change. A qualitative difference exists, however, between conventional observations and the output of a 100,000+ line computer code.

"The *Ocean*"

Even "the ocean" is a terminology in need of care. Here, "the" ocean is reserved for the real world as far as we can determine it. "An" ocean is typically one existing in a simplified system (e.g., one having a flat bottom) that is not fully representative of the real system. Similarly, "an" Atlantic can be distinguished from "the" Atlantic. "An" Earth without continents (aquaplanet) can be discussed, as opposed to "the" Earth, etc.

Feedback

The term "feedback" is widely used in discussions of climate change, but as Bates (2007) made clear, it is often confusing owing to its employment with varying, often unexplained, definitions. He pointed out that there are at least four different definitions of the term, and under some circumstances, the insistence that a feedback is "positive" under one definition can lead to behavior under another definition that is negative. Understanding may best be accomplished simply using ordinary partial derivatives and the chain rule, concepts that underlie the feedback structure.

Jargon

All fields develop specialist terminologies—they are a necessary shorthand for all types of ideas and concepts. With enough shorthand, however, the problem of jargon appears, which can grow to be a serious obstacle to interdisciplinary discourse. An example is the so-called Hovmöller diagram, which is meaningless to most scientists who would, however, understand instantly its description as a "time-longitude diagram." The term "Humboldt Current" honors an important early scientist, but the label is much less informative than "Peru-Chile Coastal Current."[2] Other examples are "Brunt-Väisälä frequency" (as opposed to "buoyancy frequency") and "4DVAR" for what physics and applied mathematics knows as the "method of Lagrange multipliers." Some labels are simply misleading. The "equilibrium theory of the tides" is not a theory of the tides but is the representation of the tidal forces. "Fast ice" is sea ice that is not moving ("stuck fast"). Many more such examples exist. A large gray area exists between a necessary condensation and the wish to honor colleagues and predecessors, and the erection of isolating barriers. Thus Rossby and Reynolds numbers are necessary and useful abbreviations. Some usages are so entrenched that avoiding them is just quixotic, as with the Sverdrup (although it is not part of the Système Internationale (SI) and should always be defined when first used).

Language use is never wholly logical, as witnessed by the failure, as mentioned in the Introduction, of the community to embrace the term "oceanology." R. B. Montgomery's (1969) proposals that the sea surface should be called the "naviface," and that oxygen concentration should be "oxty" also met with no success.

Cross-disciplinary terminology is even more difficult. In meteorology, for example, "physics" has come to exclude dynamics. Wunsch (2002a) found it necessary to provide a glossary connecting physical oceanography and meteorological tracer studies. The situation in less-related fields can be much worse: for example, the word "model" has a different

---

[2] Formal international naming conventions no longer permit the use of uninformative labels such as Lomonosov Current or Cromwell Current (both of which labels have died out of the literature).

meaning in statistics or engineering than it does in physical oceanography. "Breaking the code" can be the most difficult part of mastering another field.

Eponymy

An interesting, and only semifacetious, paper by Stigler (1980) arrives at the conclusion that almost nothing is named for the right person (including what he calls Stigler's Law of Eponymy). For those interested in the history of ideas in oceanography and climate and the vexed issues of assigning proper credit (the scientist's bread and butter), the paper makes sobering reading. Thus Rossby waves were known long before Rossby; Froude apparently never used the number named for him (Darrigol, 2005); Coriolis worked long after Laplace whose equations already included Coriolis forces, and so forth. Merton (1968) described the so-called Matthew Effect (from the eponymous gospel, King James version: "For unto every one that hath shall be given, and he shall have abundance: but from him that hath not shall be taken away even that which he hath") in which, among other symptoms, papers written by two or more authors are almost always credited to the better known one, whatever their relative contributions.

Useful Numerical Values

Higher-accuracy and precision values can be found in the references.

Earth-Ocean Geometry and Physical Parameters

Equatorial radius: 6.378×10^6 m (approximate, from Seidelmann, 2013)

Polar radius: 6.356×10^6 m

Flattening factor, f_e : $1/298.3$

1° of latitude: $110.575 + 1.110 \sin^2 \phi$ km (ϕ is geodetic lat.) ≈ 60 nautical miles

1° of longitude: $110.32 \cos \phi$ km

Area of Earth surface: 5.1×10^{14} m^2

Area of ocean: 3.618×10^{14} m^2 (Charette and Smith, 2010)

Volume of ocean: 1.332×10^{18} m^3

Mean depth of ocean: 3682 m

Volume fraction of ocean by latitude band: Figure D.1 from ECCO model.

$\Omega = 7.292 \times 10^{-5}$ rad/s

$\beta = 2\Omega/a \cos \phi = 1.983 \times 10^{-11}$ m/s (30° latitude)

$g = 9.806 - 0.026 \cos(2\phi)$ m/s^2

Astronomical Parameters

Mass of the Sun: 1.99×10^{30} kg

Ratio of mass of the Sun to Earth: 3.3×10^5

Mass of the Earth: 6.0×10^{24} kg

Ratio of mass of the Moon to mass of the Earth: 0.012

Mass of the Moon: 7.2×10^{22} kg

Mean Earth-Moon distance: 3.84×10^8 m

Mean Earth eccentricity: 0.0167 (This and other values are appropriate to 1950.)

Mean Earth obliquity: 23°26′21.4″

Rotation rate of the ecliptic: 0.4704″/year

Mean longitude of perihelion: 100°28′ (measured from ♈ on 1 January 1950)

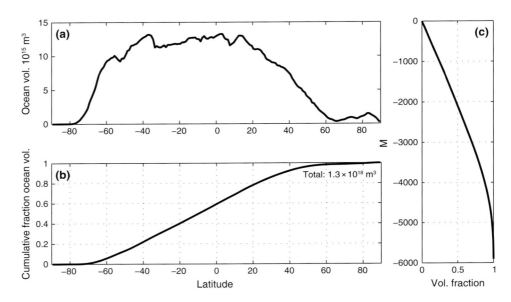

Figure D.1: (a) The ocean volume by latitude. (b) The cumulative total of the ocean volume. (c) The cumulative ocean volume by depth. Increase with depth is nearly linear until about 3500 m, when the reduced volume from bottom topography becomes apparent.

Mean Earth-Sun distance: 1.50×10^{11} m
Tropical year: 365.242 d (equinox to equinox) $\approx 3.16 \times 10^9$ s
Sidereal year: 365.256 d (fixed star to fixed star)
Anomalistic year: 365.2596 d (perihelion to perihelion)
Sidereal month (fixed star to fixed star): 27.32 mean solar days (MSD; month values are the means for 1900; They can vary by many hours.)
Nodical (or draconitic or eclipse) month: 27.21 MSD
Tropical month (passage across the equator): 27.32 MSD
Anomalistic month (passage through perigee): 27.55 MSD
Synodic month (passage through longitude of mean Sun): 27.53 MSD
Rates of change of the above can be found in Anonymous, 1961, p. 98.
Mean solar day: 86400 s
Sidereal day: 86164.09054 s
Julian century: 36525 MSD
Tidal periods are listed in Table 6.1 of the text.
Tidal amplitudes: $H_M = 0.36$ m; $H_S = 0.16$ m

Ocean Physical Properties (For 2004–2006, derived from Forget, 2010; model mean area: 3.46×10^{14} m^2, mean volume: 1.3×10^{18} m^3, mean depth: 3825 m.)

Approximate mean ocean temperature: 3.6°C
Mean ocean salinity: 34.73
Mean ocean density: 1037.6 kg/m^3
Mean potential density: 1027.5

Miscellaneous Physical Parameters

Approximate heat capacity, c_p: 4×10^3 J/kg/°C (Thorpe, 2005, p. 375)

Freezing temperature of sea water: ≈ -1.9°C

Haline density coefficient, β_S: $\approx 0.8 \times 10^{-4}$/unit salinity change (dimensionless) on the practical salinity scale $\approx 0.8 \times 10^{-4}$ g/kg. (At the time of writing, salinity has been defined by international agreement as dimensionless. A new unit, absolute salinity, whose units will again be g/kg, has recently been introduced, but for most purposes, changes in numerical values from the various definitions can be ignored relative to other errors.)

Latent heat of evaporation: 2.5×10^6 J/kg (Thorpe, 2005, p. 16)

Latent heat of fusion: 3.3×10^5 J/kg (Cuffey and Paterson, 2010)

Molecular thermal diffusivity, κ_T: $\approx 1.4 \times 10^{-7}$ m²/s

Molecular salinity diffusivity, κ_S: $\approx 1.5 \times 10^{-9}$ m²/s

Molecular kinematic viscosity, ν: $\approx 1.3 \times 10^{-6}$ m²/s (strongly temperature dependent)

(Molecular diffusivities are written as approximations because they are all partly dependent upon temperature, pressure, and salinity, but generally negligibly so for our purposes.)

Thermal expansion coefficient, α: $5-30 \times 10^{-5}$/°C (Thorpe, 2005, p. 375)

Power-of-10 Names

Giga: 10^9 (e.g., 1 GW is 1 gigawatt = 10^9 W)

Terra: 10^{12}

Peta: 10^{15}

Exa: 10^{18}

Zotta: 10^{21}

Yotta: 10^{24}

Non-SI and Miscellaneous Units

Nautical mile: 1.852 km

Knot: 1 nautical mile/hour $\approx 1'$ of latitude/hour

Fathom: 6 feet = 1.8 m

1 cm/s $\approx$ 1 km/d (1.1 cm/s more accurately)

Sverdrup (Sv): 10^6 m³/s $\approx 10^9$ kg/s (accurate to about 3.5%)

Dynamic meter: $\int_{1 \text{ meter}} g dz$

Decibar (a pressure unit): 1/10 bar = 10^4 N/m² = 10^4 Pascals $\approx$ 1 m of water pressure

Millibar: 100 Pa = 1 hectopascal (hPa)

Tritium unit (TU): concentration of atoms in a population of 10^{18} hydrogen atoms

APPENDIX E

Notation, Abbreviations, and Acronyms

The range of elements of physical oceanography, including fluid dynamics, statistics, geodesy, astronomy, control theory, etc., is so great that a unique notation is impossible. Here are listed only those symbols whose multiple definitions might cause difficulty and whose context may not make them completely clear.

$\langle \cdot \rangle$: generic average; time average, usually in a theoretical ensemble sense in time; $\langle \cdot \rangle_o$: spatial average over the ocean area.

ε: turbulent dissipation rate (Eq. 3.33); also a generic error, ε_i and a vector of error, $\boldsymbol{\varepsilon}$. Deviations from a constant of the solar orbital radius.

ϕ: latitude; ϕ_k a trigonometric phase

φ: a potential function

κ: molecular diffusion coefficient; scalar wavenumber, $\sqrt{k^2 + l^2}$

μ: Earth rigidity; Lagrange multipliers.

ρ: density; $\bar{\rho}$: a mean density, usually within the Boussinesq approximation; ρ': a perturbation density; ρ_0: a reference density; ρ_S: distance of the centers of Earth and Sun

σ: radian frequency; the standard deviation.

χ_T: the thermal variance; χ_v: the chi-square distribution with v degrees of freedom

a: radius of the Earth

C: a generic tracer; C_D: a drag coefficient; as ^{14}C: radiocarbon

S_a: the sea surface elevation measured from a satellite (Eq. 2.1); salinity, generally in parts per thousand; as a matrix, $\mathbf{S}$: a covariance

M_a: the solar mean anomaly in orbit

N: buoyancy frequency (also known as the Brunt-Väisälä frequency); the geoid elevation (written as N_g here); force unit in Newtons

$x(m\Delta t) = x_m, y(p\Delta t) = y_p$, etc., in various formulas

Some Acronyms

AABW: Antarctic Bottom Water

AAO: Antarctic Oscillation

AD: acoustic Doppler

(P)ALACE floats: (Profiling) Autonomous Lagrangian Circulation Explorer

AMO: Atlantic Multidecadal Oscillation

AMOC: Atlantic Meridional Overturning Circulation

AO: Arctic Oscillation

AOUR: Apparent oxygen utilization rate

APE: available potential energy

AVISO: Archiving, Validation and Interpretation of Satellite Oceanographic Data

BCE: before the common era (BC in the religious calendar)

BT: Bathythermograph

CDW: Circumpolar Deep Water

CE: common era (AD, anno domini, in the religious calendar)

CTD: conductivity-temperature-depth; as in an instrument called a CTD

CUEA: Coastal Upwelling Ecosystems Analysis

DORIS: Doppler Orbitography and Radio-positioning Integrated by Satellite

DWBC: deep western boundary current

ECCO: Estimating the Circulation and Climate of the Ocean

ENSO: El Niño Southern Oscillation

ENVISAT: Environmental Satellite (succeeded ERS series)

EOF: empirical orthogonal function

EOS: equation-of-state

ERS: European Remote Sensing Satellite

FFT: fast Fourier transform

FT: Fourier transform

GCM: general circulation model

GIA: glacial isostatic adjustment

GM: Garrett-Munk; as in GM Spectrum (used in some literature for the Gent-McWilliams eddy mixing parameterization)

GOCE: Gravity Field and Steady-State Ocean Circulation Explorer (a satellite). Pronounced as in Italian.

GPS: Global Positioning System

GRACE: Gravity Recovery and Climate Experiment (a satellite)

ICOADS: International Comprehensive Ocean-Atmosphere Data Set

IGY: International Geophysical Year

IOC: Intergovernmental Oceanographic Commission

IPCC: Intergovernmental Panel on Climate Change

ITCZ: intertropical convergence zone

IWEX: (Trimoored) Internal Wave Experiment

LADCP: lowered acoustic doppler current profiler

LOD: length of day

LORAN: long-range navigation

MLD: mixed-layer depth

MLM: method of Lagrange multipliers

MOC: meridional overturning circulation

MOCHA: Meridional Overturning Circulation and Heat Flux Array

MODE: Mid-Ocean Dynamics Experiment

MSD: mean solar day

NADW: North Atlantic Deep Water

NECC: North Equatorial Counter Current

NOAA: National Oceanic and Atmospheric Administration

OGCM: ocean general circulation model

PDO: Pacific Decadal Oscillation

PIRATA: Pilot Research Moored Array in the Tropical Atlantic

PSI: parametric subharmonic instability

QG: quasi-geostrophic

RAFOS: Inversion of SOFAR, implying a listening rather than transmitting system

RAPID: Rapid Climate Change Programme

RB: Rayleigh-Bénard; RB convection

RMS: root mean square

RRS: Royal Research Ship

R/V: research vessel

SA: Stommel-Arons; as in abyssal circulation theory

SAM: Southern Annular Mode

SEaOS: Southern Ocean elephant seals as Oceanographic Samplers

SI: Système Internationale (the accepted system of international units of kilograms, meters, seconds, etc.)

SLP: sea level pressure

SOFAR: sound fixing and ranging

SQG: surface quasi-geostrophic

SSH: sea-surface height

SST: sea-surface temperature

STD: salinity-temperature-depth; as in instrument called an STD

SVD: singular value decomposition

TAO: Tropical Atmosphere and Ocean

TOGA: Tropical Ocean–Global Atmosphere

TOPEX-POSEIDON (T/P): Ocean Topography Experiment—Premier Observatoire Spatial Étude Intensive Dynamique Ocean et Nivosphere, or Positioning Ocean Solid Earth Ice Dynamics Orbiting Navigator (a satellite)

VACM: vector-averaging current meter

VMCM: vector measuring current meter

WBC: western boundary current

WOCE: World Ocean Circulation Experiment

XBT: expendable bathythermograph

References

Ablain, M., A. Cazenave, G. Valladeau, and S. Guinehut (2009). A new assessment of the error budget of global mean sea level rate estimated by satellite altimetry over 1993-2008. Oc. Sci 5: 193–201.

Abraham, J. P., M. Baringer, N. L. Bindoff, T. Boyer, L. J. Cheng, J. A. Church, J. L. Conroy, et al. (2013). A review of global ocean temperature observations implications for ocean heat content estimates and climate change. Revs. Geophys. 51(3): 450–483.

Adcroft, A., and D. Marshall (1998). How slippery are piecewise-constant coastlines in numerical ocean models? Tellus A 50: 95–108.

Adcroft, A., J. R. Scott, and J. Marotzke (2001). Impact of geothermal heating on the global ocean circulation. Geophys. Res. Lett., 28: 1735–1738.

Adler, R. J. (2010). The Geometry of Random Fields. Philadelphia, Society for Industrial and Applied Mathematics.

Alcock, G. A., and D. E. Cartwright (1978). Some experiments with 'orthotides.' Geophys. J. R. Astron. Soc. 54: 681–696.

Alexander, M. A., I. Bladé, M. Newman, J. R. Lanzante, N.-C. Lau, and J. D. Scott (2002). The atmospheric bridge: The influence of ENSO teleconnections on air–sea interaction over the global oceans. J. Clim. 15(16): 2205–2231.

Alford, M. H., A. Y. Shcherbina, and M. C. Gregg (2013). Observations of near-inertial internal gravity waves radiating from a frontal jet. J. Phys. Oc. 43(6): 1225–1239.

Alford, M. H., and M. Whitmont (2007). Seasonal and spatial variability of near-inertial kinetic energy from historical moored velocity records. J. Phys. Oc. 37: 2022–2037.

Alverson, K. (2012). Vulnerability, impacts, and adaptation to sea level rise taking an ecosystem-based approach. Oceanog. 25(3): 231-235.

Anderson, D.L.T., K. Bryan, A. E. Gill, and R. C. Pacanowski (1979). Transient-response of the North-Atlantic—Some model studies. J. Geophys. Res. 84(NC8): 4795–4815.

Anonymous (1961). Explanatory Supplement to *The Astronomical Ephemeris and The American Ephemeris and Nautical Almanac.* Her Majesty's Stationery Office, London.

Anonymous (1968). Instruction Manual for Obtaining Oceanographic Data. Pub. 607. Washington DC, US Naval Oceanographic Office. Available online.

Arbic, B. K., P. St-Laurent, G. Sutherland, and C. Garrett (2007). On the resonance and influence of the tides in Ungava Bay and Hudson Strait. Geophys. Res. Lett. 34(17).

Arbic, B. K., R. B. Scott, G. R. Flierl, A. J. Morten, J. Richman, and J. F. Shriver (2012). Nonlinear cascades of surface oceanic geostrophic kinetic energy in the frequency domain. J. Phys. Oc. 42(9): 1577–1600.

Armi, L., and E. D'Asaro (1980). Flow structures of the benthic ocean. J. Geophys. Res. 85(NC1): 469–484.

Armi, L., and R. C. Millard, Jr. (1976). The bottom boundary layer of the deep ocean. J. Geophys. Res. 81(27): 4983–4990.

Armi, L., and H. Stommel (1983). 4 views of a portion of the North Atlantic sub-tropical gyre. J. Phys. Oc. 13(5): 828–857.

Baehr, J. (2010). Influence of the 26° N RAPID-MOCHA Array and Florida Current Cable Observations on the ECCO-GODAE state estimate. J. Phys. Oc. 40(5): 865–879.

Bahr, D. B., M. Dyurgerov, and M. F. Meier (2009). Sea-level rise from glaciers and ice caps: A lower bound. Geophys. Res. Lett. 36: 4.

Baines, P. (1971). The reflexion of internal/inertial waves from bumpy surfaces. J. Fluid Mech. 46(02): 273–291.

—— (1973). Generation of internal tides by flat-bump topography. Deep-Sea Res. 20(2): 179–205.

—— (1982). On internal tide generation models. Deep-Sea Res. 29: 307–338.

Baker, D. J. (1971). Density gradients in a rotating stratified fluid: Experimental evidence for a new instability. Science 172(3987): 1029–1031.

—— (1981). Ocean instruments and experimental design. In: Evolution of Physical Oceanography. Scientific Surveys in Honor of Henry Stommel. B. A. Warren and C. Wunsch, eds. Cambridge, MA, MIT Press. pp. 396–433.

—— (1990). Planet Earth: The View From Space. Cambridge, MA, Harvard University Press.

Baker, R. H. (1955). Astronomy: A Textbook For University and College Students. New York, Van Nostrand.

Baringer, M. O., and J. C. Larsen (2001). Sixteen years of Florida Current transport at 27 degrees N. Geophys. Res. Lett. 28(16): 3179–3182.

Barkan, R., K. B. Winters, and S. G. Llewellyn Smith (2013). Rotating horizontal convection. J. Fluid Mech. 723: 556–586.

Barton, E. D. (1998). Eastern boundary of the North Atlantic: Northwest Africa and Iberia. Coastal segment (18, E). Sea 11: 633–657.

Basili, V., G. Caldiera, F. McGarry, R. Pajerski, G. Page, and S. Waligora (1992). The software engineering laboratory: An operational software experience factory. Proc. 14th. Int. Conf. Software Eng., Melbourne, Australia, 11–15 May 1992. New York, ACM. pp. 370–381.

Bates, J. R. (2007). Some considerations of the concept of climate feedback. Quat. J. Roy. Met. Soc. 133(624): 545–560.

Beardsley, R. (1970). An experimental study of inertial waves in a closed cone. Stud. Appl. Math. 49: 187–196.

Becker, J. J., D. T. Sandwell, W. H. Smith, J. Braud, B. Binder, J. Depner, D. Farbo, et al. (2009). Global bathymetry and elevation data at 30 arc seconds resolution: SRTM30_PLUS. Mar. Geod. 32(4): 355–371.

Behringer, D., L. Regier, and H. Stommel (1979). Thermal feedback on wind-stress as a contributing cause of the Gulf Stream. Adv. Geophys. 37: 699–709.

Bell, T. H. (1975a). Lee waves in stratified flows with simple harmonic time dependence. J. Fluid Mech. 67(4): 705–722.

—— (1975b). Topographically generated internal waves in open ocean. J. Geophys. Res. 80(3): 320–327.

Bennett, A. F. (2006). Lagrangian Fluid Dynamics. Cambridge, Cambridge University Press.

Béranger, K., B. Barnier, S. Gulev, and M. Crepon (2006). Comparing 20 years of precipitation estimates from different sources over the world ocean. Ocean Dyn. 56(2): 104–138.

Berger, A. L. (1978). Long-term variations of caloric insolation resulting from earths orbital elements. Quat. Res. 9(2): 139–167.

Berger, A., and M. F. Loutre (1991). Insolation values for the climate of the last 10 million years. Quat. Sci. Revs. 10(4): 297–317.

Bills, B. G., and R. D. Ray (1999). Lunar orbital evolution: A synthesis of recent results. Geophys. Res. Lett. 26(19): 3045–3048.

Bindoff, N. L., and T. J. McDougall (1994). Diagnosing climate change and ocean ventilation using hydrographic data. J. Phys. Oc. 24: 1137–1152.

Bjerknes, J. (1964). Atlantic air-sea interaction. Adv. Geophys. 10: 1–82.

Bjerknes, V., J. Bjerknes, H. S. Solberg, and T. Bergeron (1933). Physikalische Hydrodynamik, mit Anwendung auf die dynamische Meteorologie. Berlin, Springer. A French translation exists.

Björck, Å. (1996). Numerical Methods for Least Squares Problems. Philadelphia, SIAM.

Blumen, W. (1972). Geostrophic adjustment. Revs. Geophys. Space Phys. 10(2): 485–528.

Blumenthal, M. B., and M. G. Briscoe (1995). Distinguishing propagating waves and standing modes: An internal wave model. J. Phys. Oc. 25(6): 1095–1115.

Bobrovich, A. V., and G. M. Reznik (1999). Planetary waves in a stratified ocean of variable depth. Part 2. Continuously stratified ocean. J. Fluid Mech. 388: 147–169.

Boccaletti, G., R. Ferrari, A. Adcroft, and J. Marshall (2005). The vertical structure of ocean heat transport. Geophys. Res. Lett. 32: L10603.

Bogden, P. S., R. E. Davis, and R. Salmon (1993). The North Atlantic circulation: Combining simplified dynamics with hydrographic data. Adv. Geophys., 51: 1–52.

Bohren, C. F., and B. A. Albrecht (1998). Atmospheric Thermodynamics. New York, Oxford University Press.

Bowditch, N. (2002). The American Practical Navigator: An Epitome of Navigation. Bethesda, MD, National Imagery and Mapping Agency.

Bower, A. S., M. S. Lozier, S. F. Gary, and C. W. Böning (2009). Interior pathways of the North Atlantic meridional overturning circulation. Nature 459(7244): 243–247.

Box, G.E.P., G. M. Jenkins, and G. C. Reinsel (2008). Time Series Analysis: Forecasting And Control. Hoboken, NJ, John Wiley.

Box, J. E., D. H. Bromwich, and L. S. Bai (2004). Greenland ice sheet surface mass balance 1991-2000: Application of Polar MM5 mesoscale model and in situ data. J. Geophys. Res. 109(D16), D16105.

Bracewell, R. N. (2000). The Fourier Transform and Its Applications. Boston, McGraw Hill.

Brambilla, E., and L. D. Talley (2006). Surface drifter exchange between the North Atlantic subtropical and subpolar gyres. J. Geophys. Res. 111(C7), C07026.

Bretherton, F. P., R. E. Davis, and C. Fandry (1976). A technique for objective analysis and design of oceanographic instruments applied to MODE-73. Deep-Sea Res. 23: 559–582.

Briscoe, M. G. (1975). Preliminary results from the trimoored Internal Wave Experiment (IWEX). J. Geophys. Res. 80(27): 3872–3884.

Broecker, W. S., H. P. Tsung, and G. Östlund (1986). The distribution of bomb tritium in the ocean. J. Geophys. Res. 91: 14331–14344.

Bromwich, D. H., and R. L. Fogt (2004). Strong trends in the skill of the ERA-40 and NCEP-NCAR reanalyses in the high and midlatitudes of the southern hemisphere, 1958fh2001. J. Clim. 17(23): 4603–4619.

Bromwich, D. H., R. L. Fogt, K. I. Hodges, and J. E. Walsh (2007). A tropospheric assessment of the ERA-40, NCEP, and JRA-25 global reanalyses in the polar regions. J. Geophys. Res.-Atmospheres 112(D10), D10111.

Bromwich, D. H., J. P. Nicolas, and A. J. Monaghan (2011). An assessment of precipitation changes over Antarctica and the Southern Ocean since 1989 in contemporary global reanalyses. J. Clim. 24: 4189–4209.

Bryden, H. L. (1977). Geostrophic comparisons from moored measurements of current and temperature during the Mid-Ocean Dynamics Experiment. Deep-Sea Res. 24: 667–681.

Bryden, H. L., and H. Stommel (1982). The origin of the Mediterranean outflow. J. Mar. Res. 40 (Suppl.): 55–71.

Bryden, H. L., and S. Imawaki (2001). Ocean heat transport. In: Ocean Circulation and Climate, G. Siedler, J. Church, and W. J. Gould, eds. San Diego, Academic Press.

Buchwald, V. T. (1968). Diffraction of Kelvin waves at a corner. J. Fluid Mech. 31: 193–205.

Buck, C. E., and A. R. Millard (2004). Tools for Constructing Chronologies. Crossing Disciplinary Boundaries. London, Springer-Verlag.

Bühler, O. (2009). Waves and Mean Flows. Cambridge, Cambridge University Press.

Bühler, O., and M. Holmes-Cerfon (2011). Decay of an internal tide due to random topography in the ocean. J. Fluid Mech. 678: 271–293.

Bumpus, D. F. (1973). A description of the circulation on the continental shelf of the east coast of the United States. Prog. Oceanog. 6: 111–157.

Byrne, D. A., A. L. Gordon, and W. F. Haxby (1995). Agulhas eddies—a synoptic view using GEOSAT ERM Data. J. Phys. Oc. 25: 902–917.

Cacchione, D. A., L. F. Pratson, and A. S. Ogston (2002). The shaping of continental slopes by internal tides. Science 296(5568): 724–727.

Cacchione, D. A., and C. Wunsch (1974). Experimental study of internal waves on a slope. J. Fluid Mech. 66: 223–239.

Cahn, A. (1945). An investigation of the free oscillations of a simple current system. J. Meteor. 2: 113–119.

Callies, J., and R. Ferrari (2013). Interpreting energy and tracer spectra of upper-ocean turbulence in the submesoscale range (1–200 km). J. Phys. Oc. 43: 2456–2474.

Calman, J. (1978). Interpretation of ocean current spectra. 1. Kinematics of three-dimensional vector time series. J. Phys. Oc. 8: 627–643.

Cardone, V. J., J. G. Greenwood, and M. A. Cane (1990). On trends in historical marine wind data. J. Clim. 3: 113–127.

Carrier, G. F. (1949). The spaghetti problem. Am. Math. Monthly 56: 669–672.

Carruthers, J. N. (1955). A plea for oceanology. Deep-Sea Res. (1953) 2: 247.

Carter, G. S., M. A. Merrifield, J. M. Becker, K. Katsumata, M. C. Gregg, D. S. Luther, M. D. Levine, et al. (2008). Energetics of M_2 barotropic-to-baroclinic tidal conversion at the Hawaiian Islands. J. Phys. Oc. 38(10): 2205–2223.

Cartwright, D. E. (1999). Tides. A Scientific History, Cambridge, Cambridge University Press.

Cartwright, D. E., and A. C. Edden (1973). Corrected tables of tidal harmonics. Geophys. J. Roy. Astron. Soc. 33: 253–264.

Cessi, P., K. Bryan, and R. Zhang (2004). Global seiching of thermocline waters between the Atlantic and the Indian-Pacific Ocean Basins. Geophys. Res. Lett. 31(4): L04302.

Chamberlin, T. C. (1890). The method of multiple working hypotheses. Science 15(366): 92–96, Reprint, Science, 148: 754–759 (1965).

Chandrasekhar, S. (1968). Hydrodynamic and Hydromagnetic Stability. Oxford, Oxford University Press.

Chapman, S., and R. S. Lindzen (1970). Atmospheric Tides. Thermal And Gravitational. Dordrecht, Reidel.

Charette, M. A., and W. H. F. Smith (2010). The volume of Earth's ocean. Oceanog. 23(2): 112–114.

Charney, J. G. (1971). Geostrophic turbulence. J. Atm. Sci. 28: 1087–1095.

Charrassin, J. B., M. Hindell, S. R. Rintoul, F. Roquet, S. Sokolov, M. Biuw, D. Consta, et al. (2008). Southern Ocean frontal structure and sea-ice formation rates revealed by elephant seals. Proc. Nat. Acad. Sci. 105(33): 11634–11639.

Chavanne, C. P., E. Firing, and F. Ascani (2012). Inertial oscillations in geostrophic flow: Is the inertial frequency shifted by $\zeta/2$ or by ζ? J. Phys. Oc. 42(5): 884–888.

Chave, A. D., D. S. Luther, and J. H. Filloux (1991). Variability of the wind stress curl over the North Pacific—implications for the oceanic response. J. Geophys. Res. 96: 18361–18379.

Chelton, D. B., R. A. DeSzoeke, M. G. Schlax, K. El Naggar, and N. Siwertz (1998). Geographical variability of the first baroclinic Rossby radius of deformation. J. Phys. Oc. 28(3): 433–460.

Chelton, D. B., J. C. Ries, B. J. Haines, L. L. Fu, and P. S. Callahan (2001). Satellite altimetry. In: Satellite Altimetry and Earth Sciences, L.-L. Fu and A. Cazenave, eds. San Diego, Academic Press. pp. 1–131.

Chelton, D. B., and M. G. Schlax (1996). Global observations of oceanic Rossby waves. Science 272(5259): 234–238.

Chelton, D. B., M. G. Schlax, and R. M. Samelson (2011). Global observations of nonlinear mesoscale eddies. Prog. Oceanog. 91: 167–216.

Chu, P. C. (2006). P-vector Inverse Method. Berlin, Springer.

Church, J. A., P. L. Woodworth, T. Aarup, and W. S. Wilson, eds. (2010). Understanding Sea-Level Rise and Variability. Chichester UK, Wiley-Blackwell.

Church, J. A., and N. J. White (2006). A twentieth century acceleration in global sea-level rise. Geophys. Res. Lett. 33(1): L01602.

Cipollini, P., D. Cromwell, P. G. Challenor, and S. Raffaglio (2001). Rossby waves detected in global ocean colour data. Geophys. Res. Lett. 28(2): 323–326.

Clarke, A. J. (2008). An Introduction to the Dynamics of El Niño and the Southern Oscillation. London, Academic Press.

Cole, S. T., D. L. Rudnick, B. A. Hodges, and J. P. Martin (2009). Observations of tidal internal wave beams at Kauai Channel, Hawaii. J. Phys. Oc. 39(2): 421–436.

Colosi, J. A., and W. Munk (2006). Tales of the venerable Honolulu tide gauge. J. Phys. Oc. 36(6): 967–996.

Compo, G. P., J. S. Whitaker, P. D. Sardeshmukh, N. Matsui, R. J. Allan, X. Yin, B. E. Gleason, et al. (2011). The Twentieth Century Reanalysis Project. Quat. J. Roy. Met. Soc. 137: 1–28.

Conrad, C. P., and B. H. Hager (1997). Spatial variations in the rate of sea level rise caused by the present-day melting of glaciers and ice sheets. Geophys. Res. Lett. 24(12): 1503–1506.

Cook, A. (1998). Edmond Halley. Charting the Heavens and the Seas, Oxford, Oxford University Press.

Cooley, J. W., and J. W. Tukey (1965). An algorithm for machine calculation of complex Fourier series. Maths. Comp. 19(90): 297–301.

Cox, C., and H. Sandstrom (1962). Coupling of internal and surface waves in water of variable depth. J. Oceanog. Soc. Japan 18: 499–513.

Craig, H. (1969). Abyssal carbon and radiocarbon in the Pacific. J. Geophys. Res. 74: 5491–5506.

Craik, A. D. D. (1985). Wave Interaction and Fluid Flows, Cambridge, Cambridge University Press.

Cuffey, K. M., and W. S. B. Paterson (2010). The Physics of Glaciers. Burlington, MA, Butterworth.

Dai, A., T. Qian, K. E. Trenberth, and J. D. Milliman (2009). Changes in continental freshwater discharge from 1948 to 2004. J. Clim. 22: 2773–2792.

Darrigol, O. (2005). Worlds of Flow: A History of Hydrodynamics from the Bernoullis to Prandtl. Oxford, Oxford University Press.

D'Asaro, E. A. (1985). The energy flux from the wind to near-inertial motions in the surface mixed layer. J. Phys. Oc. 15: 1043–1059.

—— (1995). Upper-ocean inertial currents forced by a strong storm. Part 2. Modeling. J. Phys. Oc. 25: 2937–2952.

—— (2003). Performance of autonomous Lagrangian floats. J. Atm. Oc. Tech. 20: 896–911.

D'Asaro, E. A., and M. D. Morehead (1991). Internal waves and velocity fine-structure in the arctic-ocean. J. Geophys. Res. 96(C7): 12725–12738.

Davenport, W. B. J., and W. L. Root (1958). An Introduction to the Theory of Random Signals and Noise, New York, McGraw-Hill.

Davies, J. H. (2013). Global map of solid Earth surface heat flow. Geochemistry Geophysics Geosystems 14, 4608–4622.

Davis, R. E. (1978). Estimating velocity from hydrographic data. J. Geophys. Res. 83: 5507–5509.

—— (1991). Observing the general circulation with floats. Deep-Sea Res. 38: S531–S571.

—— (1994a). Diapycnal mixing in the ocean: Equations for large-scale budgets. J. Phys. Oc. 24: 777–800.

—— (1994b). Diapycnal mixing in the ocean: The Osborn–Cox model. J. Phys. Oc. 24: 2560–2576.

—— (1996). Sampling turbulent dissipation. J. Phys. Oc. 26: 341–358.

—— (1998). Preliminary results from directly measuring mid-depth circulation in the tropical and South Pacific. J. Geophys. Res. 103: 24619–24639.

Deacon, M. (1971). Scientists and the Sea 1650–1900. A Study of Marine Science. London, Academic Press.

Decloedt, T., and D. S. Luther (2012). Spatially heterogeneous diapycnal mixing in the abyssal ocean: A comparison of two parameterizations to observations. J. Geophys. Res. 117: C11025.

Defant, A. (1950). Reality and illusion in oceanographic surveys. Adv. Geophys. 9: 120–138.

—— (1961). Physical Oceanography. Vols. 1 and 2. New York, Pergamon.

De Leon, Y., and N. Paldor (2009). Linear waves in midlatitudes on the rotating spherical earth. J. Phys. Oc. 39: 3204–3215.

Dengg, J., A. Beckmann, and R. Gerdes (1996). The Gulf Stream separation problem. In: The Warmwatersphere of the North Atlantic Ocean, W. Krauss, ed. Berlin, Gebrüder Borntraeger. pp. 253–290.

Deser, C., M. A. Alexander, and M. S. Timlin (2003). Understanding the persistence of sea surface temperature anomalies in midlatitudes. J. Clim. 16(1): 57–72.

Dewar, W.K. (1980). The Effect of Internal Waves on Neutrally Buoyant Floats and Other Near-Lagrangian Tracers. S. M. thesis, MIT.

Dewar, W. K., R. J. Bingham, R. L. Iverson, D. P. Nowacek, L. C. St. Laurent, and P. H. Wiebe (2006). Does the marine biosphere mix the ocean? Adv. Geophys. 64: 541–561.

Diaconis, P., and F. Mosteller (1989). Methods for studying coincidences. J. Am. Stat. Assoc. 84: 853–861.

Dickey, J. O., P. L. Bender, J. E. Faller, X. X. Newhall, R. L. Ricklefs, J. G. Ries, P. J. Shelus, et al. (1994). Lunar laser ranging: A continuing legacy of the Apollo Program. Science 265: 482–490.

Dietrich, G., K. Kalle, W. Krauss, and G. Siedler (1980). General Oceanography: An Introduction. 2nd ed. English translation of the 1975 German edition. New York, Wiley.

Dikii, L. A. (1966). Asymptotic behaviour of solutions to Laplace's tidal equation. Doklady Akademii Nauk SSSR 170: 67–70.

DiNezio, P. N., and G. J. Goni (2011). Direct evidence of a changing fall-rate bias in XBTs manufactured during 1986–2008. J. Atm. Oc. Tech. 28: 1569–1578.

Doherty, K. W., D. E. Frye, S. P. Liberatore, J. M. Toole (1999). A moored profiling instrument. J. Atm. Oc. Tech. 16: 1816–1829.

Doney, S. C., D. M. Glover, and W. J. Jenkins (1992). Model function of the global bomb tritium distribution in precipitation. J. Geophys. Res. 97: 5481–5492.

Donlon, C. J., P. J. Minnett, C. Gentemann, T. J. Nightingale, I. J. Barton, B. Ward, and M. J. Murray (2002). Toward improved validation of satellite sea surface skin temperature measurements for climate research. J. Clim. 15(4): 353–369.

Doodson, A. T. (1921). The harmonic development of the tide-generating potential. Proc. Roy. Soc. A 100: 305–329, Reprint, Int. Hydrog. Rev. 331: 311–335, 1954.

Douglas, B. C., Kearney, M. S., and S. R. Leatherman, eds. (2001). Sea Level Rise, History and Consequences. San Diego, Academic Press.

Drazin, P. G., and W. H. Reid (2004). Hydrodynamic Stability. Cambridge, Cambridge University Press.

Duffy, D. G. (2001). Green's Functions with Applications. Boca Raton, FL, Chapman & Hall/CRC.

Durland, T. S., and J. T. Farrar (2012). The wavenumber-frequency content of resonantly excited equatorial waves. J. Phys. Oc. 42(11): 1834–1858.

Dushaw, B. D., G. D. Egbert, P. F. Worcester, B. D. Cornuelle, B. M. Howe, and K. Metzger (1997). A TOPEX/POSEIDON global tidal model (TPXO.2) and barotropic tidal currents determined from long-range acoustic transmissions. Prog. Oceanog. 40: 337–367.

Dutkiewicz, S., M. Follows, P. Heimbach and J. Marshall (2006). Controls on ocean productivity and air-sea carbon flux: An adjoint model sensitivity study. Geophys. Res. Lett., 33: doi:10.1029/2005GL024987.

Edson, J., T. Crawford, J. Crescenti, T. Farrar, N. Frew, G. Gerbi, A. Plueddemann, et al. (2007). The coupled boundary layers and air–sea transfer experiment in low winds. Bull. Am. Met. Soc. 88: 341–356.

Edson, J. B., V. Jampana, R. A. Weller, S. P. Bigorve, A. J. Plueddemann, C. W. Faurakkm S. D. Miller, L. Mahrt, D. Vickers, and H. Hersbach (2013). On the exchange of momentum over the open ocean. J. Phys. Oc. 43: 1589–1610.

Efron, B., and R. Tibshirani (1993). An Introduction to the Bootstrap. New York, Chapman & Hall.

Egbert, G. D., and R. D. Ray (2003a). Deviation of long-period tides from equilibrium: Kinematics and geostrophy. J. Phys. Oc. 33: 822–839.

——— (2003b). semidiurnal and diurnal tidal dissipation from TOPEX/Poseidon altimetry. Geophys. Res. Lett. 30(17), 1907. doi: 1910.1029/2003GL017676.

Egbert, G. D., R. D. Ray, and B. G. Bills (2004). Numerical modeling of the global semidiurnal tide in the present day and in the last glacial maximum. J. Geophys. Res. 109(C3). doi: 10.1029/2003JC001973.

Ekman, V. W. (1905). On the influence of the earth's rotation on ocean-currents. Arkiv for Matematik, Astronomi och Fysik 2(11): 1–52.

Elgar, S., T. H. C. Herbers, V. Chandran, and R. T. Guza (1995). Higher-order spectral analysis of nonlinear ocean surface gravity waves. J. Geophys. Res. 100: 4977–4983.

Elliott, W. P., and D. J. Gaffen (1991). On the utility of radiosonde humidity archives for climate studies. Bull. Am Met. Soc. 72(10): 1507–1520.

Emile-Geay, J., and G. Madec (2009). Geothermal heating, diapycnal mixing and the abyssal circulation. Oc. Sci 5: 203–217.

Enderton, D., and J. Marshall (2009). Explorations of atmosphere–ocean–ice climates on an aquaplanet and their meridional energy transports. J. Atm. Sci. 66: 1593–1611.

Enfield, D. B., A. M. Mestas-Nuñez, and P. J. Trimble (2001). The Atlantic multidecadal oscillation and its relation to rainfall and river flows in the continental US. Geophys. Res. Lett. 28(10): 2077–2080.

Eriksen, C. C. (1978). Measurements and models of fine-structure, internal gravity-waves, and wave breaking in deep ocean. J. Geophys. Res. 83: 2989–3009.

——— (1982). Geostrophic equatorial deep jets. J. Mar. Res. 40: 143–157.

Evensen, G. (2009). Data Assimilation: The Ensemble Kalman Filter. Berlin, Springer.

Faller, A. J. (1968). Sources of energy for the ocean circulation and a theory of the mixed layer. In: Proc. 5th US Congress of Applied Mech., U. of Minn., Minneapolis, 14–17 June 1966. New York: ASME. pp. 651–672.

———— (1981). The origin and development of laboratory models and analogs of the ocean circulation. In: Evolution of Physical Oceanography: Scientific Surveys in Honor of Henry Stommel, B. A. Warren and C. Wunsch, eds. Cambridge, MA, MIT Press. pp. 462–482.

———— (2006). Letter. Physics Today 59(October): 10.

Farrell, B. F. (1989). Optimal excitation of baroclinic waves. J. Atmos. Sci. 46: 1193–1206.

Farrell, W. E. (1972). Deformation of Earth by surface loads. Revs. Geophys. Space Phys. 10(3): 761–797.

Feistel, R., D. G. Wright, K. Miyagawa, A. H. Harvey, J. Hruby, D. R. Jackett, T. J. McDougall, and W. Wagner (2008). Mutually consistent thermodynamic potentials for fluid water, ice and seawater: A new standard for oceanography. Oc. Sci 4(4): 275–291.

Feller, W. (1957). An Introduction to Probability Theory and Its Applications. 2nd ed. New York, J. Wiley.

Fenty, I., and P. Heimbach (2013). Coupled sea ice-ocean-state estimation in the Labrador Sea and Baffin Bay. J. Phys. Oc. 43(5): 884–904.

Ferrari, R., and D. L. Rudnick (2000). Thermohaline variability in the upper ocean. J. Geophys. Res. 105: 16857–16883.

Ferrari, R., and C. Wunsch (2009). Ocean circulation kinetic energy: Reservoirs, sources, and sinks. Ann. Rev. Fluid Mechs. 41: 253–282.

Ferreira, D., J. Marshall, and P. Heimbach (2005). Estimating eddy stresses by fitting dynamics to observations using a residual-mean ocean circulation model and its adjoint. J. Phys. Oc. 35: 1891–1910.

Filloux, J. H. (1971). Deep-sea tide observations from Northeastern Pacific. Deep-Sea Res. 18(2): 275–284.

Fine, R. A., J. L. Reid, and H. G. Östlund (1981). Circulation of tritium in the Pacific Ocean. J. Phys. Oc. 11(1): 3–14.

Fine, R. A., W. M. Smethie, J. L. Bullister, M. Rhein, D.-H. Min, M. J. Warner, A. Poisson, and R. F. Weiss (2008). Decadal ventilation and mixing of Indian Ocean waters. Deep-Sea Res. 55(1): 20–37.

Flagg, C. N., M. Dunn, D. P. Wang, H. T. Rossby, and R. L. Benway (2006). A study of the currents of the outer shelf and upper slope from a decade of shipboard ADCP observations in the Middle Atlantic Bight. J. Geophys. Res. 111: C06003.

Fofonoff, N. (1962). Dynamics of ocean currents. The Sea 1: 323–395.

———— (1969). Spectral characteristics of internal waves in the ocean. Deep-Sea Res. 16 (Suppl.): 59–71.

———— (1981). The Gulf Stream system. In: Evolution of Physical Oceanography: Scientific Surveys in Honor of Henry Stommel, B. A. Warren and C. Wunsch, eds. Cambridge, MA, MIT Press. pp. 112–139.

Fofonoff, N. P., and F. Webster (1971). Current Measurements in the Western Atlantic. Phil. Trans. Royal Soc. A 270(1206): 423–436.

Folland, C. K., and D. E. Parker (1995). Correction of instrumental biases in historical sea-surface temperature data. Quat. J. Roy. Met. Soc. 121: 319–367.

Forget, G. (2010). Mapping ocean observations in a dynamical framework: A 2004–06 ocean atlas. J. Phys. Oc. 40: 1201–1221.

Forget, G., and C. Wunsch (2007). Estimated global hydrographic variability. J. Phys. Oc. 37(8): 1997–2008.

Fraile-Nuez, E., F. Machín, P. Velez-Belchí, F. López-Laatzen, R. Borges, V. Benítez-Barrier, and A. Hernández-Guerra (2010). Nine years of mass transport data in the eastern boundary of the North Atlantic Subtropical Gyre. J. Geophys. Res. 115, C09009.

Frankignoul, C. and K. Hasselmann (1977). Stochastic climate models. Part II. Application to seasurface temperature, temperature anomalies and thermocline variability. Tellus 29: 289–305.

Frankignoul, C., and P. Müller (1979). Quasi-geostrophic response of an infinite β-plane ocean to stochastic forcing by the atmosphere. J. Phys. Oc. 9: 104–127.

Freeland, H. (2007). A short history of Ocean Station Papa and Line P. Prog. Oceanog. 75: 120–125.

Freeland, H. J., P. B. Rhines, and T. Rossby (1975). Statistical observations of trajectories of neutrally buoyant floats in North-Atlantic. Adv. Geophys. 33: 383–404.

Fu, L.-L. (1981). Observations and models of inertial waves in the deep ocean. Revs. Geophys., and Space Phys. 19: 141-170.

Fu, L.-L., and E. A. Cazenave (2001). Satellite Altimetry and Earth Sciences: A Handbook of Techniques and Applications. San Diego, Academic Press.

Fu, L.-L., and B. J. Haines (2013). The challenges in long-term altimetry calibration for addressing the problem of global sea level change. Adv. Space Res. 51: 1284–1300.

Fu, L.-L., T. Keffer, P. Niiler, and C. Wunsch (1982). Observations of mesoscale variability in the western North Atlantic: A comparative study. J. Mar. Res. 40: 809–848.

Fu, L.-L., and C. Ubelmann (2014). On the transition from profile altimeter to swath altimeter for observing global ocean surface topography, J. Ocean. Atm. Tech. 31: 560–568.

Fuglister, F. C. (1955). Alternative analyses of current surveys. Deep-Sea Res. 2: 213–229.

—— (1960). Atlantic Ocean Atlas of Temperature and Salinity Profiles and Data from the International Geophysical Year of 1957–1958. Woods Hole, MA, Woods Hole Oceanographic Institution.

Fukamachi, Y., S. R. Rintoul, J. A. Church, S. Aoki, S. Sokolov, M. A. Rosenberg, and M. Wakatsuchi (2010). Strong export of Antarctic Bottom Water east of the Kerguelen plateau. Nature Geosci. 3: 327–331.

Fukumori, I. (1991). Circulation about the Mediteranean Tongue: An analysis of an EOF-based ocean. Prog. in Oceanog., 27: 197–224.

Gage, K. S., and G. D. Nastrom (1985). On the spectrum of atmospheric velocity fluctuations seen by MST/ST radar and their interpretation. Radio Science 20(6): 1339–1347.

Gallagher, B. S., and W. H. Munk (1971). Tides in shallow water: Spectroscopy. Tellus 23: 346–363

Ganachaud, A. (2003a). Error budget of inverse box models: The North Atlantic. J. Atm. Oc. Tech. 20: 1641–1655.

—— (2003b). Large-scale mass transports, water mass formation, and diffusivities estimated from World Ocean Circulation Experiment (WOCE) hydrographic data. J. Geophys. Res. 108(C7), 3213.

Ganachaud, A., and C. Wunsch (2002). Oceanic nutrient and oxygen transports and bounds on export production during the World Ocean Circulation Experiment. Glob. Biogeochem. Cycles 16(4), 1057.

—— (2003). Large-scale ocean heat and freshwater transports during the World Ocean Circulation Experiment. J. Clim., 16: 696–705.

Gardiner, C. W. (1985). Handbook of Stochastic Methods for Physics, Chemistry, and the Natural Sciences. Berlin, Springer-Verlag.

Garrett, C. (1975). Tides in gulfs. Deep-Sea Res. 22: 23–36.

—— (2001). What is the "near-inertial" band and why is it different from the rest of the internal wave spectrum? J. Phys. Oc. 31: 962–971.

Garrett, C., and E. Kunze (2007). Internal tide generation in the deep ocean. Ann. Rev. Fluid Mechs. 39: 57–87.

Garrett, C., P. MacCready, and P. Rhines (1993). Boundary mixing and arrested Ekman layers: Rotating stratified flow near a sloping boundary. Ann. Rev. Fluid Mech. 25: 291–323.

Garrett., C, and W. Munk (1972). Space-time scales of internal waves. Geophys. Fl. Dyn., 3: 225–264.

Gaspar, P., Y. Gregoris, and J.-M. Lefevre (1990). A simple eddy kinetic energy model for simulations of the oceanic vertical mixing: Tests at station Papa and long-term upper ocean study site. J. Geophys. Res. 95(C9): 16179–16193.

Gebbie, G. (2007). Does eddy subduction matter in the Northeast Atlantic Ocean? J. Geophys. Res. 112 (C6), C06007.

Gebbie, G., P. Heimbach, and C. Wunsch (2006). Strategies for nested and eddy-permitting state estimation. J. Geophys. Res. 111(C10), C10073.

Gebbie, G., and P. Huybers (2011). How is the ocean filled? Geophys. Res. Lett. 38, L06604.

Gentemann, C. L., C. J. Donlon, A. Stuart-Menteth, and F. J. Wentz (2003). Diurnal signals in satellite sea surface temperature measurements. Geophys. Res. Lett. 30(3): 1140.

Gerkema, T., and V. I. Shrira (2005). Near-inertial waves on the "nontraditional" β plane. J. Geophys. Res. 110, C 1003. doi:10.1029/2004JC002519

Giering, R., and T. Kaminski (1998). Recipes for adjoint code construction. ACM Trans. Mathematical Software 24(4): 437–474.

Gill, A. E. (1982). Atmosphere-Ocean Dynamics. New York, Academic Press.

Gill, A. E., and A. J. Clarke (1974). Wind-induced upwelling, coastal currents and sea-level changes. Deep-Sea Res. 21: 325–345.

Gill, A. E., and P. P. Niiler (1973). The theory of the seasonal variability in the ocean. Deep-Sea Res. 20: 141–177.

Gill, A. E., J. S. A. Green, and A. J. Simmons (1974). Energy partition in the large-scale ocean circulation and the production of mid-ocean eddies. Deep-Sea Res. 21: 499–528.

Gille, S. T. (2004). How nonlinearities in the equation of state of seawater can confound estimates of steric sea level change. (2004) J. Geophys. Res. 109: C03005.

Gilson, J., D. Roemmich, B. Cornuelle, and L.-L. Fu (1998). Relationship of TOPEX/Poseidon altimetric height to steric height and circulation in the North Pacific. J. Geophys. Res.–Oceans 103(C12): 27947–27965.

Gladyshev, S., L. Talley, G. Kantakov, G. Khen, and M. Wakatsuchi (2003). Distribution, formation, and seasonal variability of Okhotsk Sea Mode Water. J. Geophys. Res. 108(C6), 3186.

Goff, J. A., and T. H. Jordan (1989). Stochastic modeling of seafloor morphology—a parameterized Gaussian model. Geophys. Res. Lett. 16: 45–48.

Gold, B., and C. M. Rader (1983). Digital processing of signals. Malabar, FL, Krieger.

Goldsbrough, G. R. (1933). Ocean currents produced by evaporation and precipitation. Proc. Roy. Soc. London A 141: 512–517.

Goldstein, H. (1980). Classical Mechanics. 2nd ed. Reading, MA, Addison-Wesley.

Gonella, J. (1972). A rotary-component method for analyzing meteorological and oceanographic vector time series. Deep-Sea Res. 19: 833–846.

Gordon, A. L. (1978). Deep Antarctic convection west of Maud Rise. J. Phys. Oc. 8: 600–612.

Gordon, A. L., and E. J. Molinelli (1982). Southern Ocean Atlas. Thermohaline and Chemical Distributions and the Atlas Data Set. New Delhi, Amerind.

Gordon, A. L., J. Sprintall, H. M. Van Aken, D. Susanto, S. Wijffels, R. Molcard, A. Ffield, W. Pranowo, and S. Wirasantosa (2010). The Indonesian Throughflow during 2004–2006 as observed by the INSTANT program. Dyn. Atm. Oc. 50: 115–128.

Gould, W. J. (2005). From Swallow floats to Argo: The development of neutrally buoyant floats. Deep-Sea Res. Part 2. 52(3–4): 529–543.

Gould, W. J., and E. Sambuco (1975). Effect of mooring type on measured values of ocean currents. Deep-Sea Res. 22: 55–62.

Gouretski, V. V., and K. Jancke (2001). Systematic errors as the cause for an apparent deep water property variability: Global analysis of the WOCE and historical hydrographic data. Prog. Oceanog. 48: 337–402.

Gouretski, V. V., and K. P. Koltermann (2004). WOCE Global Hydrographic Climatology: A Technical Report. Berichte des BSH 35. Hamburg, BSH.

——— (2007). How much is the ocean really warming? Geophys. Res. Lett. 34(1), L01610.

Green, R. M. (1985). Spherical Astronomy, Cambridge, Cambridge University Press.

Greenslade, D. J. M., D. B. Chelton, and M. G. Schlax (1997). The midlatitude resolution capability of sea level fields constructed from single and multiple satellite altimeter datasets. J. Atm. Oc. Tech. 14: 849–870.

Greenspan, H. P. (1963). A string problem. J. Math. Analysis Appl. 6: 339–348.

——— (1990). The Theory of Rotating Fluids. Cambridge, Cambridge University Press.

Gregg, M. C. (1987). Diapycnal mixing in the thermocline: A review. J. Geophys. Res. 92: 5249–5286.

——— (1998). Estimation and geography of diapycnal mixing in the stratified ocean. In: Physical Processes in Lakes and Oceans, J. Imberger, ed. Washington, DC, Am. Geophysical Union. pp. 305–338.

Gregg, M. C., T. B. Sanford, and D. P. Winkel (2003). Reduced mixing from the breaking of internal waves in equatorial waters. Nature 422: 513–515.

Gregg, M. C., H. E. Seim, D. B. Percival (1993). Statistics of shear and turbulent dissipation profiles in random internal wave-fields. J. Phys. Oc. 23: 1777–1799.

Griewank, A., and A. Walther (2008). Evaluating Derivatives: Principles and Techniques of Algorithmic Differentiation, 2nd ed. Philadelphia, SIAM.

Griffies, S. M. (2004). Fundamental of Ocean Climate Models, Princeton, NJ, Princeton University Press.

Griffies, S. M., and R. J. Greatbatch (2012). Physical processes that impact the evolution of global mean sea level in ocean climate models. Ocean Model. 51: 37–72.

Grignon, L., D. A. Smeed, H. L. Bryden, K. Schroeder (2010). Importance of the variability of hydrographic preconditioning for deep convection in the Gulf of Lion, NW Mediterranean. Oc. Sci 6(2): 573–586.

Gross, R. S., I. Fukumori, D. Menemenlis (2003). Atmospheric and oceanic excitation of the Earth's wobbles during 1980–2000. J. Geophys. Res 108(B8), B09405.

Groves, G. W., and E. J. Hannan (1968). Time series regression of sea level on weather. Rev. Geophys. 6: 129–174.

Groves, G. W., and R. W. Reynolds (1975). An orthogonlised convolution method of tide prediction. J. Geophys. Res. 80: 4131–4138.

Groves, G. W., and B. D. Zetler (1964). The cross spectrum of sea level at San Francisco and Honolulu. J. Mar. Res. 22: 269–275.

Haidvogel, D. B., and A. Beckmann (1999). Numerical Ocean Circulation Modeling. River Edge, NJ, Imperial College Press.

Hall, G. (1955). A plea for Oceanology. Deep-Sea Res. 2: 285–286.

Halpern, B. S., S. Walbridge, K. A. Selkoe, C. V. Kappel, F. Micheli, C. D'Agrosa, J. F. Bruno, et al. (2008). A global map of human impact on marine ecosystems. Science 319(5865): 948–952.

Hamlington, B. D., R. R. Leben, R.S. Nerem, and K.Y. Kim (2011). The effect of signal-to-noise ratio on the study of sea level trends. J. Clim. 24: 1396–1408.

Hamming, R. W. (1973). Numerical Methods for Scientists and Engineers. New York, Dover.

Harrison, J. C., ed. (1985). Earth Tides. New York, Van Nostrand Reinhold.

Hartmann, D. L. (1994). Global Physical Climate, San Diego, CA, Academic Press.

Hasselmann, K. (1970). Wave-driven inertial oscillations. Geophys. Astrophys. Fl. Dyn. 1: 463–502.

——— (1976). Stochastic climate models. Part 1. Theory. Tellus 28: 473–485.

Haubrich, R., and W. Munk (1959). The pole tide. J. Geophys. Res. 64: 2373–2388.

Haurwitz B., H. Stommel, and W. H. Munk (1959). On the thermal unrest in the ocean. In: The Atmosphere and Sea in Motion: Scientific Contributions to the Rossby Memorial Volume, B. Bolin, ed. New York, Rockefeller Institute Press. pp. 74-94.

Hazewinkel, J., F. Paparella, and W. R. Young (2012). Stressed horizontal convection. J. Fluid Mech. 692: 317–331.

Hecht, M. W., and R. D. Smith (2008). Towards a physical understanding of the North Atlantic: a review of model studies. In: Ocean Modeling in an Eddying Regime, M. W. Hecht and H. Hasumi, eds. AGU Geophysical Monograph 177. Washington, DC, AGU. pp. 213–240.

Heimbach, P., C. Hill, and R. Giering (2005). An efficient exact adjoint of the parallel MIT General Circulation Model, generated via automatic differentiation. Future Generation Computer Systems 21: 1356–1371.

Heimbach, P., C. Wunsch, R.M. Ponte, G. Forget, C. Hill, and J. Utke (2011). Timescales and regions of the sensitivity of Atlantic meridional volume and heat transport magnitudes: Toward observing system design. Deep-Sea Res. Part 2 58: 1858–1879.

Heinmiller, R. H. (1983). Instruments and methods. In: Eddies in Marine Science. A. R. Robinson, ed. Berlin, Springer-Verlag. pp. 542–567.

Heiskanen, W. (1921). Über den Einfluss der Gezeiten auf die säkuläre Acceleration des Mondes. Annales Acad. Scientiarum Fennicae 18: 1–84.

Helland-Hansen, B., and F. Nansen (1920). Temperature Variations in the North Atlantic Ocean and in the Atmosphere. Smithsonian Misc. Collect. Vol. 70, no. 4. Washington, DC, Smithsonian Institution.

Hendershott, M. (1981). Long waves and ocean tides. In: Evolution of Physical Oceanography: Scientific Surveys in Honor of Henry Stommel, B.A. Warren and C. Wunsch, eds. Cambridge, MA, MIT Press. pp. 292–341.

Hendry, R. (1977). Obsevations of the semidiurnal internal tide in the western North Atlantic Ocean. Phil. Trans. Roy. Soc. A 286: 1–24.

Hendry, R. M., and A. J. Hartling (1979). A pressure-induced direction error in nickel-coated Aanderaa current meters. Deep-Sea Res. 26: 327–335.

Henrion, M., and B. Fischhoff (1986). Assessing uncertainty in physical constants. Am J. Phys. 54: 791–798.

Hoffman, P. F., and D. P. Schrag (2002). The snowball Earth hypothesis: Testing the limits of global change. Terra Nova 14: 129–155.

Hofmann-Wellenhof, B., and H. Moritz (2006). Physical Geodesy. New York, Springer.

Hogg, N. G. (1971). Longshore current generation by obliquely incident internal waves. Geophys. Astrophys. Fluid Dyn. 2: 361–376.

―――― (1973). Preconditioning phase of MEDOC 1969. 2. Topographic effects. Deep-Sea Res. 20(5): 449–459.

―――― (1991). Mooring motion corrections revisited. J. Atm. Oc. Tech. 8: 289-295.

―――― (2001). Quantification of the deep circulation. In: Ocean Circulation and Climate, G. Siedler, J. Church and J. Gould, eds. New York, Academic Press. pp. 259–270.

Hogg, N. G., and D. E. Frye (2007). Performance of a new generation of acoustic current meters. J. Phys. Oc. 37: 148–161.

Hogg, N. G., and R.-X. Huang, eds. (1995). Collected Works of Henry M. Stommel. 3 vols. Boston, MA, American Meteorological Society.

Holbrook, W. S., I. Fer, and R. W. Schmitt (2009). Images of internal tides near the Norwegian continental slope. Geophys. Res. Lett. 36, L00D10.

Holbrook, W. S., P. Paramo, S. Pearse, and R. W. Schmitt (2003). Thermohaline fine structure in an oceanographic front from seismic reflection profiling. Science 301(5634): 821–824.

Holloway, G. (1980). Oceanic internal waves are not weak waves. J. Phys. Oc. 10: 906–914.

Holton, J. R. (1975). The Dynamic Meteorology of the Stratosphere and Mesosphere. Meteorological Monographs, Boston, MA, American Meteorological Society.

Holzer, M., and T. M. Hall (2000). Transit-time and tracer-age distributions in geophysical flows. J. Atm. Sci. 57(21): 3539–3558.

Horton, C., and W. Sturges (1979). A geostrophic comparison during MODE. Deep-Sea Res. 26A: 521–533.

Hough, S. S. (1897). On the application of harmonic analysis to the dynamical theory of the tides. Part 1. On Laplace's "Oscillations of the First Species," and on the dynamics of ocean currents. Phil. Trans. Roy. Soc. London A 189: 201–257.

Huang, R. X. (2010). Ocean Circulation: Wind-driven and Thermohaline Processes. Cambridge, Cambridge University Press.

Huang, R. X., and R. W. Schmitt (1993). The Goldsbrough-Stommel circulation of the world ocean. J. Phys. Oc. 23: 1277–1284.

Huang, R. X., and W. Wang (2003). Gravitational potential energy sinks/sources in the oceans. In: Near-Boundary Processes and Their Parameterization, Proceedings, 'Aha Huliko'a Hawaii Winter Workshop, P. Müller and D. Henderson, eds. University of Hawaii at Manoa, 21–24 January 2003. Honolulu, HI, SOEST. pp. 239–247.

Hughes, C. W., and B. A. de Cuevas (2001). Why western boundary currents in realistic oceans are inviscid: A link between form stress and bottom pressure torques. J. Phys. Oc. 31: 2871–2885.

Hughes, C. W., and S. D. P. Williams (2010). The color of sea level: Importance of spatial variations in spectral shape for assessing the significance of trends. J. Geophys. Res. 115, C10048.

Huntford, R. (1986). Shackleton. New York, Atheneum.

Huybers, P., and W. Curry (2006). Links between annual, Milankovitch and continuum temperature variability. Nature 441(7091): 329–332.

Huybers, P., and C. Wunsch (2010). Paleophysical oceanography with an emphasis on transport rates. Ann. Rev. Mar. Sci. 2: 1-34.

Ibbetson, A., and N. Phillips (1967). Some laboratory experiments on Rossby waves in a rotating annulus. Tellus 19: 81–87.

Ierley, G. R. (1990). Boundary-layers in the general ocean circulation. Ann. Rev. Fluid Mechs. 22: 111–142.

Imbrie, J., and K. P. Imbrie (1986). Ice Ages: Solving the Mystery. Cambridge, MA, Harvard University Press.

Intergovernmental Oceanographic Commission (2010). The International Thermodynamic Equation of Seawater—2010: Calculation and Use of Thermodynamic Properties. Paris, UNESCO.

Irish, J. D., and F. E. Snodgrass (1972). Quartz crystals as multipurpose oceanographic sensors. I. Pressure. Deep-Sea Res. 19: 165–169.

Ishiguro, S. (1972). Electric analogues in oceanography. Oceanog. Mar. Biol. 10: 27–96.

Ito, T., M. Woloszyn, and M. Mazloff (2010). Anthropogenic carbon dioxide transport in the Southern Ocean driven by Ekman flow. Nature 463(7277): 80–85.

Ivey, G. N., and R. I. Nokes (1989). Vertical mixing due to the breaking of critical internal waves on sloping boundaries. J. Fluid Mech. 204: 479–500.

Ivey, G. N., K. B. Winters, and I. P. D. De Silva (2000). Turbulent mixing in a sloping benthic boundary layer energized by internal waves. J. Fluid Mech. 418: 59–76.

Jackson, J. D. (1999). Classical Electrodynamics. 3rd ed. New York, Wiley.

Jacobs, S. S., H. H. Helmer, C. S. M. Doake, A. Jenkins, and R. M. Frolich (1992). Melting of ice shelves and the mass balance of Antarctica. J. of Glac. 38: 375–387.

Janssen, P. (2004). The Interaction of Ocean Waves and Wind. Cambridge, Cambridge University Press.

Jayne, S. R., and J. Marotzke (2001). The dynamics of ocean heat transport variability. Revs. Geophys. 39: 385–411.

Jayne, S. R., and R. Tokmakian (1997). Forcing and sampling of ocean general circulation models: Impact of high-frequency motions. J. Phys. Oc. 27: 1173–1179.

Jeffreys, H. (1920). Tidal friction in shallow seas. Phil. Trans. Roy. Soc. A 221: 239–264.

——— (1925). On fluid motions produced by differences of temperature and humidity. Q. J. Roy. Met. Soc., 51: 347–356.

——— (1976). The Earth: Its Origin, History, and Physical Constitution. Cambridge, Cambridge University Press.

Jenkins, G. M., and D. G. Watts (1968). Spectral Analysis And Its Applications. San Francisco, Holden-Day.

Jenkins, W. J. (1980). Tritium and He-3 in the Sargasso Sea. Adv. Geophys. 38: 533–569.

Jochum, M., and R. Murtugudde, eds. (2006). Physical Oceanography: Developments since 1950. New York, Springer.

Jones, I., and Y. Toba (2001). Wind Stress over the Ocean. Cambridge, Cambridge University Press.

Josey, S. A., E. C. Kent, and P. K. Taylor (2002). Wind stress forcing of the ocean in the SOC climatology: Comparisons with the NCEP-NCAR, ECMWF, UWM/COADS, and Hellerman and Rosenstein Datasets. J. Phys. Oc. 32: 1993–2019.

Jouzel, J., V. Masson-Delmotte, O. Cattani, G. Dreyfus, S. Falourd, G. Hoffmann, B. Minster, et al. (2007). Orbital and millennial Antarctic climate variability over the past 800,000 years. Science 317(5839): 793–796.

Joyce, T. M., D. S. Bitterman, Jr., and K. Prada (1982). Shipboard acoustic profiling of upper ocean currents. Deep-Sea Res. 29: 903–913.

Joyce, T. M., and P. Robbins (1996). The long-term hydrographic record at Bermuda. J. Clim. 9: 3121–3131.

Kadko, D. (1993). An assessment of the effect of chemical scavenging within submarine hydrothermal plumes upon ocean geochemistry. Earth and Planetary Science Letters 120: 361–374.

Kadko, D., J. Baross, and J. Alt (1995). The magnitude and global implications of hydrothermal flux. AGU Geophysical Monogr. Ser. 91: 446–466.

Kagan, B. A., and J. Sündermann (1996). Dissipation of tidal energy, paleotides, and evolution of the earth-moon system. Adv. Geophys. 38: 179–266.

Kalman, R. E. (1960). A new approach to linear filtering and prediction problems. J. Basic Eng. 82: 35–45.

Kalnay, E. (2003). Atmospheric Modeling, Data Assimilation, and Predictability. Cambridge, Cambridge University Press.

Kantha, L. H., J. S. Stewart, and S. Desai (1998). Long-period fortnightly and monthly ocean tides. J. Geophys. Res. 103: 12639–12648.

Kanzow, T., H. L. Johnson, D. P. Marshall, B. A. Cunningham, J. J. M. Hirschi, A. Mujahid, H. L. Bryden, and W. E. Johns (2009). Basinwide integrated volume transports in an eddy-filled ocean. J. Phys. Oc. 39: 3091–3110.

Kara, A. B., P. A. Rochford, and H. E. Hurlburt (2003). Mixed layer depth variability over the global ocean. J. Geophys. Res. 108(C3): 3079.

Katsumata, K., B. M. Sloyan, and S. Masuda (2013). Diapycnal and isopycnal transports in the Southern Ocean estimated by a box inverse model. J. Phys. Oc. 43: 2270–2287.

Katz, E. J. (1975). Tow spectra from MODE. J. Geophys. Res. 80: 1163–1167.

Keeling, C. D. (1998). Rewards and penalties of monitoring the Earth. Ann. Rev. Energy Environ. 23(1): 25–82.

Keffer, T. (1985). The ventilation of the worlds oceans—Maps of the potential vorticity field. J. Phys. Oc. 15(5): 509–523.

Keller, J. B., and V. C. Mow (1969). Internal wave propagation in an inhomogeneous fluid of non-uniform depth. J. Fluid Mech. 38: 365–374.

Kennedy, J. J., N. A. Rayner, R. O. Smith, D. E. Parker, and M. Saunby (2011). Reassessing biases and other uncertainties in sea surface temperature observations measured in situ since 1850. 2. Biases and homogenization. J. Geophys. Res. 116, D14104.

Kent, E. C., P. K. Taylor, B. S. Truscott, and J. S. Hopkins (1993). The accuracy of voluntary observing ships meteorological observations: Results of the VSOP-NA. J. Atm. Oc. Tech. 10: 591–608.

Kenyon, K. E. (1983). Sections along 35°N in the Pacific. Deep-Sea Res. 30: 349–370.

Killworth, P. D. (1986). A Bernoulli inverse method for determining the ocean circulation. J. Phys. Oc. 16: 2031–2051.

Killworth, P. D., and J. R. Blundell (2003). Long extratropical planetary wave propagation in the presence of slowly varying mean flow and bottom topography. Part I: The local problem. J. Phys. Oc. 33(4): 784–801.

King, B., M. Stone, H. P. Zhang, T. Gerkema, M. Marder, R. B. Scott, and H. L. Swinney (2012). Buoyancy frequency profiles and internal semidiurnal tide turning depths in the oceans. J. Geophys. Res. 117(C4): C04008.

King, M. A., R. J. Bingham, P. Moore, P. L. Whitehouse, M. J. Bentley, and G. A. Milne (2012). Lower satellite-gravimetry estimates of Antarctic sea-level contribution. Nature 491(7425): 586–590.

Klein, P., and G. Lapeyre (2009). The oceanic vertical pump induced by mesoscale and submesoscale turbulence. Ann. Rev. Mar. Sci. 1: 351–375.

Klocker, A., and T. J. McDougall, and D. R. Jackett (2009). A new method for forming approximately neutral surfaces. Oc. Sci 5: 155–172.

Knudsen, P., R. Bingham, O. Andersen, and M. H. Rio (2011). A global mean dynamic topography and ocean circulation estimation using a preliminary GOCE gravity model. J. Geod. 85: 861–879.

Köhl, A., and D. Stammer (2008). Variability of the meridional overturning in the North Atlantic from the 50-year GECCO state estimation. J. Phys. Oc. 38(9): 1913–1930.

Köhl, A., and J. Willebrand (2002). An adjoint method for the assimilation of statistical characteristics into eddy resolving ocean models. Tellus 54: 406–425.

Komen, G. J., ed. (1994). Dynamics and Modelling of Ocean Waves. Cambridge, Cambridge University Press.

Körner, T. W. (1988). Fourier Analysis. Cambridge, Cambridge University Press.

Kraus, E. B., and J. A. Businger (1994). Atmosphere-Ocean Interaction. Oxford, Oxford University Press.

Krauss, W. (1966). Interne Wellen. Berlin-Nikolassee, Borntraeger.

Krishfield, R., J. Toole, A. Proshutinsky, and M. L. Timmermans (2008). Automated ice-tethered profilers for seawater observations under pack ice in all seasons. J. Atm. Oc. Tech. 25: 2091–2105.

Ku, L. F., D. A. Greenberg, C. J. R. Garrett, and F. W. Dobson (1985). Nodal modulation of the lunar semidiurnal tide in the Bay of Fundy and Gulf of Maine. Science 230(4721): 69–71.

Kuhlbrodt, T. (2008). On Sandström's inferences from his tank experiments: A hundred years later. Tellus Series A 60: 819–836.

Kundu, P. K., and I. M. Cohen. (2008). Fluid Mechanics, 4th ed. San Diego, Academic Press.

Kunze, E. (2011). Fluid mixing by swimming organisms in the low-Reynolds-number limit. Adv. Geophys. 69: 591–601.

Kunze, E., and E. Boss (1998). A model for vortex-trapped internal waves. J. Phys. Oc. 28: 2104–2115.

Kunze, E., E. Firing, J. M. Hummon, T. K. Chereskin, A. M. Thurnherr (2006). Global abyssal mixing inferred from lowered ADCP shear and CTD strain profiles. J. Phys. Oc. 36: 1553–1576.

Kunze, E., L. K. Rosenfeld, G. S. Carter, and M. C. Gregg (2002). Internal waves in Monterey Submarine Canyon. J. Phys. Oc. 32: 1890–1913.

LaCasce, J. H. (2008). Statistics from Lagrangian observations. Prog. Oceanog. 77: 1–29.

—— (2012). Surface quasigeostrophic solutions and baroclinic modes with exponential stratification. J. Phys. Oc. 42: 569–580.

LaCasce, J. H., R. Ferrari, J. Marshall, R. Tulloch, D. Balwada, and K. Speer (2013). Float-derived isopycnal diffusivities in the DIMES experiment. J. Phys. Oc. 44: 764–780.

Lamb, H. (1932). Hydrodynamics. 6th ed. New York, Dover.

Lambeck, K. (1980). The Earth's Variable Rotation: Geophysical Causes and Consequences. Cambridge, Cambridge University Press.

Lanzante, J. R. (2005). A cautionary note on the use of error bars. J. Clim. 18: 3769–3703.

Laplace, P. S. (1839). Mécanique Céleste. Translated with a commentary by N. Bowditch. Boston, Charles C. Little and James Brown.

Large, W. G., J. C. McWilliams, and S. C. Doney (1994). Oceanic vertical mixing: A review and a model with a nonlocal boundary-layer parameterization. Rev. Geophys. 32: 363–403.

Large, W. G., and S. Pond (1981). Open ocean momentum flux measurements in moderate to strong winds. J. Phys. Oc., 11: 324–336.

Large, W. G., and S. G. Yeager (2009). The global climatology of an interannually varying air-sea flux data set. Clim Dyn. 33: 341–364.

Larsen, L. H. (1969). Oscillations of a neutrally buoyant sphere in a stratified fluid. Deep-Sea Res. 16: 587–603.

Laskar, J. E. A. (2004). A long-term numerical solution for the insolation quantities of the Earth. Astron. & Astrophys. 428: 261–285.

Lauga, E., M. Brenner, and H. Stone (2007). Microfluidics: The no-slip boundary condition. In: Springer Handbook of Experimental Fluid Mechanics, C. Tropea, A. Yarin, and J. Foss, eds. Berlin, Springer. pp. 1219–1240.

Laughton, A. S., W. J. Gould, M. J. Tucker, and H. S. J. Roe, eds. (2010). Of Seas and Ships and Scientists. Cambridge, Lutterworth Press.

Lawson, C. L., and R. J. Hanson (1995). Solving Least Squares Problems. Philadelphia, SIAM.

Le Provost, C. (2001). Ocean tides. In: Satellite Altimetry and Earth Sciences, L.-L. Fu and A. Cazenave, eds. San Diego, Academic Press. pp. 267–304.

Lea, D. J., T. W. N. Haine, M. R. Allen, and J. A. Hansen (2002). Sensitivity analysis of the climate of a chaotic ocean circulation model. Q. J. Roy. Met. Soc. 128(586): 2587–2605.

Leaman, K. D. (1976). Observations on vertical polarization and energy flux of near-inertial waves. J. Phys. Oc. 6: 894–908.

LeBlond, P. H., and L. Mysak (1978). Waves in the Ocean. Amsterdam, Elsevier.

Ledwell, J. R., E. T. Montgomery, K. L. Polzin, L. C. St. Laurent, R. W. Schmitt, and J. M. Toole (2000). Evidence for enhanced mixing over rough topography in the abyssal ocean. Nature 403(6766): 179–182.

Ledwell, J. R., A. J. Watson, and C. S. Law (1998). Mixing of a tracer in the pycnocline. J. Geophys. Res. 103: 21499–21529.

Leetmaa, A., P. Niiler, and H. Stommel (1977). Does the Sverdrup relation account for the Mid-Atlantic circulation? Adv. Geophys. 35: 1–10.

Lehrer, J. (2010). The truth wears off. New Yorker 86 (Dec, 13), p. 52.

Lenn, Y.-D., and T. K. Chereskin (2009). Observations of Ekman currents in the Southern Ocean. J. Phys. Oc. 39: 768–779.

Le Traon, P. Y., F. Nadal, and N. Ducet (1998). An improved mapping method of multisatellite altimeter data. J. Atm. Ocean Tech. 15(2): 522–534.

Levine, M. D. (2002). A modification of the Garrett–Munk internal wave spectrum. J. Phys. Oc. 32: 3166–3181.

Levine, M. D., C. A. Paulson, and J. H. Morison (1985). Internal waves in the Arctic Ocean: Comparison with lower-latitude observations. J. Phys. Oc. 15: 800–809.

Levitus, S., T. Boyer, M. E. Conkright, T. OBrien, J. Antonov, C. Stephens, L. Stathoplos, et al. (1998). NOAA Atlas NESDIS 18, World Ocean Database 1998. Vol. 1, Introduction. Washington, DC, US Government Printing Office.

Lévy, M., P. Klein, A.-M. Tréggier, D. Iovino, G. Madec, S. Masson, and K. Takahashi (2010). Modifications of gyre circulation by sub-mesoscale physics. Ocean Model. 34: 1–15.

Lien, R. C., and P. Müller (1992). Consistency relations for gravity and vortical modes in the ocean. Deep-Sea Res. A 39: 1595–1612.

Lighthill, M. J. (1958). Fourier Analysis and Generalized Functions. Cambridge, Cambridge University Press.

—— (1969). Dynamic response of Indian Ocean to onset of the southwest monsoon. Phil. Trans. Roy. Soc. A 265: 45–92.

—— (1978). Waves in Fluids. Cambridge, Cambridge University Press.

Lindzen, R. S. (1967). Planetary waves on beta-planes. J. Atmos. Sci. 95: 441–451.

Locarnini, R. A., A. Mishonov, T. P. Boyer, M. E. Conkright, T. OBrien, J. Antonov, C. Stephens, et al. (2006). World Ocean Atlas 2005. Vol. 1, Temperature. Washington, DC, US Government Printing Office.

Loder, J. W., and C. Garrett (1978). 18.6 yr cycle of sea surface temperature in shallow seas due to variations in tidal mixing. J. Geophys. Res. 83: 1967–1970.

Longuet-Higgins, M. S. (1949). The electrical and magnetic effects of tidal streams. Geophys. Suppl. Mon. Not. Roy. Astron. Soc. 5: 285–307.

—— (1964). Planetary waves on a rotating sphere. Proc. Roy. Soc. A 279: 446–473.

—— (1965a). Planetary waves on a rotating sphere. II. Proc. Roy. Soc. 284: 40–68.

—— (1965b). Some dynamical aspects of ocean currents. Q. J. Roy. Met. Soc. 91: 425–451.

—— (1968). The eigenfunctions of Laplace's tidal equations over a sphere. Phil. Trans. Roy. Soc. A 262: 511–607.

—— (1969). On the transport of mass by time-varying ocean currents. Deep-Sea Res. 16: 431–447.

Longuet-Higgins, M. S., and A. E. Gill (1967). Resonant interactions between planetary waves. Proc. Roy. Soc. London, A 299(1456): 120–144.

Longuet-Higgins, M. S., and G. S. Pond (1970). The free oscillations of fluid on a hemisphere bounded by meridians of longitude. Phil. Trans. Roy. Soc. A 266: 193–223.

Longuet-Higgins, M. S., and R. W. Stewart (1964). Radiation stress in water waves: A physical discussion with applications. Deep-Sea Res. 11: 529–562.

Longworth, H. R., and H. L. Bryden (2007). Discovery and quantification of the Atlantic meridional overturning circulation: The importance of 25°N. In: Ocean Circulation: Mechanisms and Impacts, Past and Future Changes of Meridional Overturning, A. Schmittner, J. C. H. Chiang and S. R. Hemming, eds. Geophysical Monographs 173. Washington, DC, American Geophysical Union. pp. 5–18.

Losch, M. (2008). Modeling ice shelf cavities in a z-coordinate ocean general circulation model. J. Geophys. Res. 113: C08043

Losch, M., A. Adcroft, and J.-M. Campin (2004). How sensitive are coarse general circulation models to fundamental approximations in the equations of motion? J. Phys. Oc. 34: 306–319.

Losch, M., D. Menemenlis, J. M. Campin, P. Heimbach, and C. Hill (2010). On the formulation of sea-ice models. Part 1: Effects of different solver implementations and parameterizations. Ocean Model. 33: 129–144.

Lozier, M. S. (1997). Evidence for large-scale eddy-driven gyres in the North Atlantic. Science 277(5324): 361–364.

Lozovatsky, I. D., and H. J. S. Fernando (2013). Mixing efficiency in natural flows. Phil. Trans. Royal Soc. A 371: 20120213

Lozovatsky, I. D., H. J. S. Fernando, and S. M. Shapovalo (2008). Deep-ocean mixing on the basin scale: Inference from North Atlantic transects. Deep-Sea Res. Part I 55: 1075–1089.

Lu, Y., and D. Stammer (2004). Vorticity balance in coarse-resolution global ocean simulations. J. Phys. Oc. 34: 605–622.

Lumpkin, R., and K. Speer (2007). Global ocean meridional overturning. J. Phys. Oc. 37: 2550–2562.

Lund, D. C., J. F. Adkins, and R. Ferrari (2011). Abyssal Atlantic circulation during the Last Glacial Maximum: Constraining the ratio between transport and vertical mixing. Paleoceanog. 26, PA1213.

Lupton, J. (1998). Hydrothermal helium plumes in the Pacific Ocean. J. Geophys. Res. 103(C8): 15853–15868.

Luterbacher, J., E. Xoplaki, D. Dietrich, R. Rickli, J. Jacobeit, C. Beck, D. Gyalistras, C. Schmutz, and H. Wanner (2002). Reconstruction of sea level pressure fields over the Eastern North Atlantic and Europe back to 1500. Clim Dyn. 18: 545–561.

Luther, D. S. (1980). Observations Of Long-Period Waves in the Tropical Oceans and Atmosphere. PhD diss., MIT/WHOI.

———— (1982). Evidence of a 4–6 day barotropic planetary oscillation of the Pacific Ocean. J. Phys. Oc. 12: 644–657.

Luther, D. S., J. H. Filloux, and A. D. Chave (1991). Low-frequency, motionally induced electromagnetic fields in the ocean. 2. Electric field and Eulerian current comparison. J. Geophys. Res. 96: 12797–12814.

Luther, D. S., and C. Wunsch (1975). Tidal charts of the Central Pacific. J. Phys. Oc. 5: 222–230.

Lutjeharms, J. R. E. (2007). Three decades of research on the greater Agulhas Current. Oc. Sci 3: 129–147.

Luyten, J. R., J. Pedlosky, and H. Stommel (1983). The ventilated thermocline. J. Phys. Oc. 13: 292–309.

Luyten, J. R., and H. M. Stommel (1991). Comparison of M_2 tidal currents observed by some deep moored current meters with those of the Schwiderski and Laplace models. Deep-Sea Res. 38: S573-S589.

Lvov, Y. V., K. L. Polzin, and N. Yokoyama (2012). Resonant and near-resonant internal wave interactions. J. Phys. Oc. 42: 669–691.

Lvov, Y. V., and N. Yokoyama (2009). Nonlinear wave-wave interactions in stratified flows: Direct numerical simulations. Physica D: Nonlin. Phenom. 238: 803–815.

Lyard, F., F. Lefevre, T. Letellier, and O. Francis (2006). Modelling the global ocean tides: Modern insights from FES2004. Ocean Dyn. 56: 394–415.

Maas, L. R. M. (2011). Topographies lacking tidal conversion. J. Fluid Mech. 684: 5–24.

Macdonald, A. M. (1998). The global ocean circulation: a hydrographic estimate and regional analysis. Prog. Oceanog. 41: 281–382.

Macdonald, A. M., S. Mecking, P. E. Robbins, J. M. Toole, G. C. Johnson, L. Talley, M. Cook, and S. E. Wijffels (2009). The WOCE-era 3-D Pacific Ocean circulation and heat budget. Prog. Oceanog. 82: 281–325.

MacDonald, G. J. F. (1964). Tidal friction. Rev. Geophys. 2: 467–541.

Manley, G. (1974). Central England temperatures: Monthly means 1659 to 1973. Quart. J. Royal Met. Soc. 100(425): 389–405.

Mantua, N. J., and S. R. Hare (2002). The Pacific decadal oscillation. J. Oceanogr. 58: 35–44.

Marshall, D. P., and H. L. Johnson (2013). Propagation of meridional circulation anomalies along western and eastern boundaries. J. Phys. Oc. 43: 2699–2717.

Marshall, D. P., J. R. Maddison, and P. S. Berloff (2012). A framework for parameterizing eddy potential vorticity fluxes. J. Phys. Oc. 42: 539–557.

Marshall, G. J. (2002). Trends in Antarctic geopotential height and temperature: A comparison between radiosonde and NCEP-NCAR reanalysis data. J. Clim. 15: 659–674.

Marshall, J., A. Andersson, N. Bates, W. Dewar, S. Doney, J. Edson, R. Ferrari, et al. (2009). The CLIMODE field campaign: Observing the cycle of convection and restratification over the Gulf Stream. Bull. Am. Met. Soc. 90: 1337–1350.

Marshall, J., and F. Schott (1999). Open-ocean convection: Observations, theory and models. Rev. Geophys. 37: 1–64.

Martel, F., and C. Wunsch (1993). The North Atlantic circulation in the early 1980s: An estimate from inversion of a finite difference model. J. Phys. Oc. 23: 898–924.

Martin, S. (2004). An Introduction to Satellite Remote Sensing. Cambridge, Cambridge University Press.

Maury, M. F. (1855). The Physical Geography of the Sea and Its Meteorology. New York, Harper and Bros. Reprinted by Harvard University Press, J. Leighly, ed., 1963.

Maze, G., G. Forget, M. Buckley, J. Marshall, and I. Cerovecki (2009). Using transformation and formation maps to study the role of air-sea heat fluxes in North Atlantic Eighteen Degree Water formation. J. Phys. Oc. 39: 1818–1835.

Mazloff, M. R., R. Ferrari, and T. Schneider (2013). The force balance of the Southern Ocean Meridional Overturning Circulation. J. Phys. Oc. 43: 1193–1208.

Mazloff, M. R., P. Heimbach, and C. Wunsch (2010). An eddy-permitting Southern Ocean state estimate. J. Phys. Oc. 40: 880–899.

McComas, C. H., and M. G. Briscoe (1980). Bispectra of internal waves. J. Fluid Mech. 97: 205–213.

McConnell, A. (1980). Six's thermometer: A century of use in oceanography. In: Oceanography: The Past, M. Sears and D. Merriman, eds. New York, Springer. pp. 252–265.

McDougall, T. J. (1987). Neutral surfaces. J. Phys. Oc. 17: 1950–1964.

McDougall, T. J., and C. J. R. Garrett (1992). Scalar conservation equations in a turbulent ocean. Deep-Sea Res. A 39: 1953–1966.

McIntyre, M. (1981). On the 'wave momentum' myth. J. Fluid Mech. 106: 331–347.

McKee, W. D. (1973). Internal-inertia waves in a fluid of variable depth. Proc. Cambridge Phil. Soc. 73: 205–213.

McWilliams, J. C. (1976). Maps from the Mid-Ocean Dynamics Experiment, I. Geostrophic stream function. J. Phys. Oc. 6: 810–827.

———(2008). The nature and consequences of oceanic eddies. In: Ocean Modeling in an Eddying Regime. AGU Monograph, Washington, DC. pp. 5–15.

McWilliams, J. C., and G. R. Flierl (1976). Optimal, quasi-geostrophic wave analysis of MODE array data. Deep-Sea Res. 23: 285–300.

McWilliams, J. C., and J. M. Restrepo (1999). The wave-driven ocean circulation. J. Phys. Oc. 29: 2523–2540.

Medwin, H., and J. E. Blue (2005). Sounds in the Sea: From Ocean Acoustics to Acoustical Oceanography. Cambridge, Cambridge University Press.

Meinen, C. S., M. O. Baringer, and R. F. Garcia (2010). Florida Current transport variability: An analysis of annual and longer-period signals. Deep-Sea Res. Part 1 57: 835–846.

Menard, H. W., and S. M. Smith (1966). Hypsometry of ocean basin provinces. J. Geophys. Res. 71: 4305–4325.

Menemenlis, D., C. Hill, A. Adcroft, J.-M. Campin, B. Cheng, B. Ciotti, I. Fukumori, et al. (2005a). NASA supercomputer improves prospects for ocean climate research. Eos 86(9).

Menemenlis, D., I. Fukumori, and T. Lee (2005b). Using Green's functions to calibrate an ocean general circulation model. Mon. Weather Rev. 133: 1224–1240.

Mercier, H. (1989). A study of the time-averaged circulation in the western North Atlantic by simultaneous nonlinear inversion of hydrographic and current-meter data. Deep-Sea Res. 36: 297–313.

Mercier, H., and A. C. de Verdière (1985). Space and time scales of mesoscale motions in the eastern North Atlantic. J. Phys. Oc. 15: 171–183.

Mertens, S. (2008). Guesstimation: Solving the world's problems on the back of a cocktail napkin. Science 321(5893): 1160–1160.

Merton, R. K. (1968). Matthew effect in science. Science 159(3810): 56–63.

Middleton, D. (1960). An Introduction to Statistical Communication Theory. New York, McGraw-Hill.

Miles, J. W. (1974). On Laplace's tidal equations. J. Fluid Mech. 66: 241–260.

Millard, R. C., W. B. Owens, and N. P. Fofonoff (1990). On the calculation of the Brunt-Väisälä frequency. Deep-Sea Res. Part A 37: 167–181.

Miller, G. R. (1966). The flux of tidal energy out of the deep oceans. J. Geophys. Res. 71: 2485–2489.

Miller, S., and C. Wunsch (1973). The pole tide. Nature Phys. Sci. 246: 98–102.

Millero, F. J. (1993). What is PSU? Oceanog. 6: 67.

——— (2006). Chemical Oceanography. Boca Raton, FL, CRC/Taylor and Francis.

Millero, F. J., R. Feistel, D. G. Wright, and T. J. McDougall (2008). The composition of Standard Seawater and the definition of the Reference-Composition Salinity Scale. Deep-Sea Res. Part 1 55: 50–72.

Minster, J. F., A. Cazenave, Y. V. Serafini, et al. (1999). Annual cycle in mean sea level from Topex-Poseidon and ERS-1: Inference on the global hydrological cycle. Global Planet. Change 20: 57–66.

Mitrovica, J. X., M. E. Tamisiea, J. L. Davis, and G. A. Milne (2001). Recent mass balance of polar ice sheets inferred from patterns of global sea-level change. Nature 409: 1026–1029.

Mitrovica, J. X., J. Wahr, I. Matsuyama, and A. Paulson (2005). The rotational stability of an ice-age earth. Geophys. J. Intl. 161: 491–506.

MODE Group (1978). The Mid-Ocean Dynamics Experiment. Deep-Sea Res. 25: 859–910.

Molemaker, M. J., J. C. McWilliams, and X. Capet (2010). Balanced and unbalanced routes to dissipation in an equilibrated Eady flow. J. Fluid Mech. 654: 35–63.

Molemaker, M. J., J. C. McWilliams, and I. Yavneh (2005). Baroclinic instability and loss of balance. J. Phys. Oc. 35: 1505–1517.

Monin, A. S., and A. M. Yaglom (1975). Statistical Fluid Mechanics: Mechanics of Turbulence. 2 vols. Translated from the Russian edition of 1965. Cambridge, MA, MIT Press.

Montgomery, R. B. (1969). The words naviface and oxyty. J. Mar. Res. 27(1): 161–162.

Mooers, C. N. K. (1973). A technique for the cross spectrum analysis of pairs of complex-valued time series, with emphasis on properties of polarized components and rotational invariants. Deep-Sea Res., 20: 1129–1142.

Moore, D. W. (1968). Planetary-Gravity Waves in Equatorial Ocean. PhD diss., Harvard University.

Moore, D. W., and S. G. H. Philander (1977). Modeling of the tropical oceanic circulation. The Sea 6: 319–361.

Moore, W. S. (2010). The effect of submarine groundwater discharge on the ocean. Ann. Rev. Mar. Sci. 2: 59–88.

Morgan, M. G., M. Henrion, and M. Small (1990). Uncertainty: A Guide to Dealing with Uncertainty in Quantitative Risk and Policy Analysis. Cambridge, Cambridge University Press.

Morokoff, W. J., and R. E. Caflisch (1994). Quasi-random sequences and their discrepancies. SIAM J. Sci. Comput. 15(6): 1251–1279.

Morse, P. M., and H. Feshbach (1953). Methods of Theoretical Physics. 2 vols. New York, McGraw-Hill.

Moum, J. N., D. R. Caldwell, J. D. Nash, and G. D. Gunderson (2002). Observations of boundary mixing over the continental slope. J. Phys. Oc. 32: 2113–2130.

Moum, J. N., J. M. Klymak, J. D. Nash, A. Perlin, and W. D. Smyth (2007). Energy transport by nonlinear internal waves. J. Phys. Oc. 37(7): 1968–1988.

Mowbray, D. E., and B. S. H. Rarity (1967). A theoretical and experimental investigation of phase configuration of internal waves of small amplitude in a density stratified liquid. J. Fluid Mech. 28: 1–16.

Müller, M. (2007). The free oscillations of the world ocean in the period range 8 to 165 hours including the full loading effect. Geophys. Res. Lett. 34, L05606

Müller, M., B. K. Arbic, and J. X. Mitrovica (2011). Secular trends in ocean tides: Observations and model results. J. Geophys. Res. 116, C05013.

Müller, P. (2006). The Equations of Oceanic Motions. Cambridge, Cambridge University Press.

Müller, P., G. Holloway, F. Henyey, and N. Pomphrey (1986). Nonlinear interactions among internal gravity waves. Rev. Geophys. 24: 493–536.

Müller, P., R. C. Lien, and R. Williams (1988). Estimates of potential vorticity at small scales in the ocean. J. Phys. Oc. 18: 401–416.

Müller, T. J., and G. Siedler (1992). Multi-year current time series in the eastern North Atlantic. Ocean. J. Mar. Res. 50: 63–98.

Munk, W. H. (1950). On the wind-driven ocean circulation. J. Meteor. 7: 79–93.

——— (1966). Abyssal recipes. Deep-Sea Res. 13: 707–730.

——— (1968). Once again—tidal friction. Q. J. Roy. Astron. Soc. 9: 352–375.

——— (1981). Internal waves and small-scale processes. In: Evolution of Physical Oceanography: Scientific Surveys in Honor of Henry Stommel, B. A. Warren and C. Wunsch, eds. Cambridge, MA, MIT Press. pp. 264–291.

——— (1997). Once again: Once again—tidal friction. Prog. in Oceanog. 40: 7–35.

——— (2002). Twentieth century sea level: An enigma. Proc. Nat. Acad. Sci. 99: 6550–6555.

——— (2009). An inconvenient sea truth: Spread, steepness, and skewness of surface slopes. Ann. Rev. Mar. Sci. 1: 377–415.

Munk, W., and B. Bills (2007). Tides and the climate: Some speculations. J. Phys. Oc. 37: 135–147.

Munk, W. H., and G. F. Carrier (1950). The wind-driven circulation in ocean basins of various shapes. Tellus 2: 158–167.

Munk, W. H., and D. E. Cartwright (1966). Tidal spectroscopy and prediction. Phil. Trans. Roy. Soc. A 259: 533–581.

Munk, W., M. Dzieciuch, and S. Jayne (2002). Millennial climate variability: Is there a tidal connection? J. Clim. 15: 370–385.

Munk, W., and K. Hasselmann (1964). Super-resolution of tides. In: Studies on Oceanography: A Collection of Papers Dedicated to Koji Hidaka, K. Yoshida, ed. University of Tokyo. pp. 339–344. Reprinted by the University of Washington Press, 1965.

Munk, W. H., and G. J. F. MacDonald (1960). The Rotation of the Earth: A Geophysical Discussion. Cambridge, Cambridge University Press.

Munk, W., and N. Phillips (1968). Coherence and band structure of inertial motion in the sea. Rev. Geophys. 6: 447–442.

Munk W., F. Snodgrass, and M. Wimbush (1970). Tides offshore: Transition from California coastal to deep-sea waters. Geophys. Fl. Dyn. 1: 161–235.

Munk, W., and M. Wimbush (1970). The benthic boundary layer. Sea 4: 731–758.

Munk, W., P. Worcester, and C. Wunsch (1995). Ocean Acoustic Tomography. Cambridge, Cambridge University Press.

Munk, W., and C. Wunsch (1998). Abyssal recipes II: Energetics of tidal and wind mixing. Deep-Sea Res. 45A: 1976–2009.

Murray, C. D., and S. F. Dermott (1999). Solar System Dynamics. Cambridge, Cambridge University Press

Nash, J. D., E. Kunze, J. M. Toole, and R. W. Schmitt (2004). Internal tide reflection and turbulent mixing on the continental slope. J. Phys. Oc. 34: 1117–1134.

Needler, G. (1967). A model for the thermohaline circulation in an ocean of finite depth. Adv. Geophys. 25: 329–342.

——— (1985). The absolute velocity as a function of conserved measurable quantities. Prog. Oceanog. 14: 421–429.

Niiler, P. P., A. S. Sybrandy, K. N. Bi, P. M. Poulan, and D. Bitterman (1995). Measurements of the water-following capability of holey-sock and TRISTAR drifters. Deep-Sea Res. Part 1 42: 1951–1955.

Nikurashin, M., and R. Ferrari (2010). Radiation and dissipation of internal waves generated by geostrophic motions impinging on small-scale topography. Theory. J. Phys. Oc. 40: 1055–1074.

——— (2011). Global energy conversion rate from geostrophic flows into internal lee waves in the deep ocean. Geophys. Res. Lett. 38, L08610.

Nobre, P., and J. Shukla (1996). Variations of sea surface temperature, wind stress, and rainfall over the tropical Atlantic and South America. J. Clim. 9(10): 2464–2479.

Oakey, N. S. (1988). EPSONDE: An instrument to measure turbulence in the deep ocean. IEEE J. Oceanic Engineering 13(3): 124–128.

O'Brien, J. J., ed. (1986). Advanced Physical Oceanographic Numerical Modelling. Proceedings of the NATO Advanced Study Institute, Banyuls-sur-Mer, 2–15 June 1985. Dordrecht, Reidel.

O'Connor, W. P., B. F. Chao, D. W. Zheng, and A. Y. Au (2000). Wind stress forcing of the North Sea pole tide. Geophys. J. Intl. 142: 620–630.

Olbers, D. J. (1983). Models of the oceanic internal wave field. Rev. Geophys. 21: 1567–1606.

——— (1998). Comments on "On the obscurantist physics of 'form drag' in theorizing about the Circumpolar Current." J. Phys. Oc. 28: 1647–1654.

Olbers, D., and M. Wenzel (1989). Determining diffusivities from hydrographic data by inverse methods with application to the Circumpolar Current. In: Oceanic Circulation Models: Combining Data and Dynamics, D. L. T. Anderson and J. Willebrand, eds. NATO Series. Boston, Kluwer. pp. 95–140.

Olbers, D., J. M. Wenzel, and J. Willebrand (1985). The inference of North Atlantic circulation patterns from climatological hydrographic data. Rev. Geophys. 23: 313–356.

Olbers, D., J. Willebrand, and C. Eden (2012). Ocean Dynamics. Berlin, Springer.

Olver, F., and R. Wong (2010). Asymptotic approximations. In: NIST Handbook of Mathematical Functions. Cambridge, Cambridge University Press. pp. 41–70.

Onogi, K., J. Tsutsui, H. Koide, H., M. Sakamoto, S. Kobayashi, H. Hatsushika, T. Matsumoto, et al. (2007). The JRA-25 Reanalysis. J. Met. Soc. Japan, Ser. II 85: 369–432.

Oort, A. H., and T. H. von der Haar (1976). On the observed annual cycle in the ocean-atmosphere heat balance over the northern hemisphere. J. Phys. Oc. 6: 781–800.

Oreskes, N., K. Shrader-Frechette, and K. Belitz (1994). Verification, validation, and confirmation of numerical models in the earth sciences. Science 263: 641–646.

Osborn, T. R. (1980). Estimates of the local rate of vertical duffusion from dissipation measurements. J. Phys. Oc. 10: 83–89.

Osborn, T. R., and C. S. Cox (1972). Oceanic fine structure. Geophys. Fl. Dyn. 3: 321–345.

Owens, W. B. (1991). A statistical description of the mean circulation and eddy variability in the northwestern Atlantic using SOFAR floats. Prog. Oceanog. 28: 257–303.

Owens, W. B., P. L. Richardson, W. J. Schmitz, Jr., H. T. Rossby, and D. C. Webb (1988). Nine-year trajectory of a SOFAR float in the southwestern North Atlantic. Deep-Sea Res. Part A 35: 1851–1857.

Pannella, G. (1972). Paleontological evidence on earths rotational history since early Precambrian. Astrophys. Space Sci. 16: 212–237.

Paparella, F., and W. R. Young (2002). Horizontal convection is non-turbulent. J. Fluid Mech. 466: 205–214.

Parke, M. E., R. H. Stewart, D. L. Farless, and D. E. Cartwright (1987). On the choice of orbits for an altimetric satellite to study ocean circulation and tides J. Geophys. Res. 92(C11): 11693–11707.

Passano, L. M. (1918). Plane and Spherical Trigonometry. New York, Macmillan.

Pawlowicz, R., B. Beardsley, and S. Lentz (2002). Classical tidal harmonic analysis including error estimates in MATLAB using T-TIDE. Computers & Geosci. 28: 929–937.

Peacock, S. (2010). Comment on "Glacial-interglacial circulation changes inferred from 231Pa/230Th sedimentary record in the North Atlantic region" by J.-M. Gherardi et al. Paleoceanog. 25, PA2206.

Pedlosky, J. (1965). A note on the western intensification of the oceanic circulation. Adv. Geophys. 23: 207–209.

——— (1968). An overlooked aspect of the wind-driven ocean circulation. J. Fluid Mech. 32: 809–821.

——— (1987). Geophysical Fluid Dynamics. 2nd ed. New York, Srpinger.

——— (1996). Ocean Circulation Theory. New York, Springer.

——— (2000). The transmission of Rossby waves through basin barriers. J. Phys. Oc. 30: 495–511.

Peltier, W. R. (1994). Ice age paleotopography. Science 265: 195–201.

Peltier, W. R., and C. P. Caulfield (2003). Mixing efficiency in stratified shear flows. Ann. Revs. Fluid Mech. 35: 135–167.

Percival, D. B., and A. T. Walden (1993). Spectral Analysis for Physical Applications: Multitaper and Conventional Univariate Techniques. Cambridge, Cambridge University Press.

——— (2000). Wavelet Methods for Time Series Analysis. New York, Cambridge University Press.

Perkins, H. (1972). Inertial oscillations in the Mediterranean. Deep-Sea Res. A 19: 289–296.

Peterson, R. G., L. Stramma, and G. Kortum (1996). Early concepts and charts of ocean circulation. Prog. Oceanog. 37: 1–115.

Petruncio, E. T., L. K. Rosenfeld, and J. D. Paduan (1998). Observations of the internal tide in Monterey Canyon. J. Phys. Oc. 28: 1873–1903.

Philander, S. G. H. (1978). Forced oceanic waves. Rev. Geophys. 16: 15–46.

——— (1990). El Niño, La Niña, and the Southern Oscillation. San Diego, Academic Press.

Phillips, H. E., and T. M. Joyce (2007). Bermuda's tale of two time series: Hydrostation S and BATS. J. Phys. Oc. 37: 554–571.

Phillips, O. M. (1970). On flows induced by diffusion in a stably stratified fluid. Deep-Sea Res., 17: 435–443.

——— (1977). The Dynamics of the Upper Ocean. 2nd ed. Cambridge, Cambridge University Press.

Pickart, R. S., and D. R. Watts (1990). Deep western boundary current variability at Cape Hatteras. Adv. Geophys. 48: 765–791.

Piecuch, C. G., and R. M. Ponte (2012). Importance of circulation changes to Atlantic heat storage rates on seasonal and interannual time scales. J. Clim. 25: 350–362.

——— (2013). Mechanisms of global mean steric sea level change. J. Clim. 27, 824–834.

Pillsbury, J. E. (1891). The Gulf Stream. In: Report of the Superintendent of the U. S. Coast and Geodetic Survey Showing the Progress of the Work during the Fiscal Year ending June 1890. Washington, DC, USGPO. pp. 459–620.

Pingree, R. D., and A. L. New (1991). Abyssal penetration and bottom reflection of internal tidal energy in the Bay of Biscay. J. Phys. Oc. 21: 28–39.

Pinkel, R., L. Rainville, and J. Klymak (2012). Semidiurnal baroclinic wave momentum fluxes at Kaena Ridge, Hawaii. J. Phys. Oc. 42: 1249–1269.

Platzman, G. W. (1984). Normal-modes of the world ocean. 4. Synthesis of diurnal and semidiurnal tides. J. Phys. Oc. 14(10): 1532–1550.

Platzman, G. W., G. A. Curtis, K. S. Hansen, and R. D. Slater (1981). Normal-modes of the world ocean. 2. Description of modes in the period range 8 to 80 hours. J. Phys. Oc. 11: 579–603.

Pollack, H. N., S. J. Hurrter, and J. R. Johnson (1993). Heat flow from the Earth's interior: analysis of the global data set. Revs. Geophys. 31: 267–280.

Polzin, K. L., E. Kunze, J. M. Toole, and R. W. Schmitt (2003). The partition of finescale energy into internal waves and subinertial motions. J. Phys. Oc. 33: 234–248.

Polzin, K. L., and Y. V. Lvov (2011). Toward regional characterizations of the oceanic internal wavefield. Rev. Geophys. 49, RG4003.

Pond, S., and G. L. Pickard (1983). Introductory Dynamical Oceanography. 2nd ed. Oxford, Pergamon.

Ponte, R. M. (2009). Rate of work done by atmospheric pressure on the ocean general circulation and tides. J. Phys. Oc. 39: 458–464.

Ponte, R. M., and D. Stammer (2000). Global and regional axial ocean angular momentum signals and length-of-day variations (1986-1996). J. Geophys. Res. 105: 17161–17171.

Ponte, R. M., C. Wunsch, and D. Stammer (2007). Spatial mapping of time-variable errors in TOPEX/POSEIDON and Jason-1 seasurface height measurements. J. Atm. Oc. Tech. 24: 1078–1085.

Powell, M. D., P. J. Vickery, and T. A. Reinhold (2003). Reduced drag coefficient for high wind speeds in tropical cyclones. Nature 422(6929): 279–283.

Press, W. H., S. A. Teukolsky, W. T. Vetterling, and B. P. Flannery (2007). Numerical Recipes: The Art of Scientific Computing. New York, Cambridge University Press.

Price, J. F., R. A. Weller, and R. Pinkel (1986). Diurnal cycling: observations and models of the upper ocean response to diurnal heat, cooling, and wind mixing. J. Geophys. Res. 91: 8411–8427.

Price, J. F., R. A. Weller, and R. R. Schudlich (1987). Wind-driven ocean currents and Ekman transport. Science 238(4833): 1534–1538.

Priestley, M. B. (1981). Spectral Analysis and Time Series. New York, Academic Press.

——— (1988). Non-Linear and Non-Stationary Time Series Analysis. San Diego, Academic Press.

Proudman, J. (1953). Dynamical Oceanography. New York, Wiley.

Pugh, D. T. (1987). Tides, Surges, and Mean Sea-level. New York, Wiley.

Purkey, S. G., and G. C. Johnson (2010). Warming of global abyssal and deep Southern Ocean waters between the 1990s and 2000s: Contributions to global heat and sea level rise budgets. J. Clim. 23: 6336–6351.

Rabinovich, A. B., R. N. Candella, and R. E. Thomson (2013). The open ocean energy decay of three recent trans-Pacific tsunamis. Geophys. Res. Lett. 40(12): 3157–3162.

Rainville, L., and R. A. Woodgate (2009). Observations of internal wave generation in the seasonally ice-free Arctic. Geophys. Res. Lett. 36, L23604.

Rao, D. B. (1966). Free gravitational oscillations in rotating rectangular basins. J. Fluid Mech. 25: 523–555.

Rascle, N., F. Ardhuin, P. Queffeulou, and D. Croizé-Fillon (2008). A global wave parameter database for geophysical applications. Part 1: Wave-current-turbulence interaction parameters for the open ocean based on traditional parameterizations. Ocean Model. 25: 154–171.

Ray, R. D. (2009). Secular changes in the solar semidiurnal tide of the western North Atlantic Ocean. Geophys. Res. Lett. 36, L19601.

Ray, R. D., and G. T. Mitchum (1996). Surface manifestation of internal tides generated near Hawaii. Geophys. Res. Lett. 23: 2101–2104.

Ray, R. D., and R. M. Ponte (2003). Barometric tides from ECMWF operational analyses. Annales Geophysicae 21: 1897–1910.

Reid, J. L. (1981). On the mid-depth circulation of the world ocean. In: Evolution of Physical Oceanography: Scientific Surveys in Honor of Henry Stommel, B. A. Warren and C. Wunsch, eds. Cambridge, MA, MIT Press. pp. 70–111.

——— (1982). Evidence of an effect of heat-flux from the East Pacific Rise upon the characteristics of the mid-depth waters. Geophys. Res. Lett. 9: 381–384.

Reigber, C., R. Schmidt, F. Flechtner, R. König, U. Meyer, K. H. Neumayer, P. Schwintzer, and S. Y. Zhu (2005). An Earth gravity field model complete to degree and order 150 from GRACE: EIGEN-GRACE02S. J. Geodyn. 39(1): 1–10.

Reverdin, G., F. Marin, B. Bourlès, and P. L'Herminier (2009). XBT temperature errors during French research cruises (1999–2007). J. Atm. Oc. Tech. 26: 2462–2473.

Reynolds, R. W., and D. B. Chelton (2010). Comparisons of daily sea surface temperature analyses for 2007–08. J. Clim. 23: 3545–3562.

Rhines, P. B. (1977) The dynamics of unsteady currents. The Sea 6: 189–318.

Rhines, P. B., and W. R. Young (1982). Homogenization of potential vorticity in planetary gyres. J. Fluid Mech. 122: 347–367.

Richardson, P. L. (2008). On the history of meridional overturning circulation schematic diagrams. Prog. Oceanog. 76: 466–486.

Rignot, E., I. Velicogna, M. R. van den Broeke, A. Monaghan, and J. T. M. Lenaerts (2011). Acceleration of the contribution of the Greenland and Antarctic ice sheets to sea level rise. Geophys. Res. Lett. 38, L05503.

Risien, C. M., and D. B. Chelton (2008). A global climatology of surface wind and wind stress fields from eight years of QuikSCAT scatterometer data. J. Phys. Oc. 38: 2379–2413.

Robinson, A. R., ed. (1983). Eddies in Marine Science. Berlin, Springer.

Robinson, I. S. (2004). Measuring the Oceans from Space: The Principles and Methods of Satellite Oceanography. Berlin, Springer.

Roemmich, D., J. Gilson, B. Cornuelle, and R. Weller (2001). The mean and time-varying meridional transport of heat at the tropical/subtropical boundary of the North Pacific Ocean. J. Geophys. Res. 106: 8957–8970.

Roemmich, D., W. J. Gould, and J. Gilson (2012). 135 years of global ocean warming between the *Challenger* expedition and the Argo Programme. Nature Clim. Ch. 2: 425–428.

Roemmich, D., G. C. Johnson, S. Riser, R. Davis, J. Gilson, W. B. Owens, S. L. Garzoli, C. Schmid, and M. Ignaszewski (2009). The Argo Program: Observing the global ocean with profiling floats. Oceanog. 22: 34–43.

Roemmich, D., and T. McCallister (1989). Large-scale circulation of the North Pacific Ocean. Prog. Oceanog. 22(2): 171–204.

Roemmich, D., and C. Wunsch (1985). Two Transatlantic sections: Meridional circulation and heat-flux in the sub-tropical North-Atlantic ocean. Deep-Sea Res. Part A 32: 619–664.

Roquet, F. (2013). Dynamical potential energy: A new approach to ocean energetics. J. Phys. Oc. 43: 457–476.

Roquet, F., C. Wunsch, G. Forget, P. Heimbach, C. Guinet, G. Reverdin, J.-B. Charnassin, et al. (2013). Estimates of the Southern Ocean general circulation improved by animal-borne instruments. Geophys. Res. Lett., 6176–6180.

Roquet, F., C. Wunsch, Madec, G. (2011). On the patterns of wind-power input to the ocean circulation. J. Phys. Oc. 41: 2328–2342.

Rosenberg, G. D., and S. K. Runcorn, eds. (1975). Growth Rhythms and the History of the Earth's Rotation. New York, Wiley.

Rossby, C. G. (1939). Relation between variations in the intensity of the zonal circulation of the atmosphere and the displacements of the semipermanent centers of action. Adv. Geophys. 2: 38–55.

Rossby, H. T. (1965). On thermal convection driven by non-uniform heating from below: An experimental study. Deep-Sea Res. 12: 9–16.

Rossby, T., C. Flagg, and Donahue, K. (2010). On the variability of Gulf Stream transport from seasonal to decadal timescales. Adv. Geophys. 68: 503–522.

Rossby, T., J. Price, and D. Webb (1986). The spatial and temporal evolution of a cluster of SOFAR floats in the POLYMODE Local Dynamics Experiment. J. Phys. Oc. 16: 428–442.

Rossby, T., and D. Webb (1970). Observing abyssal motions by tracking swallow floats in SOFAR channel. Deep-Sea Res. 17: 359–362.

Rubincam, D. P. (1994). Insolation in terms of Earth's orbital parameters. Theor. Appl. Climatol. 48: 195–202.

——— (2004). Black body temperature, orbital elements, the Milankovitch precession index, and the Seversmith psychroterms. Theor. Appl. Clim. 79: 111–131.

Rye, C. D., M.-J. Messias, J. R. Ledwell, A. J. Watson, A. Brousseau, and B. A. King (2012). Diapycnal diffusivities from a tracer release experiment in the deep sea, integrated over 13 years. Geophys. Res. Lett. 39(4): L04603.

Sachs, J. P. (2007). Cooling of Northwest Atlantic slope waters during the Holocene. Geophys. Res. Lett. 34(3). doi: 10.1029/2006GL028495

Sanchez, B. V. (2008). Normal modes of the global oceans: A review. Mar. Geod. 31: 181–212.

Sandström, J. W. (1908). Dynamicsche versuche mit Meerwasser. Annalen der Hydrographie under Martimen Meteorologie 36: 6–23. For an English translation, see Kuhlbrodt, 2008.

Sandwell, D. T., and W. H. F. Smith (2001). Bathymetric estimation. In: Satellite Altimetry and Earth Sciences, L.-L. Fu and A. Cazenave, eds. San Diego, Academic Press. pp. 451fh458.

Sanford, T. B., R. G. Drever, and J. H. Dunlap (1978). A velocity profiler based on principles of geomagnetic induction. Deep-Sea Res. 25: 183–196.

Sarachik, E. S., and M. A. Cane (2010). The El Niño–Southern Oscillation Phenomenon. Cambridge, Cambridge University Press.

Sarmiento, J. L., and N. Gruber (2004). Ocean Biogeochemical Dynamics, Princeton, NJ, Princeton University Press.

Schanze, J. J., and R. W. Schmitt (2013). Estimates of cabbeling in the global ocean. J. Phys. Oc. 43: 698–705.

Schanze, J. J., R. W. Schmitt, and L. L. Yu (2010). The global oceanic freshwater cycle: A state-of-the-art quantification. Adv. Geophys. 68: 569–595.

Scharffenberg, M. G., and D. Stammer (2011). Statistical parameters of the geostrophic ocean flow field estimated from the Jason-1-TOPEX/Poseidon tandem mission. J. Geophys. Res. 116: C12011.

Schiermeier, Q. (2010). Last weather ship faces closure. Nature 459: 759.

Schlee, S. (1978). On Almost Any Wind: The Saga of the Oceanographic Research Vessel *Atlantis*. Ithaca, NY, Cornell University Press.

Schlitzer, R. (1988). Modeling the nutrient and carbon cycles of the North Atlantic. 1. Circulation, mixing coefficients, and heat fluxes. J. Geophys. Res. 93: 10699–10723.

——— (2007). Assimilation of radiocarbon and chlorofluorocarbon data to constrain deep and bottom water transports in the world ocean. J. Phys. Oc. 37: 259–276.

Schmitt, R. W., J. R. Ledwell, E. T. Montgomery, K. L. Polzin, and J. M. Toole (2005). Enhanced diapycnal mixing by salt fingers in the thermocline of the tropical Atlantic. Science 308(5722): 685–688.

Schmitz, W. J. Jr., and J. R. Luyten (1991). Spectral time scales for mid-latitude eddies. Adv. Geophys. 49: 75–107.

Schott, F. A., S.-P. Xie, and J. P. McCreary, Jr. (2009). Indian Ocean circulation and climate variability. Rev. Geophys. 47(1), RG1002.

Schureman, P. (1958). Manual of Harmonic Analysis and Prediction of Tides. Spec. Pub. 98. U.S. Coast and Geodetic Survey. Washington, DC, USGPO.

Scott, R. B., B. K. Arbic, E. P. Chassignet, A. C. Coward, M. Maltrud, W. J. Merryfield, A. Srinivasan, and A. Varghese (2010). Total kinetic energy in four global eddying ocean circulation models and over 5000 current meter records. Ocean Model. 32: 157–169.

Scott, J. R., and J. Marotzke (2002). The location of diapycnal mixing and the meridional overturning circulation. J. Phys. Oc. 32(12): 3578–3595.

Seidelmann, P. K., ed. (2006). Explanatory Supplement to the Astronomical Almanac. Mill Valley, CA, University Science Books.

Seife, C. (2000). CERN's gamble shows perils, rewards of playing the odds. Science 289: 2260–2262.

Sheen, K. L., J. A. Brearley, A. C. Naveira Garabato, D. A. Smeed, S. Waterman, J. R. Ledwell, M. P. Meredith, et al. (2013). Rates and mechanisms of turbulent dissipation and mixing in the Southern Ocean: Results from the Diapycnal and Isopycnal Mixing Experiment in the Southern Ocean (DIMES). J. Geophys. Res. Oceans 118: 2774–2792.

Shepard, F. P. (1973). Submarine Geology. New York,, Harper & Row.

Shepherd, A., E. R. Ivins, G. A, V. R. Barletta, M. J. Bentley, S. Bettadpur, K. H. Briggs, et al. (2012). A reconciled estimate of ice-sheet mass balance. Science 338(6111): 1183–1189.

Shoosmith, D., P. L. Richardson, A. S. Bower, and H. T. Rossby (2005). Discrete eddies in the northern North Atlantic as observed by looping RAFOS floats. Deep-Sea Res. Part 2 52: 627–650.

Siedler, G., L. Armi, and T. J. Müller (2005). Meddies and decadal changes at the Azores Front from 1980 to 2000. Deep-Sea Res., Part 2, 52(3–4): 583–604.

Siedler, G., S. Griffies, J. Gould, and J. Church, eds. (2013). Ocean Circulation and Climate: A 21st Century Perspective. 2nd ed. New York, Academic Press.

Slobbe, D. C., F. J. Simons, and R. Klees (2012). The spherical Slepian basis as a means to obtain spectral consistency between mean sea level and the geoid. J. Geod. 86: 609–628.

Smart, W. M. (1962). Textbook on Spherical Astronomy. 5th ed. Cambridge, Cambridge University Press.

Smith, K. S. (2007). The geography of linear baroclinic instability in Earth's oceans. Adv. Geophys. 65: 655–683.

Smith, W. H. F., and D. T. Sandwell (1997). Global seafloor topography from satellite altimetry and ship depth soundings. Science 277(5334): 1956–1962.

Smyth, W. D., J. N. Moum, and D. R. Caldwell (2001). The efficiency of mixing in turbulent patches: inferences from direct simulations and microstructure observations. J. Phys. Oc. 31: 1969–1992.

Sobel, D. (1996). Longitude: The True Story of a Lone Genius Who Solved the Greatest Scientific Problem of His Time. New York, Penguin.

Somerville, R. C. J., and S. J. Hassol (2011). Communicating the science of climate change. Physics Today 64: 48–53.

Sorenson, H. W. (1970). Least-squares estimation: From Gauss to Kalman. IEEE Spectrum 7: 63–68. Reprinted in Sorenson, H. W., Ed., Kalman Filtering: Theory and Application. New York, IEEE Press, 1985. pp. 1987–1915.

Spall, M. A. (2008). Buoyancy-forced downwelling in boundary currents. J. Phys. Oc. 38: 2704–2721.

Speer, K., S. R. Rintoul, and B. Sloyan (2000). The diabatic Deacon cell. J. Phys. Oc. 30(12): 3212–3222.

Speer, K., and E. Tziperman (1992). Rates of water mass formation in the North Atlantic Ocean. J. Phys. Oc. 22: 93–104.

Spencer, R. W. (1993). Global oceanic precipitation from the MSU during 1979–91 and comparisons to other climatologies. J. Clim. 6: 1301–1326.

Stacey, F. D. (2008). Physics of the Earth. Cambridge, Cambridge University Press.

Stammer, D. (1997). On eddy characteristics, eddy mixing and mean flow properties. J. Phys. Oc. 28: 727–739.

—— (2008). Response of the global ocean to Greenland and Antarctic ice melting. J. Geophys. Res. 113 (C6), C06022.

Stammer, D., K. Ueyoshi, A. Köhl, W. G. Large, S. A. Josey, and C. Wunsch (2004). Estimating air-sea fluxes of heat, freshwater, and momentum through global ocean data assimilation. J. Geophys. Res. 109(C5), C05023.

Stammer, D., C. Wunsch, R. Giering, C. Eckert, P. Heimbach, J. Marotzke, A. Aderoft, C. N. Hill, and J. Marshall (2002). Global ocean circulation during 1992–1997, estimated from ocean observations and a general circulation model. J. Geophys. Res. 107(C9): 1.1–1.27.

—— (2003). Volume, heat, and freshwater transports of the global ocean circulation 1993–2000, estimated from a general circulation model constrained by World Ocean Circulation Experiment (WOCE) data. J. Geophys. Res. 108(C1): 7.1–7.23.

Stammer, D., C. Wunsch, and R. M. Ponte (2000). De-aliasing of global high frequency barotropic motions in altimeter observations. Geophys. Res. Lett. 27: 1175–1178.

Stansfield, K., C. Garrett, and R. Dewey (2001). The probability distribution of the Thorpe displacement within overturns in Juan de Fuca Strait. J. Phys. Oc. 31: 3421–3434.

Starr, V. P. (1968). Physics of Negative Viscosity Phenomena, New York, Mc-Graw Hill.

Stephenson, F. R. (2008). How reliable are archaic records of large solar eclipses? J. History of Astron. 39: 229–250.

Stephenson, F. R., and L. V. Morrison (1995). Long-term fluctuations in the earths rotation: 700 BC to AD 1990. Phil. Trans. Roy. Soc. London A 351: 165–202.

Stern, M. E. (1975). Ocean Circulation Physics. New York, Academic Press.

Stewart, A. L., and P. J. Dellar (2011). Cross-equatorial flow through an abyssal channel under the complete Coriolis force: Two-dimensional solutions. Oc. Model. 40(1): 87–104.

Stewart, R. H. (1985). Methods of Satellite Oceanography. Berkeley, University of California Press.

Stigler, S. M. (1980). Stigler's law of eponymy. Trans. N. Y. Acad. Sci. 39: 147–157.

Stoker, J. J. (1992). Water Waves: The Mathematical Theory with Applications. New York, Wiley.

Stommel, H. (1948). The westward intensification of wind-driven ocean currents. Trans. Am. Geophys. Un. 29: 202–206.

—— (1954). An oceanographic observatory. Research Reviews. U. S. Office of Naval Research, NAVEXOS P-510, January: 11–13.

—— (1955). Direct measurements of sub-surface currents. Deep-Sea Res. 2: 284–285.

—— (1957). A survey of ocean current theory. Deep-Sea Res. 4: 149–184.

—— (1965). The Gulf Stream: A Physical and Dynamical Description. 2nd ed. Berkeley, University of California Press.

—— (1979). Determination of water mass properties of water pumped down from the Ekman layer to the geostrophic flow below. Proc. U. S. Natl. Acad. Sci., 76: 3051–3055.

Stommel, H., and K. N. Federov (1967). Small scale structure in temperature and salinity near Timor and Mindanao. Tellus 19: 306–325.

Stommel, H., and F. Schott (1977). The beta spiral and the determination of the absolute velocity field from hydrographic station data. Deep-Sea Res. 24: 325–329.

Stommel, H., E. D. Stroup, J. L. Reid, and B. A. Warren (1973). Transpacific hydrographic sections at Lats. 43°S and 28°S: The SCORPIO Expedition-I. Preface. Deep-Sea Res. 20: 1–7.

Stone, P. H. (1978). Constraints on dynamical transports of energy on a spherical planet. Dyn. Atm. Oc. 2: 123–139.

Storch, H. v., and F. W. Zwiers (2001). Statistical Analysis in Climate Research. New York, Cambridge University Press.

Straneo, F., and P. Heimbach (2013). North Atlantic warming and the retreat of Greenland's outlet glaciers. Nature 504(7478): 36–43.

Sturges, W. (1974). Sea-level slope along continental boundaries. J. Geophys. Res. 79: 825–830.

Sturges, W., and B. G. Hong (1995). Wind forcing of the Atlantic thermocline along 32°N at low-frequencies. J. Phys. Oc. 25: 1706–1715.

Sturges, W., B. G. Hong, and A. J. Clarke (1998). Decadal wind forcing of the North Atlantic subtropical gyre. J. Phys. Oc. 28(4): 659–668.

Sullivan, P. P., and J. C. McWilliams (2010). Dynamics of winds and currents coupled to surface waves. Ann. Rev. Fluid Mechs. 42: 19–42.

Sverdrup, H. U. (1947). Wind-driven currents in a baroclinic ocean; with application to the equatorial currents of the eastern Pacific. Proc. Nat. Acad. Sci., 33: 318–326.

Sverdrup, H. U., M. W. Johnson, and R. H. Fleming (1942). The Oceans: Their Physics, Chemistry, and General Biology. New York, Prentice-Hall.

Swallow, J. C. (1955). A neutral-buoyancy float for measuring deep currents. Deep-Sea Res. 3: 74–81.

—— (1977). An attempt to test the geostrophic balance using the Minimode current measurements. In: A Voyage of Discovery: George Deacon 70th Anniversary Volume, M. Angel, ed. Oxford, Pergamon. pp. 165–176.

Szuts, Z. B., J. R. Blundell, M. P. Chidichimo, and J. Marotzke (2012). A vertical-mode decomposition to investigate low-frequency internal motion across the Atlantic at 26 degrees N. Oc. Sci. 8: 345–367.

Tachikawa, K., V. Athias, and C. Jeandel (2003). Neodymium budget in the modern ocean and paleo-oceanographic implications. J. Geophys. Res. 108(C8), 3254.

Taft, B. A. (1972). Characteristics of the flow of the Kuroshio south of Japan. In: Kuroshio: Its Physical Aspects, H. Stommel and K. Yoshida, eds. Tokyo, University of Tokyo Press. pp. 165–216.

Tai, C. K. (2006). Aliasing of sea level sampled by a single exact-repeat altimetric satellite or a coordinated constellation of satellites: Analytic aliasing formulas. J. Atm. Oc. Tech. 23: 252–267.

Tai, C. K., and L. L. Fu (2005). The 25-day-period large-scale oscillations in the Argentine Basin revisited. J. Phys. Oc. 35: 1473–1479.

Tailleux, R. (2013). Available potential energy and exergy in stratified fluids. Ann. Rev. Fluid Mechs. 45: 35–58.

Talagrand, O. (1997). Assimilation of observations, an introduction. J. Meteor. Soc. Japan 75: 191–209.

Talley, L. D. (1988). Potential vorticity distribution in the North Pacific. J. Phys. Oc. 18: 89–106.

——— (2003). Shallow, intermediate, and deep overturning components of the global heat budget. J. Phys. Oc. 33: 530–560.

——— (2007). Hydrographic Atlas of the World Ocean Circulation Experiment (WOCE). Vol. 2: Pacific Ocean, M. Sparrow, P. Chapman, and J. Gould, eds. http://www-pord.ucsd.edu/whp .mscatlas/pacific.mscindex.htm

——— (2013). Closure of the global overturning circulation through the Indian, Pacific, and Southern Oceans: Schematics and transports. Oceanog. 26: 80–97.

Talley, L. D., G. L. Pickard, and W. J. Emery (2011). Descriptive Physical Oceanography: An Introduction. Amsterdam, Academic Press.

Talley, L. D., J. L. Reid, and P. E. Robbins (2003). Data-based meridional overturning streamfunctions for the global ocean. J. Clim. 16: 3213–3226.

Tapley, B. D., S. Bettadpur, M. Watkins, and C. Reigber (2004). The gravity recovery and climate experiment: Mission overview and early results. Geophys. Res. Lett. 31(9), L09607.

Taylor, G. I. (1919). Tidal friction in the Irish Sea. Phil. Trans. Roy. Soc. A 220: 1–33.

——— (1921). Tidal oscillations in gulfs and rectangular basins. Proc. London Math. Soc., ser.2, xx: 148–181 (also in Collected Papers, Vol. 142, p.144–171).

——— (1929). Waves and tides in the atmosphere. Proc. Roy. Soc. London, A, 126(800): 169ff.

Thompson, D. W. J., J. J. Kennedy, J. M. Wallace, and P. D. Jones (2008). A large discontinuity in the mid-twentieth century in observed global-mean surface temperature. Nature 453(7195): 646–649.

Thompson, D. W. J., and J. M. Wallace (2000). Annular modes in the extratropical circulation. Part I: Month-to-month variability. J. Clim. 13: 1000–1016.

Thomson, W., and P. G. Tait (1912). Treatise on Natural Philosophy. 2 vols. Cambridge, Cambridge University Press. Reprinted as Principles of Mechanics and Dynamics, New York, Dover, 1962.

Thorpe, S. A. (2000). The effects of rotation on the nonlinear reflection of internal waves from a slope. J. Phys. Oc. 30: 1901–1909.

——— (2005). The Turbulent Ocean. Cambridge, Cambridge University Press.

——— (2010). Breaking internal waves and turbulent dissipation. Adv. Geophys. 68: 851–880.

Thurnherr, A. M., L. C. St. Laurent, K. G. Speer, J. M. Toole, and J. R. Ledwell (2005). Mixing associated with sills in a canyon on the Midocean Ridge flank. J. Phys. Oc. 35: 1370–1381.

Trefethen, L. N. (1997). Pseudospectra of linear operators. SIAM Rev. 39: 383–406.

Trenberth, K. E., and C. J. Guillemot (1998). Evaluation of the atmopsheric moisture and hydrological cycle in the NCEP/NCAR reanalyses. Clim. Dyn. 14: 213–231.

Trenberth, K. E., W. G. Large, and J. G. Olson (1989). The effective drag coefficient for evaluating wind stress over the oceans. J. Clim. 2: 1507–1516.

Tritton, D. J. (1988). Physical Fluid Dynamics. 2nd ed. Oxford, Oxford University Press.

Tsimplis, M. N., R. A. Flather, and J. M. Vassie (1994). The North Sea pole tide described through a tide-surge numerical model. Geophys. Res. Lett. 21: 449–452.

Tung, K. K., and J. S. Zhou (2013). Using data to attribute episodes of warming and cooling in instrumental records. Proc. Nat. Acad. Sci. US 110(6): 2058–2063.

Turner, J. S. (1973). Buoyancy Effects in Fluids. Cambridge, Cambridge University Press.

Tziperman, E., and A. Hecht (1987). A note on the circulation in the eastern Levantine basin by inverse methods. J. Phys. Oc. 18: 506–518.

Vallis, G. K. (2006). Atmospheric and Oceanic Fluid Dynamics: Fundamentals and Large-Scale Circulation. Cambridge, Cambridge University Press.

van Haren, H. (2008). Abrupt transitions between gyroscopic and internal gravity waves: the mid-latitude case. J. Fluid Mech. 598: 67–80.

van Haren, H., and L. Gostiaux (2010). A deep-ocean Kelvin-Helmholtz billow train. Geophys. Res. Lett. 37, L03605.

van Haren, H., and C. Millot (2004). Rectilinear and circular inertial motions in the Western Mediterranean Sea. Deep-Sea Res. 51: 1441–1455.

Vanmarcke, E. (1983). Random Fields: Analysis and Synthesis, Cambridge, MA, MIT Press.

Vanneste, J. (2013). Balance and spontaneous wave generation in geophysical flows. Ann. Rev. Fluid Mechs. 45: 147–152.

Velicogna, I. (2009). Increasing rates of ice mass loss from the Greenland and Antarctic ice sheets revealed by GRACE. Geophys. Res. Lett. 36, L19503.

Veronis, G. (1969). On theoretical models of the thermocline circulation. Deep-Sea Res. 16 (suppl.) 301–323.

——— (1981). Dynamics of large-scale ocean circulation. In: Evolution of Physical Oceanography: Scientific Surveys in Honor of Henry Stommel, B. A. Warren and C. Wunsch, eds. Cambridge, MA, MIT Press. pp. 140–183.

Veronis, G., and H. Stommel (1956). The action of variable wind stresses on a stratified ocean Adv. Geophys. 15: 43–75.

Vignudelli, S., A. G. Kostianoy, P. Cipollini, and P. Benveniste, eds. (2010). Coastal Altimetry. Berlin, Springer.

Vinnichenko, N. K., and J. A. Dutton (1969). Empirical studies of atmospheric structure and spectra in free atmosphere. Radio Sci. 4: 1115–1126.

Vinogradov, S. V., R. M. Ponte, P. Heimbach, and C. Wunsch (2008). The mean seasonal cycle in sea level estimated from a data-constrained general circulation model. J. Geophys. Res. 113(C3), C03032.

Von Arx, W. S. (1962). An Introduction to Physical Oceanography. Reading, MA, Addison-Wesley.

Voorhis, A. D. (1970). Measuring vertical vorticity in the Sea. Deep-Sea Res. 17: 1019–1023.

Waelbroeck, C., A. Paul, M. Kucera, A. Rosell-Melé, M. Weinelt, R. Schneider, A. C. Mix, et al. (2009). Constraints on the magnitude and patterns of ocean cooling at the Last Glacial Maximum. Nature Geoscience 2: 127–132.

Wang, D. P., C. N. Flagg, K. Donohue, and H. T. Rossby (2010). Wavenumber spectrum in the Gulf Stream from shipboard ADCP observations and comparison with altimetry measurements. J. Phys. Oc. 40(4): 840–844.

Wang, J., G. R. Flierl, J. H. LaCasce, J. L. McClean, and A. Mahadevan (2013). Reconstructing the ocean's interior from surface data. J. Phys. Oc. 43: 1611–1626.

Wang, W. Q., A. Köhl, and D. Stammer (2010). Estimates of global ocean volume transports during 1960 through 2001. Geophys. Res. Lett. 37.

Warren, B. A. (1981). Deep circulation of the world ocean. In: Evolution of Physical Oceanography: Scientific Surveys in Honor of Henry Stommel, B. A. Warren and C. Wunsch, eds. Cambridge, MA, MIT Press. pp. 6–41.

——— (1983). Why is no deep water formed in the North Pacific? Adv. Geophys. 41: 327–347.

——— (1999). Approximating the energy transport across oceanic sections. J. Geophys. Res. 104: 7915–7920.

———, ed. (2006). Historical introduction: Oceanography of the general circulation to the middle of the twentieth century. In: Physical Oceanography: Developments since 1950, M. Jochum and R. Murtugudde, eds. New York, Springer. pp. 1–14.

——— (2008). Nansen-bottle stations at the Woods Hole Oceanographic Institution. Deep-Sea Res. Part I 55: 379–395.

Warren, B. A., T. Whitworth, and J. H. LaCasce (2002). Forced resonant undulation in the deep Mascarene Basin. Deep-Sea Res. Part 2 49: 1513–1526.

Wearn, R. B., and N. G. Larson (1982). Measurements of the sensitivities and drift of digiquartz pressure sensors. Deep-Sea Res. Part A 29: 111–134.

Weaver, A., and P. Courtier (2001). Correlation modelling on the sphere using a generalized diffusion equation. Q. J. Roy. Met. Soc. 127: 1815–1846.

Webb, D. J. (1974). Green's function and tidal prediction. Rev. Geophys. 12: 103–116.

Webb, D. J., and B. A. de Cuevas (2003). The region of large sea surface height variability in the southeast Pacific Ocean. J. Phys. Oc. 33: 1044–1056.

Weis, P., M. Thomas, and J. Sundermann (2008). Broad frequency tidal dynamics simulated by a high-resolution global ocean tide model forced by ephemerides. J. Geophys. Res. 113(C10),. C10029.

Welander, P. (1983). Determination of the pressure along a closed hydrographic section. Part 1. The ideal case. J. Phys. Oc. 13: 797–803.

Wells, J. W. (1963). Coral growth and geochronometry. Nature 197(487): 948–950.

Wessel, P., D. T. Sandwell, and S. S. Kim (2010). The global seamount census. Oceanog. 23: 24–33.

Whalen, C. B., L. D. Talley, and J. A. MacKinnon (2012). Spatial and temporal variability of global ocean mixing inferred from Argo profiles. Geophys. Res. Letts. 39, L18612, 10.1029/2012gl053196.

Whittaker, E. T., and G. N. Watson (1996). A Course of Modern Analysis. Cambridge, Cambridge University Press.

Williams, G. E. (2000). Geological constraints on the Precambrian history of Earth's rotation and the Moon's orbit. Rev. Geophys.38: 37–59.

Williams, R. G., and M. Follows (2011). Ocean Dynamics and the Carbon Cycle: Principles and Mechanisms. Cambridge, Cambridge University Press.

Winters, K. B., P. Bouruet-Aubertot, and T. Gerkema (2011). Critical reflection and abyssal trapping of near-inertial waves on a beta-plane. J. Fluid Mech. 684: 111–136.

Winters, K. B., P. N. Lombard, J J. Riley, and E. A. D'Asaro (1995). Available potential energy and mixing in density-stratified fluids. J. Fluid Mech. 289: 115–128.

Winters, K. B., and W. R. Young (2009). Available potential energy and buoyancy variance in horizontal convection. J. Fluid Mech. 629: 221–230.

Woloszyn, M., M. Mazloff, and T. Ito (2011). Testing an eddy-permitting model of the Southern Ocean carbon cycle against observations. Ocean Model. 39: 170–182.

Woods, J. (1968). Wave-induced shear instability in the summer thermocline. J. Fluid Mech. 32: 791–800.

Woodworth, P. L., S. A. Windle, and J. M. Vassie (1995). Departures from the local inverse barometer model at periods of 5 days in the central South Atlantic. J. Geophys. Res. 100: 18281–18290.

Wöppelmann, G., N. Pouvreau, A. Coulomb, B. Simon, and P. L. Woodworth (2008). Tide gauge datum continuity at Brest since 1711: France's longest sea-level record. Geophys. Res. Lett. 35, Art No. L22605.

Wortham, C. (2012). A Multi-Dimensional Spectral Description of Ocean Variability with Applications. PhD diss. MIT/WHOI Joint Program in Physical Oceanography.

Wortham, C., J. Callies, and M. G. Scharffenberg (2014). Asymmetries between wavenumber spectra of along- and across-track velocity from tandem mission altimetry. J. Phys. Oc. 44(4): 1151–1160.

Wortham, C., and C. Wunsch (2014). A multi-dimensional spectral description of ocean variability. J. Phys. Oc. 44: 944–966.

Worthington, L. V. (1976). On the North Atlantic Circulation. Baltimore, Johns Hopkins University Press.

——— (1981). The water masses of the world ocean: some results of a fine-scale census. In: Evolution of Physical Oceanography. Scientific Surveys in Honor of Henry Stommel, B. A. Warren and C. Wunsch, eds. Cambridge, MA, MIT Press. pp. 42–69.

Wright, C. J., R. B. Scott, D. Furnival, P. Ailliot, and F. Vermet (2012). Global observations of ocean-bottom subinertial current dissipation. J. Phys. Oc. 43: 402–417.

Wunsch, C. (1971). Note on some Reynolds stress effects of internal waves on slopes. Deep-Sea Res. 18: 583–592.

——— (1972a). Bermuda sea level in relation to tides, weather, and baroclinic fluctuations. Rev. Geophys. 10: 1–49.

——— (1972b). The spectrum from two years to two minutes of temperature fluctuations in the main thermocline at Bermuda. Deep-Sea Res. 19: 577–593.

——— (1975). Internal tides in the ocean. Rev. Geophys. Space Phys. 13: 167–182.

——— (1977). Determining the general circulation of the oceans: A preliminary discussion. Science 196: 871–875.

——— (1981). Low frequency variability of the sea. In: Scientific Surveys in Honor of Henry Stommel, B. A. Warren and C. Wunsch, eds. Cambridge, MA, MIT Press. pp. 342–374.

——— (1989). Sampling characteristics of satellite orbits. J. Atmo. and Oceanic Tech. 6: 891–907.

——— (1994). Dynamically consistent hydrography and absolute velocity in the eastern North Atlantic Ocean. J. Geophys. Res. 99: 14071–14090.

——— (1996). The Ocean Circulation Inverse Problem. New York, Cambridge University Press.

——— (1997). The vertical partition of oceanic horizontal kinetic energy. J. Phys. Oc. 27: 1770–1794.

——— (2000). On sharp spectral lines in the climate record and the millennial peak. Paleoceanog. 15: 417–424.

——— (2002a). Oceanic age and transient tracers: Analytical and numerical solutions. J. Geophys. Res. 107(C6), 3048

—— (2002b). What is the thermohaline circulation? Science 298: 1180–1181.

—— (2002c). Ocean observations and the climate forecast problem. In, Meteorology at the Millennium. R. P. Pearce, ed. London, Academic Press. pp. 217–224.

—— (2005). Thermohaline loops, Stommel box models, and the Sandström theorem. Tellus A 57(1): 84–99.

—— (2006a). Discrete Inverse and State Estimation Problems: With Geophysical Fluid Applications. Cambridge, Cambridge University Press.

—— (2006b). Towards the World Ocean Circulation Experiment and a bit of aftermath. In: Physical Oceanography: Developments Since 1950, M. Jochum and R. Murtugudde, eds. New York, Springer. pp. 181–201.

—— (2010). Towards a mid-latitude ocean frequency-wavenumber spectral density and trend determination. J. Phys. Oc. 40: 2264–2281.

—— (2011). The decadal mean ocean circulation and Sverdrup balance. J. Mar. Res. 69: 417–434.

—— (2013). Baroclinic motions and energetics as measured by altimeters. J. Atm. Ocean Tech. 30(1): 140–150.

Wunsch, C., and J. Dahlen (1970). Preliminary results of internal wave measurements in the main thermocline at Bermuda. J. Geophys. Res. 75: 5889–5908.

Wunsch, C., and A. E. Gill (1976). Observations of equatorially trapped waves in Pacific sea level variations. Deep-Sea Res. 23: 371–390.

Wunsch, C., D. B. Haidvogel, M. Iskandarani, and R. Hughes (1997). Dynamics of long-period tides. Prog. in Oceanography (D. Cartwright Volume) 40: 81–108.

Wunsch, C., and P. Heimbach (2007). Practical global oceanic state estimation. Physica D-Nonlin. Phenom. 230: 197–208.

—— (2008). How long to ocean tracer and proxy equilibrium? Q. Sci. Rev. 27: 1639–1653.

—— (2009). The global zonally integrated ocean circulation, 1992–2006: Seasonal and decadal variability. J. Phys. Oc. 39(2): 351–368.

—— (2013a). Two decades of the Atlantic meridional overturning circulation: Anatomy, variations, extremes, prediction, and overcoming its limitations. J. Clim. 26: 7167–7186.

—— (2013b). Dynamically and kinematically consistent global ocean circulation state estimates with land and sea ice. In: Ocean Circulation and Climate, J.C.G. Siedler, S. M. Griffies, J. Gould, and J. A. Church, eds. 2nd ed. Amsterdam, Elsevier. pp. 553–579.

—— (2014). Bidecadal thermal changes in the abyssal ocean and the observational challenge. J. Phys. Oc. 44, 2013–2030.

Wunsch, C., R. M. Ponte, and P. Heimbach (2007). Decadal trends in sea level patterns: 1993–2004. J. Clim. 20: 5889–5911.

Wunsch, C., and D. Stammer (1995). The global frequency-wavenumber spectrum of oceanic variability estimated from TOPEX/POSEIDON altimetric measurements. J. Geophys. Res. 100: 24895–24910.

—— (1997). Atmospheric loading and the oceanic "inverted barometer" effect. Rev. Geophys. 35: 79–107.

—— (1998). Satellite altimetry, the marine geoid and the oceanic general circulation. Ann. Rev. Earth Plan. Sci. 26: 219–254.

Wüst, G. (1924). Stratification and Circulation in the Antillean-Caribbean Basins. New York, Columbia University Press.

—— (1978). Stratosphere of the Atlantic Ocean. New Delhi, Amerind. Available from the US Dept. of Commerce, National Technical Information Service.

Wüst, G., and A. Defant (1993). Atlas of the Stratification and Circulation of the Atlantic Ocean. New Delhi, Amerind. Published for the Division of Oceanic Sciences, National Science Foundation.

Wyrtki, K., E. B. Bennett, and D. J. Rochford (1971). Oceanographic Atlas of the International Indian Ocean Expedition. Washington, DC, USPO.

Xie, P. P., and P. A. Arkin (1997). Global precipitation: A 17-year monthly analysis based on gauge observations, satellite estimates, and numerical model outputs. Bull. Am. Met. Soc. 78: 2539–2558.

Xu, Y. S., L.-L. Fu, and R. Tulloch (2011a). The global characteristics of the wavenumber spectrum of ocean surface wind. J. Phys. Oc. 41: 1576–1582.

Xu, C., X. D. Shang, and R. X. Huang (2011b). Estimate of eddy energy generation/dissipation rate in the world ocean from altimetry data. Ocean Dyn. 61: 525–541.

Xu, Y. S., and L. L. Fu (2012). The effects of altimeter instrument noise on the estimation of the wavenumber spectrum of sea surface height. J. Phys. Oc. 42: 2229–2233.

Young, W. R. (2010). Dynamic enthalpy, conservative temperature, and the seawater Boussinesq approximation. J. Phys. Oc. 40: 394–400.

Zektser, I. S., and R. G. Dzhamalov (2007). Submarine Groundwater. Boca Raton, FL, CRC Press.

Zhai, X. M., H. L. Johnson, and D. P. Marshall (2010). Significant sink of ocean-eddy energy near western boundaries. Nature Geosci. 3: 608–612.

Zhai, X. M., H. L. Johnson, D. P. Marshall, and C. Wunsch (2012). On the wind power input to the ocean general circulation. J. Phys. Oc. 42: 1357–1365.

Zhai, X. M., and C. Wunsch (2013). On the variability of wind power input to the oceans with a focus on the subpolar North Atlantic. J. Clim. 26: 3892–3903.

Zhao, Z., M. H. Alford, R.-C. Lien, M. C. Gregg, and G. S. Carter (2012). Internal tides and mixing in a submarine canyon with time-varying stratification. J. Phys. Oc. 42: 2121–2142.

Zhao, Z. X., M. H. Alford, J. A. MacKinnon, and R. Pinkel (2010). Long-range propagation of the semidiurnal internal tide from the Hawaiian Ridge. J. Phys. Oc. 40: 713–736.

Zika, J. D., T. J. McDougall, and B. M. Sloyan (2010). A tracer-contour inverse method for estimating ocean circulation and mixing. J. Phys. Oc. 40(1): 26–47.

Index